THE ALCOHOL TEXTBOOK
3rd Edition

A reference for the beverage, fuel and industrial alcohol industries

Front cover

Typical pot stills, the heart of Scotch and Irish whiskies

The Alcohol Textbook

3rd Edition

A reference for the beverage, fuel and industrial alcohol industries

K. Jacques, PhD
T.P. Lyons, PhD
D.R. Kelsall

Nottingham University Press
Manor Farm, Main Street, Thrumpton
Nottingham, NG11 0AX, United Kingdom

NOTTINGHAM

Published by Nottingham Universtiy Press (2nd Edition) 1995
Third edition published 1999

ISBN 1-897676-735

The Alcohol Alphabet
© John Murtagh

Typeset by Nottingham University Press, Nottingham
Printed and bound by Redwood Books, Trowbridge, Wiltshire

Contents

Foreword

This book actually began in 1970 when, as a recent graduate of brewing, I went to work for a distillery and became involved in designing a distillery for both malt and grain-type whisky. To my amazement, unlike in the brewing sector there was no reference text one could turn to for information on the cooking, fermentation, whisky maturation or even the distillation steps. The science of alcohol production was an art hidden away in the old distiller's brain or in some obscure notebook. Blind thermometers and magic recipes abounded, and the whole process was kept secret. To this day many old distillers hold the mash bills, their cuts to whisky collection and even their maturation cycles or barrel recipes like state treasures.

Meanwhile, like the author in 1970, many distillers hunger for more information. This book, the third such effort, seeks to remove some of the mystery; and while respecting proprietary information, satisfy that hunger. It is intended to be suitable for reading at all levels of our industry. To make the technical language understandable, a glossary of terms used in the industry is included at the end of the book.

The nature of alcohol production is a blend of chemical engineering, microbiology, botany, biochemistry and art. The objective of this book is to show how these disciplines are interwoven, and hopefully it will make the search for information for some distillers easier. It would have done so for me had it been available in 1970.

T. Pearse Lyons
Alltech Inc.
September, 1999

Chapter 1

Thinking outside the box
Ethanol production in the next millennium: processors of raw materials, not just ethanol producers

T.P. Lyons
Alltech Inc., Nicholasville, Kentucky, USA

Shifting perspectives on our industry

As our industry moves into the next millennium, it – like all industries - must make changes. The 21st century will not be business as usual, but rather business at the speed of light. Enterprises, whether they be distilleries or not, must be capable of responding quickly to avail themselves of new opportunities, new challenges and new technology.

Few industries are as conservative or as protective as the distilling industry. This is exemplified by the rules dictating that only whisky produced in Scotland is Scotch, or that tequila can only be produced in certain areas of Mexico. Even our whole approach to sugar extraction and fermentation is conservative.

Distillers cling to the age-old processes for good reasons where marketing is concerned, but for poor reasons where technology and processes are concerned. Many plants run 5-10% ethanol in the fermenter, yet yeast can tolerate as high as 23% ethanol. Most distillers consider alcohol to be their primary product rather than one of a number of products. The future, however, will be for companies who think 'outside the box' and who look to changing their business perspectives. They are processors of grain, not alcohol producers. They must focus on increasing yeast's performance in the fermenter, so it can produce ethanol at its maximum rate. Antinutritional factors such as phytic acid and mycotoxins must be considered. Just because a process was designed around certain levels of backset does not mean that it remains the best system.

Thinking outside the box also means using spreadsheets to objectively measure effects of change on productivity and profitability. 'What if?' should be the by-word. Whether we are using molasses or grain, new uses of so-called by-products must be found. It is not sufficient to simply sell distillers grains as a commodity, nor to accept spent molasses as a waste needing to be got rid of. The future is for those who find or develop new markets for these materials. We must think of ourselves as processors of grain, sugar beets, cane, etc., not just alcohol producers.

The potential for ethanol

All of this requires education of operators, marketers and management as to the possibilities. All of this will require more and more research. It is disheartening to note how little research is currently underway in our industry. When one considers the growth of the industry over the last few years we cannot but be impressed (Figure 1). In the United States, ethanol production has grown to over 1.6 billion gallons annually. In 1998, nearly 10 billion bushels of corn were harvested (>250 million metric tons). Over 5%, or 500 million bushels were converted into ethanol. This is about the same percentage of the nation's corn crop used to produce high fructose syrups (Andreas, 1999).

When one considers that the oxygenate MTBE (methyl tertiary butyl ether) is expected to be either substantially reduced or eliminated from gasoline, then our industry has an excellent future. In the United States demand for finished motor fuel is around 120 billion gallons annually (Miller, personal communication). If MTBE is removed, then there will be a 4% shortfall to be replaced. Ethanol, with an octane of 113 compared to the octane booster MTBE of 109, is one of the few ways that a gasoline producer can achieve this high octane rating. Furthermore, high octane gasoline always commands a higher price at the pump.

Furthermore, the US Clean Air Act insists that an average of 2% oxygen be present in reformulated gasoline (RFG), which represents about 30% of the US gasoline market. Pure ethanol, contributing 35% oxygen, could clearly provide this 2% oxygen if it were used at 5% of the RFG. However if that were to happen, ethanol production would not be 1.4 billion gallons, but rather 3-4 billion gallons (Miller, personal communication).

Nor is alcohol production limited to the US. In 1998 some 31 billion liters of ethanol were produced and the vast majority were produced by fermentation - 93% or 29 billion liters (7.8 billion US gallons). The major feedstocks used to product this alcohol were sugar crops such as cane and beet sugars, accounting for 60% of the major ethanol producers, i.e. Brazil, and corn in the US and Europe (Table 1). The total ethanol capacity of Russia is estimated to be 2.5 billion liters with beverage alcohol accounting for 60% of this. Political changes are also ocurring that will alter these figure dramatically. In Brazil the highest production of ethanol was in 1997 when output topped 16 billion liters. While a surplus existed in 1998, the government has now introduced 'green fleets' and blending of anhydrous alcohol (5%) into diesel fuel to take care of this. In Europe there has been much discussion about Bio-Ethanol; but only recently have there been tangible results. The long-term goal is a 12% share for renewable fuels with biofuels playing a major role. Despite concessions, progress has been slow. France has been leading the way. A French law dating back to 1987 allows the blending of 3-15% organic oxygenated compounds (3% for ethanol, 15% for ethers such as ETBE). In November of 1996, a draft bill on clean air makes use of oxygenated components mandatory by 2000; and the powerful French farm lobby are pushing for ethanol to be included in gasoline at 2%. This would require almost 0.5 billion liters. Seven self-contained distilleries with a combined capacity of 82 million liters/day and 16 integrated sugar

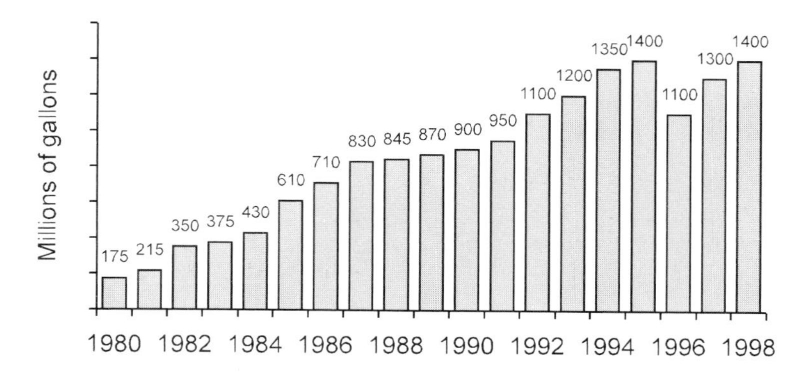

Figure 1. Fuel ethanol production in the US 1980-1998 (from Energy Information Administration and Renewable Fuels Association).

and alcohol complexes will be easily able to supply that amount.

Table 1. Geographic distribution of the majority of fermentation alcohol (fuel, industrial and potable).[1]

Country	*Production (billion liters)*	*Raw material*
Brazil	14.0	Sugarcane, beets
United States	5.3	Cereal grains (mostly corn)
Europe	4.3	Cereal grains, beets
Russia	2.5	Cereal grains, beets
Total world production	28.0	

[1]Berg, 1999.

The future of ethanol is thus very positive and does not depend on subsidies. A 1997 study by Dr. Michael Evans of the prestigious Kellogg School of Management at Northwestern University concluded that the net effect of ethanol production was a reduction in the US federal budget deficit (Table 2). This is brought about by increased tax revenues due to increases in personal income from wages and salaries, higher farm income, more corporate tax revenue and declining unemployment compensation payments.

Table 2. Impact of ethanol production on the US federal budget[1].

Revenue source	*Million US$*
Personal income taxes from wages and salaries	532
Personal income taxes, farm income	675
Social security taxes	1608
Decline in unemployment benefits	561
Corporate income taxes	846
Less the ethanol tax incentive	-648
Net savings to the federal budget	$3,574

[1](adapted from Evans, 1999)

The alcohol production process

The alcohol process has a number of steps that enable us to differentiate among the various types of spirits (Figure 2). In the first step starch-containing raw materials must be liquefied so that dextrins and subsequently fermentable sugars can be obtained. Other raw materials such as molasses and whey have sugars pre-formed or their extraction may be taken care of in the 'cooking' process such as agave. The sugar will in all cases be passed over to the workhorse of all distilleries – yeast. This microorganism, the world's most heavily used microbe, converts sugars to ethanol and caron dioxide. The ethanol content, which can be as much as 23% but in practice is 10-13, is then concentrated. At this point the chemical engineer takes over from the biochemist as the beer moves toward distillation. Even with the disillation some classification of spirits can be made. The rums and malt whiskies from pot stills; gin, vodka, fuel and industrial alcohol from continuous stills.

Further classifications of the process occur when we decide whether to mature the distilled product. As noted in the chapter on American whisky, bourbon must by law be aged for a minimum of one year in charred white oak barrels. This opens up an opportunity for the Scots and the Irish, and indeed for the rum manufacturers, who use barrels that once carried bourbon and use them for a further three years to store and mature their own Irish whiskey, malt whisky, or Scotch grain whisky. The rum and tequila manufacturers, unlike the Irish and Scotch whisky producers, also age their product in barrels, but not for such a long period. Rum is commonly matured for 1-2 years; although many fine rums are aged for up to 12 years. While the bulk of the tequila sold is not aged; regulations defining aged tequilas call for Reposado to be aged at least one month while Añejo must be aged for at least 12 months.

Recovery of by-products

As we review our overall process, it is obviously critical that we maximize utilization of by-products. In the production of spirits, assuming the feedstock is corn, the major cost factors are the raw material, energy, enzymes, processing

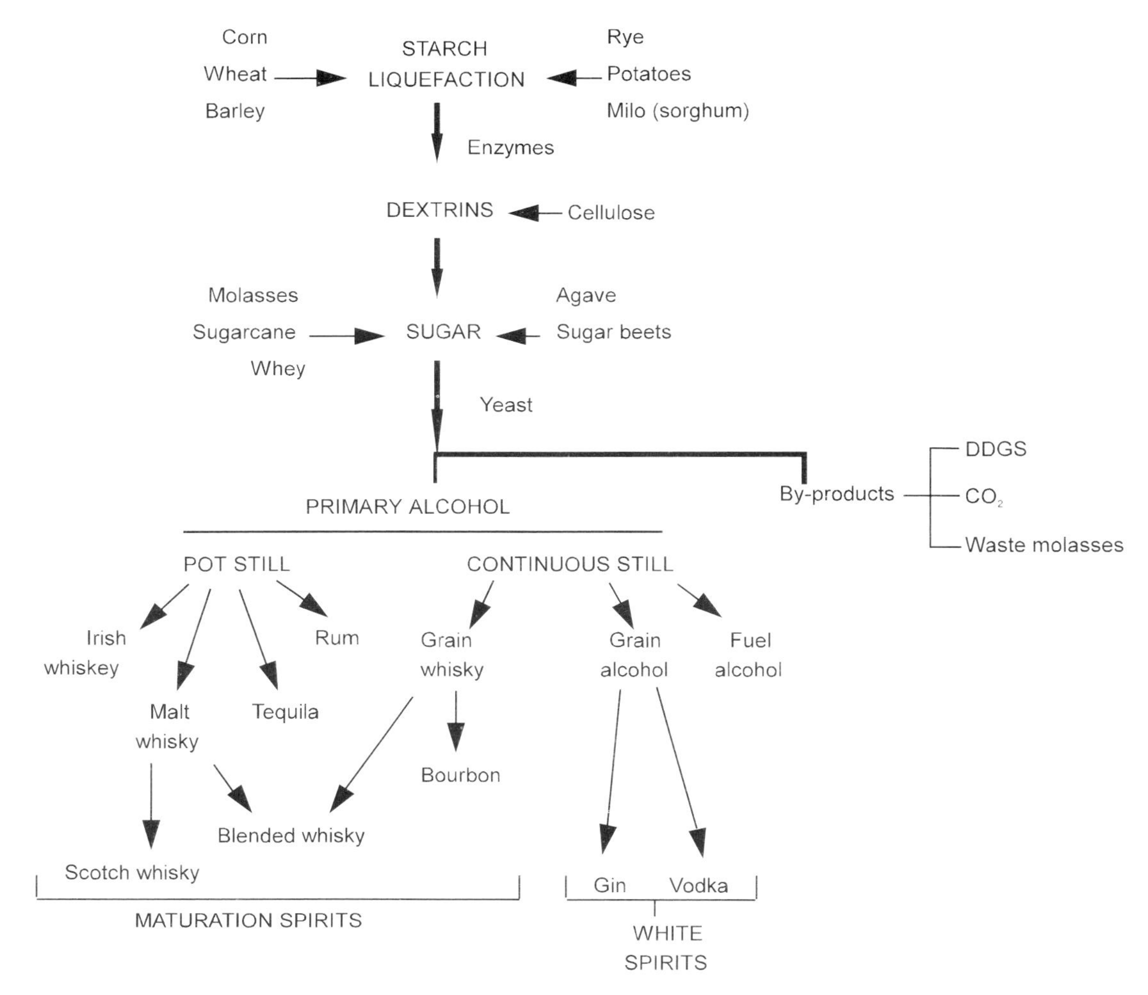

Figure 2. Beverage alcohol feedstocks and products.

chemical cost (water treatment, etc.) and labor. Much of these costs can be offset with 17 lbs of distillers grains coming from each bushel of corn; but we must maximize the return on this essential product. At the time this book was printed, distillers grains were selling for around $80/ton while corn, at under $2 per bushel, is at an all time low – as is distillers grain. At 4 cents per lb ($80/ton), this 17 lbs of DDGS gives us a credit of 68 cents. This 68 cents is set against the $2 bushel of corn or the 3-4 cents for enzymes. If on the other hand, new or novel products could be made from distillers grains, (i.e. improved bypass proteins for cattle, pre-hydrolyzed DDGS for fiber use in human applicatoins), the financial impact would be substantial. The future will see many new products built around DDGS. This is the subject of a separate chapter in this volume.

Therefore, as we move forward to improve the value of distillers grains and derived products, a continual emphasis on yield (while still very important), would not be as critical. Perhaps in the future distillers grains and other derived products will, and indeed should, be the single most important raw material from a distillery. After all, they are rich and valuable proteins, a commodity in very short supply across the world.

The chapters in this book we hope will help you realize the complexity and at the same time the simplicity of the process of converting sugars to ethanol. If we make your search for information just a little easier, we will have achieved our objective. If we can encourage more research, then we will be well-rewarded.

References

Andreas, M. 1999. A global perspective. In: The 1999 International Fuel Ethanol Workshop and Trade Show. Cedar Rapids, Iowa, June 22-25.

Berg, C. 1999. World ethanol production and trade to 2000 and beyond. World Ethanol Report.

Evans, M.K. 1997. The economic impact of of the demand for ethanol. Report to Midwestern Governors Conference. Lombard, IL, February. (cited by the Renewable Fuels Association).

Chapter 2

Grain dry milling and cooking for alcohol production: designing for 23% ethanol and maximum yield

D.R. Kelsall and T.P. Lyons
Alltech Inc., Nicholasville, Kentucky, US

This chapter deals with the milling and cooking stages of alcohol production from whole cereals. In brief, in this process the whole cereal, normally corn (maize), is ground in a mill to a fine particle size and mixed with liquid, usually a mixture of water and backset stillage. This slurry is then treated with a liquefying enzyme to hydrolyze the cereal to dextrins, which are oligosaccharides. The hydrolysis of starch with the liquefying enzyme, called α-amylase, is helped along by cooking the mash at an appropriate temperature to break down the granular structure of the starch and cause it to gelatinize. Finally the dextrins produced in the cooking process are further hydrolyzed to glucose in a saccharification process using the exoenzyme glucoamylase and another enzyme (Rhizozyme™) that may be added to the yeast propagation tank or the fermenter. These separate stages of milling, cooking and saccharification will be explained in more detail.

Milling

The incoming cereal is usually inspected upon receipt. The distiller will check the grain for bushel weight, moisture content, mold infestation and general appearance. If the cereal complies with the quality control standards, it will be unloaded into silos in preparation for milling.

The purpose of milling is to break up the cereal grains to as small a particle size as possible in order to facilitate subsequent penetration of water in the cooking process. A wide variety of milling equipment is available to grind the whole cereal to a meal. Normally, most distilleries use hammer mills, although some may use roller mills, particularly for small cereal grains.

Hammer mills

With a hammermill, the cereal grain is fed into a grinding chamber in which a number of hammers rotate at high speed. The collision of the hammers with the grain causes a breakdown to a meal. The mill outlet contains a retention screen that holds back larger particles until they are broken down further so that there will be a known maximum particle size in the meal. The screens are normally in the size range of 1/8-3/16 in.

A size distribution test or 'sieve analysis' of the meal should be conducted regularly. Sieve analysis shows whether the hammermill screens are in good order and whether the mill is correctly adjusted. Table 1 shows a typical sieve analysis for corn. The two largest screens retain only 11% of the particles while the quantity passing through the 60 mesh screen is also fairly low at 7%. For efficient processing of the cereal starch into alcohol, the particles should be as fine as possible. However a compromise must be made such that the particles are not too fine and cause balling in the slurry tank or problems in the by-product recovery process.

Table 1. Typical results of a sieve analysis of corn meal.

Screen size	*Hole size (in)*	*Corn on screen (%)*
12	0.0661	3.0
16	0.0469	8.0
20	0.0331	36.0
30	0.0234	20.0
40	0.0165	14.0
60	0.0098	12.0
Through 60		7.0

Fineness of the grind is a significant factor in the final alcohol yield. It is possible to obtain a 5-10% difference in yield between a fine and a coarse meal. Table 2 shows the typical alcohol yield from various cereals. It can be seen that the normal yield from corn is 2.65 gallons of anhydrous ethanol per bushel (56 lbs). However, the yield with coarsely ground corn may drop to 2.45 gallons per bushel, a reduction in yield of 7.5%. This is a highly significant reduction and would have serious economic consequences for any distiller.

Table 2. Typical alcohol yields from various cereals.

Cereal	Yield[a] (US gallons of anhydrous alcohol/bushel)
Fine grind corn, 3/16 in.	2.65
Coarse grind corn, 5/16 in.	2.45
Milo	2.60
Barley	2.50
Rye	2.40

[a]Note that a distiller's bushel is always a measure of weight. It is always 56 lb, regardless of the type of grain.

It is recommended that a sieve analysis of the meal be done at least once per shift. The distiller should set specifications for the percentage of particles on each sieve; and when the measured quantity falls outside of these specifications the mill should be adjusted. Normally the hammers in a hammer mill are turned every 15 days, depending on usage; and every 60 days a decision should be made as to whether or not to replace the hammers and screens.

Since sieve analysis is critical, a case can also be made for recycling to the hammer mill any grain not ground sufficiently fine. Fineness of the grind also has an important bearing on centrifugation of the stillage post-distillation. A finer grind may yield more solubles and hence place a greater load on the evaporator. However since the key is to maximize yield, dry house considerations, while important, cannot override yield considerations.

Roller mills

Some distillers use roller mills (e.g. malt whisky producers), particularly where cereals containing substantial quantities of husk material are used. In a roller mill, the cereal is nipped as it passes through the rollers, thus exerting a compressive force. In certain cases, the rollers operate at different speeds so that a shearing force can be applied. The roller surfaces are usually grooved to aid in the shearing and disintegration. Figure 1 shows the general configuration of a roller mill.

In Scotland the solids in whisky mash, made entirely from malted barley, are usually removed by using a brewery-type lauter tun, which is a vessel with a perforated bottom like a large colander. In this case, a roller mill should be used as the shear force allows the husk to be separated with minimal damage. The husk then acts as the filter bed in the lauter tun for the efficient separation of solids and liquid.

Cooking

Cooking is the entire process beginning with mixing the grain meal with water (and possibly

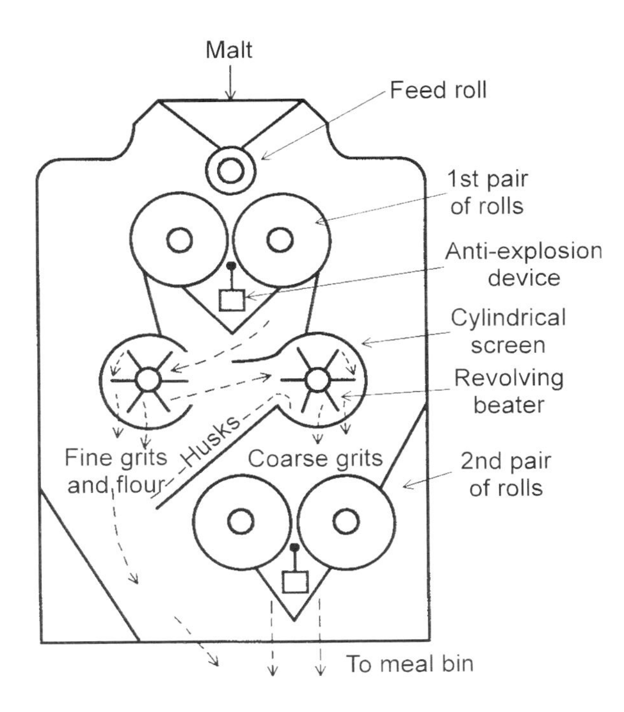

Figure 1. Roller mill.

backset stillage) through to delivery of a mash ready for fermentation. Figure 2 shows the components that make up a typical milling and cooking system. This schematic diagram could represent the processes involved in beverage, industrial or fuel alcohol production, except that nowadays only the whisky distillers use malt as a source of liquefying and saccharifying enzymes. All other alcohol producers use microbial enzyme preparations. The key to cooking is to simply liquefy the starch so it can be pumped.

Hydrolysis of starch

Before discussing cooking operations, it is necessary to consider the biochemical processes involved. The source of alcohol from cereal grains is the glucose polymer known as starch. The purpose of cooking and saccharification is to achieve hydrolysis of starch into fermentable sugars. Hydrolysis normally involves use of the

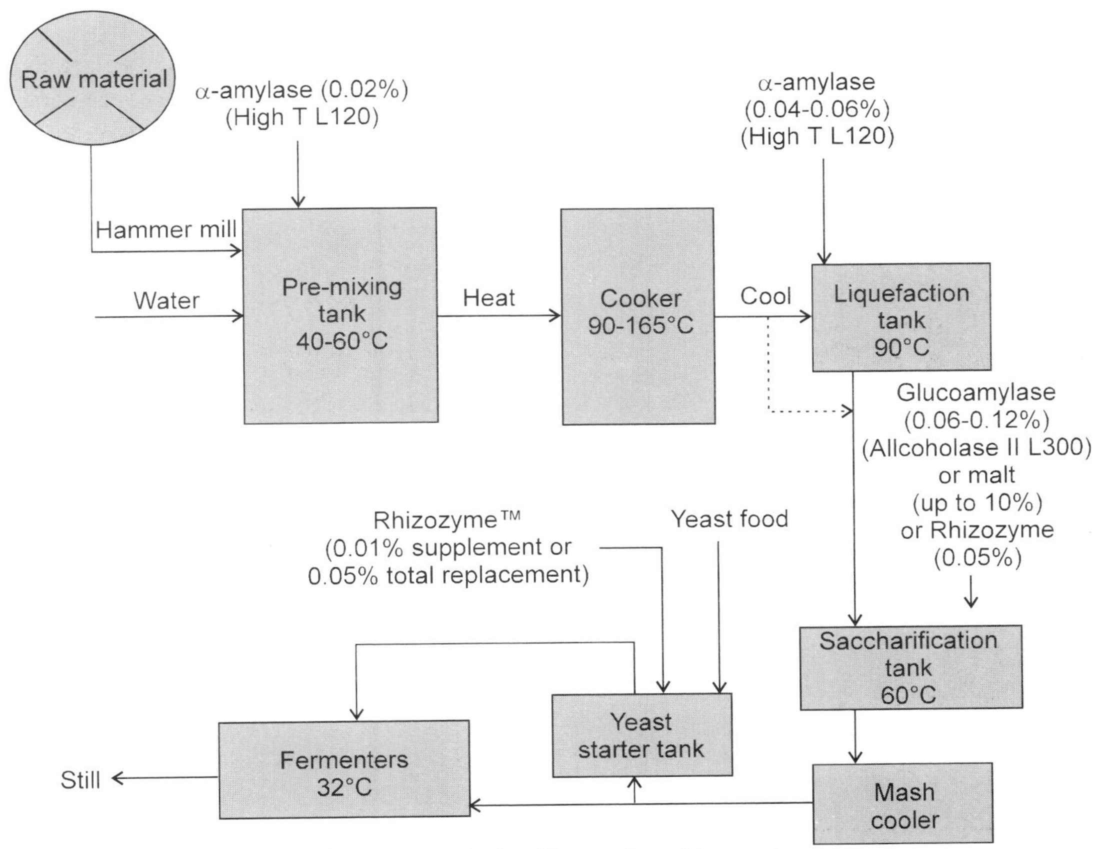

Figure 2. Typical milling and cooking system.

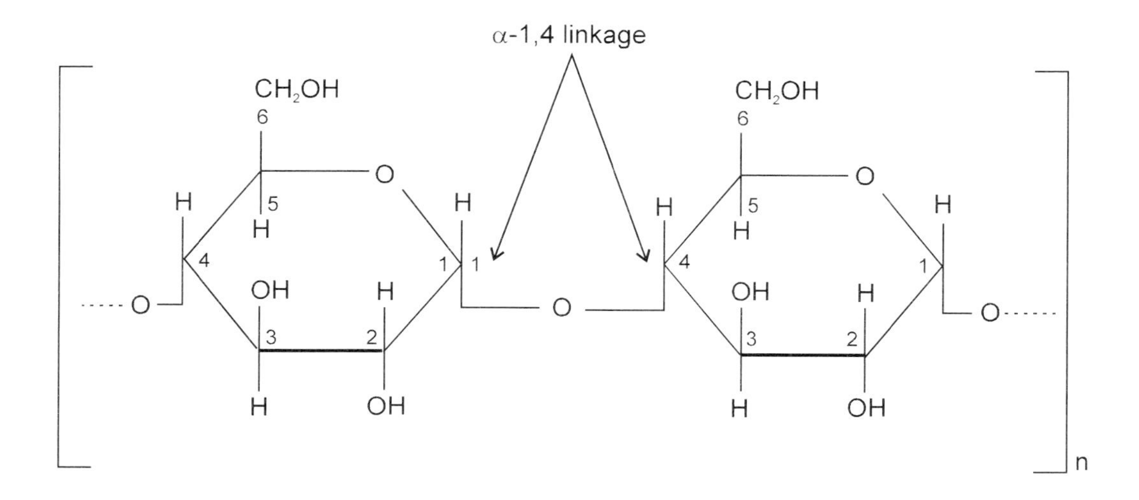

Figure 3. Chemical structure of amylose. The polymer can have several thousand α-1,4 glucosidic linkages.

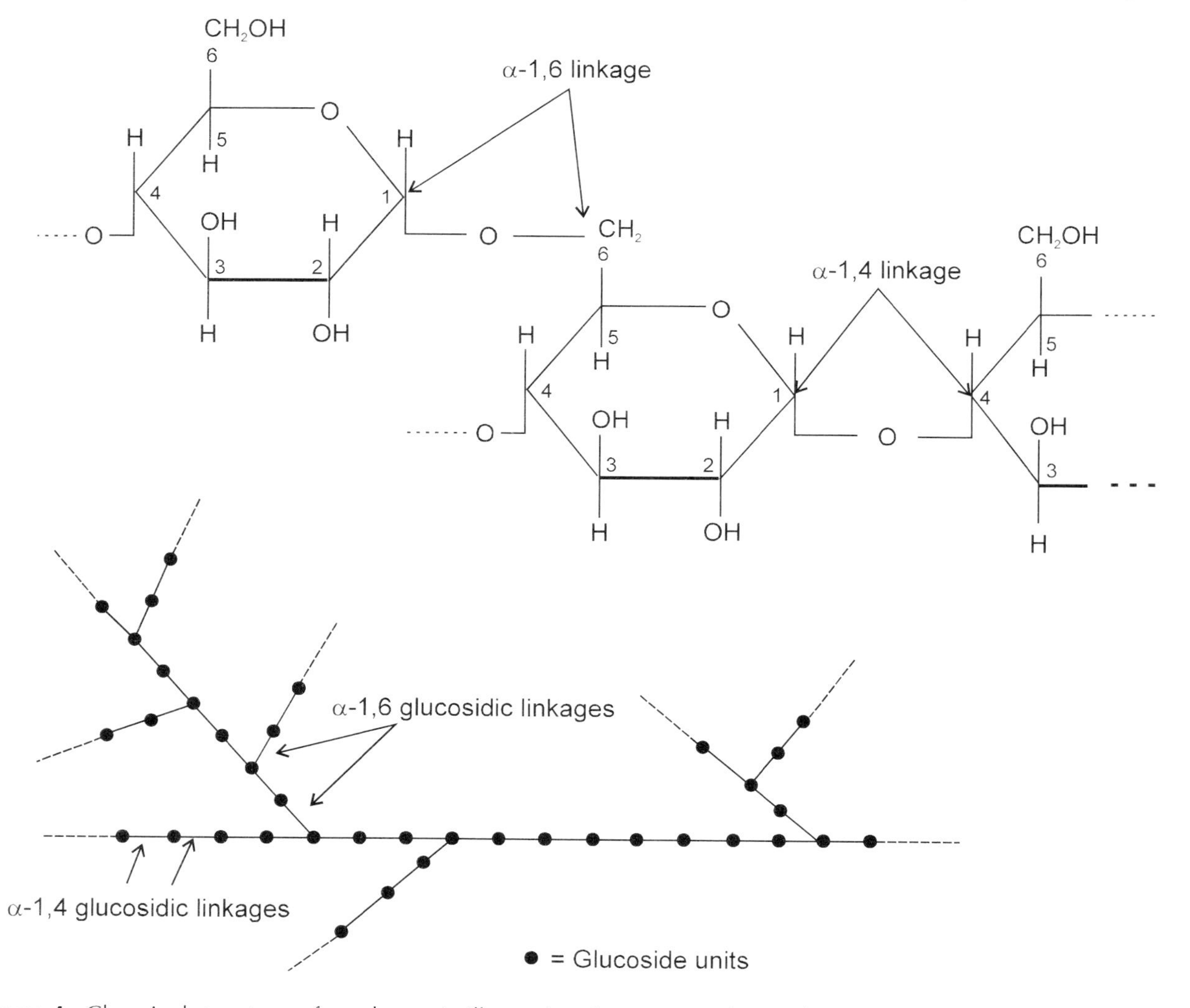

Figure 4. Chemical structure of amylopectin illustrating the α-1,4 and α-1-6 bonds and the general structure of the molecule.

endoenzyme α-amylase followed by the exoenzyme glucoamylase (amylo-glucosidase, to produce glucose. However, when malt is used as a source of both α-amylase and the exoenzyme ß-amylase, the fermentable sugar produced is maltose, a dimer made up of two glucose units.

Starch exists in two forms. One form is the straight-chained amylose, where the glucose units are linked by α-1,4 glucosidic linkages (Figure 3). The amylose content of corn is about 10% of the total starch; and the amylose chain length can be up to 1,000 glucose units. The other form of starch is called amylopectin, which represents about 90% of the starch in corn. Amylopectin has a branched structure (Figure 4). It has the same α-1,4 glucosidic linkages as in amylose, but also has branches connected by α-1,6 linkages. The number of glucose units in amylopectin can be as high as 10,000. The α-amylase enzyme used for the initial liquefaction or hydrolysis of the starch acts randomly on the α-1,4 glucosidic linkages but will not break the α-1,6 linkages of amylopectin. Corn, wheat and milo have similar levels of starch (Table 3). That is why they are the most commonly used cereals in the alcohol distilling industry.

Table 3. Starch content of various cereal grains.

Raw material	*Starch (%)*
Corn	60-68
Wheat	60-65
Oats	50-53
Barley	55-65
Milo	60-65
Potato	10-25
Cassava	25-30
Rye	60-63
Rice (polished)	70-72
Sorghum (millet)	75-80

In order for the α-amylase to bring about hydrolysis of the starch to dextrins, the granular structure of the starch must first be broken down in the process known as gelatinization. When the slurry of meal and water are cooked, the starch granules start to adsorb water and swell. They gradually lose their crystalline structure until they become large, gel-filled sacs that tend to fill all of the available space and break with agitation and abrasion.

The peak of gelatinization is also the point of maximum viscosity of a mash. Figures 5, 6 and 7 show the progressive gelatinization of cornstarch, as viewed on a microscopic hot stage. In Figure 5 the granules are quite distinct and separate from the surrounding liquid. In Figure 6 these same granules have swollen in size and some of the liquid has entered the granules. Figure 7 shows the granules as indistinct entities in which the liquid has entered to expand them considerably.

Gelatinization temperatures vary for the different cereals (Table 4). Some distillers consider it important for the slurrying temperature of the meal to be below the temperature of gelatinization. This avoids coating of grain particles with an impervious layer of gelatinized starch that prevents the enzymes from penetrating to the starch granules and leads to incomplete conversion. Many distillers, however, go to the other extreme and slurry at temperatures as high as 90 °C. At these temper-atures starch gelatinizes almost immediately and with adequate agitation there is no increase in viscosity and no loss of yield.

Table 4. Temperature range for the gelatinization of cereal starches.

Cereal	*Gelatinization range (°C)*
Barley	52-59
Wheat	58-64
Rye	57-70
Corn (maize)	62-72
High amylose corn	67- >80
Rice	68-77
Sorghum	68-77

Figure 8 shows the hydrolysis reaction of the α-1,4 glucosidic linkage and represents the breakdown of starch to the less viscous dextrins. Dextrins are oligosaccharides resulting from the

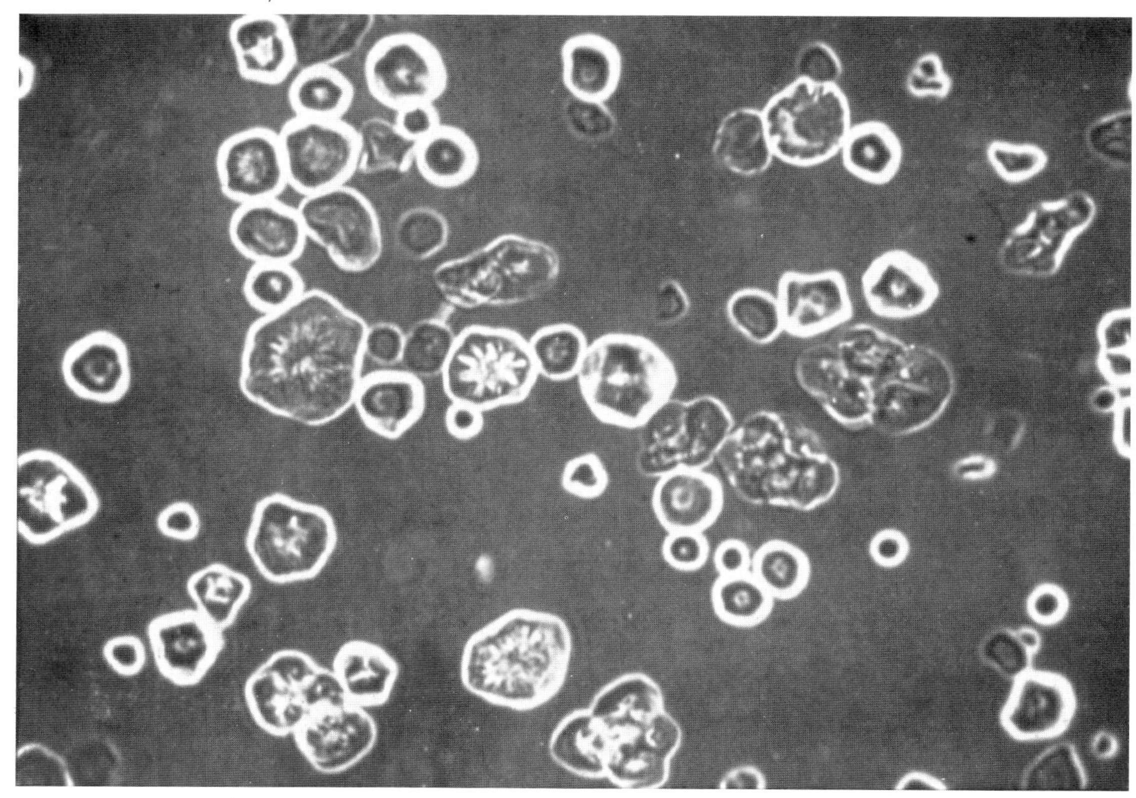

Figure 5. Gelatinization of cornstarch. Group of granules viewed on microscope hot stage at 67 °C under normal illumination.

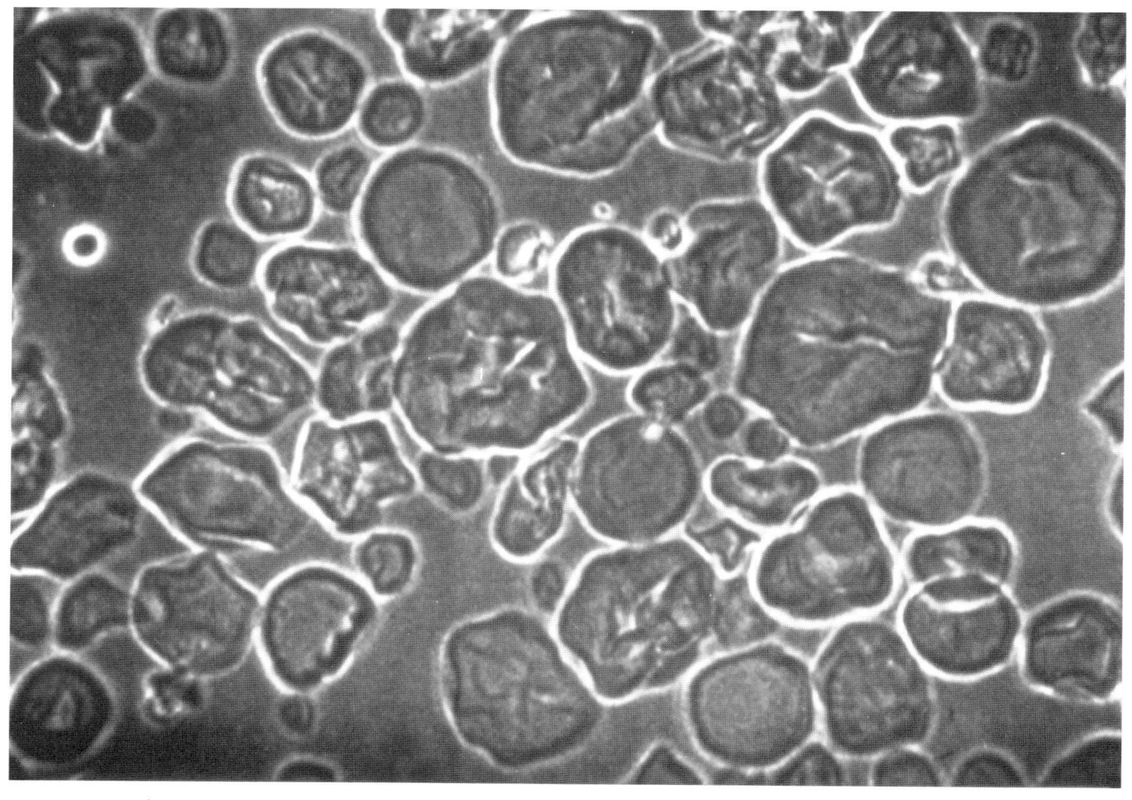

Figure 6. Gelatinization of cornstarch. Same group of granules as in Figure 5, viewed on microscope hot stage at 75 °C under normal illumination.

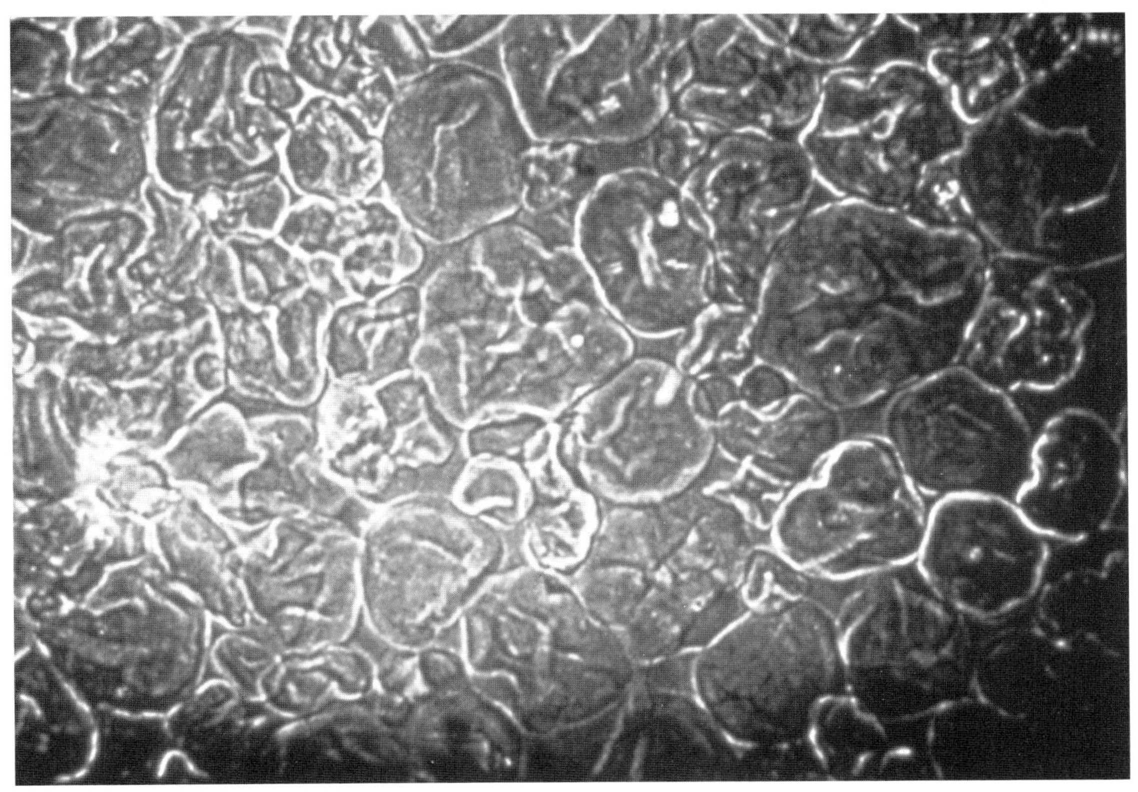

Figure 7. Gelatinization of cornstarch. Same group of granules as in Figure 6, viewed on microscope hot stage at 85 °C under normal illumination.

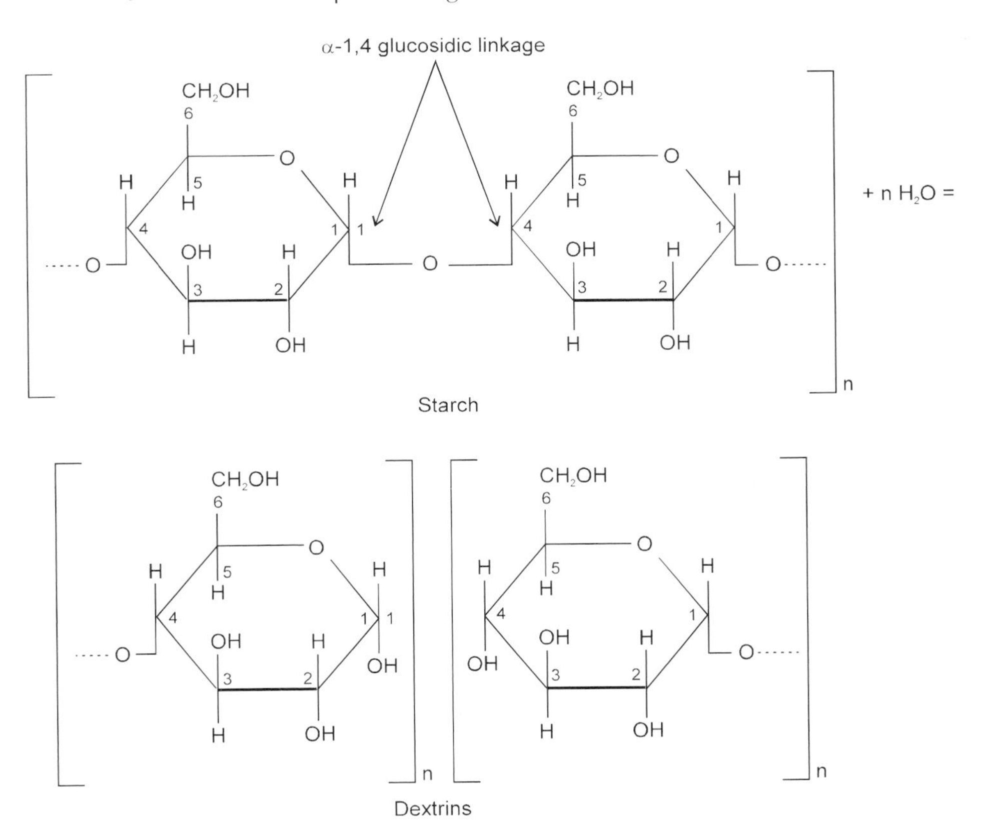

Figure 8. Hydrolysis of the α-1,4 glucosidic linkage in starch.

hydrolysis of starch using the endoenzyme α-amylase. α-amylase works randomly and rapidly to catalyze hydrolysis of the starch molecule. The dextrins will be of varying chain lengths. However, the shorter the chain length the less work remaining for the exoenzyme glucoamylase, which releases single glucose molecules by hydrolyzing successive α-1,4 linkages beginning at the non-reducing end of the dextrin chain. Glucoamylase also hydrolyzes α-1,6 branch linkages, but at a much slower rate.

Premixing, cooking and liquefaction

In considering all the different processes that make up cooking, it should first be explained that there are a variety of types of both batch and continuous cooking systems. For a batch system there is usually only one tank, which serves as slurrying, cooking and liquefaction vessels. Live steam jets are typically installed in the vessel to bring the mash to boiling temperature along with cooling coils to cool the mash for liquefaction. Figure 9 shows a typical batch cooking system.

In the batch cooking system, a weighed quantity of meal is mixed into the vessel with a known quantity of water and backset stillage. These constituents of the mash are mixed in simultaneously to ensure thorough mixing. The quantity of liquid mixed with the meal will determine the eventual alcohol content of the fermented mash. When a distiller refers to a '25 gallon beer', it means 25 gallons of liquid per bushel of cereal. For example, for a corn distillery with an alcohol yield of 2.5 gallons of absolute alcohol per bushel, the 25 gallons of liquid would contain 2.5 gallons of alcohol. Therefore it would

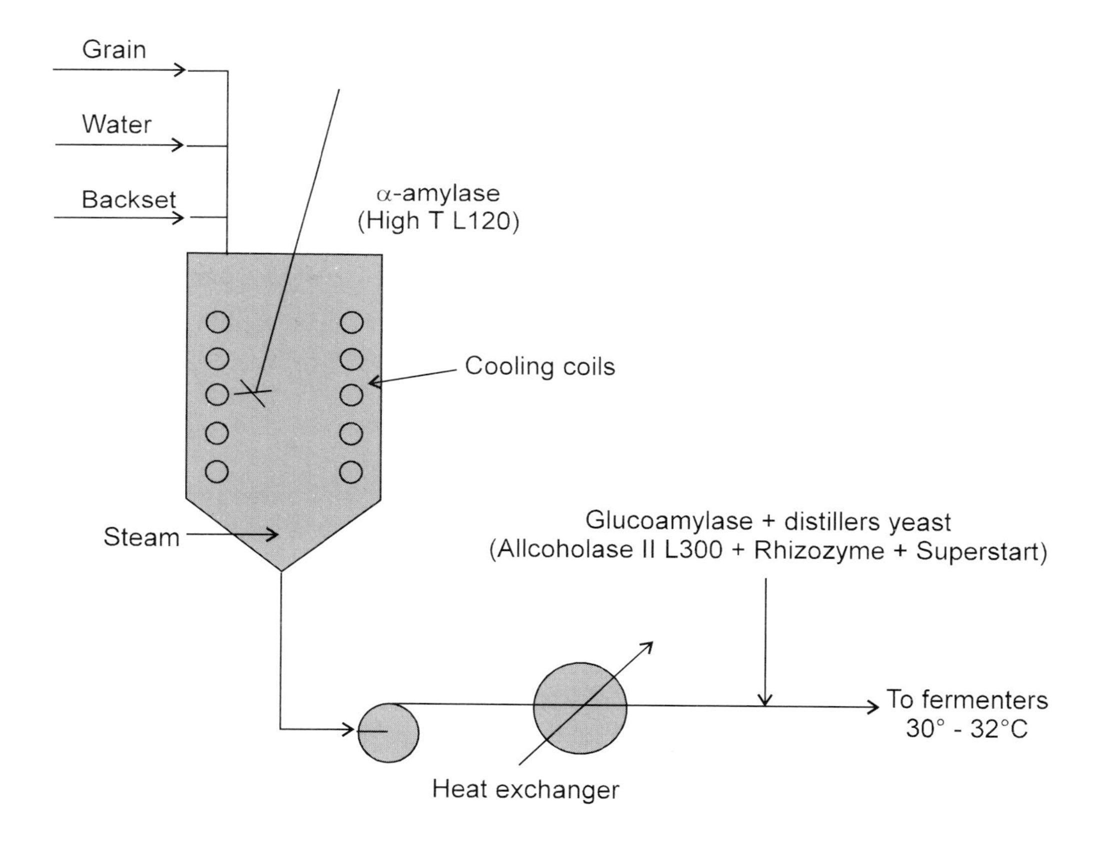

Figure 9. Batch cooking system.

contain 10% alcohol by volume. Using the distillery alcohol yield, the distiller can determine the quantity of cereals and liquid to use. Most distilleries operate with beers in the 10-15% alcohol range, although some beverage plants run at alcohol levels as low as 8%.

In the batch cooking system, a small quantity of α-amylase is added at the beginning (0.02% w/w of cereal) to facilitate agitation in the high viscosity stage at gelatinization. After boiling, usually for 30-60 minutes, the mash is cooled to 75-90 °C and the second addition of α-amylase made (0.04-0.06% w/w cereal). Liquefaction then takes place, usually over a holding period of 45-90 minutes. The mash should always be checked at this stage to make certain that no starch remains. Starch produces a blue or purple color with iodine. Mash should not be transferred from the liquefaction hold until it is 'starch-negative'.

The pH range for efficient α-amylase usage is 6.0-6.5. Therefore, mash pH should be controlled in this range from the first enzyme addition until the end of liquefaction. The glucoamylase enzyme has a lower pH range (4.0-5.5), so after liquefaction the pH of the mash should be adjusted with either sulfuric acid or backset stillage, or a combination of the two.

The quantity of backset stillage as a percentage of the total liquid varies from 10 to 50%. On one hand, the backset stillage supplies nutrients essential for yeast growth. However too much backset stillage can result in the oversupply of certain minerals and ions that suppress good fermentation. Especially noteworthy are the sodium and lactate ions. Sodium concentrations above 500 ppm or lactate above 0.8% inhibit yeast growth and can slow or possibly stop the fermentation prematurely. Overuse of backset must be avoided to prevent serious fermentation problems.

In the continuous cooking process (Figure 10) meal, water and backset stillage are continually fed into a premix tank at a temperature just below that of gelatinization. The mash is pumped continuously through a jet cooker, where the temperature is instantly raised to 120 °C. It then

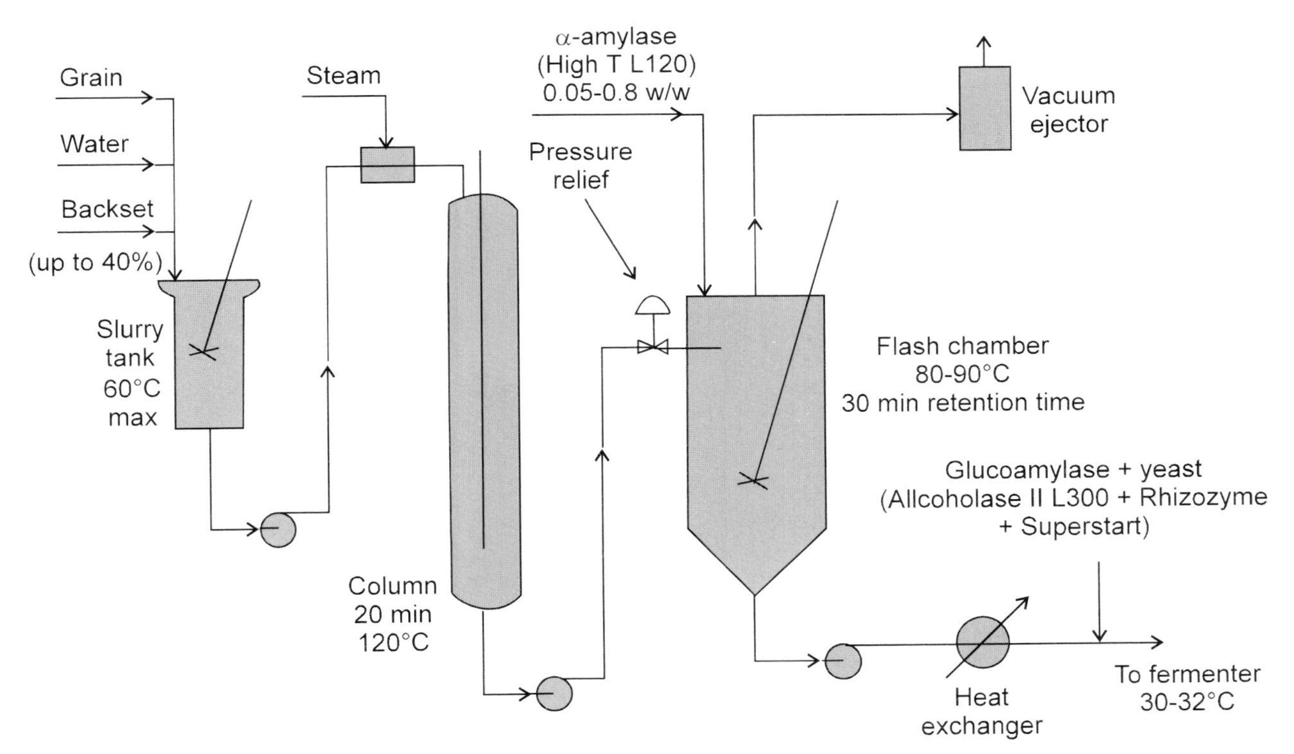

Figure 10. Continuous columnar cooking system.

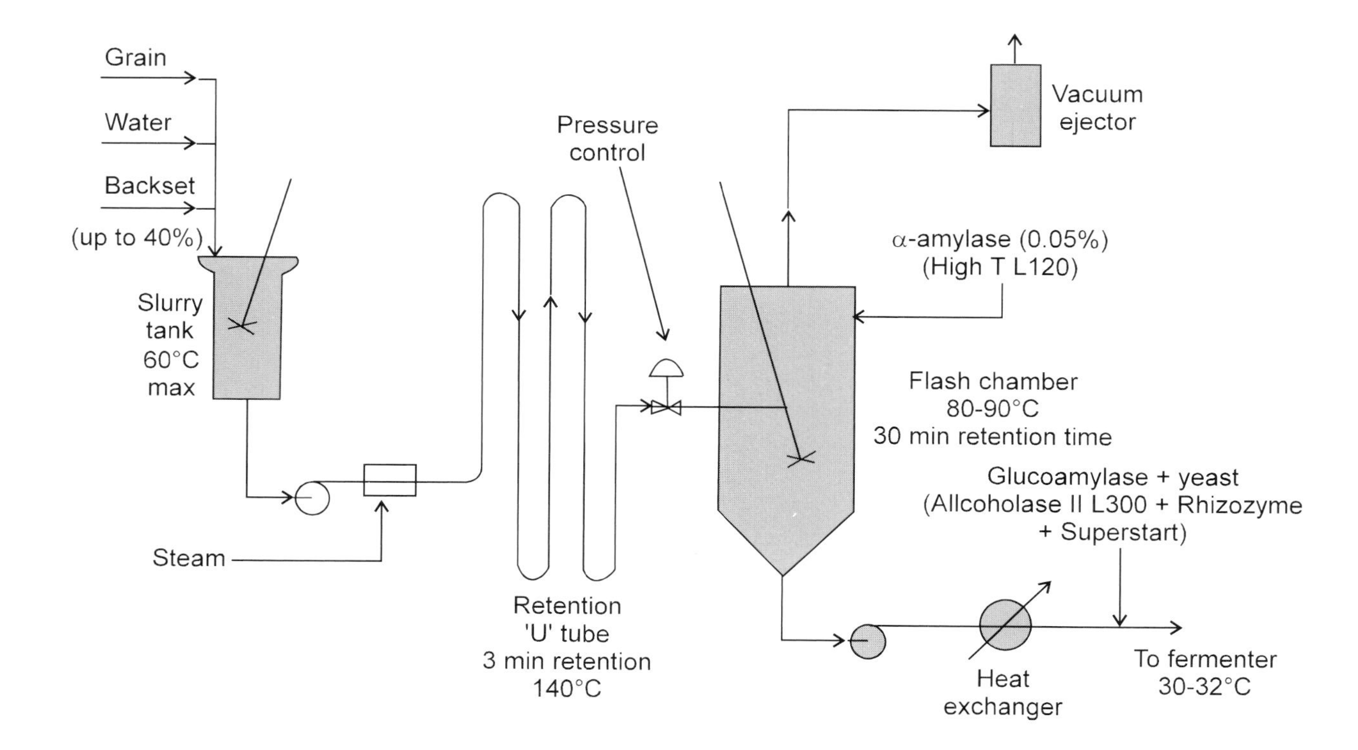

Figure 11. High-temperature, short-time, continuous U-tube cooking system.

passes into the top of a vertical column. With plug flow, the mash moves down the column in about 20 minutes and passes into the flash chamber for liquefaction at 80-90 °C. High temperature-tolerant α-amylase is added at 0.05-0.08% w/w cereal to bring about liquefaction. The retention time in the liquefaction/flash chamber is 30 minutes. The pH from slurrying through to the liquefaction vessel must be controlled within the 6.0-6.5 range. The greatest advantage of this system is that no enzyme is needed in the slurrying stage, leading to significant savings in enzyme usage. From the liquefaction chamber, the mash is pumped through a heat exchanger to be cooled for saccharification or fermentation.

The continuous U tube cooking system (Figure 11) differs from the columnar cooking system in that the jet cooker heats the mash to 120-140°C prior to being transferred through a continuous U tube. The retention time in the U tube is only three minutes, after which it is flashed into the liquefaction vessel at 80-90°C and the enzyme is added (high temperature-tolerant α-amylase 0.05-0.08% w/w cereal). The residence time in the liquefaction vessel is a minimum of 30 minutes.

The main advantage of this system is the relatively short residence period in the U tube. If properly designed there is no need to add any α-amylase enzyme in the slurrying stage. However, because of the relatively narrow diameter of the tubes, some distillers add a small amount of enzyme to the slurry tank to guarantee a free flow.

The relative heat requirements of the three cooking systems can be seen in Table 5. Surprisingly, the batch system is the most energy-efficient. Batch systems also generally use less enzyme than the other systems, possibly due to the difficulty of accurate dosing and good mixing with the continuous systems. The main disadvantage of the batch system compared to the continuous system is the poor utilization or productivity per unit of time. The temperature-

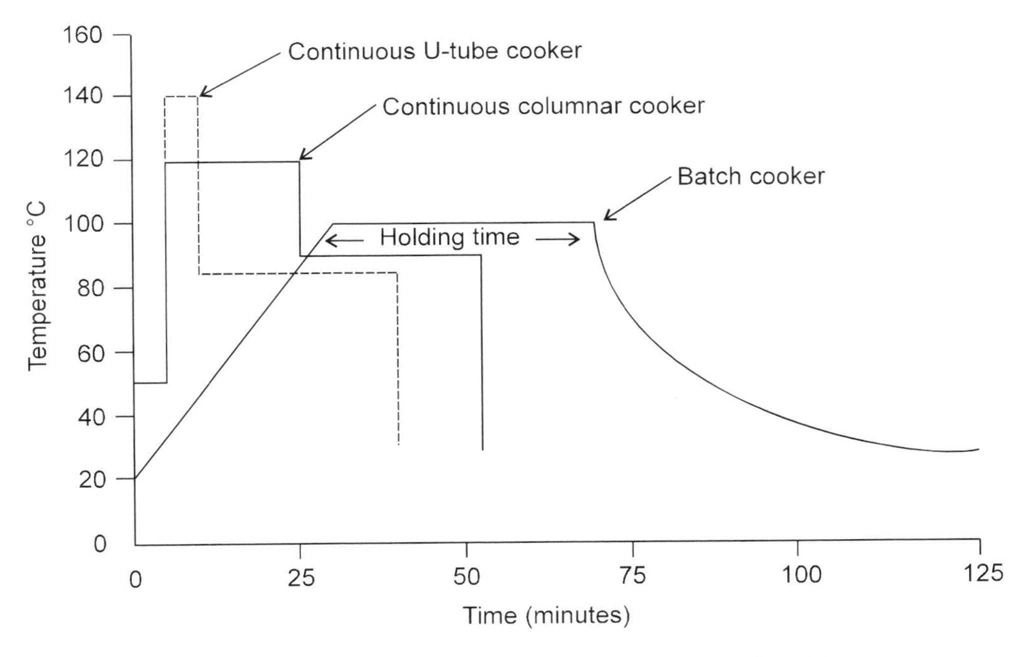

Figure 12. Temperature-time sequences in various types of cooking systems.

time sequences for the three systems shown in Figure 12 demonstrate how much more efficiently time is used in continuous systems compared to the batch system.

Table 5. Relative heat requirements of cooking systems.

Batch	1
Continuous columnar	1.18
Continuous U-tube	1.37

In the continuous systems, the flow diagrams show steam addition to raise the mash temperature. This temperature increase is brought about instantaneously by a jet cooker or 'hydroheater' as shown in Figure 13.

One purpose of the cooking process is to cleave the hydrogen bonds that link the starch molecules, thus breaking the granular structure and converting it to a colloidal suspension. Another factor in the breakdown of starch is the mechanical energy put into the mash via agitation of the different vessels in which the cooking process takes place. Well-designed agitation is very important in a cooking system; and the problem is intensified when plug flow is also desired.

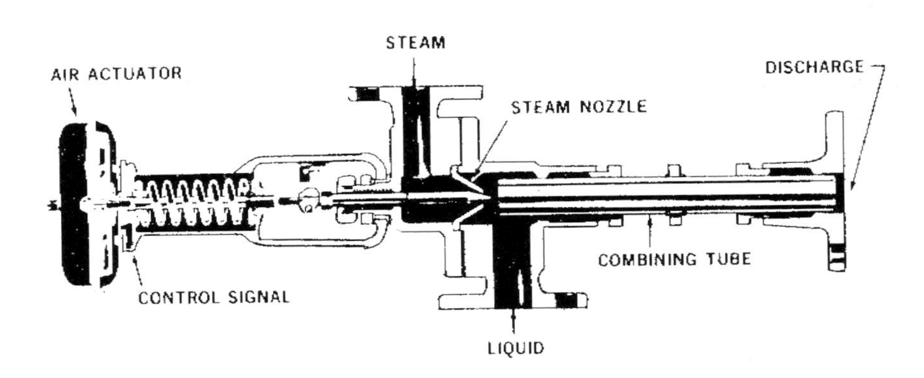

Figure 13. Automatic hydroheater.

Mash viscosities give an indication of the relative ease or difficulty with which some cereals are liquefied. Figure 14 compares viscosity against temperature for corn and waxy maize (amioca) and demonstrates the difference in viscosity profiles.

All of the cooking systems described require the addition of enzymes at least for the liquefaction stage where most of the hydrolysis takes place. Many distilleries now use a high temperature-tolerant α-amylase produced by the microorganism *Bacillus lichenformis*. The optimum pH range for this enzyme is between 6.0 and 6.5, although it shows good stability up to pH 8.5 (Figure 15) while the optimum temperature range is 88°C-93°C (Figure 16). Typically, this type of enzyme would be used at between 0.06% and 0.08% by weight of cereal. Where it is necessary to add some α-amylase enzyme to the slurrying vessel, the dosage rate may be slightly higher.

The reaction time for enzyme-catalyzed reactions is directly proportional to the concentration of enzyme. Consequently, distillers wishing to minimize the quantity of enzyme used should design equipment to have long residence times to allow the reactions to be completed with the minimal dosage of enzyme.

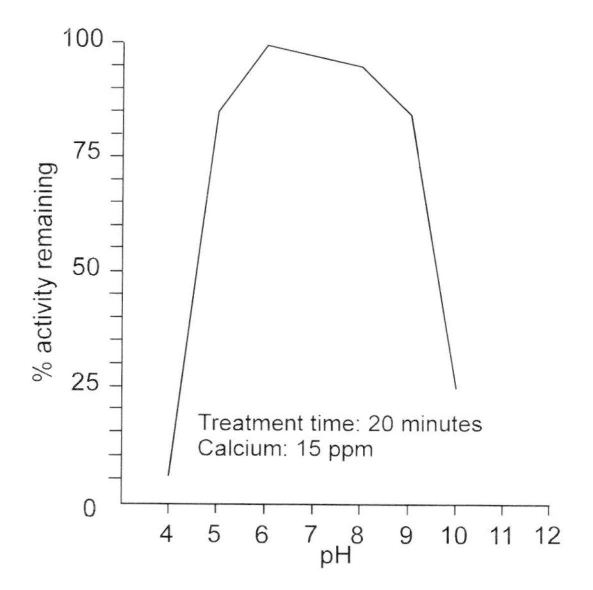

Figure 15. Effect of pH on the activity of α-amylase.

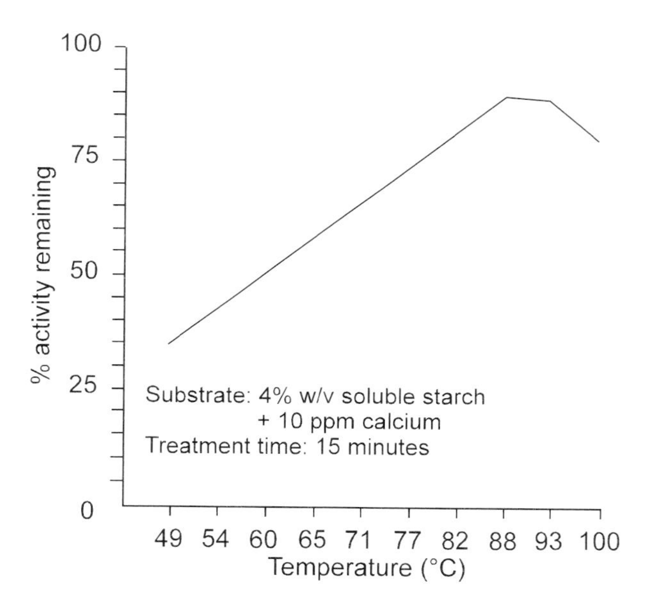

Figure 16. Activity of high temperature-tolerant α-amylase in relation to temperature.

Saccharification - yes or no?

Saccharification of distillery mashes is a somewhat controversial subject. Over the last ten years many distillers have changed from saccharifying mash in a dedicated saccharifi-cation vessel (or sacc' tank) to adding the saccharifying enzyme directly to the fermenter in a process referred to as Simultaneous Saccharification and Fermentation. Saccharifi-cation in a separate

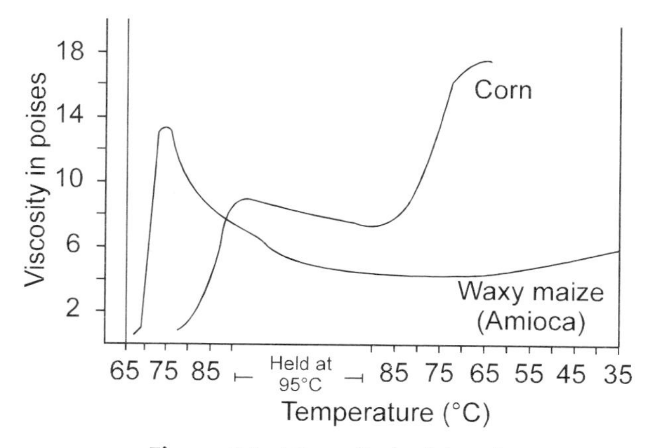

Figure 14. Viscosity build up in cooking corn meal.

vessel is still the practice in some distilleries, particularly for beverage production.

A further complication occurs if Rhizozyme™, a surface culture enzyme, is used. Rhizozyme™ has an optimum pH of 3.5-4.5 and an optimum temperature of 30-35°C. As such, it is a glucoamylase more suited to a distillery fermenter (Table 6). Rhizozyme™ has particularly found favor in SSF situations because the enzyme can work at near its optimum in the fermenter. The enzyme also contains side activities that assist in releasing more carbohydrate and protein (Table 6).

Table 6. Advantages of Rhizozyme™ over conventional glucoamylase.

	Conventional glucoamylase	*Rhizozyme™*
Temperature optimum, °C	60+	30+
pH optimum	4.5-6	3.5-4.5
Amylase activity,		
SKB units	None	50,000
Cellulase activity,		
CMC-ase units	None	2500
Amylopectinase, AP units	None	5000

If saccharification step is used

Mash from the liquefaction vessel is cooled, usually to 60-65°C, and transferred to a liquefaction vessel where the glucoamylase (amyloglucosidase) enzyme is added. This exoenzyme starts hydrolyzing the dextrins from the non-reducing end of the molecule and progressively, though slowly compared to endoenzymes, releases glucose. The saccharifying process is usually carried out with a residence time of between 45 and 90 minutes and the glucoamylase is added at 0.12% by weight of cereal used. Some distillers actually measure the quantity of glucose produced by measuring the dextrose equivalent (DE) of the mash. A DE of 100 represents pure glucose, while zero represents the absence of glucose. This test is rarely used nowadays, as many distillers have high performance liquid chromatography systems that can measure sugars directly. Recent experience, however, shows that DE, provided it is above 10, is of no concern. In focusing on 23% ethanol, the key in any cooking and liquefaction process is to liquefy, i.e., lower viscosity, so that mash can be pumped through heat exchanger to the fermenter (for SSF) or to the saccharification tank.

The functional characteristics of glucoamylase, which is usually prepared from the microorganism *Aspergillus niger*, can be seen in Figures 17-20. Two parameters, temperature and pH, dictate how enzymes can be used. While liquefaction is carried out at a pH of 6.0-6.5 and a temperature of 90°C, this is not at all acceptable for saccharification. The pH must be in the 4-5 range for saccharification; and the optimum temperature for the glucoamylase activity is 75°C. The mash, therefore, must be acidified with either sulfuric acid or backset stillage or both before addition of the glucoamylase. Temperature must also be adjusted. As mentioned previously, normal mash saccharification temperature is 60-65°C; although for microbiological reasons 70-75°C would be preferable. Lactobacillus can survive at 60°C; and frequent infection of saccharification systems has caused many distillers to change to saccharifying in the fermenter as described below.

If no saccharification step is used

If no saccharification step is planned, the liquefied mash is simply cooled from 90° C through a heat exchanger and transferred to the fermenter. A portion of the liquefied mash is diverted to a yeast starter tank where yeast, glucoamylase and the RhizozymeTM is added. Conventionaly glucoamylase (L300) is added at 0.08% with 0.01% Rhizozyme™ recommended as a supplement. Rhizozyme™ alone can be added at 0.05%.

Raw materials

Corn

Corn contains 60-68% starch and is certainly the most widely used cereal in dry milling operations. It is easy to process from cooking through fermentation. A bushel of corn weighs 56 lbs and generally contains approximately 32 lbs of starch, which is present in the endosperm portion of the kernel in the form of granules. When hydrolyzed, this starch yields about 36 lbs of glucose. (The weight increases as water is taken up in the hydrolysis process.)

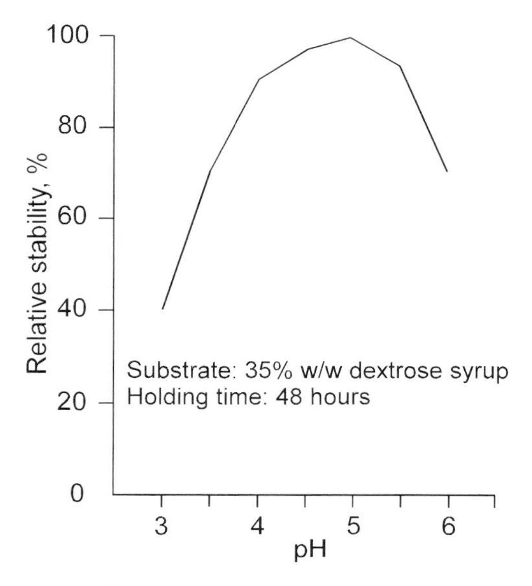

Figure 17. Influence of pH on the stability of glucoamylase

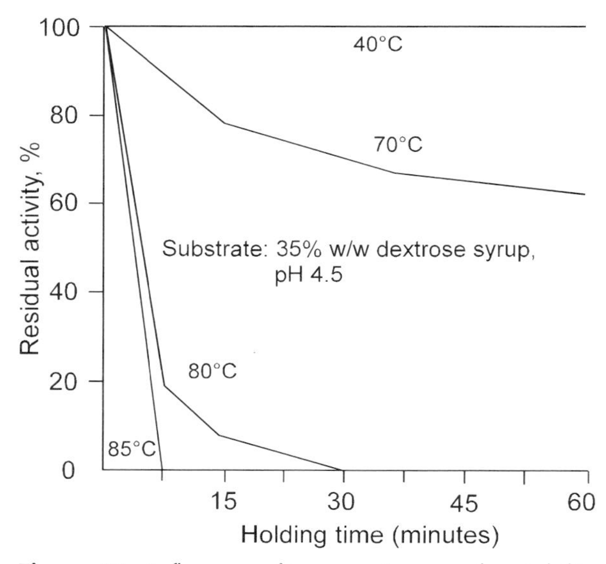

Figure 19. Influence of temperature on the stability of glucoamylase.

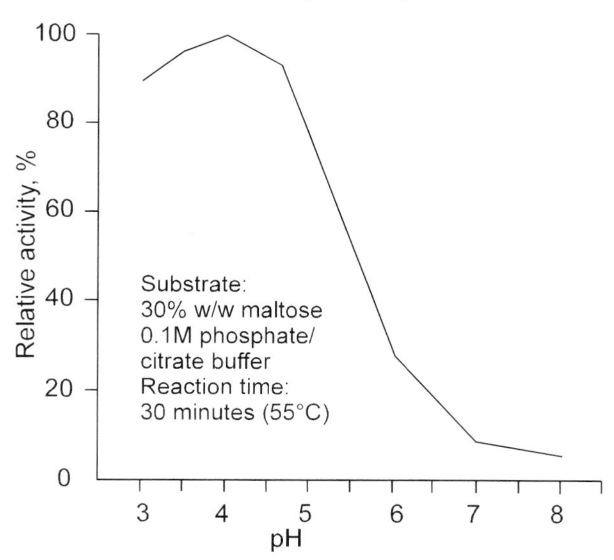

Figure 18. Influence of pH on the activity of glucoamylase.

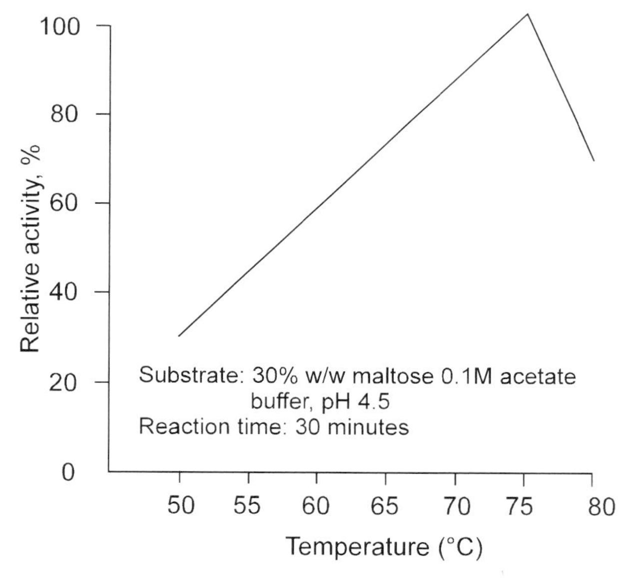

Figure 20. Influence of temperature on the activity of glucoamylase.

Barley

Barley is used in many countries as a raw material for alcohol production both in its ungerminated form and in its germinated state when it is called barley malt. This cereal contains starch at levels of 55-65% of dry weight. The starch is easily processed using methods similar to those used for corn, although barley is much more abrasive to equipment than corn due to its high fiber content.

Both the two row and six row varieties of barley contain high levels (1-4%) of the polysaccharide gum known as betaglucan. This is a very viscous gum and can lead to processing problems unless the mash is treated with a betaglucanase enzyme to hydrolyze the gum to glucose.

Alcohol yields from barley are slightly lower than for corn, normally 2.2-2.3 US gallons of anhydrous alcohol per 56 lb bushel.

Milo (millet or grain sorghum)

Milo has a smaller kernel than corn but yields about the same quantity of alcohol per bushel. The cereal is treated in the same way as corn. Occasionally foaming problems occur during fermentation; and the distillers dried grain has a slightly different color. As milo is purchased at a lower price per bushel than corn, the alcohol may be produced at a lower raw material cost. An important feature of milo fermentation is the formation of a crusty head above the liquid, which requires constant agitation to break up. For this reason, some distillers who use milo regularly install an extra agitator just below the fermenter's fill level.

Rye

Rye contains almost as much starch as corn, and is used for alcohol production in rye-producing regions. The alcohol yield is 2.4 gallons per bushel. It is an unusual cereal in that ungerminated rye contains a high level of α-amylase and the mash can almost be liquefied without the addition of α-amylase from external sources. During cooking, a hold at 65-70°C allows these enzymes to work. Normally microbial α-amylase preparations are added at 0.03% w/w rye. The fermenting mashes have a tendency to foam; and rye also contains gums that lead to serious viscosity problems. Treatment of rye mashes with betaglucanase enzyme helps reduce the viscosity, but the problem is not as easy to solve as with barley. The bitter taste of rye results in a distillers dried grain with a different character to that of corn, but the plants in North America that use rye exclusively have no difficulty in selling this by-product.

Wheat

While vast quantities of wheat are grown in the US and Canada, little has been directly used in distilleries because it generally tends to be more expensive than corn. The by-product wheat gluten may be extracted before converting the starch to alcohol as shown in the wheat gluten processing system in Figure 21.

Depending on the end use of gluten, it can be extracted by washing the wheat starch with water or by dissolving in ammonium hydroxide. Water is used when the gluten is intended for bread baking, while ammonia is used when the gluten is intended for use as a protein supplement or is to be processed further. When the gluten is removed, the resulting product may be deficient in free amino acid nitrogen and yeast food may need to be added to the mash to ensure a satisfactory fermentation.

Recommendations

Given the goals and the variable involved, which cooking system should be chosen? A comparison of systems used in four distilleries demonstrates the diversity of approach possible, yet points out many similarities (Table 7). It would appear that there are three schools of thought:

a) A long liquefaction period (1-2 hrs at 90+ ^{0}C) prior to the high temperature jet cooker.
b) A short liquefaction prior to, but a long liquefaction period (1-2 hrs at 90+ ^{0}C) after the high temperature jet cooker.
c) No saccharification step.

From the experience of the authors, all the systems described in Table 7 work; however the saccharification step is used to suit the glucoamylase. Furthermore it runs the risk of infection and over-production of glucose (detrimental to the yeast). A long liquefaction period (pre- or post-jet cooking), no saccharification and addition of glucoamylase plus Rhizozyme™ to yeast starter or fermenter would be recommended, and is now in use in many distilleries. However, if the objective is to 'spoon-feed' the yeast with glucose and hence achieve high alcohol levels, saccharification should not be used.

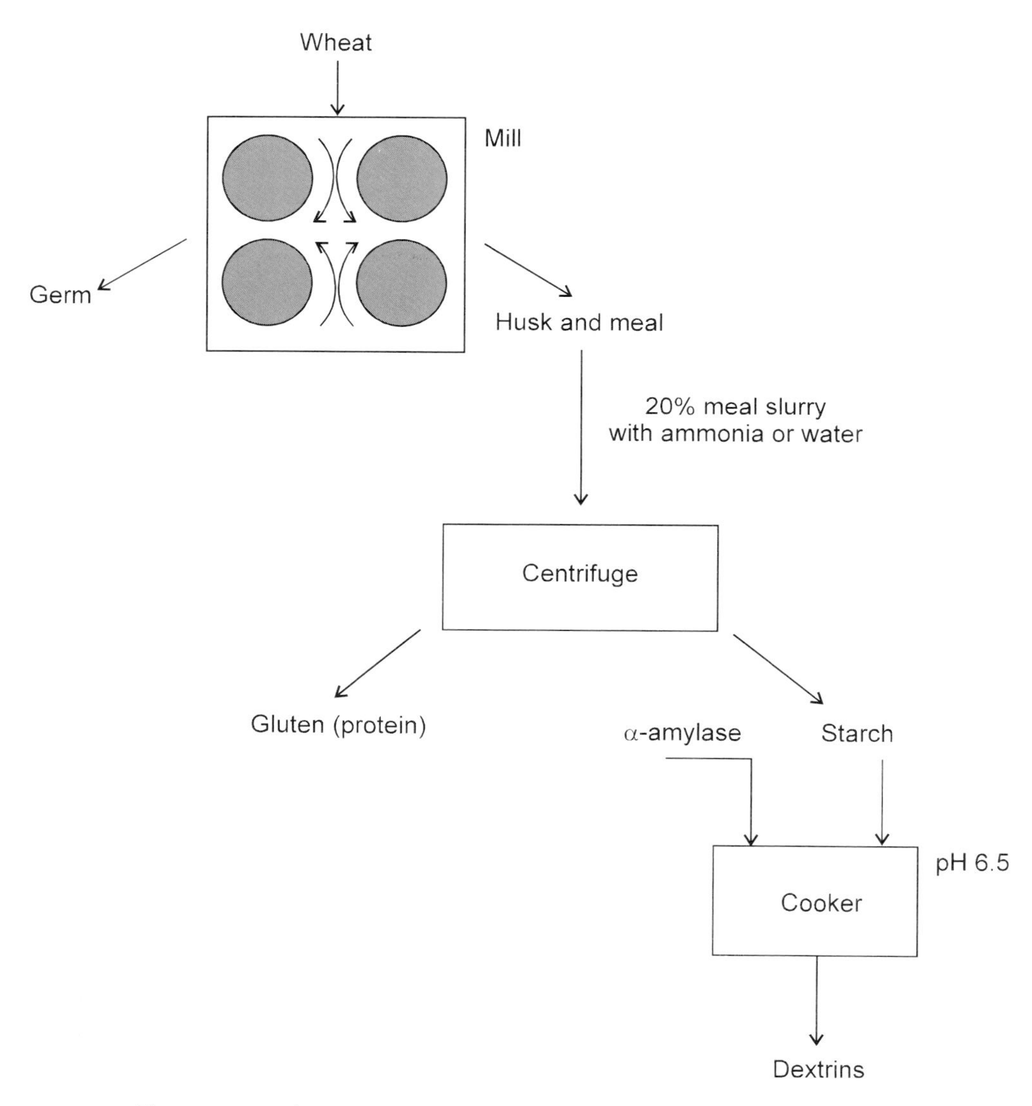

Figure 21. Wheat processing including gluten separation.

Table 7. Variations in the cooking process used by different distilleries.

Steps and functions	*Distillery 1*	*Distillery 2*	*Distillery 3*	*Distillery 4*
Slurry · Mix corn with water · Hydrate starch · Adjust pH	140°F	140°F - Add 1/3 α-amylase (0.03%)	135°F - Adjust pH to 6.15 - Add 50% α-amylase	140°F Adjust pH to 5.8-6.0 Add lime
Pre-liquefaction · Not always carried out · ⅓-½ low or high temperature enzyme · Temperature near point where starch gels · Time: 6 minutes to 2 hrs	180°F - Add 50% α-amylase - (0.04% High T L120) - 30 minutes	190°F	195°F - 6 minutes	No pre-liquefaction
Liquefaction · Typically jet cooker · Short time/high temperature under pressure	220°F for 5 minutes, release vacuum	220°F for 6 minutes, release vacuum	220°F for 6 minutes, release vacuum	220°F release vacuum
Post-liquefaction · Release vacuum, temperature drops · Add ⅓-½ enzyme · Typically 180-190°C · Time: 1-2 hrs	190°F - Add 50% α-amylase (0.04% High T L120) - Cool	185°F - Add 2/3 α-amylase (0.06%) - Cool	185°F - 2 hrs - Add 50% α-amylase - Cool	185°F - 2 hrs - Cool
Saccharification · Generate sugars: is it necessary? · Temperature: 140°F · Add glucoamylase from Aspergillus (not Rhizozyme)	140-150°F - Add glucoamylase (0.07% Allcoholase L300)	No saccharification step	No saccharification step	No saccharification step
Fermentation · Typically 90°F · Add glucoamylase (Aspergillus or Rhizozyme)	90°F Add 0.01% Rhizozyme	90°F Add glucoamylase (0.12% Allcoholase II L300)	90°F Add glucoamylase (0.08% Allcoholase II L400) Add 0.01% Rhizozyme	90°F Add protease Add glucoamylase (0.08% Allcoholase II L300) Add 0.01% Rhizozyme

Chapter 3

Management of fermentations in the production of alcohol: moving toward 23% ethanol

D.R. Kelsall and T.P. Lyons
Alltech Inc., Nicholasville, Kentucky, USA

Introduction

Fermentation is the critical step in a distillery. It is here that yeast converts sugar to ethyl alcohol; and it is here that contaminating microorganisms find an opportunity to divert the process to lactic acid, glycerol, acetic acid, etc. The fermentation step (and to a much lesser degree the cooking process providing the DE is kept low) is also where yeast must be 'coached' to produce 23% ethanol. Fermentation will also have a direct bearing on downstream processing. If sugar levels remaining in beer are too high, evaporative capacity may be impaired or syrup contents increased. Both can lead to distillers grains being more difficult to dry and altered in color. Truly the fermentation step is the heart of the distillery.

When reflecting on the overall flow diagram of the distillery (Figure 1) the central role of yeast is obvious. The process of converting sugars to ethanol takes between 10 and 48 hrs during which heat is generated. The various factors affecting the efficiency of that process need to be considered. In a separate chapter the chemistry of ethanol production by yeast will be covered in detail. In this chapter we will restrict ourselves to a discussion of managing the stress factors affecting ability of yeast to produce ethanol.

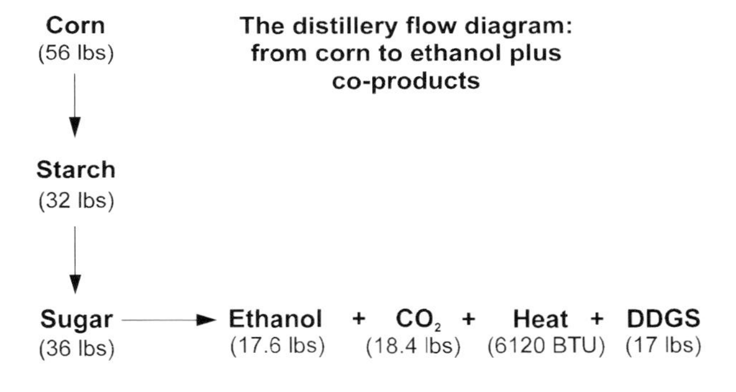

Figure 1. Distillery product flow.

Stress factors affecting alcohol yield

Temperature: the first stress factor

Inability to precisely control the temperature of fermentation is possibly the biggest and most commonly encountered factor or problem affecting alcohol yield. Almost all distilleries suffer from some degree of inadequate design, and probably the most common fault is underestimation of the amount of energy released during fermentation. Consider the general chemical equation for the conversion of glucose to alcohol as illustrated in Figure 1.

Generation of the 6,000 BTU for about 17 lbs of ethanol largely takes place between hours 10 and 30 of fermentation. In other words, the fermenter cooling system should be designed to cool 17,000 BTU per 50 lbs of ethanol over a 20 hr period. Usually this amount of cooling is not designed into the system and the fermenters become overheated.

The theoretical temperature rise for a 25 gallon beer (gallons of water per bushel of grain) is 22°C per bushel (Figure 2). The optimum temperature of fermentation when fermenting with yeast *(Saccharomyces cerevisiae)* is 32°C and the optimum temperature of reproduction for the yeast is 28°C. It can also be seen (in Figure 2) that without adequate cooling, a fermenter set at 32°C would inevitably increase in temperature. Every degree in temperature above 32°C will have a tendency to depress fermentation, as the yeast cannot handle these higher temperatures. Furthermore, the increase in temperature favors the growth of lactobacilli, the microorganism that competes with yeast for glucose.

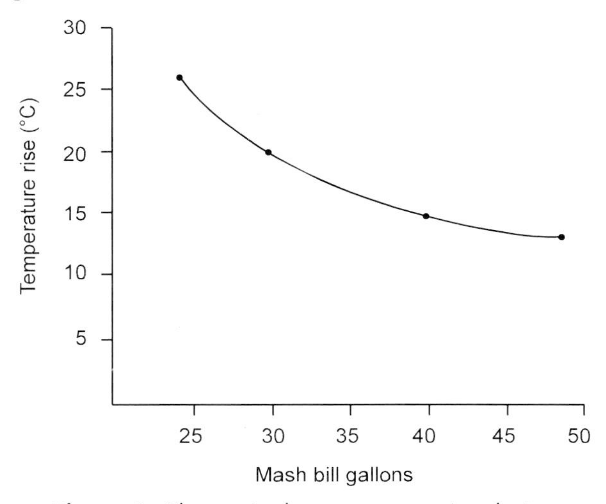

Figure 2. Theoretical temperature rise during fermentation.

Combining enzyme activities to control sugar delivery to yeast

How can we best control fermentation temperatures? The easiest way to reduce fermenter temperature would be to reduce the sugar level going in and therefore reduce yeast growth, however less alcohol would be produced. Since the objective is more, not less alcohol, this is not an option. However, sugar level, or more importantly sugar type, is important.

Yeast must be maintained in a growth (budding) phase, since their ability to produce alcohol is more than 30 times greater during growth than in non-growth mode. Yeast however can respond to high glucose levels in two ways. Cell growth is either inhibited by high glucose levels or yeast may grow rapidly and then stop. In either case there is an almost 'OD' or over-dosing effect on fermentation efficiency or possibly a spike in temperature. Glucose must be 'spoon-fed' to the yeast; and for this reason it is recommended that glucoamylases be used both sparingly and in conjunction with an enzyme produced in surface culture called Rhizozyme™. The combination provides ideally balanced sugar delivery to maintain good yeast growth and maximum alcohol productivity while avoiding high temperature peaks. Glucoamylase (Allcoholase II L300) provides glucose at the start of fermentation while the Rhizozyme™, with pH and temperature optimums more closely aligned to fermentation conditions, continues to work during fermentation. The net result is a steady, slow release of glucose and a steady formation of alcohol by yeast (Figure 3).

Equipment for heat control

Despite the sequential feeding of glucose, heat is generated during fermentation and must be removed. Four types of heat exchangers are typically used; and these are illustrated later in this chapter when fermentation design is discussed.

a) *Internal cooling coils.* These are the least desirable option because they are practically impossible to clean and sterilize.
b) *Internal cooling panels.* Like cooling coils, panels are suspended in the fermentation vessel. Also like cooling coils, cooling is adequate but sterilization may be difficult.

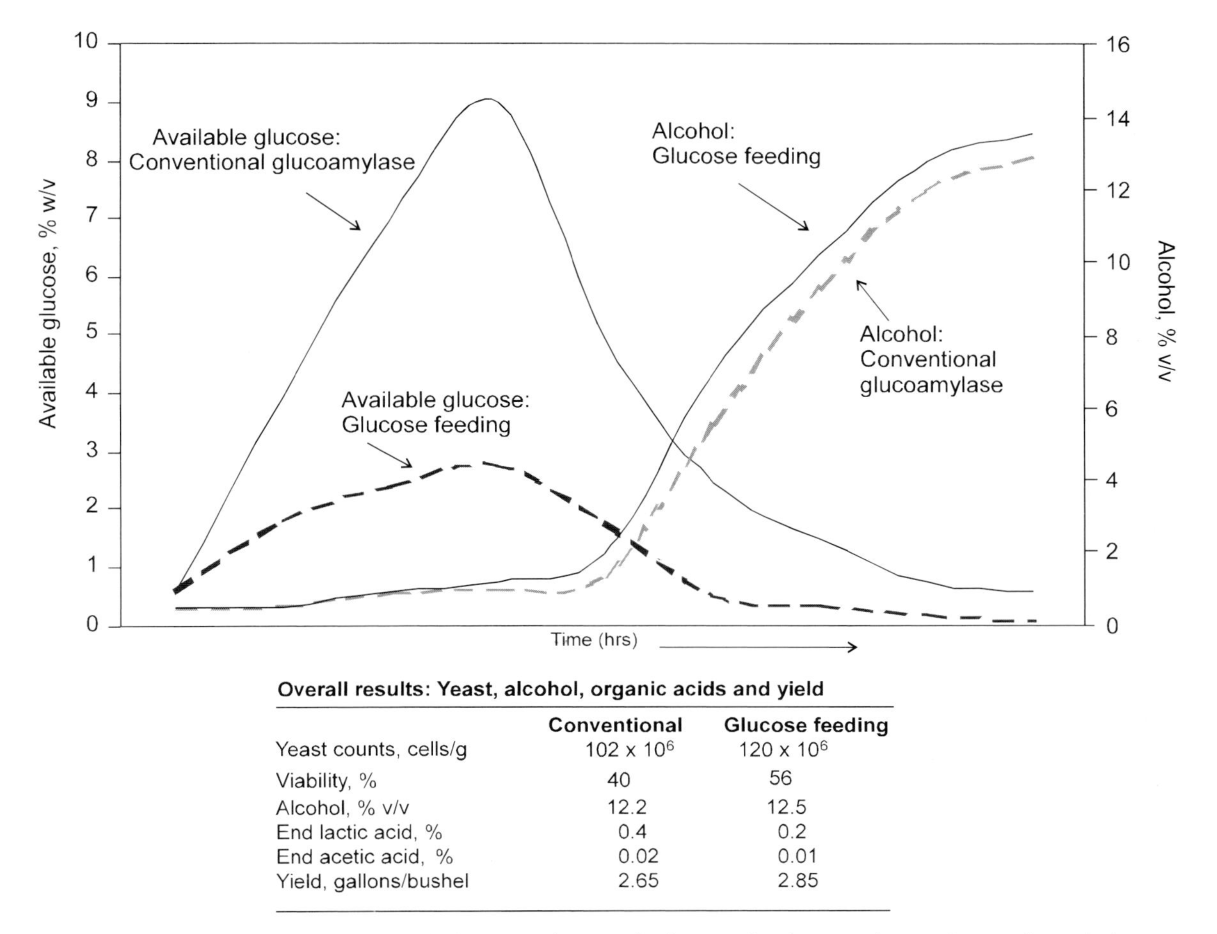

Overall results: Yeast, alcohol, organic acids and yield

	Conventional	Glucose feeding
Yeast counts, cells/g	102 x 10^6	120 x 10^6
Viability, %	40	56
Alcohol, % v/v	12.2	12.5
End lactic acid, %	0.4	0.2
End acetic acid, %	0.02	0.01
Yield, gallons/bushel	2.65	2.85

Figure 3. Comparison of conventional glucoamylase and 'glucose feeding' with a surface-cultured Rhizopus enzyme on glucose and alcohol levels during fermentation (Allcoholase II L400 added at 0.1% (conventional) vs Rhizozyme™ (glucose feeding) added at 0.05%; both enzymes added to the fermenter).

d) *External cooling jacket.* Cooling jackets, often dimpled for added contact with the fermentation, can cover either part of the vessel or the entire vessel. Recirculating coolant through the jacket allows fermentation temperatures to be monitored. If jackets are spaced out over the fermenter sides, then cooling can commence as the fermenter is filled.

e) *Heat exchanger with recirculation.* Contents of the fermenter are pumped through a heat exchanger (typically spiral or plate-and-frame) and returned to the fermenter. This has the added advantage of acting as a means of keeping the fermenter agitated; and nutrients (sterol donors, peptides, oxygen, etc.) can be introduced at critical times. It is suggested that optimum control of temperature can be obtained by placing spiral heat exchangers at early stages in a continuous fermenter where heat generation is high; while plate-and-frame exchangers can be placed at later stages where alcohol production has subsided.

Of the four types, the cooling jacket is best for mimimizing infection, but the heat exchanger is the most effective at heat control. As the distiller moves toward 23% ethanol, heat recovery and maintenance of low temperatures will become even more important as temperature rises of 6-8^0 will become the norm. A word of caution: some distilleries, particularly in continuous operations, share external heat exchangers bet-ween fermenters in order to economize. Based on the experience of many existing distilleries using this sharing concept, infection can easily be transferred among fermenters.

Infection: the second stress factor

The various types of infection are described in detail elsewhere in this book, however several points are relevant in this context. Lactobacilli consume glucose to produce lactic acid, which is the second major factor affecting the yield of alcohol in fermentation (Table 1). Yeast do produce some organic acids during fermentation, but concentrations are relatively low compared to those produced by lactobacilli and other contaminating bacteria. As a general rule of thumb, if the titratable acidity of an uninoculated mash is X, then the titratable acidity of the fermented beer is usually about 1.5X. This is a general rule and seems to work quite well in practice, although it is not based on known scientific principles. The 1.5X represents the titratable acidity of the finished beer when there is no significant contamination from lactobacilli. When lactobacilli are active, the generation of lactic acid substantially increases the titratable acidity and often the high acid content will cause the yeast fermentation to stop or dramatically slow down.

Practical means of controlling infection

The most fundamental way to control infection is to control management (Table 2). Fermenters must be cleaned on a regular basis; and designs must avoid sharing of heat exchangers and dead-legs in piping. Fermenters should be designed to facilitate complete emptying. Scaling will occur regardless of the fermenter design, and should be removed using either a scale inhibitor (such as Scale-Ban) or a descaling chemical. Fermentations must be fast; and yeast should be added either in a form that will quickly rehydrate or preferrably one that has been 'started' in a pre-starter tank. Saccharification tanks can be source of infection as lactobacillus strains can grow and multiply at the temperatures used in this process. The goal is to make conditions as suitable for yeast growth as possible (without jeopardizing alcohol yield) and as unsuitable as possible for infectious microorganisms. By restricting glucose release in fermentation through use of Rhizozyme™ and lower levels of glucose-producing glucoamylase (typically 50-70%), the yeast uses the sugar as it is becomes available and less is available for lactic and acetic acid-producing bacteria. Equally, when yeast growth begins to tail off (typically at 20-24 hrs) sterol

Table 2. Practical means of controlling infection.

Avoid deadlegs in lines
Avoid sharing fermenter heat exchangers
Proper cleaning programs
Periodic use of descaling chemicals
Control fermentation temperatures
Avoid the saccharification step in cooking
Maximize yeast growth (budding)
Use yeast pre-starters
Use sterol donors and peptides as yeast foods in yeast propagator
Antimicrobials
Allpen (against Gram-positive organisms)
Lactoside (against Gram-postive and Gram-negative organisms)

Table 1. Inhibitory concentrations and typical levels of organic acids.

Acid	*Inhibitory level (%)*	*Typical level at the end of fermentation (%)*	*Comment*
Lactic	0.8	0.2-0.4	Most common acid produced by lactobacilli
Acetic	0.05	0.014-0.02	Produced by Acetobacter and Lactobacillus
Succinic	>0.3	0.1	Normal by-product of yeast

donors and oxygen, which at that stage have become limiting, will revive activity by ensuring that fatty acids are available for yeast reproduction.

Regardless of precautions, infection invariably becomes a problem on occasion. To tip the balance and maintain it in favor of yeast, some antimicrobial factors may need to be considered. Two products that have stood the test of time are Allpen and Lactoside. The former is an anti-Gram-negative factor effective at 1-2 ppm (1-2 g/1000 liters of beer) and Lactoside, a broader range antimicrobial used at 2-3 ppm (2-3 g/1000 liters of beer). In both cases the active ingredients are destroyed during distillation and no carryover occurs in finished products.

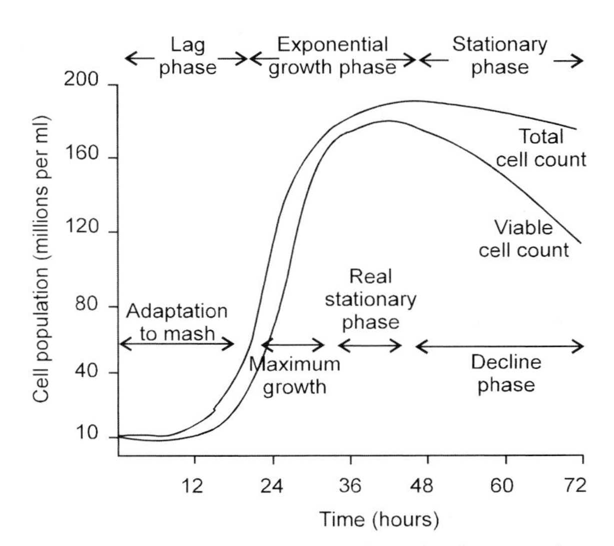

Figure 4. Typical yeast growth in distillery mash.

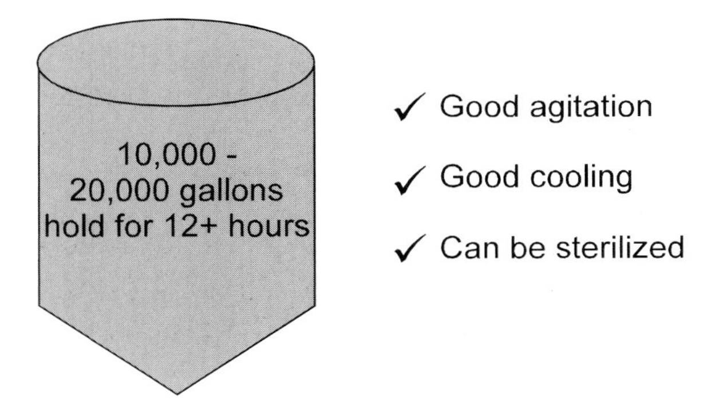

Figure 5. Yeast pre-starter tanks. A key factor in maximizing yeast cell numbers and improving alcohol levels.

Alcohol levels: a third stress factor

High alcohol beers have a tendency to quit fermenting. Understanding how alcohol levels act to stress yeast requires a working knowledge of yeast metabolism and growth. It has been shown that when yeast are in the reproductive phase they produce alcohol over 30 times faster than when not reproducing. Figure 4 shows typical yeast growth patterns or phases during fermentation. The first phase is the so-called lag phase, during which the yeast adapt to the fermenter environment. During this period there is little or no yeast growth and consequently little or no alcohol production. This period can last from 4 to as long as 12 hrs. Ways of managing and reducing the length of the lag phase are shown in Figure 5 in reference to yeast propagation and pre-starter tanks.

The second phase is the exponential growth phase. This is the most important phase where nearly all of the alcohol is produced. There is a limited time in which the yeast stay in this phase; and this is the factor limiting the quantity of alcohol produced during a given fermentation. The length of time during which the yeast can remain reproducing depends on the nutrition available in the fermenter.

The three major factors that significantly affect alcohol yield are therefore temperature, acidity of the mash and the alcohol level produced during fermentation. Yeast are tremendously resilient microorganisms and usually produce efficiently despite all the negative conditions imposed. Yeast fermentation will yield well if the temperature is a few degrees high or if the acidity is a little high indicating a mild infection. Beers containing up to 13% alcohol v/v can be produced without slowing fermentation. However, when changes in two or more of these factors happen simultaneously, serious losses in yield usually occur. For example, if the temperature is 35°C in a mash contaminated with 10 million CFU lactobacilli/ml, once the alcohol level reaches 8% it is very probable that the fermentation would stop only at two-thirds complete, with a corresponding 33% loss of yield. By feeding the glucose to the yeast as described before and by adding fresh mash to existing fermentations, yeast can be 'coached' to produce up to 17% ethanol in batch ferm-enters, 20+% in continuous fermentations.

Yeast is the powerhouse of any distillery; and without 'happy yeast' both alcohol percentages in the fermenter and alcohol yield will drop (Table 3). Prefermenters are a way to ensure good cell numbers in the fermenter. At the start of batch fermentation there should be a minimum of 50 millions cells/ml which should increase to 150-200 million cells/ml at the height of fermentation and drop to 100 million cells/ml at the end. Continuous fermenters should maintain cell numbers of 150 million from a high of 250-300 million cells/ml. Yeast viability as measured by methlylene blue stain should be 95% in the prefermenter and 90% at the start of fermentation. There will be a drop to around 50% viability at the end of fermentation. In a cascade continuous fermentation system achieving a final 17-18% ethanol yield, viability was seen to drop to 30% in the last fermenter.

Table 3. Keys to a good yeast prefermenter.

Keep Brix to 10-20°
Good agitation
Use active dried yeast
Use peptide-based yeast food
Achieve minimum of 200 ppm amino acids
Add Rhizozyme™ to 'spoon-feed' yeast glucose
Hold for 10-12 hrs with agitation
Recharge with yeast either:
every 12 hrs
every week (with yeast food every 12 hrs)
Achieve 300-500 million cells/ml before transfer to main fermenter

A typical charge for a continuous yeast prefermenter is shown in Table 4. The mash was sterilized and cooled to 90°F. Yeast was added with good agitation along with yeast food and saccharifying enzyme (Rhizozyme™). The mash was held for 12 hrs at which stage a cell count in excess of 500 million was achieved. 50% of the mash was transferred to the fermenter and a further charge of mash, yeast peptide food and Rhiozyme was added. The procedure was repeated and 12 hrs later a full charge transferred. At the end of one week the pre-fermenter was emptied, cleaned and sterilized. In this way the distillery is able to minimize yeast purchases while mazimumizing yeast counts and alcohol yields. In such a system yeast is essentially eliminated as a cost factor (<0.1 cents per gallon).

Table 4. Typical charge for yeast prefermenter for a 500,000 gallon fermenter.

Prefermenter size	20,000 gals
Mash	Direct from jet cooker: 27% solids
Active yeast	400 lbs Superstart
Peptide yeast food	20 lbs AYFP
Rhizozyme™	l lb
Yeast cell numbers expected	400-500 million
Time	12 hrs

Mycotoxins: a stress factor

In 1985 the Food and Agricultural Organization (FAO) of the United Nations estimated that at least 25% of the world's grain supply is contaminated with mycotoxins. The type and degree of contamination depend on many factors including conditions during growth and harvest and geographic area. Toxins such as zearalenone and vomitoxin (deoxynivalenol) produced by growth of Fusarium molds in stored grain tend to be the mycotoxins of concern in the more temperate climates of North America and Europe while aflatoxin, produced by various species of Aspergillus, predominates in tropical climates. However a wide range of toxins and mixtures of toxins are usually present (Figure 6, Table 5) and all are of concern when in the animal feed and human food chain. To a distiller, the main concern with mycotoxins is whether they pass through into the DDGS. The FDA have set definitive limits on aflatoxin levels for interstate grain shipments; and the increasing knowledge of the variety of mycotoxins and extent of contamination possible have heightened concern in all of the animal feed industry.

Table 5. Major mycotoxins affecting cereal grains and oilseeds, toxic syndromes and fungal sources.

Toxin	*Toxic syndrome in animals and/or humans*	*Fungal source*
***Aspergillus* toxins (primary)** Aflatoxins B_1, B_2, G_1 and G_2 (M_1 and M_2 are contaminants in milk	Hepatotoxin; carcinogenic, decreased growth, performance and immunity	*A. flavus* and *A. parasiticus*
Ochratoxin	Nephrotoxin	*A. ochraceus and Penicillium viridicatum*
***Penicillium* toxins** primarily luteoskyrin	Tremors, convulsions	*P. islandicum*
Patulin	Hemorrhagic, possibly carcinogenic	*P. urticae, P. expansum, P. claviforme* and *A. clavatus*
Rubratoxin	Liver damage, hemorrhage	*P. rubrum*
Citrinin	Kidney damage	*P. citrinum*
***Fusarium* toxins**		
Zearalenone	Estrogenic syndrome	*F. graminearum, F. tricinctum*
Vomitoxin (deoxynivalenol)	Emetic or feed refusal	*F. graminearum*
Other trichothecenes (T-2, HT-2, MAS, DAS)	Digestive tract toxins	*F. tricinctum, F. graminearum, F. poae, F. equiseti, F. lateritium, F. porotrichoides*
Fumonisin B_1, B_2	Leucoencephalomalacia	*F. moniliforme*
Ergot toxins		
Ergopeptines	Nervous disorders	*Claviceps purpurea*
Ergovaline	Vasoconstriction in extremities	*Acremonium coenophlalum*

Of more recent concern to the distiller has been the finding that mycotoxins can also affect yeast growth (Table 6). Perhaps presence of these toxins may provide at least part of the reason behind unexplained unfinished fermentations. Since mycotoxins are not destroyed by heating, it is unlikely that the cooking step would remove or inactivate them. In animals many of the mycotoxins damage protein metabolism causing reduced growth and increased disease susceptibility. Research into ways of nutritionally compensating for mycotoxin presence in animals feeds led to discovery that esterified glucomannans have characteristics that make them very specific adsorbents for certain mycotoxins. For example, about 80% of the zearalenone in solution can be bound and inactivated by adding esterified glucomannans. These products are also beginning to be used successfully in the fermentation industry.

Table 6. Effects of mycotoxins on yeast growth.

Mycotoxin	*Level required to inhibit yeast growth (ppm)*
Zearalenone	50
Vomitoxin	100
Fumonisin	10

Figure 6. Primary mycotoxin concerned in various regions of the world.

Novel stress factors

Phytic acid: an antinutritional factor

Phytic acid, the form in which 60-80% of the phosphorus is stored in cereal grains, is a stress factor for yeast that may be converted into a nutritional benefit. The phytic acid molecule has a high phosphorus content (28.2%) and large chelating potential. In addition to holding phosphorus in tightly bound form, phytic acid can form a wide variety of insoluble salts with di- and trivalent cations including calcium and zinc, copper, cobalt, manganese, iron and magnesium (Figure 7). This binding potentially renders these minerals unavailable in biological systems; and makes phytin content a familiar 'antinutritional factor' when cereal grains are fed to livestock. Animal nutritionists either compensate for the presence of phytin with extra-nutritional levels of zinc, calcium and phosphorus or add microbial phytases to diets for pigs and poultry.

Phytin may also have an antinutritional effect on yeast. Calcium ion concentration affects enzyme activity; and minerals such as zinc are needed for yeast growth. Phytic acid is also

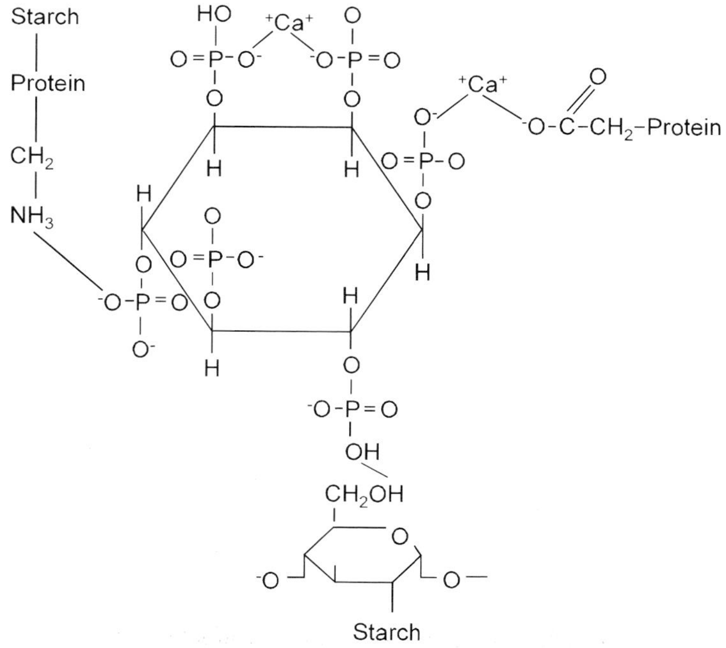

Figure 7. Stucture of the phytate (myoinositol hexakis phosphate) molecule and the possible bonds formed.

known to inhibit proteolytic enzymes; and phytate-protein or phytate-mineral-protein complexes may reduce the breakdown of protein to amino acids required by yeasts for growth. Starch is also known to be complexed by phytate. Finally, inositol, a limiting nutrient for yeast growth, forms a part of the phytin molecule.

Use of an enzyme called Phytozyme during the cooking process or as a fermenter addition stimulates phytin breakdown and increases starch and protein breakdown. Furthermore, the phytin is converted to two nutrients for yeast: inositol and phosphate. In effect, a negative has been converted to a positive. The net effect is as much as 0.3 gallons per bushel increased yield.

Summary: stress factors affecting yeast activity

As we 'coach' yeast to produce 23% ethanol, it becomes of paramount importance that we both understand the stress factors involved and design equipment to allow us to control them (Figure 8). Critical stress factors such as temperature, alcohol levels and infection, while understood individually, combine to produce unique conditions in every fermentation. Such conditions will vary with feedstocks used, but also with changes in the composition of individual feedstocks due to season or climatic variables. An understanding of other feedstock-related stress factors such as mycotoxins or phytin content will also become more important as we move toward higher ethanol production.

Fermentation systems used in distilleries

Before considering the specific management tools available to the distiller, it is necessary to consider the differences among the fermentation systems available. Over the last 30 years the brewing and distilling industries have developed new fermenting systems; and the distilling industry has largely converted to rapid batch fermentation with cylindro-conical or sloping

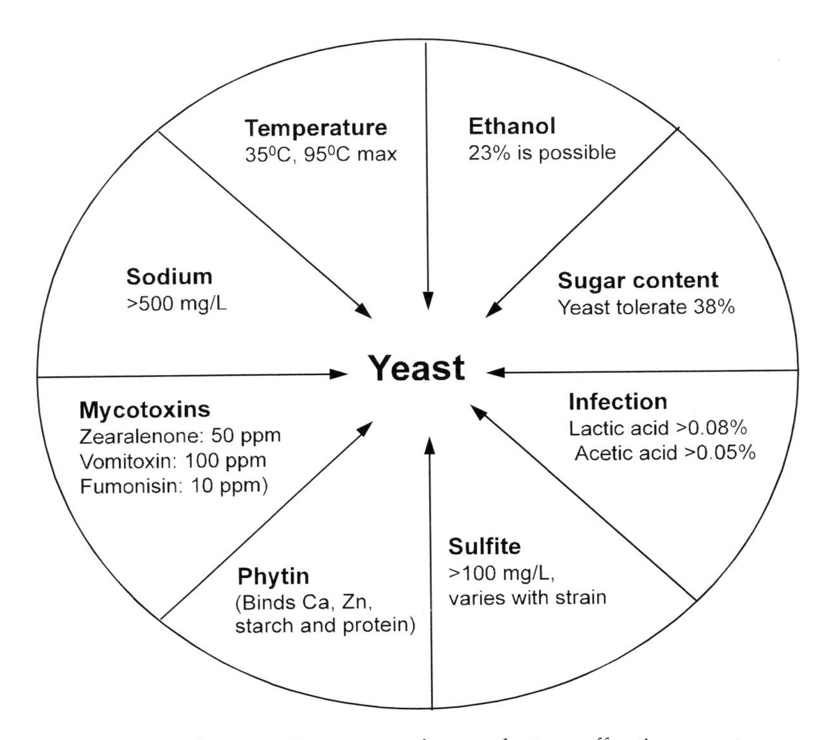

Figure 8. Summary of stress factors affecting yeast.

bottom fermenters, or to cascade continuous fermentation using cylindro-conical fermenters.

Batch fermentation

A typical cylindro-conical fermenter with skirt support is illustrated in Figure 9. These fermenters are usually designed to ferment from 30,000 gallons to 500,000 gallons. Fermenters need to be fabricated on site if the diameter and length exceed a size that can be safely transported by road.

Different cooling systems are available. Many distilleries use chilled water to cool the fermenters. From a microbiological standpoint

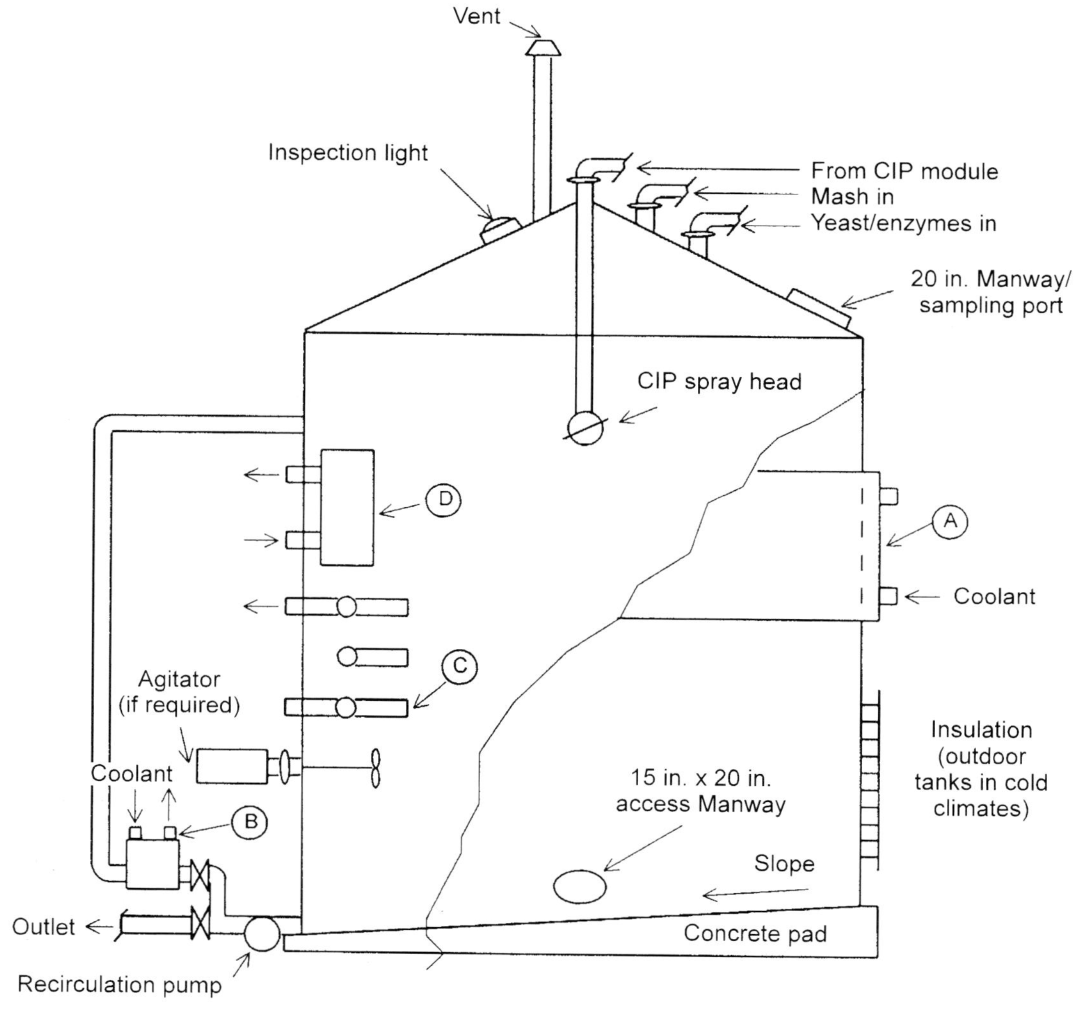

Figure 9. Alcohol fermenter with sloping bottom (Alltech/Bishopric, 1981).

the most desirable cooling system by far uses external cooling jackets. The least desirable option employs internal cooling coils because they are practically impossible to clean and sterilize from the top-mounted CIP (clean-in-place) spray nozzles. The external recirculation heat exchanger is frequently used in distilleries and is sometimes shared with other fermenters. This is a poor option as there are times when more than one fermenter needs to be cooled. The main disadvantage of this system when shared among fermenters is bacterial cross-contamination. With such a system, the microbiological condition of all fermenters will be determined by that of the worst-contaminated fermenter. Even when these recirculating heat exchangers are not shared, they are difficult to clean. Velocities of at least 7 ft^3/sec must be obtained to ensure turbulent flow through the tubes.

The agitator is required particularly at the start and end of fermentation. Frequently the agitators are badly designed and provide poor mixing. A folding action is required to ensure proper mixing of the mash solids and to ensure an even temperature throughout the fermenter.

Carbon dioxide is removed through the vent. Fermenters must be fitted with pressure relief valves and vacuum breakers to avoid serious accidents. The carbon dioxide is frequently collected and sold. In any case the carbon dioxide should be scrubbed to remove alcohol, which is returned to the beer well.

The lag phase is a great opportunity for lactobacilli bacteria to become established. Bacteria under optimum growth conditions can reproduce every 20-30 minutes; however yeast, which are much larger microorganisms, can only reproduce every 3 hrs. A single bacterium reproducing every 20 minutes would have a population of 256 in 3 hrs. Some yeast strains have lag phases as long as 12 hrs, during which time the lactobacilli can become heavily established. Therefore, it is very important to choose a yeast strain with a very short lag phase and to use a pre-fermenter so that the main fermentation starts immediately after the yeast are added.

There is a choice of materials to use in the manufacture of fermenters. Stainless steel is generally the best choice as it is easier to clean and sterilize. It also lasts much longer than mild steel or lined vessels; and when considered over the expected lifetime of a fermenter it is by far the most economical option. If chloride levels are high, then special stainless steels should be considered to avoid stress corrosion.

It is important that the slope of the bottom be sufficient for the mash to run out when discharging. This type of fermenter is much better suited for use with clear or semi-clear mashes than with whole cereal mashes.

Fermenter cleaning is accomplished with clean-in-place (CIP) equipment. CIP sprayheads are high-pressure devices (100-120 psi) which usually have automated cleaning cycles. A typical cleaning cycle would be:

Pre-rinse with water	10 minutes
Detergent circulation	20 minutes
Post-rinse with water	10 minutes
Sterilization	10 minutes

The detergent used would be based on caustic soda, normally with added wetting agent, antifoam and de-scaling agent. The caustic strength should normally be in the 3-5% range. Ideally, the detergent should be hot (80-90°C). The detergent and rinse waters should be continually drawn off from the fermenter to prevent accumulation of liquid during the cycle. The detergent and post-rinse should at least be recirculated and continually made up to strength. Chlorine dioxide and iodophors make ideal sterilizing agents, but many distillers still use steam to sterilize the fermenters. This is time-consuming and is probably not as effective as chemical sterilization.

It is important to control the build up of beerstone (calcium magnesium phosphate and calcium oxalate). Bacteria can penetrate beerstone, which is insoluble in straight alkaline solutions. Beerstone thus protects the bacteria from detergent and sterilant. EDTA-based chelating agents added to the detergent help dissolve beerstone and will control accumulation. The level of chelating agent needed can be calculated from the calcium level in the mash.

Continuous fermentation

The most successful continuous fermentation system used in distilling is the cascade system (Figure 10). The system illustrated in Figure 10 has twin fermenters and a beer well where the yeast are recycled. Yeast can only be recycled when clear mashes are used. The yeast can then be centrifuged, washed and reused. Typically, yeast can be washed using phosphoric acid at a pH of 2.2-2.4 and a holding time of 90 minutes. These conditions are sufficient to kill the bacteria without doing permanent damage to the yeast. Chlorine dioxide at 40-50 ppm is an alternative to acid washing, and seems to be gentler on the yeast.

The system in Figure 10 has two fermenters whereas most cascade plants have five fermenters and a pre-fermenter. The majority of these plants in the US use whole mash fed into the first two fermenters. Both of these fermenters are aerated continuously with sterile air and are cooled with external coolers. The pre-fermenter feeds yeast into both of these fermenters equally. The mash has been previously saccharified and enters the fermenters at a pH of 4.0 or slightly less. Usually these systems, if yeast stress factors are taken into account, produce 15-17% beers in 24-30 hrs although clear mash systems operate in 10-14 hrs. Each fermenter is individually cooled and agitated either with air or carbon dioxide or mechanically. The clear mash systems are capable of maintaining a fixed yeast count although they have the potential to recycle.

Typically yeast cell counts in these fermenters are slightly higher than in batch systems, with counts in the range of 180-220 million cells/ml for the first two fermenters and slightly less as the mash proceeds through the system. The viability and percentage of budding cells decrease from one fermenter to the next. Typically, the beer entering the beer well could be down to 120 million cells/ml with a viability of around 30%. Alcohol levels could be 10-11% in Fermenters 1 and 2, 12-13% in Fermenter 3 and up to 17% in Fermenters 4 and 5.

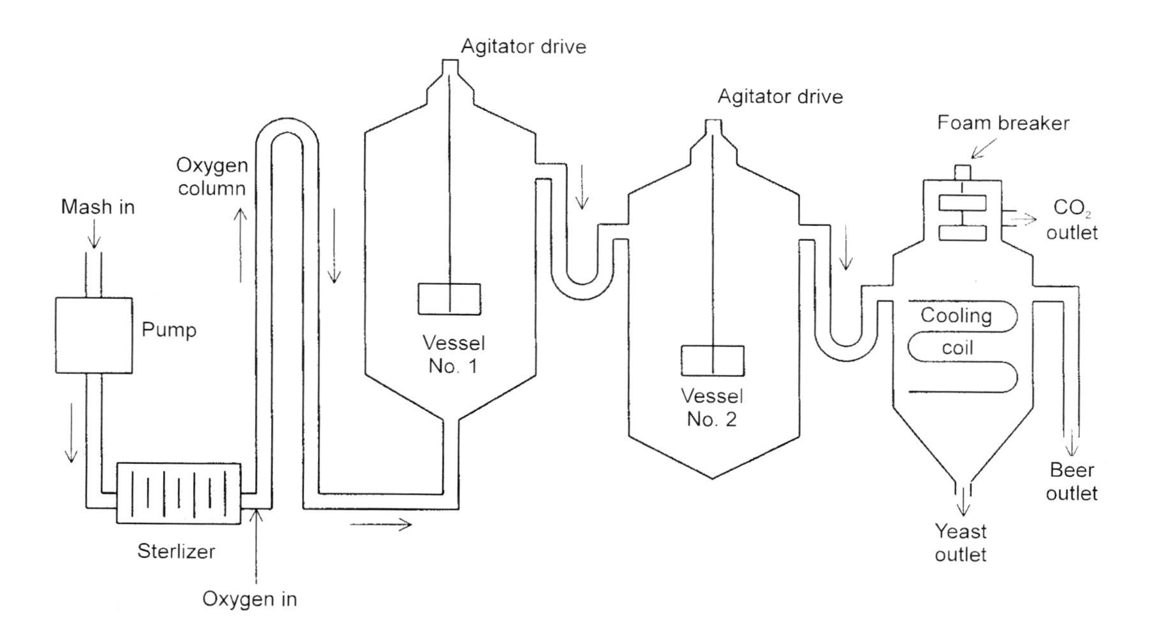

Figure 10. Twin-vessel continuous stirred fermenter.

Use of HPLC to monitor the presence of sugars reveals some interesting preferences of yeast for different sugars. The carbohydrate uptake during distillers mash fermentation is shown in Figure 11. It can be seen that yeast metabolize sucrose most rapidly, followed by glucose, fructose and maltose, while maltotriose is metabolized slowly or not at all. It is interesting to observe that continuous systems fermenting in less than 24 hrs usually have residual maltose, as this does not allow yeast enough time to metabolize this sugar fully.

There are other types of continuous fermentation systems that were used mainly by brewers in the 1970s and 1980s. The inclined-tube continuous and continuous tower systems are perhaps the best known (Figures 12 and 13). Neither of these systems has been used extensively for commercial alcohol production, but they were fashionable to brewers for a short time.

The main advantages of continuous fermentation (and they are very important advantages) are rapid throughput and the fact that these fermenters can be run for very long periods without stoppage. Cleaning costs are therefore practically eliminated; and vessel utilization is superior to any batch system. These continuous systems are frequently used for 12 months without stopping and are only cleaned when the plant has its annual shutdown.

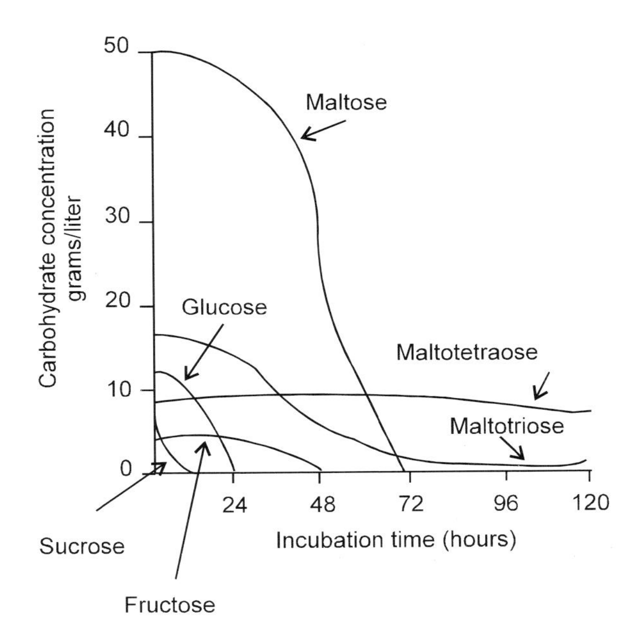

Figure 11. Carbohydrate uptake during mash fermentation.

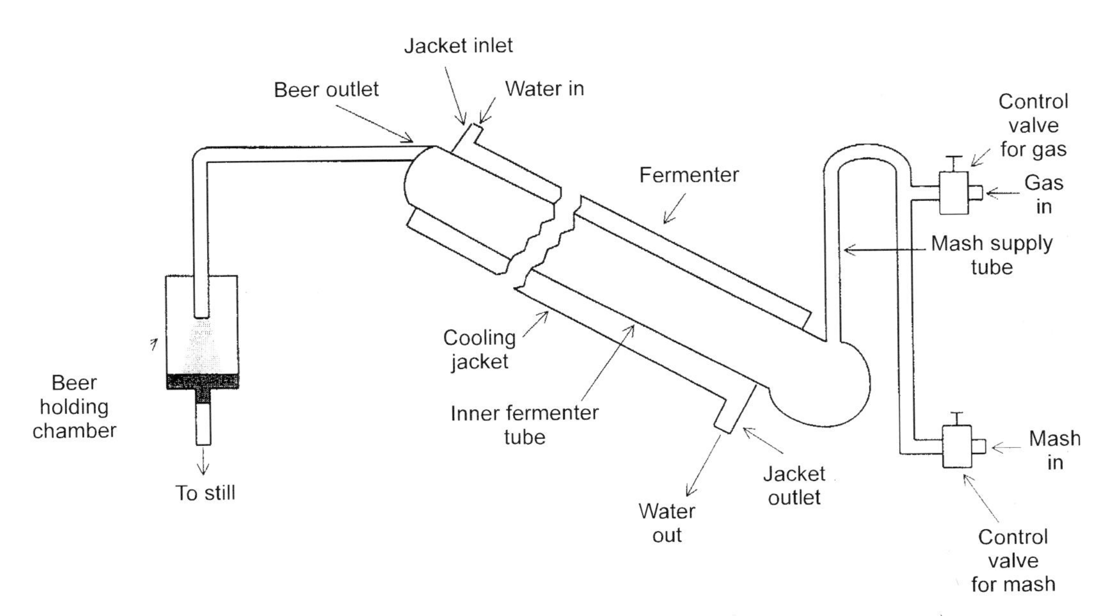

Figure 12. Inclined-tube continuous fermenter (US patent 3407069).

The most serious problem encountered with the cascade system is the occasional infection. The fermenters can be infected through non-sterile air injected into the first two fermenters. In this case, acetic acid bacteria, which convert the ethanol to acetic acid, can be introduced. This can be detected by a vinegar odor, or more accurately on an HPLC. There is also a possibility of infection by lactobacilli that usually propagate in contaminated mash coolers or saccharification tanks.

Conclusions

Understanding the stress factors that affect yeast and the fermentation equipment will help us move to the theoretical 23% ethanol that yeast are capable of producing. It is critical, however, that we push the positives: 'spoon-feed' sugar to yeast, provide yeast with peptides to ensure high cell numbers in early stages of fermentation. Control temperature and overcome the negatives. These negatives, infection, lactic acid, acetic acid and phytic acid and mycotoxins all prevent yeast and therefore the distillery from achieving potential yields.

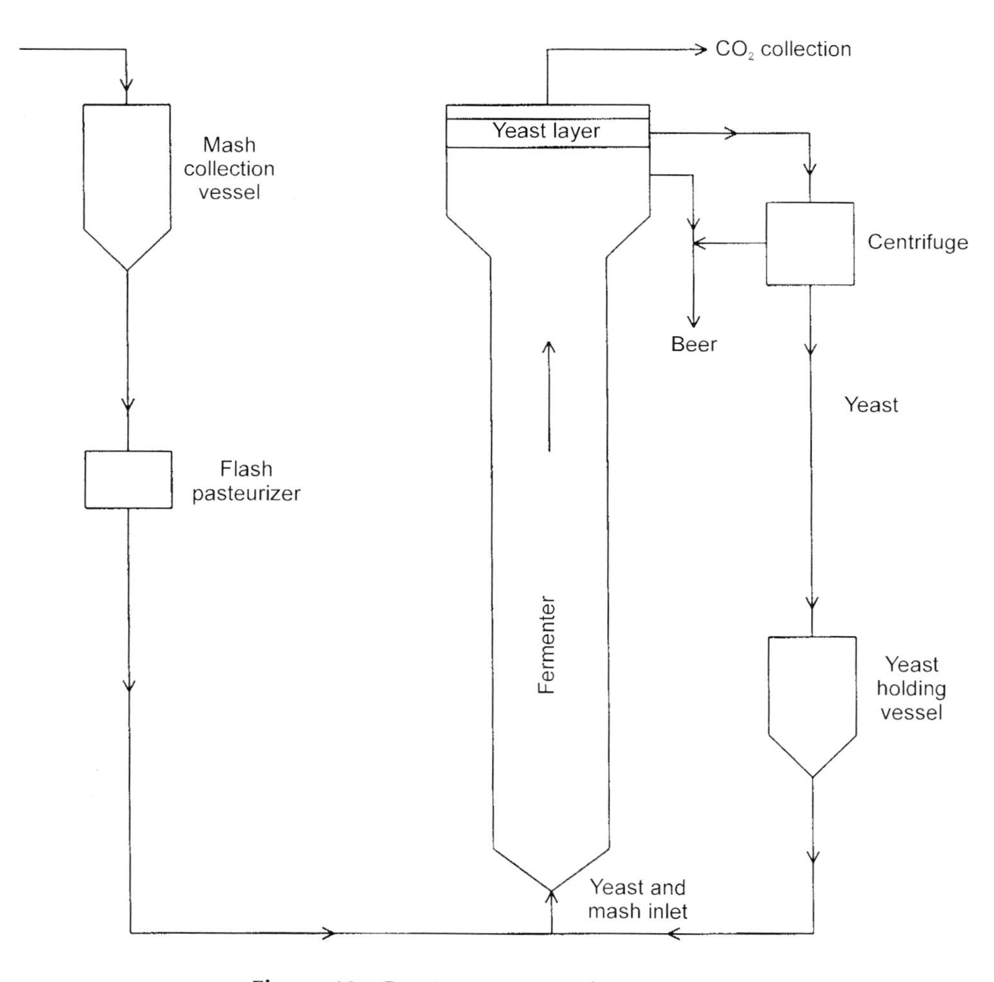

Figure 13. Continuous tower fermenter.

Chapter 4

The wet milling process: the basis for corn wet milling alcohol production

C.R. Keim
Sarasota, Florida, USA

Introduction

Broadly speaking, processes for making alcohol by fermentation include steps for 1) preparing the feedstock, 2) fermenting simple sugars, 3) alcohol recovery and often 4) recovery of residual non-alcohol materials. The feedstock may already contain sugars as in the case of molasses, sugar cane juice or whey; or it may contain sugar polymers such as cellulose, hemicellulose or starch, which can be depolymerized to fermentable sugars.

This chapter focuses on grain as the feedstock (and therefore starch as the sugar source) and concerns wet milling (as opposed to dry milling) as the process for preparing grain for fermentation. Use of the term 'wet milling' in the context of preparing grain for alcohol fermentation is of recent origin. It seems to have been first used in the late 1970s or early 1980s as a means to differentiate between two distinct processes for people entering the new and explosively-growing fuel ethanol industry. In the corn wet milling industry, alcohol could be made from standard products such as purified starch and a range of sugars and syrups. In contrast, conventional distilleries made alcohol from whole grain after it was processed in a hammer or roller mill in 'dry' condition.

In addition to wet milling and the regular distillery dry milling process, there are other processes for preparing grain prior to fermentation. One process uses specific flours made from wheat by the classic flour milling process. Flour is mixed with water to form either a dough or 'batter' after which the elastic hydrated gluten is recovered. Alcohol can then be made from the remainder, preceded or not by starch purification.

Another process is that practiced in the US by the venerable corn dry milling industry. The process resembles wheat flour milling except the starch-rich endosperm fraction is intentionally not ground entirely into flour. Instead, it is separated by screening and other means into various size fractions as indicated in Table 1. Annual US production of these corn fractions is in the region of 6 billion lbs (2.7 million metric tonnes). Some 30% of this, primarily coarse and regular grits, is used to replace part of the malt in brewery fermentations (Watson and Ramstad, 1987). These grits contain approximately 78% starch (on an 88% dry basis), so their use for making distilled alcohol has been suggested from time to time.

However the economics for using grits as a feedstock have never proven favorable.

Table 1. Products of corn dry milling[1].

US screen size		Product
Through	*On*	
3.5	6	Flaking grits
10	15	Coarse grits
15	30	Regular grits
30	60	Corn meal
40	80	Corn cones
60	325	Corn flour

[1]Watson and Ramstad, 1987.

The history of wet milling

The wet milling process was first introduced in the mid-1800s (Kerr, 1950) to produce cornstarch sufficiently pure for use not only in the laundry, but also in the kitchen for making puddings and custards and for thickening gravies. Meanwhile, an industry focusing on making substitutes for cane sugar from different starches by treatment with acid had been growing since the blockades of the Napoleonic wars. By the early 1900s at least, production of corn (or glucose) syrup had been joined to the wet milling starch process. In the US corn syrup began to be used in candy-making and in the home as a replacement for other syrups. In the 1920s pure crystallized dextrose was successfully produced (Newkirk, 1923). For several decades afterwards industry growth was slow but steady. Increases primarily resulted from industrial applications for a broadening line of regular corn syrups. This included introduction of enzymes to replace acid in at least part of the conversion from starch to syrup. However, sales of such products from starch were always limited by a low level of sweetness in comparison to cane and beet sugars and 'invert syrups' made by hydrolysis to their component simple sugars dextrose (glucose) and levulose (fructose).

Sales of starch-derived products changed dramatically during the 1970s with the commercialization of high-fructose corn syrup (HFCS), which had sweetness equal to syrups made from sucrose. The volume, measured in pounds of dry HFCS used per capita, rose from 1.4 lbs in 1973 (Corn Refiners Association, 1977) to 56 lbs in 1993/94 (USDA, 1994a) - an amount equal to 86% of the sucrose used. In that year the industry consumed some 660 million bushels (USDA, 1994b) (16.8 million metric tonnes) of corn in making fermentable sugars for use as sweeteners.

The first significant use of a wet milling plant for alcohol production began after the end of prohibition when the Standard Brands Company bought an existing plant in Clinton, Iowa and began making the grain neutral spirits required for its Fleischmann brand of beverages. Later, an interesting change from the dry milling to the wet milling process took place in stages at the Grain Processing Corporation (GPC) plant in Muscatine, Iowa. This plant was originally a dry milling unit built to produce alcohol for the World War II Rubber Reserve program. After the War it became a major supplier of industrial and beverage alcohols. In the 1950s GPC added facilities to recover the solubles and the oil-containing germ from the corn, employing the first steps of the wet milling process while continuing to use the remainder of the corn for fermentation in the existing war-time equipment. At a later time, equipment was added to take out part of the protein and some clean starch. Ultimately they completed the change to wet milling (see process description below). Eventually GPG became the largest and lowest-cost producer of neutral spirit, to the extent that a great many distillers purchased spirit from GPC instead of producing it themselves.

When the petroleum crisis of the 1970s set off a rush to build plants to make fuel alcohol from grain, it was widely expected that production would come primarily from plants using the dry milling process. Instead, we find the wet milling process now dominating the industry. Wet milling got a fast start when Archer Daniels Midland (ADM) added alcohol facilities to its then under-used HFCS plant in Decatur, Illinois. Steady expansion of those facilities followed, with acquisition and expansion of the Clinton plant and addition of alcohol to the product line at the Cedar Rapids, Iowa starch and corn sweetener plant. Meanwhile, a joint venture of Texaco and Corn Products Company (CPC) (traditional petroleum and corn wet milling companies, respectively) modified an aging starch and dextrose plant in Pekin, Illinois to produce fuel alcohol exclusively.

In retrospect, it can be said that the dry milling-type distilleries built with the financial support of the US government in the late 1970s and early 1980s were generally financially unsuccessful and required from modest to extensive rebuilding to operate. On the other hand, most of the existing wet milling plants added alcohol to their product lines and have been successful in the fuel alcohol market without direct government support.

The wet milling industry

Wet milling plants using corn as the raw material developed in countries other than the US also in the mid-1800s, especially in England, Scotland and Canada. Today there are plants in nearly every country in the world. The grain source is still predominately corn, although some plants in Europe have been built for, or converted to, processing wheat. The US has the largest wet milling industry, with a grind in 1994 of at least 1,200 million bushels (30.5 million metric tonnes).

The wet milling process is divided into two distinct sections. The first is the millhouse, which produces a slurry consisting of refined starch plus various by-products from corn. The other part consists of 'finishing' departments that process starch slurry from the millhouse into a myriad of starch, sweetener and fermentation products (Figure 1).

The corn kernel

A corn kernel is made up physically of four major parts (Figure 2). The largest, which represents about 83% of kernel dry weight, is the endosperm, containing both starch and protein. The protein is more concentrated in the hard starch area where a protein matrix encloses particles of starch, while in the upper part of the endosperm there is less protein structure and the starch particles are much easier to separate.

The germ is the corn plant embryo. It contains about 1/3 fat. It is an elongated structure on one side of the kernel and it makes up 10-12% of the dry weight. Covering the entire kernel is the pericarp, more usually known as the hull or bran. It is composed almost entirely of cellulose and accounts for about 5-6% of the weight. The other 1% is the 'tip cap' a small broken tube that originally connected the kernel to the cob.

Chemical composition of the corn kernel may vary within wide limits. The averages shown in Table 2 are representative of commercial corn available in the late 1980s.

Table 2. Chemical composition of corn (% of dry matter).

	Range	*Average*
Starch	61.0 - 78.0	71.7
Protein	6.0 - 12.0	9.5
Fat	3.1 - 5.7	4.3
Ash (oxide)	1.1 - 3.9	1.4
Cellulose[a]	3.3 - 4.3	3.3
Pentosans[b]	5.8 - 6.6	6.2
Sugars[c]	1.0 - 3.0	2.6
Other		1.0

[a]plus lignin
[b]as xylose
[c]as glucose

In the millhouse, the kernel is separated into five well-defined streams: a slurry of prime starch and four other streams used to produce valuable by-products. The process streams are 1) steepwater containing the solubles, 2) germ, containing most of the oil, 3) fiber, and 4) gluten, containing a large part of the insoluble protein. (Corn gluten is significantly different in composition and physical properties from the wheat gluten mentioned earlier, and the two should not be confused).

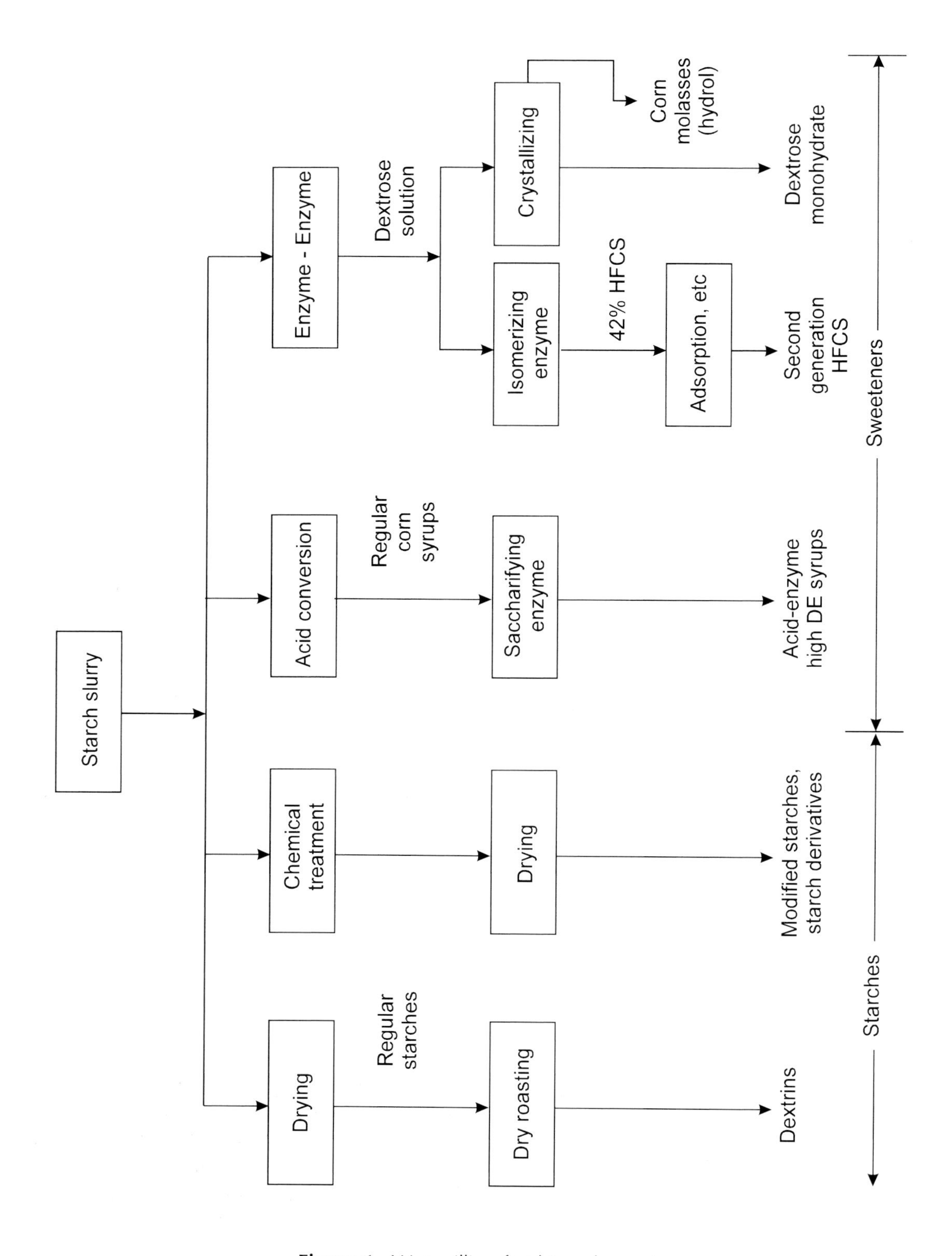

Figure 1. Wet milling finishing channels.

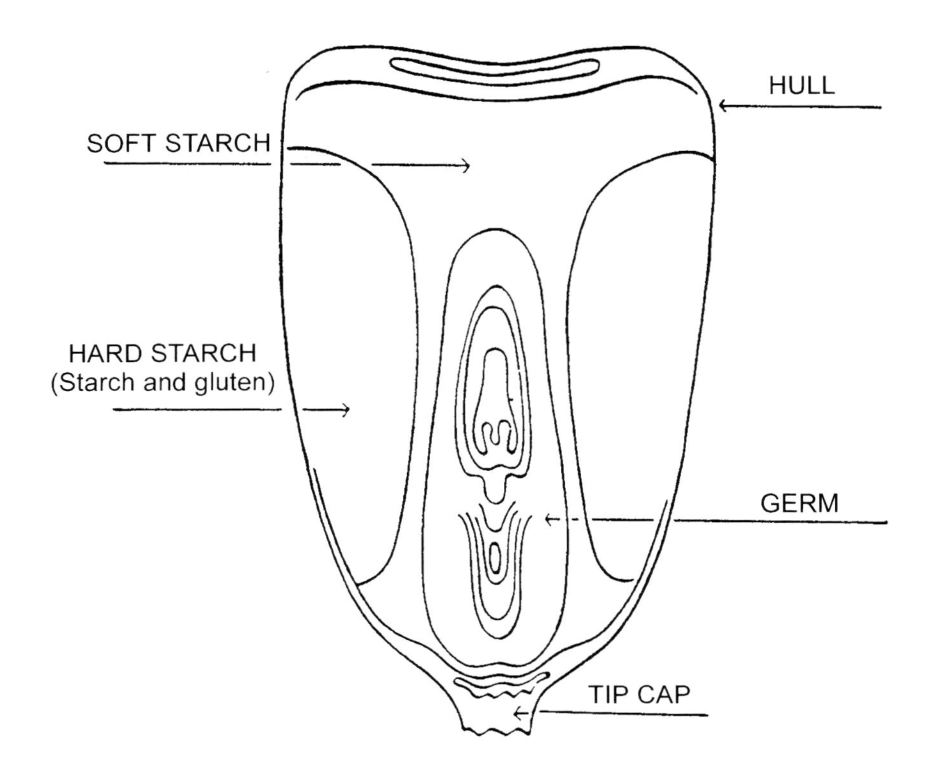

Figure 2. Structure of the corn kernel.

The usual by-products are unrefined corn oil expelled or extracted from the germ, corn gluten meal (60% protein) made by drying the gluten stream, and corn gluten feed (21% protein) made by mixing and drying the germ residue, fiber and steepwater streams. Yields of these by-products vary with corn composition and among facilities, but the yields in Table 3 are those used by the USDA (1994b).

Table 3. Wet milling yields listed by the USDA.

Product	*Moisture (%)*	*lbs/bu*	*kg/tonne*
Crude corn oil	-	1.55	27.7
Corn gluten feed	10	13.5	241.1
Corn gluten meal	10	2.65	47.3
Starch	0	31.5	562.5

The by-product oil, gluten feed and gluten meal are crucial to the economics of wet milling inasmuch as they produce revenue, which is on an average some 60% or more of the cost of the corn itself. This contrasts with the dry milling alcohol production process where the return from by-product sales is more usually on the order of 45% of the cost of the corn. The difference amounts to about $0.15 per gallon of alcohol in the US.

The wet milling process

General

The millhouse produces starch using a series of steps beginning with a lengthy soaking (steeping) where solubles are removed followed by recovery of the germ, fiber (bran) and gluten (protein). This leaves impure starch, which in the last millhouse step is washed in a countercurrent of fresh water to produce a 99% pure product.

The flow of water is actually countercurrent throughout the entire process (Figure 3). After starch-washing it moves stepwise toward the steeps, accumulating solubles as it goes. While small amounts of water leave with the various by-product streams, most of the water goes to steep the incoming corn, after which it is removed, carrying a significant amount of solubles with it.

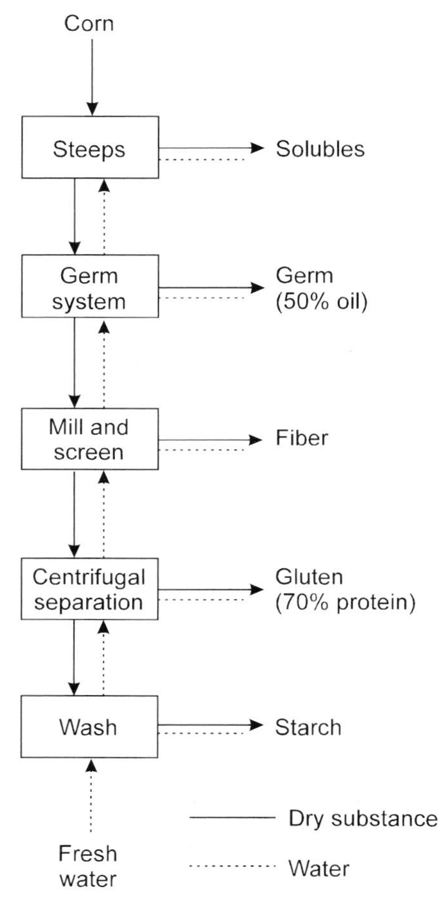

Figure 3. Water flow in corn wet milling.

Corn receiving and storing

Shelled corn arrives at the plant in rail cars or trucks (both provided with hopper bottoms), barges and sometimes ships. In the plant the corn is weighed, sampled, cleaned, elevated and stored in bins or silos. Depending upon location and availability of transport, a plant may store large amounts of corn or as little as only a few days' grind. Broken grain removed by the cleaning process may be sold as such or is more typically ground and mixed into corn gluten feed.

Steeping

Clean corn is first soaked in tanks for 20-40 hrs with steep acid containing about 1600 ppm sulfur dioxide (SO_2) at a temperature of about 125 ^{0}F (52 ^{0}C) (Figure 4). The tanks are filled with corn in sequence and are discharged after steeping in the same order. Steeping swells the kernels, bringing them to about 45% moisture, and the action of the SO_2 toughens the outer hull and germ while softening the protein structures. As a result, it is easier to remove intact germs and hulls and to separate starch and protein. The steep acid moves countercurrently through the steeps, being used first on the 'oldest' corn and finally on the newest corn put in the steep tanks. On removal, the steep acid contains about 7% dry substance, representing some 6.5% of the dry weight of the incoming corn and is referred to as 'light' steepwater. The solids, which total about 28%, are high in nitrogen (about 47% calculated as protein) and contain some 16% ash plus lactic acid and residual sugar.

Light steepwater also contains various unidentified growth factors and <100 ppm SO_2. Light steepwater is concentrated to about 45% dry matter in multi-effect or vapor recompression evaporators. It is then called 'heavy steepwater' and is sold under the name of 'condensed fermented corn extractives' for use as a component of liquid cattle feeds. More usually heavy steepwater is mixed in the plant with fiber (see below) and often other streams such as germ meal, corn screenings and refinery residue (protein and fat). When dried together these form corn gluten feed (21% protein), which is widely used as an ingredient in compounded cattle feeds. A large part of corn gluten feed is exported to Europe.

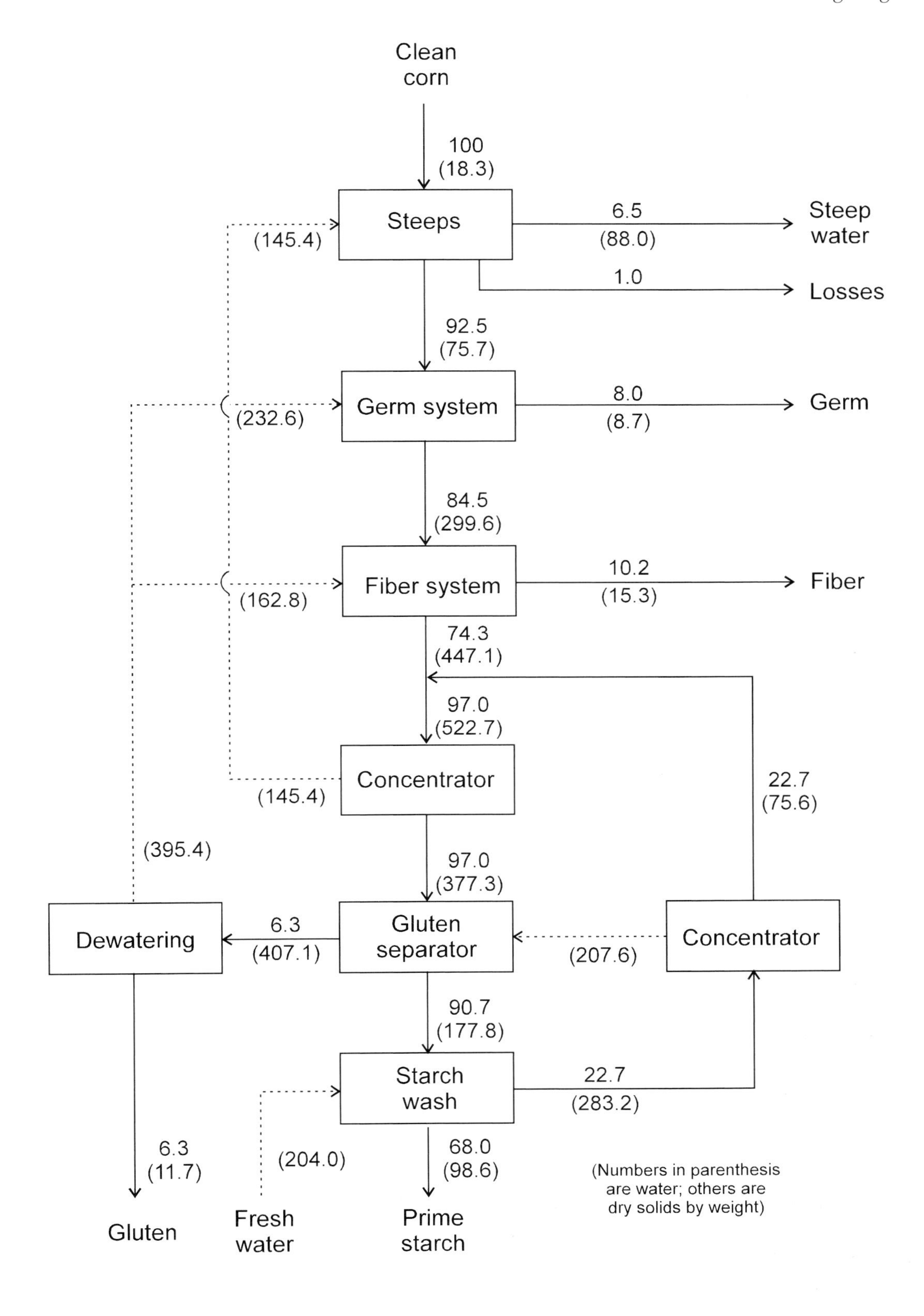

Figure 4. Corn wet milling block flow diagram.

Germ system

Steeped corn next passes through mills made of rotating toothed discs set far enough apart to ensure that kernels are torn rather than pulverized. This action frees the intact germ from the rest of the kernel. Germ is separated from the slurry in hydrocyclones owing to the fact that the oil content makes the germ lighter than the rest of the slurry. Germ moves to the center of the hydrocyclone and is withdrawn. After two stages of milling and separation, the germ is washed countercurrently in three steps, after which it is mechanically dewatered and dried in rotary steam tube dryers. At this point the germ contains about 50% oil along with fiber, protein and starch. Whole germ is subsequently processed by mechanical expelling and/or solvent extraction to recover the oil, which is the high value component. Either oil recovery is done at the wet milling plant or the whole germ may be sold to others for processing. The residual 'cake' from expelling, or 'flakes' from extracting, are sold as cattle feed ingredients or mixed into corn gluten feed. There is a commercial market for the crude corn oil, which is refined for use in frying, baking, salad dressings and other commercial and domestic food uses.

Fiber system

After removal of the germ, the remaining material is a slurry of starch, protein and fiber with particles of various shapes and sizes. At this point, washing screens are used to recover the fiber pieces to which large amounts of starch and protein are adhered, while fine particles of starch and protein free of fiber advance to the next process step. The fiber remaining behind is milled with closely-set discs to dislodge as much starch and protein as possible. After milling, the fiber is countercurrently washed through several stages of screens, mechanically dewatered and partially dried, then mixed with heavy steepwater and other components. This material is dried together to produce corn gluten feed.

Gluten separation

The remaining thin slurry of starch and protein is next thickened by removal of water using a disc-nozzle centrifugal machine followed by a similar machine that separates the lighter protein from the starch by a centrifugal force of 3,000 x g. Protein content is adjusted to around 68% at discharge so that it will be above 60% when dried to a commercial moisture of about 10%.

The stream of gluten is thickened by another centrifugation, dewatered on a rotary vacuum filter and dried. The product is corn gluten meal (60% protein). The xanthophyll oil that gives yellow corn its color is concentrated in this fraction. Used primarily in poultry feed formulations, it provides not only high protein, but also yellow color to the skin and egg yolks, which is desirable in many markets.

Starch washing

Starch slurry coming from the gluten separator contains impurities, primarily about 2% insoluble and nearly 1% soluble protein, which must be greatly reduced to produce starch of acceptable quality. This is accomplished by countercurrent washing with fresh water through a series of up to 15 stages of hydrocyclones. Impurities are concentrated by centrifugation and returned to the process stream after the fiber system, while the washed starch slurry is collected as the main product. The washed starch is in a slurry containing about 40% dry matter with the dry material containing only about 0.3% total protein and less than 0.01% soluble protein (van Beynum and Roels, 1985). On analysis very little oil or fat will be noted, but as much as 0.6% 'hidden' fat held in the spirals of the starch molecules may actually be present. This fat is released when the starch is hydrolyzed to form sugars.

Alcohol production in wet mills

In order to make alcohol from this starch, it must first be converted into low molecular weight fermentable sugars, of which dextrose (glucose) and maltose are important examples. The reaction is called hydrolysis since water combines with the starch, coincidentally producing a weight gain:

$$(C_6H_{10}O_5)_n + nH_2O \rightarrow nC_6H_{12}O_6$$

Starch	Water	Glucose
(162)	(18)	(180)

Hydroysis can be achieved by the action of acids or of suitable enzymes. In the dry milling distillery process where whole grain is used, hydrolysis and subsequent fermentation take place in the presence of large amounts of other materials such as germ fiber and protein. In contrast, these other materials are absent when clean wet milling starch is used and the fermentation broth is said to be a 'clear substrate'. This is important, because with a clear substrate yeast can be recovered at the end of the fermentation and re-used for subsequent fermentations instead of being lost in the stillage. The advantages of the clear substrate are:

a) Fermentation starts rapidly because of high yeast concentrations.
b) High yeast concentrations tend to keep unwanted organisms from flourishing.
c) In cascade fermentation clear substrate results in a high alcohol concentration in the first fermenter, which substantially decreases formation of glycerol in favor of more alcohol yield.
d) The need to grow large amounts of new yeast during fermentation is reduced, saving as much as 4% of the sugars for alcohol instead of yeast growth.

The yeast recovery process involves passing the fermented beer through a centrifuge on its way from the fermenter to the first distillation column. The yeast is concentrated into a 'cream' which is acid treated to eliminate bacteria and returned to the process.

Summary

The process known as grain wet milling has been practiced for more than a century to obtain pure food grade starch used both as such and for conversion by hydrolysis into fermentable sugars and syrups. Some of these sugars and syrups have been used historically to make beverage alcohol - both outside and in the wet milling plant.

When the demand for fuel alcohol first arose in the US, the corn wet millers were thus already very efficiently producing large amounts of fermentable sugars. All that was required to produce alcohol was to add facilities for fermentation, distillation and alcohol dehydration - everything else was already in place. ADM, which had considerable excess HFCS capacity at the time, did this on a large scale and became the principal alcohol producer in the US. A different approach was taken by the A.E. Staley Company (Tate and Lyle) who designed their new HFCS plant for large alcohol capacity in addition to syrup production. This offered two advantages. First, the starch volume required for alcohol increased the millhouse size and allowed the economy of scale. Secondly, alcohol production rate could be adjusted to balance the seasonality of HFCS sales and thereby the millhouse could be kept operating steadily at full production. Another plant (Pekin Energy) discontinued starch and sweetener production completely and devoted its entire grind to alcohol. The success of this operation demonstrated that it is not inherently necessary for wet milling alcohol plants to make other products to be viable.

At present almost all wet milling plants use corn (maize) although some wheat is being used in Europe. The wet milling process obtains pure starch by successively removing other fractions of the kernel, arriving at a somewhat impure starch, which is then washed clean with fresh water prior to further processing. The non-starch fractions are customarily sold in the form of unrefined corn oil, corn gluten feed (21% crude protein), and corn gluten meal (60% crude protein). Together these by-products usually return a value equal to at least 60% of the cost of corn. This results in a significantly lower raw material cost when compared with the dry milling process for making alcohol.

Due to the lower raw material cost and to the economies of scale of the wet milling plants, a great majority of the grain neutral spirits and fuel alcohol produced in the US is made by the wet milling process.

References

Watson, S.A. and P.E. Ramstad. 1987. Corn chemistry and technology. American Association of Cereal Chemists, Inc. St. Paul, MN. p. 357.

Kerr, R.W. 1950. Chemistry and industry of starch. Academic Press, New York. p 29.

Newkirk, W.B. 1923. Method of making grape sugar. U.S. Patent 1, 471,347.

Corn Refiners Association. 1977. Updating the corn sweetener revolution. In: Corn Annual. Washington, D.C.

USDA. 1994a. Sugar and Sweetener. Situation and Outlook Report. Economic Research Service p.17.

USDA. 1994b. Sugar and Sweetener. Situation and Outlook Report. Economic Research Service. p. 70.

van Beynum, G.M.A. and J.A. Roels. 1985. Starch conversion technology. Marcel Dekker Inc. New York and Basel, p. 56.

Chapter 5

Alcohol production by *Saccharomyces cerevisiae*: a yeast primer

W.M. Ingledew
Applied Microbiology and Food Science Department, University of Saskatchewan, Saskatoon, SK, Canada

Introduction

Saccharomyces cerevisiae remains the most exploited microorganism known to industry and is still the primary microorganism used for the production of virtually all potable and industrial ethanol. Ethanol is, quantitatively and economically the world's premier biotechnological commodity (Walker, 1998), and is produced at 24 billion L/year (Dixon, 1999). The potable alcohol industry produces brewing, winery and distillery products destined for human consumption, while the non-potable alcohol industry manufactures ethanol for fuel and industrial purposes. In North America, the approximate production of pure (200 proof) alcohol is: 1.51×10^9 L/yr (brewing), 6.84×10^8 L/yr (distilling), 2.1×10^8 L/yr (winery), and 6.0×10^9 L/year (fuel and industrial alcohol) (Anonymous, 1977; Anonymous, 1999 with Canadian figures added). This is a total production of ~8.4×10^9 L of pure alcohol per year.

This chapter will concentrate on fermentation aspects of ethanol production and on related considerations of the yeast fermentation. Other sections of this book will deal in more detail with substrate preparation and processing.

Composition of mash

In this chapter, we will assume corn is the substrate for fermentation. In passing, the major differences in wet and dry milling of corn will be outlined to set the scene for discussions (in a generic way) on the microbiological and biochemical aspects of the process.

The products of a corn wet milling plant are starch, corn oil, gluten feed, corn syrup, ethanol, yeast, carbon dioxide, and corn steep liquor. The starch slurry made in a wet milling facility is produced at 38-42% dry solids. All or a portion of this slurry can be made available for ethanol production. Excess or lower grade starch is fermented to ethanol as a secondary end product.

Table 1 compares the products of wet and dry milling of corn. Fermentation by the dry milling process yields the end products ethanol, spent grain, thin stillage and carbon dioxide (CO_2). Dry milling is extensively carried out in North America for potable beverages and for fuel purposes. An example of inputs and outputs in a dry milling fuel ethanol operation is given in Table 2.

Table 1. Products from corn mash.

Dry milling[a]
ethanol
carbon dioxide
distillers dried grains
thin stillage
Wet milling[a]
gluten feed
syrups
corn oil
starch
ethanol
yeast
carbon dioxide
corn steep liquor

[a]lesser amounts of higher alcohols, aldehydes, and esters are also produced

Table 2. Inputs and outputs - corn dry milling plant.

Yearly inputs [a]	
Corn	203,000 tonnes
Active dry yeasts	180-300 tonnes
Yeast foods	100-400 tonnes
Steam load (via coal)	1.7×10^9 MJ
Electricity	6.1×10^7 kWh
Gasoline denaturant	4,000 L
Yearly outputs [b]	
Ethanol (76,000 kL)	62,000 tonnes
Distillers dried grains with solubles	65,000 tonnes
Carbon dioxide	60,000 tonnes
Aldehydes, fusel oils, esters, ketones	23 tonnes

[a]Calculated from data provided by Kentucky Agriculture Energy Corp. and energy inputs by Hussain (1988).
[b]Other inputs include microbial enzymes, antifoams, penicillin, sanitizers, cleaning agents, chemicals, and labor.

After milling, the unit operations in dry and wet milled corn processing include gelatinization of starch, enzymatic liquefaction and saccharification to fermentable sugars and fermentation of sugars by yeast to ethanol and carbon dioxide. Subsequent steps include concentration and purification of ethanol and processing of the stillage. Detailed information on all these steps is covered in this book in other locations, and has been reviewed by Lewis (1996).

The net cost of ethanol is directly related to the cost of substrate. Between 1981 and 1986, the net cost for corn (cost of corn minus the cost recovered in DDG by-product) converted to ethanol in a dry milling plant ranged from $0.13 to $0.235/L. The average was $0.143/L for the 1980-1994 period. Since that time the cost has fluctuated widely. The industry has little control over substrate costs; and the only other ways total production costs can be reduced are with a) technological advances (like distillation costs and dewatering by molecular sieve), b) increasing ethanol yield, c) increasing carbon dioxide (CO_2) or DDG value, or d) lowering labor, capital and overhead costs. There is not much flexibility in these options. For wet milling, although the net cost of corn would be less than in dry milling due to the high value of end products produced (oil, starch, corn gluten feed, corn gluten meal and corn steep liquor), other costs remain high. See Lewis (1996) for further economic considerations.

Fermentation - The 'Black Box'

Alcohol production is a multidisciplinary process based on the chemistry, biochemistry and microbiology of the raw materials, the yeast fermentation, and the downstream processing technologies used to separate the ethanol, thin stillage and spent grains (Figure 1). Operators and managers of alcohol plants are not usually trained microbiologists and biochemists. This is one reason why the science of the fermentation is not as well known as the science and engineering of grain processing and that of liquid handling and post-fermentation processing. The intent of this chapter is to review aspects of the scientific use of yeast in fermentations where alcohol is the major product. In the past, alcohol production has been an 'art'. Slowly over the last 150 years, we have come closer to describing the science of the alcohol fermentation (understanding the black box - Figure 1). The hope is that technologies will eventually be detailed with fewer microbiological and engineering problems and therefore fewer operating difficulties.

Plant engineering trouble spots

Much of the information presented in this article refers to the operation of batch fermentation systems. Continuous fermentation, however, is carried out by many fuel alcohol plants in an attempt to improve productivity, eliminate down time (filling, emptying and cleaning), reduce capital and labor costs and simplify control. Continuous systems are said to work best for low value - high volume products like ethanol. In such a system working optimally, a steady state yeast population is maintained with fresh medium (substrates) added, and with product removed at the same rate. Commonly, 4-stage or 5-stage fermenter trains are used. Ideally, these fermentation trains are designed to operate over many months without constant shutdowns to clean or decontaminate as required in batch fermentation systems. In reality, many problems have been experienced in continuous operations; and workers constantly battle the effects of bacterial contamination that invariably result in levels of lactic acid, acetic acid and other factors sufficient to stress and inhibit the growth and routine production of ethanol by the yeast (see Figure 2). The onset of these problems in some cases is rapid, and consequently they are difficult to manage. The sources of the problem are not easily located, and considerable job stress results. As a result, some continuous plants have reverted to batch operation at the fermentation stage. A 'cleansing' of the system occurs as each fermenter is processed, cleaned, and restarted with new yeast; each batch is individually controlled. In this way, the loss of alcohol (and sugar) is reduced significantly and fewer long-term concerns are faced by workers. Output in most cases is not unduly compromised, although additional tanks are usually needed to maintain productivity.

Microbial concerns with continuous (and batch) fermentation can occur at any point in the system after mash temperatures are reduced to less than 65°C. For this reason, microbial problems can exist in the saccharification tanks (if used), transfer lines, yeast propagation tanks, surge (supply) tanks, fermenters, beer wells, and any other vessel, pump, external heat exchanger,

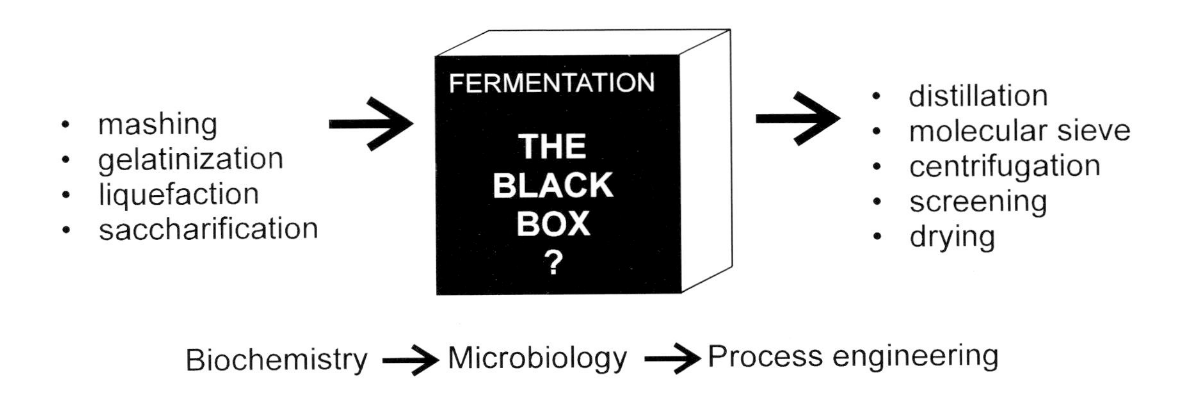

Figure 1. The 'Black Box' of fermentation, the least understood segment of the alcohol production process.

or the air supply. In addition, all liquid streams and tanks used to store well water or recycled waters entering the process and all solid additions (if made) to fermenters can be a source of bacterial or wild yeast contamination, and should be monitored.

Two of these sites deserve special mention. Saccharification tanks have always been a microbiological hazard because some lactic acid bacteria are able to grow at or near the temperatures used for saccharification of dextrins (near 60°C). As these bacteria propagate, they increase in number, compete for nutrients (prior to and after yeast addition), and form lactic acid and acetic acid that stress the yeast. Secondly, yeast propagation tanks (conditioning tanks), which are often 'engineered' to aerobically cultivate distillery yeasts, can be microbiologically unsound. In these tanks, yeasts only grow aerobically to high yield (cell mass/L) when sugar levels are kept below 0.1% w/v, adequate nutrients are present in the mash, and aeration is more than 1 volume/volume medium/min. These conditions are never met in fuel alcohol plants, breweries or wineries due to the equipment used. For this reason, such tanks at best are anaerobic propagators with cell yields the same as expected in fermenters. Small volumes of oxygen (from air) are needed in order to supply approximately 8-20 ppm of O_2 to the mash as yeasts grow, but this does not lead to aerobic growth. However, if such propagators are used in continuous mode, they can become bacterial propagators whenever the generation time of the contaminating bacterium is shorter than the generation time of the yeast. Then, a continuous inoculum of the 'selected' bacterium takes place. In similar fashion, antibiotic-resistant bacteria can be selected in continuous propagators as mutants form and are no longer inhibited by the antibiotics added to these tanks. Recycled viable yeast can similarly be a source of recycled harmful and antibiotic-resistant bacteria - all leading to reduced ethanol yields or reduced yeast viability and vigor. All such areas and incoming streams should be monitored in a quality control program in order to reduce ethanol losses in the plant. Every molecule of

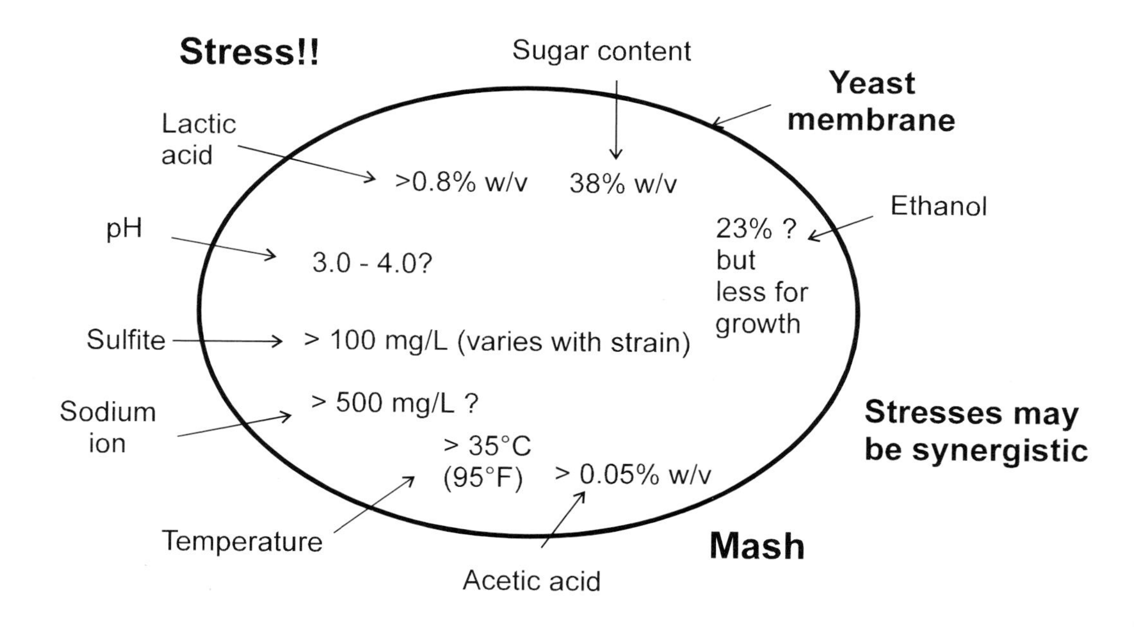

Figure 2. Typical stress factors known to affect yeast fermentations.

lactic acid made in a fermenter means the loss of a molecule of ethanol! Some of the bacteria isolated from commercial sources by our lab are resistant to 10-15% v/v ethanol and grow faster than yeast in commercial grain mashes at pH values as low as 3.6-3.8.

Another area on which we need to focus is the process flow of an alcohol plant where chemicals are added and a number of waste streams are recycled to eventually form part of the overall composition of the fermentation mash. Some of these waste streams include nutrients that aid the fermentation - for example, light steep water, corn steep liquor (Stanbury *et al.*, 1995), or corn steep powder (Table 3), stillage/backset, aqueous ammonia and enzymes. Much of the nitrogen in corn steep liquor (powder) is in peptide and protein form, not degradable by yeast. Free amino nitrogen (FAN) is not indicated, but we have determined that ~16 mg FAN is present in each gram of corn steep powder (Thomas and Ingledew, unpublished data). The free amino acid profile indicates that alanine, proline, leucine and tryptophan are the most prevalent free amino acids (Table 4). This is very different from the profile of amino acids in *hydrolyzed* protein from either corn steep liquor or powder. Corn steep powder contains ~0.115 g of free amino acids per gram of dry material. Protein and the majority of peptides are not usable by yeast. However, corn steep products provide some usable phosphorus, sulfur, other ions and vitamins. The exact amounts have not been determined. Lactic acid levels are also not indicated.

Thin stillage (Table 5) has also been analyzed (Jones and Ingledew, 1994d). Values vary widely depending on the source and the degree of problems experienced by each of the suppliers (one sample, for example, contained 33.1 g/L of glucose indicating a stuck or incomplete fermentation). Thin stillage also contains yeasts, heat killed in the still, which are a source of vitamins, minerals, carbohydrates, fats and other cellular components. Significant variations are also seen in the lactic acid levels, which range from 1.7% (a level inhibitory to yeast) down to

Table 3. Proximate composition (g/100 g dry weight) of corn steep powder made from corn steep liquor[a].

	Traders[b]	*Marcor*[b]
Dry matter	95	
Nitrogen[c]	7.5-7.7	
Protein	47-488	
Fat	0.4	
Carbohydrate	0	
Fiber	0	
Ash	17	
Calcium	0.06	0.3
Magnesium	1.5	1.5
Phosphorus	3.3	3.9
Available phosphorus	1.1	
Potassium	4.5	5.2
Sulfur	0.58	
Biotin	0	
Choline	0.00056	
Niacin	0.000016	
Pantothenate	0.000003	
Pyridoxine	0.000002	
Riboflavin	0.000001	
Thiamin	0.000001	
Arginine[d]	3.3	2.3
Cystine[d]	1.9	1.7
Glycine[d]	4.5-5.1	2.3
Histidine[d]	2.8	1.6
Lysine[d]	2.5-3.4	1.7
Methionine[d]	1.9-2.1	1.1
Phenylalanine[d]	3.2-4.4	1.7
Threonine[d]	3.7-4.0	1.9
Tryptophan[d]	0-0.2	0.1
Tyrosine[d]	2.2-3.4	1.1
Valine[d]	4.7-5.8	2.6
Alanine[d]	7.4	3.7
Aspartic acid[d]	5.7	2.9
Glutamic acid[d]	13.9	7.3
Proline[d]	7.8	4.2
Serine[d]	4.1	2.2
Leucine[d]	8.2-11.3	4.2
Isoleucine[d]	2.8-3.6	1.6

[a]Trader's Protein, Memphis, TN, and Marcor Development Corp., Hackensack NJ.
[b]Reproducibility of results and methods used are unknown.
[c]Total crude protein x 0.16 (a typical protein contains 16% N).
[d]Amino acids expressed as % of crude total protein (after hydrolysis), not as free amino acids.

Table 4. Free amino acid composition of corn steep powder.*

Component	*µmoles/g powder*	*g amino acid /g powder*
Arginine	11.04	.00198
Cystine	4.68	.00112
Glycine + asparagine[a]	44.43	.00334
Histidine	0.32	.00005
Lysine	9.89	.00145
Methionine	17.94	.00268
Phenylalanine	36.06	.00596
Threonine	32.41	.00386
Tryptophan	97.58	.01929
Tyrosine	33.21	.00602
Valine	60.57	.00710
Alanine	193.78	.01726
Aspartic acid	7.21	.00096
Glutamic acid	42.78	.00067
Proline	122.83	.01414
Serine	41.85	.00440
Leucine	112.10	.01470
Isoleucine	28.97	.00380
TOTAL		.1150

*Thomas and Ingledew, unpublished data; [a] calculated from unresolved peak as glycine.

Table 5. Proximate composition of corn and wheat thin stillage (backset).*[a]

Parameter	*A (Wheat)*	*B (Wheat)*	*C (Wheat)*	*D (Corn)*	*E (Corn)*	*F (Corn)*
Total solids, % w/v						
Solubles	4.6	6.3	3.4	4.5	2.8	2.4
Insolubles	8.0	7.8	2.3	3.2	2.0	2.0
Total N, % w/v	0.18	0.11	0.13	0.026	0.053	0.54
Crude protein, % w/v[b]	1.1	0.7	0.8	0.2	0.3	3.4
FAN, mg/L	96	79	51	118	75	88
Glucose, g/L[c]	1.6	33.1	1.0	0.7	0.6	0.2
Lactic acid, g/L	11	17	1.5	7	11	6
pH	4.2	3.6	4.7	4.7	3.7	4.0
Minerals, mg/L						
Phosphorous	1762	815	726	1310	1051	924
Potassium	2643	705	938	2049	1567	1378
Sulphur	344	309	574	195	399	169
Calcium	178	52	224	43	96	38
Magnesium	721	204	431	552	513	437
Copper	0.7	1.3	0.1	0.1	0.1	0.1
Iron	6.8	3.8	6.8	10.0	2.7	2.2
Manganese	9.5	6.7	5.1	2.2	5.5	1.7
Zinc	7.6	5.6	3.7	4.9	4.9	6.4

[a]Backset samples were centrifuged upon receipt, and insolubles were re-suspended to the same volume using sterile distilled water. With the exception of total solids, all data were collected from the soluble fraction. Unfractionated backsets were analyzed for minerals

[b]Total N x 6.25

[c]Glucose was determined after treatment with excess glucoamylase to convert residual dextrins to fermentable sugars.

*Jones and Ingledew, 1994d by permission.

0.6% w/v. After addition of thin stillage and its dilution in mash, significant amounts of lactic acid will be present in the fermenter. This can be calculated from analytical data and addition rates of the stillage into the mash. Well-fermented mashes lead to backset similar to sample F (Table 5), which has low sugar, low nitrogen, low lactic acid and therefore may have low value as a nutrient source for yeasts in the fermenter. Too much emphasis may have been placed on the nutritive value of thin stillage (backset) to the alcohol fermentation. The major reason for its use is for water recycling/pollution control, in spite of the expectation prevalent in the industry that backset is a good source of nutrients.

Other processing streams in an alcohol plant are also reused in an attempt to reduce water costs and effluent charges, but these definitely do not contribute significantly to the fermentation. These include process condensate water (which may include distillation column water, evaporator condensate water and flash condensate), stripper bottoms from the still, and tank wash waters. In our corn wet milling example, the overall effect of using all such streams outlined above is that the final concentration of the starch is only about 1/2 to 1/3 of its original concentration of 40% w/v. This significantly reduces the possible alcohol concentration that could have been achieved without such extensive dilution (see later information on very high gravity fermentation). However, wet milled starch after gelatinization and liquefaction is virtually free from insoluble materials, and therefore this substrate stream has advantages for both continuous fermentation and distillation. It should be emphasized that the fermenter is a sensitive biological system, not a repository for recycled watery factory wastes. Inhibitory levels of sodium ion, sulfite, lactic acid, and acetic acid increase to inhibitory levels during recycling of these waste streams, and all lead to yeast stress (Figure 2). Fermentation must be run under favorable physiological conditions in order to ensure it will complete in a predictable and consistent fashion.

Fermentation biochemistry

Yeast fermentation proceeds as outlined in Figure 3. Sugar is converted to ethanol, carbon dioxide and yeast biomass as well as much smaller quantities of minor end products such as glycerol, fusel oils, aldehydes and ketones. It is the biochemical conversion of sugar to end products that this article will deal with at length. This discussion may help in understanding the 'black box' of fermentation.

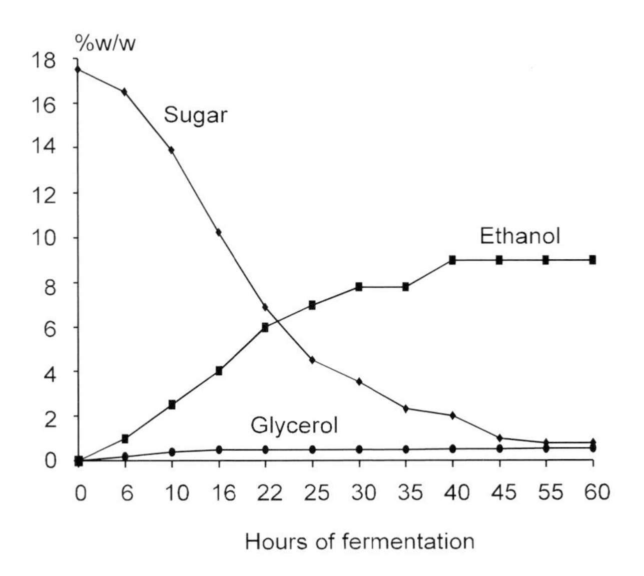

Figure 3. Progress of a typical fermentation for the spirit industry. Note that neither the alcohol nor the sugar concentrations in this figure are typical for the fuel alcohol industry.

Yeast growth

The development of the baking yeast industry from both sales and technology standpoints is an interesting story. This industry propagates the largest amount of a single type of microbial biomass in all the fermentation industry. The annual production of dry bakers yeast was about 82,000 tonnes in 1990 (Reed and Nagodawithana, 1991). The technological aspects of their production and use have been well documented (Rose and Yijayalakshmi, 1993; Reed and Nagodawithana,1991).

Yeast can be supplied in one of three forms. Yeasts grown aerobically can be collected and used directly as a slurry (yeast cream), a 14-16 % w/v suspension of yeast in water or in the medium or diluted medium in which it was grown. Yeast slurries are relatively hardy, but have to be kept at 4°C to maintain viability. They must be transported in refrigerated tanker trucks and stored at or near 4°C. The yeasts can be washed with phosphoric acid to reduce any bacterial contamination just prior to use using established procedures detailed in the brewing literature (Simpson and Hammond, 1990).

To prepare compressed yeasts, slurried yeast at 4°C with added salt is filtered in a cloth press or by a rotary drum vacuum filter. The resultant yeast is 27-33% dry matter and has a consistency similar to that of butter. It is then extruded into blocks and stored at 4°C where it may lose 5-10% viability per week. It is best used within 30 days; and must be shipped and stored at refrigerated temperatures.

Active dry yeast (ADY) is made by extruding compressed yeast in spaghetti-like strands which are placed in an appropriate dryer to remove moisture (Bayrock and Ingledew, 1997a,b). The tiny cylindrical pellets of dry yeast produced in air lift (fluidized bed) dryers are relatively stable especially if stored under inert nitrogen gas. They require no refrigeration (but may benefit from storage at 10°C or below), and only lose ~1% activity per month under these conditions. This product is very useful to small bakeries and those not located near the source for shipping of slurry or compressed product. For home baking, active dry yeast is a miracle product. These preparations may have as many as $2.2\text{-}2.5 \times 10^{10}$ viable yeast cells per gram. The traditional technique of saving a portion of leavened bread dough for the next batch may soon be a long forgotten part of the art of making bread owing to development of ADY.

Cream yeast and compressed yeast are at times used in fuel alcohol distilleries. However, the technology of active dry yeast production for baking has now been expanded to drying specialty yeasts for brewing, enology and industrial alcohol production. These are used as direct inocula for the corresponding fermentations, and have reduced significantly the need for yeast propagation, expensive storage, microbiological expertise and microbiological surveillance at alcohol plants. Predictability of fermentation has been significantly increased. The major impact of these products has been in the smaller companies. In larger plants, active dry yeasts or selected company cultures are sometimes propagated in stages of increasing volume or 'conditioned' prior to use. During this time a small increase in cell number occurs (anaerobically), preparing the yeast in advance for the medium used in fermentation and lowering slightly the cost of yeast procurement. However, the availability of a shelf-stable dried microbial inoculum, where each gram contains $\sim 2.2 \times 10^{10}$ viable yeasts, is a major biotechnological development. Active dry yeasts are easily used in the distillery, fuel alcohol and winery industries to inoculate fermenters to recommended values of 1.0-2.0 million cells/mL per °Plato[1] of wort or mash (Casey and Ingledew, 1985). When rehydrated and inoculated, active dry yeasts produced under aerobic conditions will initiate fermentation with no apparent difficulties. Other microbes are similarly being dried for the fermentation industry and for probiotics.

In an alcohol plant, mash is fermented under anaerobic conditions in spite of the very small amount of air that should be supplied at or near the time of initial inoculation. Air addition can be carried out by 'rousing' the cooled mash to give 8-20 ppm O_2 using air or bottled oxygen or helped by 'splash filling' of the fermenter usually just prior to adding yeast. This amount of oxygen, *if* used aerobically, would only be enough to allow respiration of 0.008% w/v glucose. It therefore has no effect on sugar utilization; and the oxygen is probably all used anaerobically due to the excessive amounts of glucose present in the mash. Oxygen will remain in the fermenter headspace until used or until it is replaced by metabolically-produced CO_2. Oxygenation at the above low levels is a useful practice and has been

[1]grams extract (as sucrose) per 100 g solution as determined by hydrometer or density meter.

carried out for many years by the brewing industry. Recent research in this laboratory has shown that the most effective time to ensure oxygen availability is when the yeasts begin to actively grow (O'Connor-Cox and Ingledew, 1990; Yokoyama and Ingledew, 1997). Yeasts require this small amount of oxygen to synthesize membrane sterols and unsaturated fatty acids they are unable to make in the complete absence of oxygen (Kirsop, 1982). In addition to the synthesis of sterols and unsaturated fatty acids, oxygen is required for a number of hitherto poorly understood functions (Thomas *et al.*, 1998). When oxygen is unavailable, available sterols and unsaturated fatty acids are partitioned between mother and daughter yeast cells to ensure cell membrane integrity. In time, these compounds are diluted below a threshold value required for growth. Cell multiplication ceases, therefore reducing ability of the cells to mediate or complete fermentation and making them more susceptible to the wide range of external stresses (temperature, ethanol, salts, acids, sugars) shown in Figure 2. It is not yet known if these ions and chemical stress agents act synergistically when more than one is present.

Sugar utilization

Yeasts are considered facultative anaerobes; ie., they are capable of growing in the presence of oxygen or in its absence. When oxygen is sufficient and substrate level is kept low, little or no ethanol is made by the growing yeast. All sugars are utilized for energy production and to make new cells. When oxygen is reduced and/or glucose levels exceed 0.1% w/v, ethanol is made. Yeasts then ferment sugars derived from any of a number of possible substrates (in our case, enzymatically produced glucose from corn starch) to make alcohol even in the presence of the small amount of oxygen in the mash. The yeast cell mass produced under such 'anaerobic' conditions is less than 5% of the weight of sugar

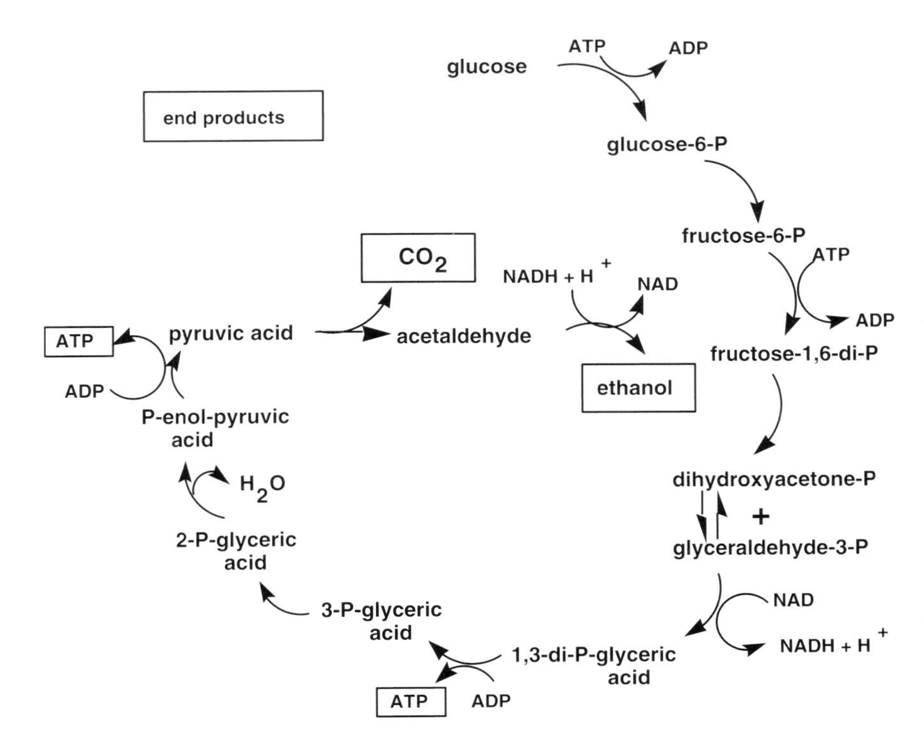

Summary reactions: Glucose + 2ADP + $2P_i$ + $2NAD^+$ → 2 Pyruvate + 2ATP + 2NADH + $2H^+$

2 Pyruvate + 2ADH $^+$ $2H^+$ → 2 Ethanol + $2CO_2$ + 2NAD

Figure 4. The Embden-Meyerof-Parnas (glycolytic) pathway for sugar utilization.

metabolized, and only one tenth of the cell yield that could be realized under aeration in non-repressing concentrations of glucose. (For more detailed information, the reader is directed to the review by Fiechter and Seghezzi, 1992). Yeasts use the glycolytic or Embden-Meyerof-Parnas pathway (Figure 4) to break down or dissimilate sugars into energy and intermediates the yeast requires for growth. In the process, substantial amounts of ethanol and CO_2, the major end products, as well as heat, are produced. This pathway is mediated completely by enzymes. That is, after glucose has entered the yeast cell, enzymes under physiological conditions catalyze the reactions shown in Figure 4, changing the glucose to chemical intermediates at a very fast rate. Eventually, the six carbons of most of the glucose molecules are converted to two 3-carbon molecules of pyruvic acid. Some of the pyruvic acid and other intermediates can be used by the yeast cell through a variety of metabolic pathways as 'building blocks' for new yeast cells. Some glycerol is also made from the intermediate, dihydroxyacetone, but most pyruvic acid is immediately converted to ethanol and CO_2. The pathway also yields the energy the yeasts require to carry out the limited amount of growth that occurs anaerobically. The yeast's objective is to grow and survive, and therefore one cannot divorce alcohol synthesis from cell growth and energy production! Growth and fermentation reactions are tightly coupled.

The glycolytic pathway operates in the presence or the absence of air (oxygen) to convert glucose to pyruvic acid, energy and reduced nicotinamide adenine dinucleotide ($NADH + H^+$).

Heat is also given off in the reaction, although much of the energy liberated from the biochemical steps is conserved by the yeast and stored as adenosine triphosphate (ATP) for use in biosynthetic reactions.

Ethanol production

Yeasts are not very tolerant of high concentrations of acidic end products like pyruvic acid. Through evolution, they have developed a way to 'detoxify' this acidic end product by converting the pyruvic acid into CO_2 and ethanol - both of which are easily excreted. In doing so, NADH formed in glycolysis is re-oxidized to NAD as acetaldehyde is converted to ethanol. The NAD is then available to participate again in glycolysis. This provides for oxidation/reduction balance in the cells. In this way, the yeasts are able to continue to grow and metabolize sugar. The 2-step reaction leading to alcohol can be written:

$$\underset{\text{pyruvate}}{CH_3\text{–}C{=}O\text{–}COOH} \xrightarrow[\text{pyruvate decarboxylase}]{\nearrow CO_2} \underset{\text{acetaldehyde}}{CH_3\text{–}CHO} \xrightarrow[\text{alcohol dehyrogenase}]{NADH + H^+ \nearrow NAD} \underset{\text{ethanol}}{CH_3\text{–}CH_2OH}$$

Ethanol yield

The conversion of starch to alcohol then becomes as shown in Figure 5 (with theoretical yields indicated). The theoretical ethanol yield based on moisture-free starch from grain can be calculated at 56.7% w/w; and the yield of ethanol based on glucose generated from starch would be 51.1% w/w.

A calculation done to determine the 'in-plant' yield is normally based on starch content of corn, but must be based on starch content in moisture-free corn. One tonne of corn at 13.2% moisture contains 868 kg corn. If the starch content was 71.8% (dry weight basis), one tonne of corn (13.2% wet weight) would contain 623 kg starch. On hydrolysis, the net weight of monomers is greater than the weight of the starch due to the hydrolysis of the chemical bonds in starch creating a total of 623 x 180/162 = 692 kg of potential glucose. The theoretical yield of ethanol from glucose is 51.1% (above), therefore 353 kg of ethanol (or 448 L, as the density of alcohol is 0.789 g/mL) can potentially be made. If one knows the actual amount of ethanol made by the plant per tonne of corn, the efficiency can be calculated by the formula in Figure 6. The

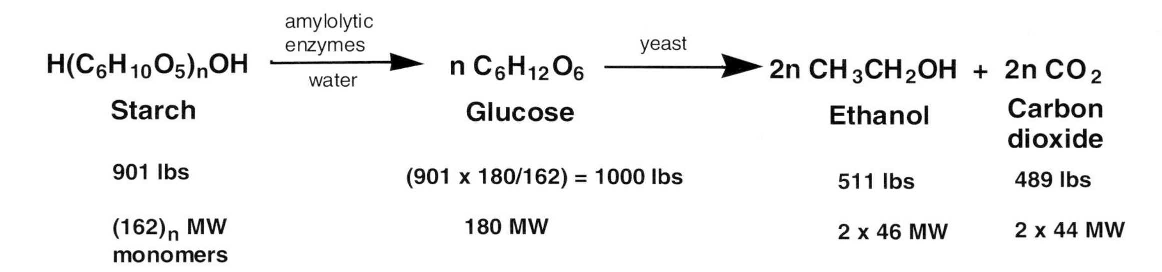

Figure 5. The conversion of starch to alcohol.

conversion efficiency in most alcohol plants is at best 90-93% due to cell growth, production of minor end products of metabolism and less than complete enzyme hydrolysis of starch. If so, one would expect 403 to 417 L of ethanol per tonne of moist corn. If a plant exceeds these calculated outputs, it can be concluded that the corn had less than 13.2% moisture, the starch content was more than 71.8%, or that small increases in efficiency over the 90-93% range were realized. Ideally, moisture content and starch content should be known for every delivery of corn.

Conversion of the above figures to non-metric equivalents is based on the assumption of a 56 lb bushel of corn and the constant factor of 1/3.79 to convert a liter to a US gallon. One then obtains what for some are the more familiar yields of 2.69-2.79 US gallons of alcohol per bushel corn. Interestingly, ethanol yields in the 2.45-2.5 gal/bushel range are common in industry. Attention to microbiological problems and the more refined technologies proposed by some alcohol plant designers may lead to the production of the calculated higher yields of ~2.69-2.79 gal/bushel.

Fermentation efficiency

The efficiency of alcohol production can be expressed as shown in Figure 6. As industrial efficiencies vary between 90 and 93% of theoretical, there is little room for further improvement in the ability of yeast to make alcohol. Unfortunately, fermentation efficiencies can rarely be calculated as easily as shown because it is not normal for plant operators to know the exact starch content, the moisture of the grain used, or even the consistency of such values in the total grain shipment. Therefore, it is not possible to calculate accurately the glucose yield produced by the action of amylase enzymes. The presence of insoluble grain particles and their contribution to the volume of mash in the fermenter also affects the calculation (Thomas *et al.*, 1996a).

Factors affecting alcohol yield

A number of factors can reduce yields of alcohol and lower productivity of a plant. These factors must be reduced or eliminated to minimize

$$\text{\% fermentation efficiency} = \frac{\text{actual weight of alcohol produced}}{\text{theoretical weight of alcohol from produced glucose}} \times 100$$

Figure 6. The efficiency of alcohol production.

alcohol losses and maximize both yield and profits. When ethanol yield is increased, distillers dried grains with solubles (DDGS) production obviously decreases, but Lewis (1997) has shown that because ethanol is worth more per pound than DDG(S), this increase in ethanol is 'value added' technology.

Losses in mashing

Decreased alcohol yield due to unconverted starch in mashing results from poor milling (whole and part kernels found), less than complete enzyme hydrolysis, retrogradation of starch, or starch 'blinding' by protein. In such cases, a measurable quantity of unhydrolysed starch is found in stillage. This is rarely checked in industry. Poor enzymatic conversion can also occur due to a lack of plug flow in continuous cooking and in saccharification tanks, or due to the presence of lumps of starch which are not homogeneous and not available to enzymatic attack. Starch can easily be overestimated when grain is dark or when distressed grains or screenings are used.

Cell growth lowers ethanol yield

When yeast is put into any suitable microbiological medium under favorable temperatures and pH, it grows and metabolizes until substrates (nutrients) are exhausted, or lethal levels of toxic metabolic end products accumulate. Growth of *Saccharomyces cerevisiae* under anaerobic conditions is tightly coupled to production of ethanol. Moreover, growing yeasts may produce ethanol and carbon dioxide at a rate ~33 times higher than that observed with resting (non-growing) cells (Kirsop, 1982). To maximize ethanol productivity, therefore, we must expect some substrate to be used for cell mass production. By keeping cells longer in the growth phase, more carbohydrate is efficiently converted to alcohol, fermentation rates are maintained, and the ferments are more predictably completed.

Yeasts channel glucose through a number of metabolic pathways to synthesize building blocks for growth. There is a continual 'bleed' of intermediates from the Embden-Meyerof-Parnas or glycolysis pathway and other metabolic routes in order to make amino acids, proteins, lipids, carbohydrate polymers, and nucleic acids for new cells. The citric acid (tricarboxylic acid or Krebs) cycle (Figure 7) in conjunction with oxidative phosphorylation produces an excess of energy when adequate oxygen is available and when glucose is present at non-repressing concentrations of less than 0.1% w/v. However, when oxygen is limiting or when glucose repression is in effect (as in the alcohol fermentation), oxidative phosphorylation does not operate or contribute to the energy requirements of the cell. In fact, the citric acid cycle, normally the most important pathway for energy and intermediate synthesis in an aerobically grown cell, only functions during fermentation in sections in a biosynthetic role (Fraenkel, 1982) for the manufacture of a few key chemical precursors - not at all for energy production. The cycle is interrupted at α-ketoglutarate oxidation; only a small amount of synthesis in the normal (oxidative) direction may take place (Gancedo and Serrano, 1989). Oxaloacetic acid (made from pyruvic acid by CO_2 fixation) is used to re-initiate the cycle; and this intermediate is also shunted 'backwards' to make malic acid, fumaric acid, and succinic acid.

The hexose monophosphate pathway also functions anaerobically. It consumes glucose in order to form pentose (5-carbon) sugars for nucleic acid biosynthesis and tetraose (4-carbon) sugar for aromatic amino acid formation (Sols *et al.*, 1971; Gancedo and Serrano, 1989).

Obviously, if glucose is used as a carbon source for cell growth, it is no longer available to make ethanol. As growth and ethanol production are coupled in *Saccharomyces* yeasts, some potential ethanol must be sacrificed to obtain enough growing yeast cells to catalyze the remaining sugar in an efficient manner to ethanol. Yang and coworkers (1982) have calculated that as much as 3.1% of the initial substrate present was converted to cell mass in normal gravity corn mash fermentations. Harrison and Graham (1970) reported the production of 1.9-2.1 g of cells/L.

The following example (Ingledew, 1993) shows how cell growth influences ethanol yield. If the weight of inoculated yeast increases from 0.5 mg/mL (~2 x 10^7 cells/mL) to 5 mg/mL (~2 x 10^8 cells/mL) during growth in a normal fermentation, we would have a net synthesis of ~4.5 mg cells/mL. Assuming a formula for yeast[2] of $C_{3.72}H_{6.11}O_{1.95}N_{0.61}$ (Harrison, 1967; Nagodawithana, 1986), and therefore that approximately 49.3% of the cell is carbon, then the net 4.5 mg cells/mL that were made would contain 2.2 mg/mL of carbon. Assuming that all carbon comes from supplied glucose, 2.2 mg/mL x 180 ÷ [6 x 12] = 5.5 mg of glucose/mL would be needed to grow these yeasts. If these cells were *not* grown, this amount of glucose might be fermented to provide nearly 0.27 g of ethanol/100 mL. By a similar calculation, it can be shown that stimulation of fermentation by the addition of both yeast food and the small amounts of required oxygen could result in a total cell yield of 12 mg/mL of yeast and could lead to reductions in fermentable glucose of about 14.5 mg/mL and therefore a potential loss of 0.73% w/v ethanol. A yield of 0.46% more ethanol would then be sacrificed to achieve the greater mass of actively fermenting yeasts, to enhance the rate of the fermentation, and to ensure that a high sugar (high gravity) fermentation finishes. The real loss of ethanol, however, is much less than calculated above due to other nutrients in mash and in yeast foods added - and the fact that all cell carbon does not come from glucose. Whatever the real loss of ethanol, it must be balanced against the benefits of much more rapid, trouble-free batch fermentations yielding higher alcohol content per unit volume. Ethanol cannot be produced without significant cell growth. High concentrations of sugar ferment

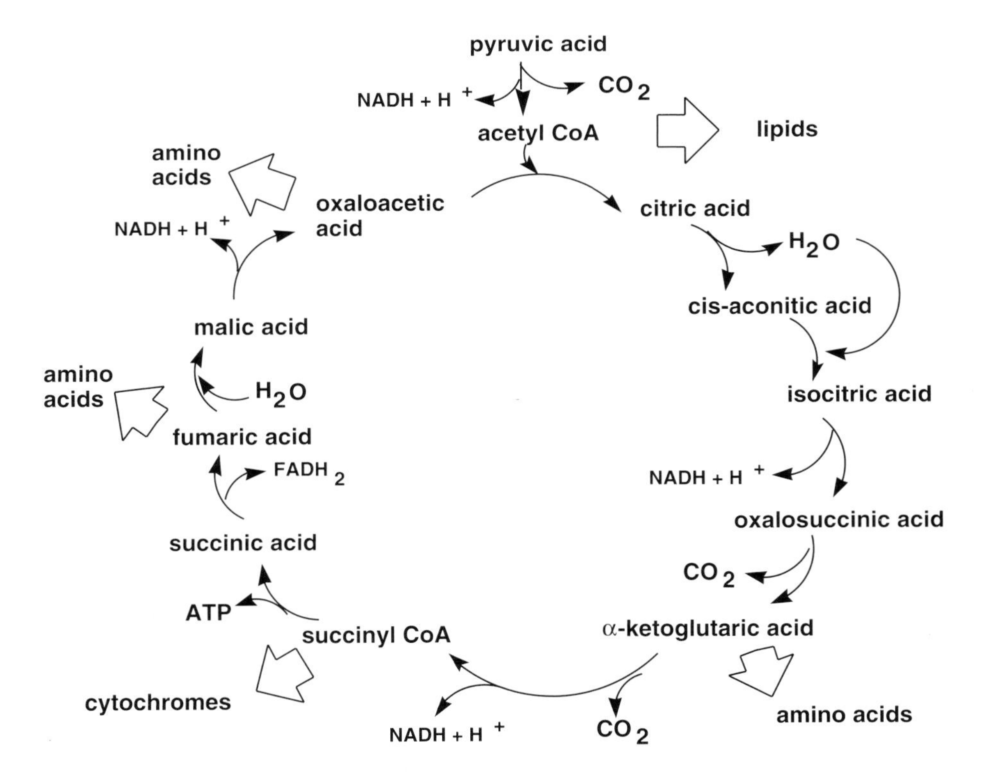

Figure 7. The citric acid (Kreb's or TCA) cycle for synthesis of intermediates and energy under aerobic conditions.

[2] a calculated formula from aerobically-grown bakers yeast. It satisfies for estimation purposes.

slowly or stick when growth is not stimulated. A slight reduction of ethanol yield due to cell growth must therefore be tolerated.

'Minor' end products of metabolism lower ethanol yields

Organic acids

Succinic acid and glycerol (see below) are two secondary end products of the alcohol fermentation. Pyruvic, malic, fumaric, oxaloacetic, citric, α-ketoglutaric, glutamic, propionic, lactic and acetic acids are also produced during alcoholic fermentation, and small amounts remain in the 'beer'. Some of these organic acids are made via functioning enzymes through the limited operation of the citric acid cycle. In potable beverages, many of the organic acids can affect flavor; and they also are formed into esters. Another group of organic acids is made by fatty acid biosynthetic pathways leading to saturated and unsaturated fatty acids used in membrane lipids (Berry and Watson, 1987). It should be pointed out however that unless oxygen is present in small quantities when growth begins, desaturation steps do not proceed and unsaturated fatty acids and sterols cannot be synthesized by the cells. Contaminating bacteria also alter the normal profile of organic acids. Increases in lactic and acetic acids especially are observed.

Acidic end products can inhibit the growth of yeast if produced in excess or if recycled and built up in backset. In this laboratory, it has been shown that concentrations of lactic acid over 0.8% and of acetic acid over 0.05% affect the growth of yeasts. Such levels are easily produced by the five Lactobacillus species we have isolated from commercial alcohol plants and used to discover the significance of bacterial contamination on the ethanol fermentation (Narendranath *et al.*, 1997). Lactobacilli are very ethanol tolerant and are capable of very rapid growth in distillery mashes. Lactic acid is the main end product of their metabolism. In addition, acids (by lowering pH) may affect continuing glucoamylase activity on residual dextrins in the fermenter. Levels of acids will vary from 0.5-1.4 g/L depending on the nitrogen source and can rise dramatically (to over 15 g/L) when bacterial contaminants or wild yeasts are also present (Narendranath *et al.*, 1997). When bacteria are present at high levels, ethanol yield can be reduced by 3.8-7.6%. The major reasons for reduced yeast growth and viability are the production of lactic acid and the competition of the bacteria for essential growth factors in the fermenting medium. Even a 1% decrease in ethanol is a significant loss due to the small profit margin in this industry (Makanjuola *et al.*, 1992; Narendranath *et al.*, 1997). A 1% loss in a small plant producing 12 x 10^6 L of ethanol/year is a loss of >$35,000/year. Losses of up to 5% in isolated incidents are not uncommon.

Glycerol

Higher than normal mash pH (Table 6), increased osmotic stress, lower flux of pyruvate due to utilization of glycolytic intermediates for biosynthesis (intermediates subsequent to the step in the pathway producing reduced NAD), or addition of sulfite into the fermentation can all stimulate the conversion of dihydroxyacetone phosphate (an intermediate in glycolysis) to glycerol (below). Glycerol production results in re-oxidation of accumulated NADH generated in the glycolytic pathway, and detracts from ethanol yields. Yang *et al.* (1982) reported glycerol levels in normal corn mash fermentations to be as high as 7.2 % w/w of the initial sugar consumed (approximately 1.4 g/100 mL if 20% sugar is used). Most alcohol fermentations contain glycerol at values of about 1.0 % w/v.

$CH_2O{\sim}P$ | $C{=}O$ | CH_2OH (glycerol phosphate) $\xrightarrow[\text{glycerol phosphate dehydrogenase}]{NADH + H^+ \rightarrow NAD}$ $CH_2O{\sim}P$ | $CHOH$ | CH_2OH (dihdroxy-acetone phosphate) $\xrightarrow{\text{phosphatase}}$ CH_2OH | $C{=}O$ | CH_2OH + P (glycerol)

glycerol phosphate **dihdroxy-acetone phosphate** **glycerol**

Table 6. The influence of pH on production of glycerol and organic acids*.

Product[a]	*pH 3.0*	*pH 4.0*	*pH 5.0*	*pH 6.0*	*pH 7.0*
Ethanol	171	177	173	161	150
Carbon dioxide	181	190	188	177	161
Glycerol	6.2	6.6	7.8	16.2	22.2
Acetic acid	0.5	0.7	0.8	4.0	8.7
Lactic acid	0.8	0.4	0.5	1.6	1.9

*adapted from Neish and Blackwood, 1951 by permission.
[a]millimoles of product per millimole glucose fermented.

Interestingly, it is now considered that the degradation of glycerol by lactic acid bacteria results in the production of acrolein (see below), which reacts with phenolic groups in the mash to form bitter compounds. Acrolein has a very peppery, lacrimating and irritating effect on workers operating a still. Not all lactic bacteria will degrade glycerol, but it would seem that mashes containing high numbers of the contaminating lactic acid bacteria able to degrade glycerol will contain higher amounts of acrolein; and problems at the still may then occur. Certain lactic bacteria degrade glycerol using a dehydratase enzyme forming 3-hydroxy-propionaldehyde, which spontaneously degrades (acidic conditions) to acrolein in storage or when heated (Sobolov and Smiley, 1960).

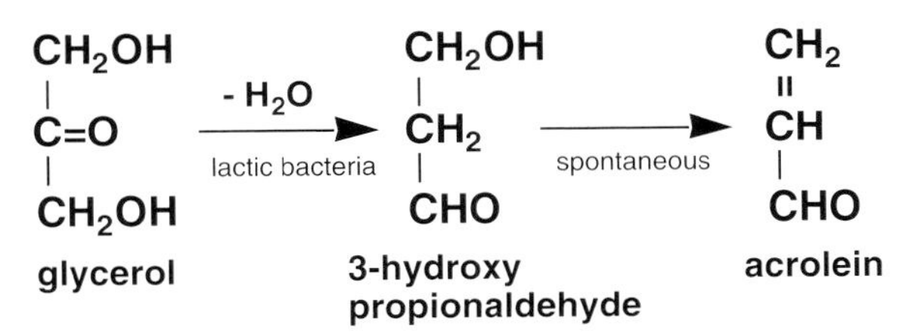

Higher alcohols

From a quantitative point of view, the most important higher alcohols (fusel oils) are n-propanol, amyl alcohol, isoamyl alcohol, isobutanol and phenethyl alcohol. These are made directly by anabolic (synthetic) pathways from keto acids, or by catabolic pathways directly from amino acids (Erlich pathway). In the latter mechanism, a specific amino acid is transaminated to the corresponding α-keto acid. The keto acid is then decarboxylated to form an aldehyde which in turn is reduced to an alcohol by alcohol dehydrogenase, the same enzyme that catalyses the conversion of acetaldehyde to ethanol. A typical reaction leading to the formation of isoamyl alcohol (3-methyl butanol) is shown in Figure 8. Yang *et al.* (1982) have reported fusel oil levels as high as 0.12% w/w of the initial substrate in normal corn mash fermentations (approximately 0.024 g/100 mL from 20% sugar). The level and composition of yeast-assimilable nitrogen in wort will influence the formation of higher alcohols. Low levels of usable N result in less yeast growth and increased yields of higher alcohols. If assimilable free amino nitrogen is increased in the medium, the anabolic pathway becomes less dominant and the Erlich pathway takes over. If ammonium ion or urea is the main source of nitrogen supplied, all higher alcohols are made via the anabolic (synthetic) pathway.

Levels of higher alcohols produced in batch fermentation systems vary with: yeast strain (levels produced by *Saccharomyces cerevisiae* are higher than that produced by *Saccharomyces carlsbergensis),* temperature (higher temperature results in higher alcohols), aeration and agitation (increased amounts lead to production of the higher alcohols) and wort composition (N sources and readily metabolizable sugars)

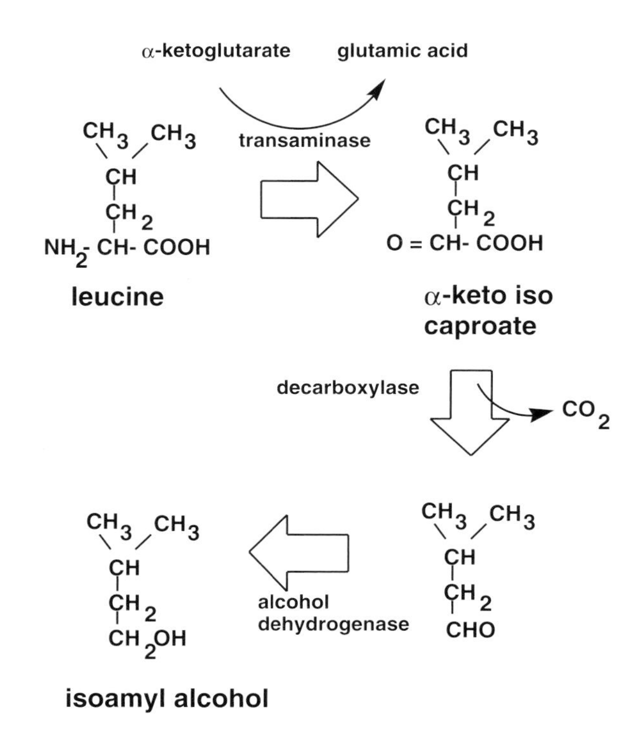

Figure 8. Conversion of leucine to isoamyl alcohol as an example of higher alcohol production.

Esters

Many esters are made by yeast during fermentation. The predominant esters, derived from the most abundant alcohols (ethanol, then isoamyl alcohol, then propanol) and acids (acetate), lead to esters such as ethyl acetate (Figure 9). The importance of each ester relates to its concentration and its flavor threshold, if applicable (Ingledew, 1979; Berry and Watson, 1987). Ester concentrations are generally quite low (less than 175 mg/L), but may increase in fermentations as a result of factors such as availability of the coenzyme derivative of the acid (acetyl CoA predominates), temperature, presence of the alcohol (ethanol in excess, others in limited supply), yeast strain variability, oxygenation (low aeration throughout fermentation suppresses ester formation), addition of unsaturated fatty acids (i.e., grain pressings), mash composition (more esters when gravity was increased) and the presence of increased α-amino nitrogen.

Other compounds

A number of aldehydes such as acetaldehyde (produced at about 1 L/1000 L ethanol or approximately 120 µL/L of medium if 20% sugar was used), ketones such as diacetyl (and related compounds acetoin and 2,3-butanediol) and volatile sulfur compounds (diethyl and dimethyl sulfide) are formed but are normally in low concentrations. They are more of a concern as flavor and aroma problems when present in alcoholic beverages. Diacetyl, for example, has an aroma threshold value of approximately 100 µg/L, while the sulfur compounds, diethyl sulfide and dimethyl sulfide, approximate 0.4 µg/L and 50µg/L, respectively. Information on the

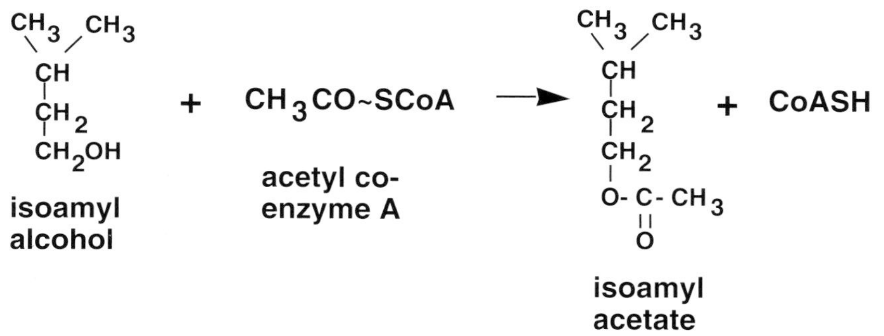

$$(CH_3)_2CH{-}CH_2{-}CH_2OH + CH_3CO{\sim}SCoA \longrightarrow (CH_3)_2CH{-}CH_2{-}CH_2{-}O{-}C(=O){-}CH_3 + CoASH$$

isoamyl alcohol + acetyl co-enzyme A → isoamyl acetate + CoASH

Figure 9. Formation of isoamyl acetate from isoamyl alcohol and acetic acid (coenzyme form).

synthesis of these compounds is given in Engan (1981) and Berry and Watson (1987).

$CH_3COCOCH_3$ diacetyl	$CH_3CHOHCOCH_3$ acetoin
$CH_3CHOHCHOHCH_3$ 2,3-butanediol	CH_3CHO acetaldehyde
$CH_3CH_2SCH_2CH_3$ diethyl sulfide	CH_3SCH_3 dimethyl sulfide

Infections will lower ethanol yields

Although the numbers must normally be very high ($>10^6$/mL), bacteria interfere with yeast metabolism through competition for nutrients and production of end products such as acetic and lactic acids that inhibit yeast growth. Bacterial metabolism can also subtract significantly from ethanol yield. This is because the stoichiometric conversion of glucose to an end product such as lactic acid through bacterial action results in a stoichiometric reduction of the amount of ethanol capable of being produced from glucose by yeast. Bacteria, or more likely their by-products (like lactic acid), can also be recycled in backset and therefore can infect or chemically alter the contents of every fermenter in the plant. Volatile acids (for example, acetic acid) can be recycled in process condensate water; and even at low concentrations can have a startling effect on yeast viability.

Figure 10 illustrates the various by-products developed by the groups of microorganisms that are a problem in distilleries and which might at times develop in fermentation tanks or in spent grains. The mixed acid bacteria including coliforms are only a problem in mash above pH 5.0 as they are not tolerant to the lower pH values in fermenting mash. Butyric acid bacteria are found mostly in wet grain piles that have been allowed to 'incubate' in farmyards or processing plant areas.

Not shown in Figure 10 is the hetero-fermentative degradation of glucose by certain strains of lactic acid bacteria. Heterolactic bacteria degrade glucose (and pentose); and some homofermentative bacteria degrade pentose through an alternate pathway that converts glucose or pentose to the 5-carbon sugar xylulose, which is enzymatically converted to one molecule of acetyl phosphate and one molecule of glyceraldehyde (Axelsson, 1998). Depending on strain, the acetyl phosphate can be converted to acetic acid (the source of volatile acidity in fuel alcohol production?) or ethanol; and the glyceraldehyde is converted into lactic acid. These lactic bacteria may be solely responsible for the acetic acid production seen in a number of fuel alcohol situations. In addition, aerobic Acetobacter and Gluconobacter bacteria can be present in unique situations such as in propagation tanks where ethanol is present and when the head space of the vessel contains air. In this special case, ethanol may be converted to acetic acid by these vinegar bacteria.

Stuck (sluggish) fermentations affect ethanol yield

The fifth major factor to affect alcohol yields is the stuck or sluggish fermentation. Stuck fermentations are commonplace in very high gravity brewing above 16° Plato, and in winery fermentations of white grape juices in the 20-24° Brix range, but are also a problem at lower levels of sugar such as in high gravity brewing at 13-16° Plato, and in alcohol distilleries. Sluggish or stuck fermentations occur when the rate of sugar utilization becomes extremely slow or protracted, especially toward the end of fermentation.

Work on stuck fermentations is rooted in the basic science explaining ethanol tolerance and the osmotic tolerance of yeasts which make ethanol. Much of this science is not fully understood; new discoveries frequently appear in the scientific literature. However, practical information about alcohol has already been exploited to some degree in brewing, in enology, and in a limited number of fuel or potable alcohol distilleries where production of more than 14% v/v ethanol is now a reality.

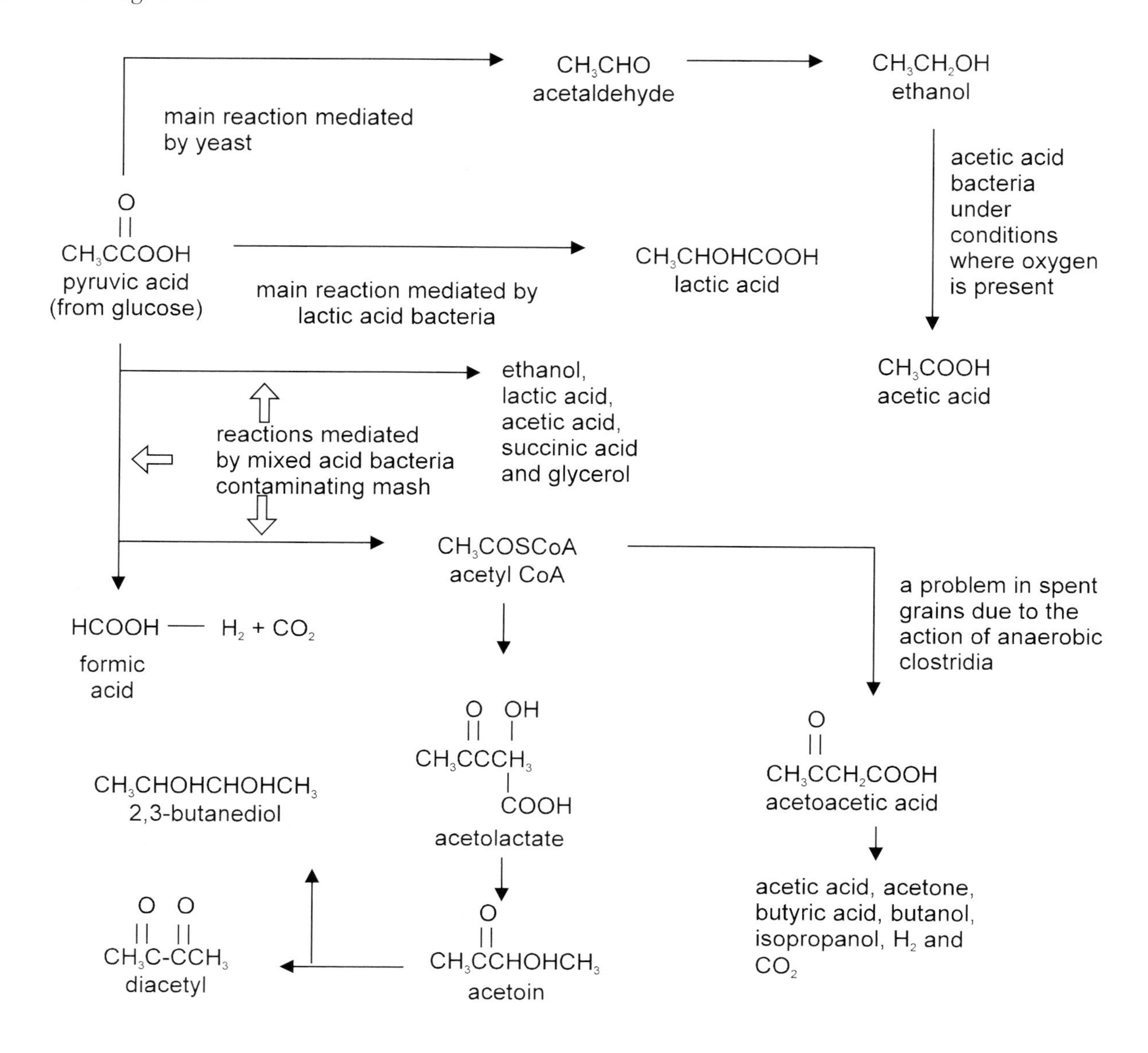

Figure 10. The key role of pyruvic acid (from glycolysis) in the production of spoilage end products in fermentations contaminated with bacteria.

Information on avoidance of stuck fermentation is of value even if a particular plant does not wish to attempt the fermentation of high amounts of sugar to make higher concentrations of alcohol than is normal in that alcohol plant. The technology is designed to keep the yeast cells robust and able to metabolize sugar, and growing actively as long as possible. Nutrition is the secret to the 'black box' of Figure 1 that scientists often refer to because of the biological difficulties they do not clearly understand.

As indicated previously in this article, yeast cell growth and energy production cannot be uncoupled from ethanol production. As a result, some carbohydrate in mash must be budgeted for use in yeast growth instead of for alcohol yield. As yeast growth increases, more catalytic activity is present in the mash to convert sugar to ethanol. Cell growth relies on a number of nutrients in addition to the sugar, and therefore it is also important to consider substrates other than sugar that are available in corn or wheat mashes. When sugars are high but other nutrients are limiting (as would be the case when candy waste or isolated corn starch slurry is fermented), stuck and sluggish fermentations result. Such a medium is not nutritionally balanced. Research in this laboratory has shown that in brewing and enology, stuck and sluggish fermentation of high and very high gravity worts or juices are always caused by inadequate levels of yeast nutrients such as assimilable nitrogen and small amounts of oxygen. This leads to a slowing or cessation of yeast growth. Nutrient deficiency also affects the ethanol tolerance of the yeast (Figure 11).

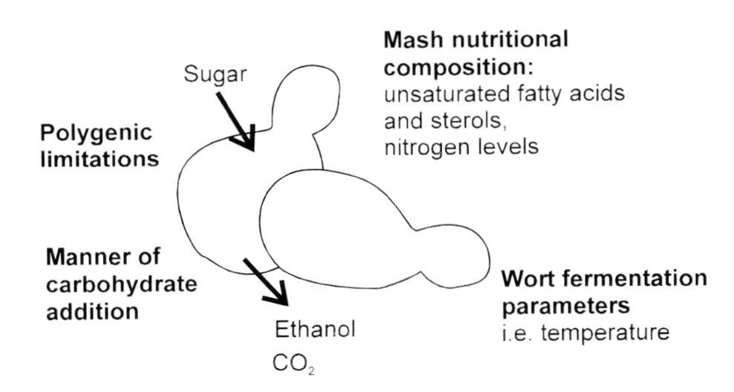

Figure 11. Factors that dictate ethanol tolerance of Saccharomyces yeasts (Casey and Ingledew, 1985, by permission).

Nitrogenous constituents

Usable or assimilable nitrogen becomes problematic in wort, juice or mash fermentation media. Yeasts are only able to utilize low molecular weight nitrogenous materials such as inorganic ammonium ion, urea, amino acids and small peptides. Yeasts used in industrial and potable alcohol manufacture are not proteolytic. They cannot derive assimilable nitrogen for growth from protein or the breakdown of peptides much larger than tripeptides (Patterson and Ingledew, 1999; Ingledew and Patterson, 1999). Levels of ammonium ion, urea and free α-amino nitrogen (FAN) in the medium can therefore be growth limiting. This is certainly the case in brewer's worts and in varietal grape juices such as Chardonnay. Starvation for nitrogen, a common finding in fermentation systems, results in catabolism of amino acids. In this case, basic amino acids are converted on entry to their N-free carbon skeletons. Deaminated carbon skeletons of hydrophobic amino acids are then excreted as α-keto acids and the corresponding fusel oils. Starvation for nitrogen also leads to a small increase in protein degradation inside the cell (Cooper, 1982). Proteinases have been characterized from yeast in spite of the fact that Saccharomyces yeasts appear unable to metabolize exogenously supplied proteins or peptides in the medium. Proteolysis, therefore, is intracellular unless lysis of intact cells occurs.

In brewing, traditional gravity worts of approximately 12° Plato normally contain 100-250 mg/L of free amino-nitrogen (FAN) depending on the amount of adjunct (a carbohydrate replacing malt) used. Yeast growth increases almost linearly with FAN levels up to 100 mg/L (Pierce, 1987). Very high amounts of FAN (up to 878 mg/L) have been shown by some to have no additional effect on yeast growth, but to markedly stimulate the fermentation rate (Vos and Gray, 1979; Monk *et al.*, 1986). In all-malt worts (and some grape juices), the FAN level exceeds that recommended for growth (>150 mg/L). Stuck fermentations, however, are very common with juices from some grapes (Chardonnay) grown under certain conditions in particular vineyards. Such grapes may have less than 40 mg of FAN per liter (Ingledew and Kunkee, 1985) and are most susceptible to 'sticking'. Nitrogenous sources become increasingly important as the initial sugar level is raised or the purity of the sugar source increases.

If we assume that the formula proposed for the major elements in yeast cell mass of $C_{3.72}H_{6.11}O_{1.95}N_{0.61}$ (Harrison, 1967) is correct, it can be calculated that the nitrogen content of a cell is of the order of 9% w/w. Although this is generally true for aerobically-grown yeast, 9% is a significant overestimation for yeast from anaerobic fermentations where 'natural' nitrogen limitation can result in levels of carbohydrates at times exceeding 65% (Patel and Ingledew, 1973) and therefore nitrogen contents below 5%. The level of protein inside a yeast is a function of the concentration of usable nitrogen (ammonium salts, urea, amino acids, di- or tripeptides and nitrogen derived from nucleic acid breakdown).

Ammonium ion is a preferred nitrogen source for yeast cultivation in laboratory studies. It appears to be utilized by all genera of yeasts, and is usually supplied as ammonia or as the sulfate or phosphate salt (Suomalainen and Oura, 1971). Urea is also utilized by most yeasts although biotin and other growth factors may then be required. Urea, which is not a normal constituent of mashes, is easily broken down to two molecules of ammonium ion and one molecule of carbon dioxide. Unfortunately, its use as a yeast food in potable beverage manufacture is now banned in many countries because it leads to the production of small amounts of urethane (ethyl carbamate), a suspected carcinogen (Ingledew *et al.*, 1987a).

Free amino acids are the main sources of nitrogen in mashes, worts and grape juices. Not all amino acids are well used by Saccharomyces (Suomalainen and Oura, 1971). For example, the Schultz and Pomper survey (1948) showed that alanine, arginine, asparagine, aspartic acid, glutamic acid, leucine and valine promoted good growth of yeast when used as sole nitrogen sources. Isoleucine, methionine, phenylalanine, serine and tryptophan allowed good to moderate growth, while cystine, glycine, histidine, lysine, proline and threonine were poor sources of nitrogen. A more quoted study by Pierce (1987) showed the order (Groups A to D) of utilization of amino acids when all were present. Cooper (1982) determined maximum specific growth rates (μ_{max}) for yeast cells using a mixture of amino acids as nitrogen source, but noted that the exact order of use of amino acids was strain dependent. Although some variance in the utility of single amino acids exists, a balanced nitrogen source (mixture of amino acids) is obviously more efficient in providing nitrogen than a single source. In this case, some amino acids do not undergo transamination, but instead are incorporated directly into protein. Cooper has suggested that good nitrogen sources should undergo rapid transport, should be converted rapidly to glutamate or ammonium ion, and should be non-toxic.

Fermentation problems arise when starch slurries or other relatively pure carbohydrates devoid of yeast nutrients are used for fuel alcohol production. Such sources of nutritionally inadequate fermentable substrate will not support the growth necessary to ensure that enough yeasts grow in the ferment to catalyze complete conversion of sugars to alcohol in a reasonable time frame. A good fermentation rate depends on the addition of corn steep water (or liquor), ammonia and other yeast foods that permit some yeast growth. Molasses-based media can also be deficient in yeast-usable nitrogen and then must be supplemented with inorganic ammonium ion, urea or free amino acids. They are usually also deficient in biotin, sulfur and phosphorus. Moreover, intentional dilution of nutrients by, for example, the addition of a sugar syrup (compounds with no nutritive value other than the sugar content) serves to dilute all non-carbohydrate components while increasing the mash specific gravity. This leads to a nitrogen deficiency that manifests itself in sluggish or stuck fermentations (O'Connor-Cox and Ingledew, 1989). Proof of this phenomenon was provided in reports on commercial malt extracts for home brewing and microbreweries that were shown to ferment poorly due to high levels of added adjuncts (leading to reduced levels of nitrogenous nutrients) (Paik *et al.*, 1991).

The fact that growing yeasts produce alcohol much faster than non-growing (resting) yeasts is probably because yeast cells do not take up and ferment sugars unless there is a need for energy. Non-growing (resting) cells ferment only enough sugar to produce energy for cell maintenance. As energy production, growth and ethanol production are tightly coupled, all efforts therefore should be made to keep yeast under conditions which do not lead to low growth rates or to death. Attention to the nutritive status of mash will prevent premature termination of fermentation. This is especially important when high carbohydrate (high gravity) fermentations are carried out. Whereas very high gravity technology may lead to production of alcohol at lower cost, ignorance of the role of assimilable nitrogen and the need for oxygen or 'oxygenated growth factors' for sterol and unsaturated fatty acid synthesis would compromise this scientific advance, leading to incomplete or inadequate fermentations. It is too late to provide nutrients or new yeast after fermentations become stuck or sluggish.

Yeast foods and fermentation supplements designed to provide nutrients and to make up for nutritional deficiencies of mashes or fermentation media have been used for many years to enhance and ensure predictable fermentation times and product yields. These products were not always designed for yeast fermentations, nor were they specific for grain mash fermentations limited in usable nitrogenous constituents. They were not produced with the complete understanding of the deficiencies in mashes and the requirements of the particular microbes used; and many were found to be poor sources of utilizable nitrogen for the alcohol fermentation (Ingledew *et al.*, 1986). Some contained large amounts of protein (unavailable to yeast); and one supplier provided analytical data showing the amino acid composition of the food as if an acid hydrolysis had been carried out in the yeast food production process.

In much of the work published by this lab group, yeast extract (a water soluble yeast autolysate) was used when a good source of nitrogen was needed. A typical yeast extract preparation contains 28-45 mg FAN per gram, along with approximately 250 mg of protein and less than 1 mg of ammonium ion. Yeast extract is a natural choice as a supplement of usable nitrogen; and it also contains a wide array of vitamins, minerals and other growth factors. In every case in our laboratory, however, the prime function of yeast extract was to serve as a source of assimilable nitrogen in very high gravity and normal gravity fermentations where deficiency in nitrogen was always demonstrable.

Yeast extract is normally expensive, although it is now available for this industry in a technical grade. In distillery mashes where the initial carbohydrate levels are high, the deficiency in nitrogen can be overcome prior to fermentation by the use of such yeast foods and/or the addition of proteolytic enzymes able to create usable FAN from unusable but soluble/insoluble proteins or peptides. Figure 12 shows the benefit of proteases to the alcohol fermentation (Thomas and Ingledew, 1990; Jones and Ingledew, 1994a). This work conducted a cost analysis of proteases per liter of ethanol produced, the enhancement of the rate of ethanol production, and the optimum dosage of one selected protease from the standpoint of FAN release, time taken for full fermentation and enzyme cost per liter. It was also proven that a gross excess of protease had no effect on glucoamylase activity - an essential requirement of the enzyme due to the simultaneous saccharification and fermentation procedure. If bacteriological problems exist in a plant, protease addition may be made after yeast addition so that liberated amino acids are consumed by yeasts (in high numbers) rather than by the bacteria (Jones and Ingledew, 1994a). We have been investigating the use of proteases since the late 1980s in distillery alcohol production. We published on the use of a protease in such an application in 1990 (Thomas and Ingledew, 1990), and can say that commercial proteases were already in use at that time and sold by more than one enzyme company. In spite of that prior art, a patent was filed in 1991 by an enzyme company in Indiana (Lantero and Fish, 1993) which claimed novel use of proteases added with glucoamylase enzyme and also the production of ethanol when a "higher dry solids mash level" is present in the fermenter, and higher ethanol levels are obtained. Proteases may also increase the yield of glucose made from starch in the mashing process by increasing the availability of the starch to amylases. Through enzyme-assisted mash protein degradation, a number of

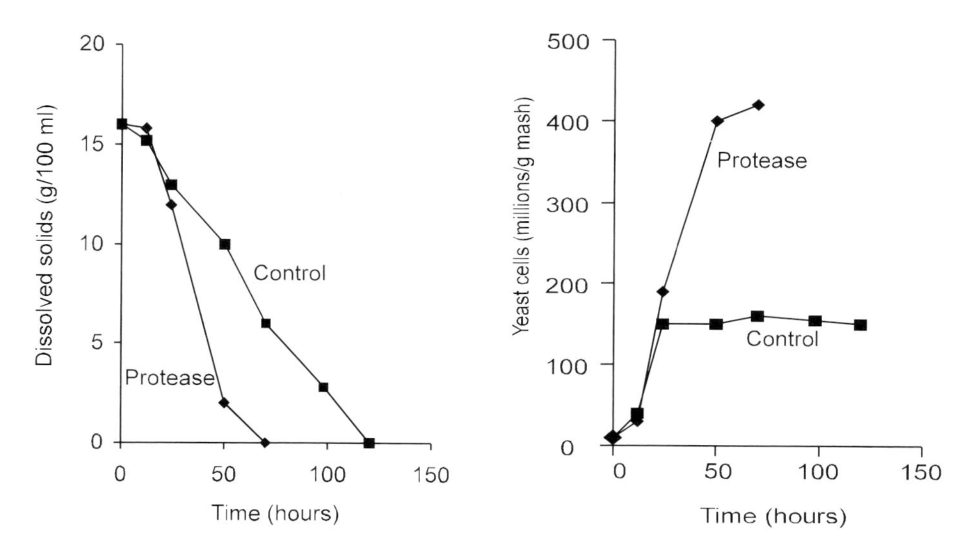

Figure 12. The benefit of protease addition to the alcohol fermentation (Thomas and Ingledew, 1990; by permission).

fermentation improvements are possible. These include increased fermentation rates, improved yeast growth, more consistent fermentations, increased alcohol yields, increased carbohydrate fermented, increased saccharification, reduced carbohydrates in thin stillage, lower viscosities, and higher protein in thin stillage. Other improvements in the process were evaporator performance, lower recycle levels, reduced foaming, reduced fermentation time, increased ethanol output and a more consistent plant operation (Lewis, 1996; 1997).

Yeast extract promotes growth of yeast as well as an increased production rate (but only slightly the amount) of alcohol; and we continued to use it in experiments in spite of its cost. With lab wheat mashes containing 35% dissolved solids, addition of yeast extract reduced the total fermentation time from eight to three days and a final ethanol yield of 17.1% v/v was obtained (Thomas and Ingledew, 1990). With yeast foods, regardless of which of the good nitrogenous source were used, values of ~21-23% ethanol were achieved when carbohydrate levels were increased to over 37% dissolved solids (Thomas and Ingledew, 1992a; Thomas *et al.*, 1994a, Jones and Ingledew, 1994c). These levels of alcohol were produced by normal commercial active dry yeasts without the need for genetic selection or preconditioning. Yeast extracts, special N-containing yeast foods and urea were all able to reduce the fermentation time of a mash containing as much as 36 g of solids per 100 g from 9 days (control) to 4 days, producing 21% alcohol (v/v) in each case (20% in the control) (Jones and Ingledew, 1994c).

Basic science studies have indicated a qualitative difference among amino acids with regard to their suitability to serve as sources of nitrogen in wheat mashes for yeast. Very high gravity (VHG) wheat mashes in excess of 20 g or more dissolved solids per 100 mL, although containing only about 70 mg FAN/L, do ferment to completion (albeit somewhat more slowly than normal gravity mashes; Thomas and Ingledew, 1990). With only this level of nitrogen, stuck or sluggish fermentations had been expected. This led to a study of a number of amino acids including the 'right' kind of amino acids, and their effects on fermentation rate. It has now been shown that some amino acids such as glycine and lysine are readily taken up by yeast but are inhibitory to growth and fermentation only under nitrogen limiting conditions. When usable nitrogenous substances (other amino acids) are in excess, the lysine and glycine are not inhibitory and may even be stimulatory. It is interesting to compare the free amino acid content to the total content of amino acids after mash protein is hydrolyzed. The two are rather different in content. Importantly, the concentration of the inhibitory amino acids (like lysine) in free form appears to be rather low in wheat, perhaps explaining why wheat mash ferments so well (Thomas and Ingledew, 1992b).

Oxygen requirements

Oxygen in small amounts is required by Saccharomyces yeasts for the synthesis of sterols and unsaturated fatty acids, both critical to the function and integrity of yeast cell membranes. Saccharomyces yeasts are unable to ferment sugars and undergo anaerobic growth over an extended number of generations without a source of oxygen or pre-formed lipid precursors of the necessary cell membrane constituents. Ergosterol and Tween 80™ (a source of oleic acid, an unsaturated fatty acid) are useful precursors in lab studies to show that oxygen *per se* is not a requirement. Unfortunately, in many mashes and worts, lipids are in sub-optimal supply and oxygen is not available at optimal levels due to industrial practices. Moreover, oxygen solubility is lower in mashes poised at >35°C that contain increased amounts of solids. When oxygen is depleted, yeast cells will continue to grow and 'dilute out', for a time, the required fatty acids and sterols by sharing them with new daughter cells. Eventually, they become limiting and cell growth stops. Rousing with air or oxygen is routinely practiced in brewing to provide oxygen. Interestingly, although a wide range of oxygen requirements has been demonstrated among yeasts, the amount of oxygen appears to be not as critical as the timing of the supply (O'Connor-Cox and Ingledew, 1990; Yokoyama and Ingledew, 1997). Most effectively (and obviously), oxygen should be supplied during the growth and division phases of yeast. In distillery practice, it may be sufficient to splash fill tankage just prior to fermentation, especially if active dry (aerobically grown) yeasts are used. Experience indicates, however, that clear substrates without insoluble lipid materials present may require more dissolved oxygen. Brewers, for example, will oxygenate worts between 16 and 22° Plato to saturation using bottled oxygen. Such worts have been filtered and are almost devoid of insolubles. The amount of oxygen needed by yeast growing anaerobically is small but important. Deprivation leads to serious difficulties in fermentation.

Sterols in aerobically-grown yeasts may reach 1% total dry weight, but in anaerobic growth with no usable sterols in the medium they dilute out to a value less than 0.1% (David and Kirsop, 1973). Reproductive growth eventually ceases when the limiting value of sterols is reached. Rapid uptake of both nitrogen sources and carbohydrates are then affected. This has a quick impact on the rate of ethanol production. The limiting level of unsaturated fatty acids is 0.5% (Haukeli and Lie, 1979).

In 1985, Ingledew and Kunkee demonstrated that Ruby Cabernet and Chardonnay grape juice fermentations became sluggish and eventually stuck whenever oxygen access was denied during juice preparation and fermentation. If such juice was also low in yeast-utilizable nitrogenous nutrients, the effect on fermentation was even more evident. The set of experiments in Figure 13 graphically illustrates the effect of oxygen and nitrogenous nutrients on fermentation rate. All five experiments used the same juice. Additions to the juices are noted.

Addition of very small amounts of air, even if only in the head space, and excess FAN resulted in very rapid fermentations. Seven day end-fermentations of 25° Brix juice at 14°C rather than 25 to 30 day sluggish control fermentations (not shown) were then possible. Furthermore, it was evident that ergosterol and oleic acid additions could replace the need for oxygen. This proved that oxygen is mainly needed for production of sterols and unsaturated fatty acids for membranes; and is only necessary until higher viable yeast cell numbers (~3 x 10^8 /mL) are

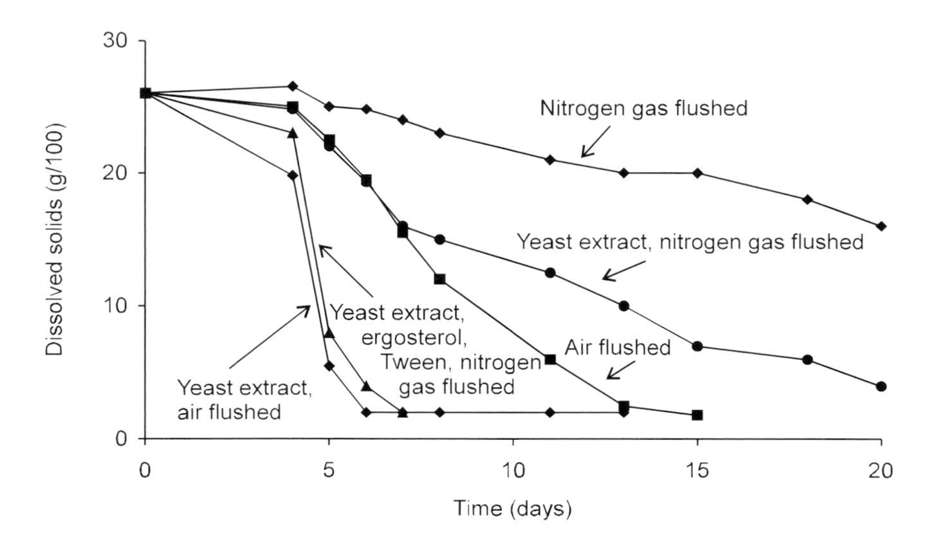

Figure 13. Fermentation of grape juice showing the importance of nutrition to the fermentation rate and predictability of end fermentation (Ingledew and Kunkee, 1985, by permission).

reached - numbers capable of converting high sugar concentrations to alcohol in a practical time frame. However, it is also important to point out that other requirements for oxygen in yeast growing anaerobically can be demonstrated (O'Connor-Cox *et al.*, 1996; Thomas *et al.*, 1998).

Active dry yeasts (ADY) which have been grown in the presence of oxygen (containing a high 'credit' of sterols and unsaturated fatty acids) have been recommended to small brewers, winemakers and alcohol distilleries because of the above phenomena. As these yeasts are grown under conditions of high aeration, their unsaturated fatty acid and sterol content is much higher than in yeasts propagated or conditioned 'in house' in an alcoholic fermentation plant. It is likely that active dry yeasts will complete a fermentation without the same oxygen (or sterol/fatty acid) requirement in the fermentation broth. These yeasts are never reused in wine or distillery practices as they cannot be easily harvested. However, the above consideration of both yeasts and yeast nutrients in wineries has resulted in more rapid fermentations of white wine. This research provided industry with the opportunity to use tankage more than once per harvest period for white wine manufacture.

Many distilleries and fuel alcohol plants already use ADY. Aeration through 'splash filling' of batch fermenters or circulation of mash through heat exchangers for cooling purposes is probably adequate when ADY is used. Aeration at the heat exchanger should be done on the cold side of the exchanger to ensure maximum solubility of oxygen on injection. This may be more important in fermentations done at ~35°C than for white wine production and brewing, which are fermented near 14°C.

Minerals

Depending on the composition of mashes, mineral nutrition could also be important. Phosphorus and sulfur are two important minerals necessary for yeast growth. Phosphorus is taken up in the monobasic form ($H_2PO_4^-$) and can be added to the medium as the acid or in the form of ammonium, sodium or potassium salts. Phosphorus is essential for metabolic reactions in sugar metabolism (sugar phosphates), in lipid synthesis, in the cell membrane and cell wall and for production of nucleic acids in the cell. Yeasts also store polyphosphates as an energy reserve and use this material during times of starvation. A rule of thumb is to be sure that available phosphorus is present at approximately 1-2% of the cell dry weight per unit volume of medium.

Sulfur (given as sulfate ion) is required in yeasts for synthesis of the sulfur-containing amino acids, methionine and cysteine/cystine, and for vitamins for the coenzymes used in metabolism (acetyl coenzyme A for example). Sulfur is required at approximately 0.3-0.5% of the weight of expected yeast mass per volume of medium.

Kosaric *et al.* (1987) have listed the optimum concentrations of inorganic ions for growth of yeast in a continuous stirred tank reactor. Berry and Brown (1987) and Matthews and Webb (1991) have documented the concentrations and function of minerals required for growth of Saccharomyces yeasts. Macro elements (potassium, magnesium, calcium, iron, zinc, and manganese) were all required in concentrations between 0.1 and 1.0 mM. Micro elements (cobalt, boron, cadmium, chromium, copper, iodine, molybdenum, nickel and vanadium) were required between 0.1 and 100 µM. Minerals such as silver, arsenic, barium, mercury, lithium, nickel, osmium, lead, selenium and tellurium can be inhibitory when concentrations exceed 100 µM. Deficiencies in minerals have rarely been described in fermentation media with the possible exceptions of zinc and calcium which, at times, have been reported to stimulate fermentation rates. A proper ratio between calcium and magnesium positively influences fermentation rates (Walker *et al.*, 1996). For yeast, it is also instructive to examine ion concentrations in yeast extract. These levels no doubt resemble those present in whole cells used to make the extract, although they may not be optimal as yeasts often accumulate minerals to very high levels, a phenomenon exploited in the animal feed and neutraceutical industries (Ingledew, 1999).

Vitamins

The growth factors commonly required by Saccharomyces yeasts include biotin, pantothenic acid and inositol. Others of concern may include thiamin, nicotinic acid and pyridoxine. Their importance to yeast metabolism is summarized by Soumalainen and Oura (1971) and Berry and Brown (1987). It is important to note that most literature on this subject is concerned with the aerobic propagation of bakers' yeast. Growth in anaerobic fermentations is only approximately one tenth of that in aerobic growth, and therefore concentrations listed for propagation media can be reduced considerably for the alcohol fermentation. In fact mashes usually contain sufficient amounts, and vitamins are not directly added (unless the yeast food used contains them). Active dry yeasts used for alcohol production only multiply a few generations, and vitamin deficiencies may never be encountered unless yeast recycling is practiced.

Nutritional deficiencies

When considering the effects of any nutrient deficiency, it is worthwhile to reflect on the fact that in traditional batch fermentations, 3 to 10-fold increases in yeast cell numbers occur. Yeasts are normally inoculated at 5×10^6 to 2×10^7 viable cells per mL (approximately 2 lb of active dry yeasts per 1000 US gallons or 1 kg/4000 L are required). New yeast made in the fermentation medium would then be produced and would weigh from 6 to 20 lb (in 1000 gallons). As the nitrogen content of a yeast can be 3-8% w/w depending on glycogen levels, a suitable nitrogen-containing nutrient is needed to provide this level of nitrogen. For 1000 gallons, (where n = the number of N atoms in the nitrogenous yeast food supplied) assuming the mid range value of nitrogen in yeast of 5.5 and of new cell growth at 13 lb dry weight per 1000 gallons (real values should be ascertained in-plant), the amount of nitrogen to be added can be calculated (Figure 14).

In a similar fashion, the phosphate and sulfate ions required by the yeast can be calculated along with a number of other nutrients at much lower concentrations. We must not forget that the mash is really a microbiological medium. If the right

$$\left(\frac{5.5 \times 13\ \text{lb}}{100} - \begin{matrix}\text{usable N}\\ \text{in mash}^{3}\\ \text{per 1000 gal}\end{matrix}\right) \times \frac{\text{F.W. of Nutrient}}{\text{M.W. of N (14)} \times n} = \begin{matrix}\text{required amount of nutrient}\\ \text{chosen as N source per 1000 gal}\end{matrix}$$

nitrogen needed — *nitrogen available* — *nitrogen to be added*

Figure 14. Calculation of the amount of nitrogen to be added.

nutrients are not supplied, the yeasts will neither grow well nor ferment quickly. Under such conditions, the maximum ethanol content dictated by sugar levels (added or derived) may not be achieved.

The composition of a microbe provides a clue as to the nutrition they require to grow, but wide ranges of values for cellular components of baker's and brewer's yeasts are known. Table 7 is a compilation of values from industrial preparations of two major Saccharomyces yeast preparations available in quantity (Ingledew, 1999). Although scientists might quibble over the magnitude of the values reported above, the importance of these compositional analyses is in assessment of the utility of such yeast when it can be harvested for potential industrial use. Certainly, the constituents of yeasts vary extensively among strains as well as among yeast genera. Less appreciated are the changes to composition caused by the environment and nutritive status of the growth media used in production of the yeast strain (Casey *et al.*, 1984). Not only can the concentration of the major components of yeast (carbohydrate, protein and nucleic acid) be influenced, but so can mineral content, the levels of enzymes made by the yeast (Thomas et al., 1996b), and the amounts of end products harvested from a medium (Casey *et al.*, 1983; O'Connor-Cox *et al.*, 1991).

To demonstrate the versatility of Saccharomyces yeasts and the breadth of industrial opportunities available, all that is needed is to show both the historical development and current usage of these yeasts in industry. The yeast strains used in these industries are different, and the handling procedures are not the same. In brewing, for example, worts are inoculated with from 5×10^6 to 2×10^7 yeasts/mL. There is a 3-8 fold growth of the yeast anaerobically as it makes alcohol from grain starch-derived sugars. This provides the brewer with a crop of yeast that is harvested, chilled and reused for the next batch as an inoculum. Because there is an increase in cell number in each fermenter, there is always a purging of the oldest or least desirable of the yeast crops, which are usually incorporated into animal feed (via the spent grains from the process).

In winery and distillery technology, the yeasts are not reused in fermentation. Presumably this is only due to the inability to collect them from the rest of the insoluble materials on tank bottoms and the lower viability of harvested yeasts caused by the lower pH and higher alcohol levels of the final ferment. In such fermentations, the peak yeast cell numbers in the fermenters can exceed 1.90×10^8 yeasts/mL. This figure will become relevant below.

North American fuel alcohol production now totals some 7 billion liters per year. On average, the concentration of alcohol attained in fermenters is about 12%. Therefore ~60 billion liters of mash are fermented in North America. If peak yeast levels in fermenters average 1.9×10^8/mL for each of the 60 billion L of mash in fermentation, over 1.14×10^{22} yeasts would have been made. If we assume that 1 gram of yeast before any stresses are applied equates to approximately 4.87×10^{10} cells (K.C. Thomas, unpublished data; Bayrock and Ingledew, 1997a,b)[4], then 2.34×10^{11} grams of excess yeast or 234,000 tonnes of yeasts would be available in North America. At a conservative protein

[3] can be estimated by conducting total nitrogen analysis before and after an unsupplemented fermentation.

[4] The total yeast cell number in compressed yeast is ~4.87×10^{10}/gram dry weight. The best ADY preparations are ~2.5×10^{10} cells/gram dry weight. This indicates that even though the dryer is gentle to the yeast, ~50% of the yeasts do not survive the dehydration process (Bayrock and Ingledew, 1997). Our assumption is that one gram of any *Saccharomyces* yeast would contain at least 4.87×10^{10} yeast cells/gram dry weight (viable and dead cells).

Table 7. Approximate composition (g/kg dry yeast) of industrially produced bakers and brewers yeasts.[1]

Component in the cell	*Bakers yeasts*[2]	*Brewers yeasts*[3]
Carbohydrate	180-440	390-600
Protein	380-590	370-420
Ash	45-75	73-81
Nucleic acid (DNA and RNA)	52-95	39-43
Lipids	40-50	
P	10-19	14-20
S	3.9	
K	20-21	17
Na	0.12-0.3	0.7
Ca	0.6-0.75	01.3
Fe	0.02-0.03	0.1
Mg	1.3-1.65	2.3
Co	0.008	.0002
Cu	0.008	.033
Mn	0.0059	.0057
Zn	0.170-0.197	.0387
Cr	0.0005-0.0025	
Mn	0.008	
Ni	0.003	
Sn	0.003	
Mo	0.00004	
Li	0.000017	
V	0.00004	
Se	0.005	
Pantothenate (Coenzyme A)	0.065-0.10	0.110-0.120
Choline (membranes)	2.71-4.00	3.80-4.55
Thiamine (Vit B_1)	0.090 -0.165	0.092-0.150
Riboflavin (Vit B_2)	0.040-0.100	0.035-0.045
Nicotinic acid/niacin (NAD)	0.30-0.585	0.450
Pyridoxine (Vit B_6)	0.020-0.040	0.043-0.050
Biotin (biocytin)	0.0006-0.0013	0.001
p-aminobenzoate (folic)	0.160	0.005
Inositol (phospholipids)	3.0	3.9-5.0
Folic acid (1-C transfer)	0.013-0.015	0.010

[1]Malony, 1998; Reed and Nagodawithana, 1991; Ingledew *et al.*, 1977; Patel and Ingledew, 1973; Peppler, 1970
[2]Aerobically grown
[3]Anaerobically grown

content of 38%, 88,920 tonnes of yeast protein would be distributed into the spent grains. Moreover, large amounts of growth factors, vitamins and beneficial minerals would be present. If production of 1 kL of ethanol leads to 0.86 tonnes of distillers dried grains (DDG) (Lewis, 1997), then 7 x 10^6 kL of ethanol would lead to 6 x 10^6 tonnes of DDG. The contribution of yeast biomass to the weight of what would become distillers dried grains with solubles (DDGS) would be at least 3.9 %. The contribution of the yeast protein to the protein content of DDGS (of 28%) would be at least 5.3% (Ingledew, 1999). The contribution of spent yeasts to overall vitamin content of DDG feed might be even higher as many of the vitamins from grain would have been scavenged by the yeast population during fermentation. Mash insolubles and thin stillage are collected after distillation, a process that leads to yeast death and denaturation of proteins. We believe that thin stillage and spent grains from the alcohol fermentation industry benefit animals more than generally realized due to the nutrient content of yeast cells and the fact that a good portion of the yeast protein could be 'bypass' protein for the animal (Ingledew, 1999).

Some interesting calculations in DDG and ethanol co-products have been provided by Lewis (1996) in a mass balance consideration based on the premise that every additional kg of ethanol results in a loss of 2 kg of DDG. Net revenue from the process, of course, depends on cyclical commodity prices for ethanol and DDG, but in general 0.1% w/v increase in a 100,000 gallon (378,540 L) fermenter leads to an additional $50 (US). If the gain in alcohol was the result of addition of extra enzymes (amylases, cellulase, proteases), it is obvious that the additional costs of enzyme must not exceed the increased revenue. Similar calculations can be done for determining cost benefit of addition of yeast foods or higher yeast levels (inoculated) which lead to shorter total fermentation time (increased rates of fermentation), increase predictability of fermentation and eliminate stuck/sluggish fermentations.

Very high gravity fermentation

The technology reviewed below, very high gravity (VHG) fermentation, was developed from a knowledge of nutrition (discussed above) and the effect of nutrients on increased tolerance of yeast cells to ethanol. Tolerance to ethanol has been reviewed in depth (Casey and Ingledew, 1986).

As mentioned earlier, as wort carbohydrate levels in the brewing industry were increased in high gravity (HG) brewing, stuck and sluggish fermentations sometimes occurred. As a result, most breweries imposed an upper limit of 15 or 16°P over which they would not manufacture wort. Many argued at that time that brewers yeasts were only tolerant to 7-8% v/v alcohol, and that high levels of carbohydrate were also an osmotic problem for these yeasts. Work done by Hayashida and Ohta (1981) and Kalmokoff (Kalmokoff and Ingledew, 1985) had indicated, however, that if brewers yeasts were grown and measured in the same way as distillers and sake yeasts, differences in ethanol tolerance were not as great as previously thought. This led to studies that capitalized on prior information provided by Kirsop (Kirsop, 1982). The elimination of stuck and sluggish fermentations in an industrial setting was, in fact, very simple.

In California, it was shown that stuck fermentation problems could be avoided by a simple measurement of 'must' free amino nitrogen levels and by preventing complete removal of oxygen from juice made with green grapes so that yeast had at least a minimal access to the trace amounts of oxygen they require (Ingledew and Kunkee, 1985). The economic ramifications of this work led us to exploit the technology in the more traditional distillation industry; and led to the description of VHG fermentation technology and the influences of nutrition on ethanol tolerance. We have now, in fact, been able to shatter dogma regarding alcohol tolerance of yeast by producing up to 23.8% v/v alcohol in batch fermentation with all carbohydrate provided at the start of fermentation. No conditioning or genetic manipulation of yeast is necessary. The yeasts used are commercially available, active dry yeasts. Yeasts tolerate these ethanol levels moderately well at the fermentation temperatures employed. Higher yeast levels would likely reduce fermentation times even further.

To use VHG technology effectively for fuel or distillery alcohol production, alterations must be made in the production process (Thomas *et al.*, 1996a). Crucial to this process are:

i) *Methods to increase carbohydrate concentration in batch cooking:* by recycling the first clear mash to use as makeup liquid for a second mash; by adding a concentrated source of sugar like molasses; by developing a suitable jet cooking regimen to recycle some prepared mash; or handling grain/water slurries of higher solids content (up to 360-380 g dissolved solids/L).

ii) *Removal of the insoluble grain residues after mashing to reduce viscosity, avoiding the expansion of viscous grain-water mixtures by the CO_2 produced in fermentation.* Also avoided is the handling of grain throughout fermentation and distillation. This pre-fermentation separation also prevents or reduces microbial proliferation in spent grain by lactic bacteria, and may provide for alternate use of this grain as a food grade product. This technology change will require engineering modification to most plants - a centrifuge, screw press, rotary drum vacuum filter, or mash filter will be required to remove the grain solids in suspension prior to fermentation (rather than post-distillation).

iii) *Rinsing and removal of dextrin from grain* as it is removed by separation technology.

iv) *Additions of yeast foods and oxygen* to allow multiplication of yeast to the higher numbers needed to convert the higher concentration of sugars to alcohol without cessation of fermentation.

v) *Increased cooling capacity in fermentation.* Due to the increased sugar concentration, more heat is provided per unit volume (fermentation is an exothermic process).

vi) *Assessment of the total process to ensure that plant retrofit or new design is possible* without limiting steps resulting from plant engineering.

The advantages of this technology are obvious. VHG technology leads to:

- reduction in water used per L of ethanol produced,
- increased plant capacity or reduction in capital costs/L ethanol in new plants,
- more alcohol produced per unit of labor
- more alcohol produced per unit of provided services
- fewer contaminating (yield reducing) microorganisms in the process
- increases in alcohol from 5-7% v/v to 9-12% in brewing (16% in the lab) or from 7-10% v/v to 18% or more in distilling/ fuel alcohol (23% in the lab)
- a decrease in stuck fermentations due to understanding of the yeasts' ability to make alcohol
- availability of increased fermenter space (~ 30%) due to removal of grain residuals
- possible special use or recycling of spent yeast due to the reduced grain solids in thin stillage (less or no need for DDG processing from stillage)
- economy at the still (less energy per L of ethanol is required over the range of 14-18% alcohol and there is less water to remove)
- liquid-liquid rather than solid-liquid-liquid extraction in the still
- possible food use for grain removed pre-fermentation.

Increasing throughput and yield should, in fact, be an operating strategy. This may best be achieved by investment in capital or in raw materials (enzymes, yeast nutrients, yeast) (Lewis, 1997).

A number of achievements in VHG fermentations can be related. Under simulated industrial conditions (in the laboratory and at a commercial pilot plant facility), normal active dry yeast manufactured for fuel alcohol purposes will produce 16-23% v/v alcohol at 20°C from mashes prepared with 300-370 g of sugars per liter. This is achieved in batch fermentation without the need for incremental feeding or more complicated fermentation techniques (unpublished data). Procedures for preparing VHG mashes have been established (Thomas *et al.*, 1996a). VHG fuel alcohol manufacture is carried out in the lab by adding gelatinized, liquefied (high temperature α-amylase treated) wheat extract which is clarified, lyophilized and added back to a normal mash in order to increase carbohydrate content. A method whereby a similar extract can be made industrially has been developed for 'in plant' production of mashes by batch cooking technology. Supplementation of mash with FAN (in the form of yeast extract, amino acids or other nitrogenous sources) resulted in an increased rate of alcohol production (a faster fermentation), and an increased yield of alcohol. The beneficial use of proteolytic enzymes has also been demonstrated. Optimization of the conditions for very high gravity alcohol production has been carried out by concentrating on the sugar content of the medium, the available nitrogen content of yeast foods, the inoculation rates in the fermenters, and the effect of temperature (Jones and Ingledew, 1994b; Wang *et al.*, 1999a). Figures 15 and 16 show the startling effect of adding an inexpensive nitrogen source on the rate of fermentation (to completion) and the elimination of stuck fermentations at the higher sugar concentrations. This is evidenced by the incremental ethanol increases attained as sugar levels increase when usable nitrogenous material is present. It was also seen that the time required to complete fermentation decreased with increases in temperature from 17-33°C. The optimum temperature for fermentation in VHG mashes was found to be 27°C (Wang *et al.*, 1999a; Jones and Ingledew, 1994b). This is significantly below the temperatures used in industry at sugar levels of approximately 23° Brix, but higher than the 20°C used in much of our earlier research. As expected, increases in dissolved solids in the mash resulted in longer fermentation times, regardless of the temperature chosen.

The considerable savings of water resulting from the use of VHG technology with the concomitant higher sugar/dextrin concentration and increased mash viscosity result in the need to alter thinking regarding the completeness of starch conversion to fermentable sugars, the influence of ß-glucans, pentosans and/or proteins on viscosity, the binding of water into polymers, the need for free water in gelatinization, hydrolysis of starches/polymers and sugar dissolution and the water activity of the resultant fermentation medium.

The result of the above considerations has been the generation of a number of formulae to determine the amounts of grain and water required to prepare mashes with predetermined concentrations of dissolved solids and insoluble materials (Table 8), the quantity of the liquid portion of the mash before and after fermentation and the concentration of ethanol in fermented mash. These formulae, unlike those in current use, take into account changes in weight and volume of the mash during fermentation (Thomas *et al.*, 1996a). As Table 9 shows, the formulae allow comparative efficiencies to be calculated for normal (16° Plato) vs. VHG (31°P) mashes and provide graphic evidence of water savings. It should also be noted that removal of solids (with rinse) increases dramatically the amount of ethanol that can be made in a given fermenter in VHG (or even normal) gravity conditions (Thomas *et al.*, 1996a). Additionally, the use of enriched starch fractions from grain using milling techniques (Wang *et al.*, 1997; Wang *et al.*, 1999b), if economic, also increases productivity.

Successful implementation of VHG fermentation technology requires considerations regarding the ability to clarify and rinse dextrin from the grain prior to fermentation, and a certain number of alterations in the process through mashing, cooling, fermentation, and distillation. A number of engineering changes have been addressed as the work has progressed; but except for grain separation experimentation done at lab and pilot plant levels, the published work does not yet address the types of equipment nor the designs or sizing of such instrumentation. For separation, decanter centrifuges and basket centrifuges have been tested; and rotary drum vacuum filters have been proposed. Increased milling capacity, cooling capacity and distillation capacity along with the ability to dewater and press DDG so that recycling of stillage and perhaps yeast can be attempted, have all been addressed. Technology

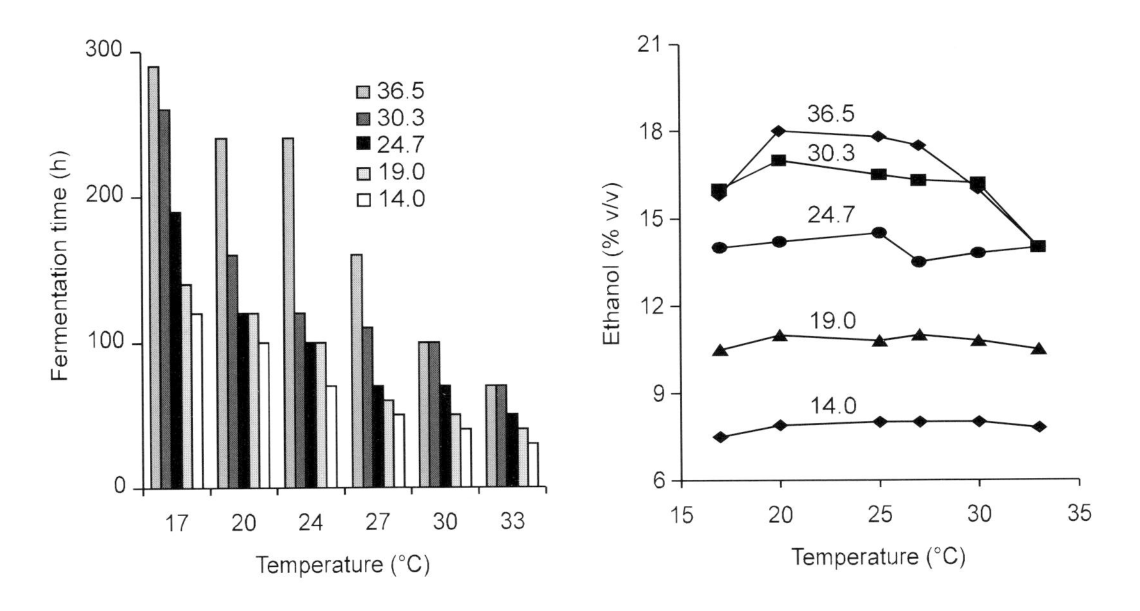

Figure 15. Effects of temperature and initial substrate concentration on the time taken to end ferment unsupplemented wheat mash (left) and on ethanol yield (right) with no urea added. Wheat mashes initially contained 14.0 to 36.5 g of dissolved solids per 100 mL (Jones and Ingledew, 1994b by permission).

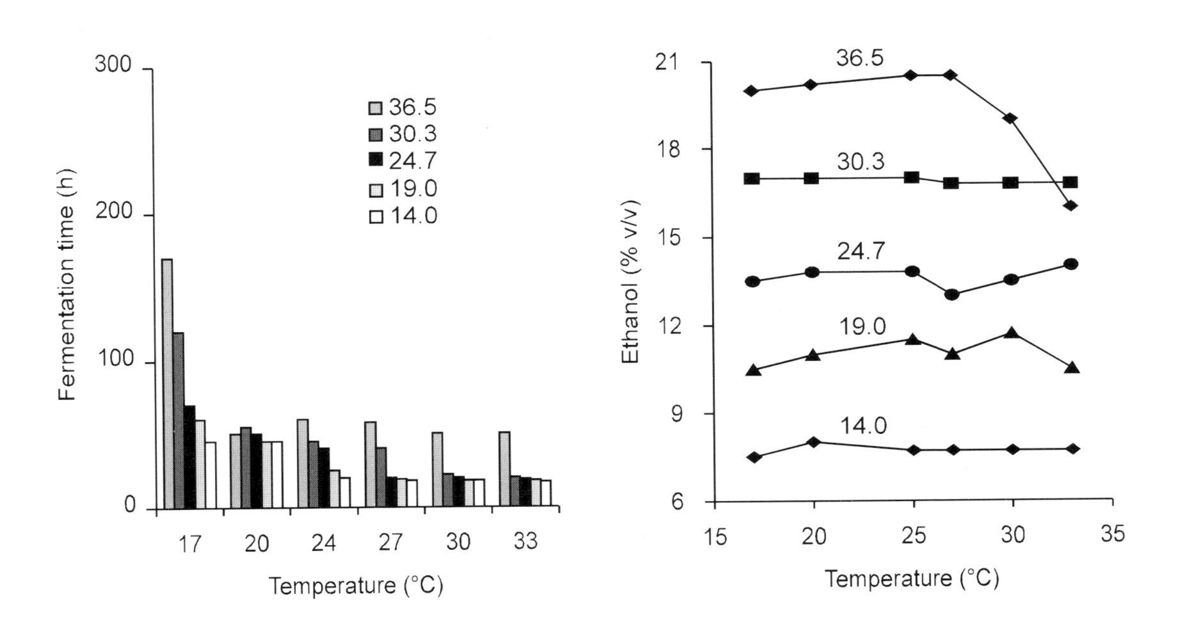

Figure 16. Effects of temperature and initial substrate concentration on the time taken to end ferment wheat mashes supplemented with 16 mM urea (left) and on ethanol yield (right). Wheat mashes initially contained 14.0 to 36.5 g of dissolved solids per 100 mL (Jones and Ingledew, 1994b by permission).

Table 8. Derivation of an equation to calculate the weight of grain required to make a given weight of mash with a specified dissolved solids concentration.*

weight of mash = weight of grain + weight of water added **(M = G + W)**
weight of starch in grain = weight of grain x kg of starch per kg of grain **(= GS)**
weight of glucose obtainable from grain = weight of starch in grain x 1.111 **(= 1.111 GS)**
weight of insolubles in mash = weight of grain x kg of non-starch solids per kg grain = **GR or G(1 - S - N)**
weight of the liquid portion of mash = weight of mash - weight of insolubles in mash (GR) **(= M - GR)**

$$\text{concentration of glucose liquid portion (C)} = \frac{\text{weight of glucose}}{\text{weight of liquid portion}} = \frac{\mathbf{GS\ 1.111}}{\mathbf{M - GR}}$$

$$\text{on rearranging, } G^{a} = \frac{\mathbf{M\ C}}{\mathbf{1.111S + RC}}$$

*From Thomas *et al.*, 1996, with permission.

[a] If a fraction of the insolubles is removed, formula is changed to: $G = \frac{MC}{1.111S + FRC}$ where *F* is the fraction left in the mash.

Table 9. Efficiencies and calculations of weights of wheat required to prepare 100,000 kg of normal and VHG wheat mashes with and without insoluble materials (Thomas *et al.*, 1996 by permission).

Mash type	*Insolubles removed (%)*	*Weight of wheat (kg)*	*Fermentation time (hours)*[a]	*Throughput rate (kg wheat/h)*	*Comparative efficiency (%)*
NG[b] (16°P)	0	25,000	60	417	100
NG (16°P)	100	27,275	60	455	109
VHG[c] (31°P)	0	44,558	80	557	134
VHG (31°P)	50	48,349	80	604	145
VHG (31°P)	100	52,846	80	661	159

[a] based on fermentation temperature of 20°C and 1.0 x 10^8 cells/g dissolved solids in the mash.
[b] NG = normal gravity.
[c] VHG = very high gravity.

development for continuous jet cooking is needed.

The brewing industry has embraced VHG technology. A number of companies are now making beer in concentrate which is diluted at the end of the aging process to the desired alcohol concentration using the volume of 'de-aerated' brewing water that could have been added to the process at its inception as done in normal brewing. Beer containing up to 16.2% v/v ethanol has been made in the lab and up to 11.2 % v/v (limit by choice not restrictions) in a commercial pilot plant (McCaig *et al.*, 1992). This success helped to lead the brewing industry to commercial fermentation of worts as high as 25° Plato, a level previously considered impossible to reproducibly ferment because of stuck or sluggish fermentations. Widespread commercial success and continued acceptance by industry will depend on our understanding of the biochemistry of the yeast cell, the regulation of its metabolism and the tolerance of yeasts to ethanol and sugar. This will allow us to maximize yields and enhance plant productivity for fuel alcohol production - lowering the overall production costs and making alcohol even more competitive as an octane enhancer or fuel extender.

VHG technology has also been used in the wine industry to prevent, for example, stuck fermentations of high sugar Chardonnay juices, some of which were shown to contain less than 40 mg of FAN per liter (Ingledew and Kunkee, 1985). In the brewing industry VHG technology has been used to expose the poor fermentability of canned and bulk malt extracts (many of which were adulterated with carbohydrate adjuncts that diluted the normal nutrient spectrum found in an all-malt wort as most of these preparations were supposed to be; Paik *et al.*, 1991). In another application, the beneficial effects of yeast hulls used in enology to suppress stuck fermentations were also proven through supplementation of one of the two groups of nutrients required by yeasts, namely the sterols and unsaturated fatty acids (Munoz and Ingledew, 1990). In proving that VHG technology has merit for industry, it has been shown that much of the dogma respecting alcohol and osmotic tolerance of yeast must be reconsidered.

Other biological factors of importance

Yeast concentration and vitality

An alternate and effective method to increase fermentation rate is to increase the inoculation level of a fermentation. When a very large number of cells in the range of 8×10^7 to 7×10^8/mL (O'Connor-Cox and Ingledew, 1991; Thomas and Ingledew, 1992a) is used, a vast amount of sugar is fermented even though little to no net growth is observed. In fact, the slower fermentation rate of these non-growing (resting) cells as carried out for cell maintenance purposes approaches the rate achieved at more normal yeasting levels (5×10^6/mL per ° Plato) where the smaller number of cells are actively growing and metabolizing at the much faster rate described by Kirsop (1982). For this reason, very high inoculation rates will often overcome stuck and sluggish fermentations and obviate the need for increased assimilable nitrogen. Unfortunately, it is not usually cost effective to increase cell number with active dry yeast (and they are not very active on initial rehydration), so this method of stimulating fermentation works only when recycling of viable yeast is possible. A side benefit from such large inoculations is that ethanol production efficiency increases somewhat at the same time. That is, lesser amounts of sugar are diverted to new biomass production.

High pitching rates increase the rate of fermentation not only when the amount of utilizable nitrogen is adequate, but more importantly even when the supply of assimilable nitrogen is low or growth limiting. Although the mechanism of stimulation at high pitching rates is not clearly understood, two possibilities have been suggested. First, although no increase in viable cell count was observed, there was an increase in the biomass and in the catalytic rate of glucose conversion to alcohol. There was cell growth in terms of cell enlargement even though there was no net increase in the number of viable cells. Alternatively, some cells equal to the number of newly formed cells might have lost viability, with the result that there was no measurable change in the viable counts. Moreover, although few have been detected by methylene blue, dead cells might have lysed and released nutrients stimulatory to the fermentation.

Whatever the mechanism, stimulation of fermentation at high pitching rates has important practical applications. By increasing pitching rate, one can stimulate the overall rate of fermentation and avoid stuck or sluggish fermentations. This is especially beneficial when it is not allowed by law to adequately supplement the 'wort' or 'must' with growth-stimulatory nutrients in the form of yeast foods - or when it is too expensive. A simple test for FAN is predictive of possibilities for stuck fermentations (Ingledew and Kunkee, 1985) although FAN requirements in musts, worts and mashes differ extensively; and the minor oxygen requirements of fermenting yeast must not be ignored. Moreover, only 'usable' FAN is the true indication of availability of nitrogenous substances to yeast. Ammonium ion or urea is important to measure as well when this pertains. Above all, it is not possible to correct nutritional problems *after* fermentation has stuck. Advanced knowledge of usable FAN and ammonium contents is the key to prevention of fermentation disasters.

Such high levels of inoculation are probably not practical. However, pitching yeast slurry can be concentrated and recycled, or the pitching level can be somewhat decreased from the higher rate used in this work. Active dry yeasts would certainly be too expensive to provide such high levels of inoculated yeast. However, any increase over the recommended pitching rate would help to alleviate the problems of nitrogen limitation and slow fermentations. Most of the work published by this lab has been done without large numbers of added yeasts as it was not practical to do so.

Osmoprotectants

As part of our attempt to define the best nitrogen source(s) available to stimulate VHG and normal fermentations, the amino acid arginine was found most effective in promoting fermentation (and also in relieving lysine-induced inhibition). Because it appeared to be better than ammonium salts or yeast extract, it was suggested that arginine may play a role above and beyond serving as a source of nitrogen. We have now demonstrated conversion of arginine to proline during alcohologenesis, and a subsequent excretion of proline to the medium. Additions of proline (which cannot be used as a sole source of nitrogen in true fermentation due to the lack of oxygen and need for an oxidase enzyme - see Ingledew *et al.*, 1987b) also stimulated the rate of uptake and sugar fermentation. We believe that proline may serve to protect yeasts from osmotic stress. It has a similar role in plants and some bacteria. It is also of interest to note that grape juice and grain fermentations are very high in proline (Ingledew and Kunkee, 1985; Thomas and Ingledew, 1992b), a fact that may partly explain the ability of yeast to ferment high sugar (VHG) mashes with little apparent osmotic stress or alcohol intolerance. The influence of osmoprotectants in maintaining high yeast viability under fermentation conditions has clearly been demonstrated (Thomas *et al.*, 1994). Without additions, the viability of the yeast dropped significantly at the end of the fermentation.

Yeast hulls

Yeast cell walls surround the cell and are responsible for the rigidity of the cell shape and the protection of the cell membrane, which functions for the diffusion and/or transport of nutrients in and products out of the cell. The cell wall may account for up to 20% of the cell weight and is composed of glucan, mannan, and smaller amounts of chitin (80-90%), protein (6-13%) and lipids (2-5.8%) (Fleet, 1991). Structural proteins are bound to the polysaccharides. The unsaturated fatty acids found are likely part of the cell membrane still adhering to the yeast cell wall. Yeast hulls are the ghosts or residuals (insolubles) from the production of yeast extract. They have a verified role as an additive in wine making where they stimulate stuck and sluggish fermentations. Although the mechanism of action of the hulls has been in dispute, the ghosts appear to act as nutrient sources for the oxygen substitutes needed by the yeast for sterol and unsaturated fatty acid components in newly synthesized membranes (daughter cells) (Munoz and Ingledew, 1990). We believe that the use of hulls in wine fermentations is only justified if oxygen input to juice is considered detrimental to color or flavor. This may not be the case (Long and Lindblom, 1986).

Conclusions

Scientists are getting a lot closer to the understanding of many of the problems encountered in the 'black box' of fermentation. In time, we hope fermentation will be so well understood that industry will be able to follow a 'recipe' related to the substrates and conditions imposed on the yeast that will result in more successful fermentation at higher alcohol concentrations than are now imagined. Above all, it is important to reduce fermentation problems so that the economics of the alcohol production industry are sound and unchallenged.

Acknowledgments

The author would like to acknowledge the Natural Sciences and Engineering Research Council, Western Grains Research Foundation, the Agriculture Development Fund and Value Added Crops programs of the Saskatchewan Government, and the College of Agriculture Research Trust Fund - all of which supported this work over some years. The efforts of a number of technicians, graduate students and postdoctoral fellows along with Dr. K. C. Thomas are obvious in the development of this research program.

Literature cited

Anonymous, 1997. World drink trends 1997. International beverage consumption and production trends. NTC Publications Ltd. pp. 134-139.

Anonymous, 1999. Ethanol production hits all-time peak; industry group sees new applications. Milling & Baking News, March 9. pp. 21-22.

Axelsson, L. 1998. Lactic acid bacteria: classification and physiology. In: Lactic Acid Bacteria. Microbiology and Functional Aspects. Edited by: S. Salminen and A. von Wright. Marcel Dekker, Inc., N.Y. pp 1-72.

Bayrock, D. and W.M. Ingledew. 1997a. Fluidized bed drying of baker's yeast: moisture levels, drying rates, and viability changes during drying. Food Research International, 30(6): 407 -415.

Bayrock, D. and W.M. Ingledew. 1997b. Mechanism of viability loss during fluidized bed drying of baker's yeast. Food Research International, 30(6): 417 - 425.

Berry, D.R. and C. Brown. 1987. Physiology of yeast growth. In: Yeast Biotechnology. Edited by: D.R. Berry, I. Russell, G.G. Stewart. Allen and Unwin, Winchester Mass.

Berry, D.R. and D.C. Watson. 1987. Production of organoleptic compounds. In: Yeast Biotechnology. Edited by: D.R. Berry, I. Russell, G.G. Stewart. Allen and Unwin, Winchester Mass.

Casey, G.P. and W.M. Ingledew. 1985. Re-evaluation of alcohol synthesis and tolerance in brewers' yeast. Journal of the American Society of Brewing Chemists, 43:75-83.

Casey, G.P. and W.M. Ingledew. 1986. Ethanol tolerance in yeast. CRC Critical Reviews in Microbiology, 13: 219-280.

Casey, G.P., C.A. Magnus and W.M. Ingledew, 1983. High gravity brewing: nutrient enhanced production of high concent-rations of ethanol by brewing yeast. Biotech-nology Letters 5 (6): 429-434.

Casey, G.P., C.A. Magnus and W.M. Ingledew. 1984. High gravity brewing: effects of nutrition on yeast composition, fermentative ability, and alcohol production. Applied and Environmental Microbiology. 48: 639-646.

Cooper, T.G. 1982. Nitrogen metabolism in *Saccharomyces cerevisiae*. In: The Molecular Biology of the Yeast *Saccharomyces*. Metabolism and Expression. Edited by: J.N. Strathern, E.W. Jones and J.R. Broach. Cold Spring Harbor Laboratory. Cold Spring Harbor, New York.

David, M.H. and B.H. Kirsop. 1973. Correlation between oxygen requirements and the products of sterol synthesis in strains of *Saccharomyces cerevisiae*. Journal of General Microbiology, 77, 529 -531.

Dixon, B. 1999. Yeasts: rising stars in biotechnology. American Society for Microbiology News, 65(1): 2-3.

Engan, S. 1981. Beer composition: volatile substances. In: Brewing Sciences, Volume 2. Edited by: J.R.A. Pollock. Academic Press. New York.

Fiechter, A. and W. Seghezzi. 1992. Regulation of glucose metabolism in growing yeast cells. Journal of Biotechnology, 27: 27-45.

Fleet, G.H. 1991. Cell walls. In: The Yeasts Vol 4. Cell Organelles. Edited by: A.H. Rose, and J.S. Harrison. Academic Press, London.

Fraenkel, D.G. 1982. Carbohydrate metabolism. In: The Molecular Biology of the Yeast *Saccharomyces*. Metabolism and Expression. Edited by: J.N. Strathern, E.W. Jones, and J.R. Broach. Cold Spring Harbor Laboratory. Cold Spring Harbor, New York.

Gancedo, C. and R. Serrano. 1989. Energy-yielding metabolism. In: The Yeasts Vol 3. Metabolism and Physiology of Yeasts. Second Edition. Edited by: A.H. Rose and J.S. Harrison. Academic Press. New York.

Harrison, J.S. 1967. Aspects of commercial yeast production. Process Biochemistry 2(3): 41-46.

Harrison, J.S. 1971. Yeast production. Progress in Industrial Microbiology, 10:129-177.

Harrison, J.S. and J.C.J. Graham. 1970. Yeast in distillery practice. In: The Yeasts Vol 3, Yeast Technology Edited by: A.H. Rose and J.S. Harrison. Academic Press, New York.

Haukeli, A.D. and S. Lie. 1979. Yeast growth and metabolic changes during brewery fermentations. Proceedings of the Congress of the European Brewing Convention, 17: 461-473.

Hayashida, S. and K. Ohta. 1981. Formation of high concentrations of alcohol by various yeasts. Journal of the Institute of Brewing, 87: 42-44.

Husain, H. 1988. Engineering aspects of ethanol production. Proceedings - Fuel Alcohol Production and Opportunity Conference, Saskatoon, March, p74.

Ingledew, W.M. 1979. Bacterial contaminants - their effect on beer - a review. Journal of the American Society of Brewing Chemists, 37: 145-150.

Ingledew, W.M. 1993. Yeasts for production of fuel alcohol. In: The Yeasts Vol 5. Yeast Technology. Second Edition. Edited by: A.H. Rose and J.S. Harrison. Academic Press. New York.

Ingledew, W.M. 1999. Yeast - could you build a business on this bug? In: Under the Microscope. Focal Points for the New Millennium. Biotechnology in the Feed Industry. (T.P. Lyons and K.A. Jacques, eds). Nottingham University Press, Nottingham, UK. pp. 27-50.

Ingledew, W.M. and R.E. Kunkee, 1985. Factors influencing sluggish fermentations of grape juice. American Journal of Enology and Viticulture, 36, 65-76.

Ingledew, W.M., L.A. Langille, G.S. Menegazzi and C.H. Mok. 1977. Spent brewers yeast - analysis, improvement, and heat processing considerations. Master Brewers Association of the Americas Technical Quarterly, 14: 231-237.

Ingledew, W.M., C.A. Magnus and J.R. Patterson. 1987a. Yeast foods and ethyl carbamate formation in wine. American Journal of Enology and Viticulture, 38: 332-335.

Ingledew, W.M., C.A. Magnus and F.W. Sosulski. 1987b. Influence of oxygen on proline utilization during the wine fermentation. American Journal of Enology and Viticulture, 38:246-248.

Ingledew, W. M. and C. A. Patterson. 1999. Effect of nitrogen source and concentration on the uptake of peptides by a lager yeast in continuous culture. Journal of the American Society of Brewing Chemists, 57:9 -17.

Ingledew, W.M., F.W. Sosulski and C.A. Magnus. 1986. An assessment of yeast foods and their utility in brewing and enology. Journal of the American Society of Brewing Chemists, 44:166-171.

Jones, A.M. and W.M. Ingledew. 1994a. Fuel alcohol production: assessment of selected commercial proteases for very high gravity wheat mash fermentation. Enzyme and Microbial Technology, 16:683-687.

Jones, A.M. and W.M. Ingledew. 1994b. Fuel alcohol production: optimization of temperature for efficient very-high-gravity fermentation. Applied and Environmental Microbiology, 60:1048-1051.

Jones, A.M. and W.M. Ingledew. 1994c. Fuel alcohol production: appraisal of nitrogenous yeast foods for very high gravity wheat mash fermentation. Process Biochemistry, 29: 483-488.

Jones, A.M. and W.M. Ingledew. 1994d. Fermentation of very high gravity wheat mash prepared using fresh yeast autolysate. Bioresource Technology, 50: 97-101.

Kalmokoff, M.L. and W.M. Ingledew. 1985. Evaluation of ethanol tolerance in selected *Saccharomyces* strains. Journal of the American Society of Brewing Chemists, 43 (4), 189-196.

Kirsop, B.H. 1982. Developments in beer fermentation. Topics in Enzyme and Fermentation Technology, 6: 79 -131.

Kosaric, N., F. Adalbert, H. Sahm, O. Goebel and D. Mayer. 1987. Ethanol. In: Ullmann's Encyclopedia of Industrial Chemistry. Fifth edition. Edited by: W. Gerhartz. Weinheim, New York, NY. Volume A9: 587-653.

Lantero, O.J. and J.J. Fish. 1993. Process for producing ethanol. U.S. Patent #5,231,017. July 27.

Lewis, S.M. 1996. Fermentation alcohol. In: Industrial Enzymology (2nd Edition). Edited by: T. Godfrey and S. West. Macmillan Press Ltd. London UK.

Lewis, S.M. 1997. Value added enzyme applications for fuel alcohol production. Fuel Alcohol Workshop, Omaha NB, June 25-27.

Long, Z. and B. Lindblom. 1986. Juice oxidation experiments. Wines and Vines, 67(11): 44-49.

Makanjuola, D.B., A. Tymon, and D.G. Springham. 1992. Some effects of lactic acid bacteria on laboratory scale yeast fermentations. Enzyme and Microbial Technology, 14: 351-357.

Maloney, D. 1998. Yeasts. In: Kirk-Othmer Encyclopedia of Chemical Technology. 4th Edition. (J. I. Kroschwitz and M. Howe-Grant, eds).John Wiley & Sons, New York, 761-788.

Matthews, J.M. and C. Webb. 1991. Culture systems. In: *Saccharomyces* . Biotechnology Handbooks. 4. Edited by: M.F. Tuite and S.G. Oliver. Plenum Publishing Corp. New York.

McCaig, R., J. McKee, E.A. Pfisterer, D.W. Hysert, E. Munoz, and W.M. Ingledew. 1992. Very high gravity brewing - laboratory and pilot plant trials. Journal of the American Society of Brewing Chemists, 50, 18-26.

Monk, P.R., D. Heok and B.M. Freeman. 1986. Amino acid metabolism by yeast. Proceedings of the Sixth Australian Wine Industry Technical Conference, July pp 129-133.

Munoz, E. and W.M. Ingledew. 1990. Yeast hulls in wine fermentations - a review. Journal of Wine Research, 1: 197-209.

Nagodawithana, T.W. 1986. Yeasts. Their role in modified cereal fermentations. Advances in Cereal Science and Technology, 8:15-104.

Narendranath, N., S.H. Hynes, K.C. Thomas and W.M. Ingledew. 1997. Effects of lactobacilli on yeast-catalyzed ethanol fermentation. Applied and Environmental Microbiology, 63(11): 4158-4163.

Neish, A.C. and A.C. Blackwood. 1951. Dissimilation of glucose by yeast at poised hydrogen ion concentrations. Canadian Journal of Technology, 29: 123-129.

O'Connor-Cox, E.S.C. and W.M. Ingledew. 1989. Wort nitrogenous sources - their use by brewing yeasts: a review. Journal of the American Society of Brewing Chemists, 47:102-108.

O'Connor-Cox, E.S.C. and W.M. Ingledew. 1990. Effect of the timing of oxygenation on very high gravity brewing fermentations. Journal of the American Society of Brewing Chemists, 48: 26-32.

O'Connor-Cox, E.S.C. and W.M. Ingledew. 1991. Alleviation of the effects of nitrogen limitation in high gravity worts through increased inoculation rates. Journal of Industrial Microbiology, 7: 89-96.

O'Connor-Cox, E.S.C., E.J. Lodolo and B.C. Axcell. 1996. Mitochondrial relevance to yeast fermentation performance: a review. Journal of the Institute of Brewing, 102: 19-25.

O'Connor-Cox, E.S.C., J. Paik and W.M. Ingledew. 1991. Improved ethanol yields through supplementation with excess assimilable nitrogen. Journal of Industrial Microbiology, 8: 45-52.

Paik, J., N.H. Low and W.M. Ingledew. 1991. Malt extract: relationship of chemical composition to fermentability. Journal of the American Society of Brewing Chemists, 49: 8-13.

Patel, G.B., and W.M. Ingledew. 1973. Internal carbohydrates of *Saccharomyces carlsbergensis* during commercial lager brewing. Journal of the Institute of Brewing, 79: 392-396.

Patterson, C.A. and W.M. Ingledew. 1999. Utilization of peptides by a lager brewing yeast. Journal of the American Society of Brewing Chemists, 57: 1-8.

Peppler, H.J. 1970. Food yeasts. In: The Yeasts, Vol 3. Yeast Technology. Academic Press, New York, 421-461.

Pierce, J.S. 1987. The role of nitrogen in brewing. Journal of the Institute of Brewing, 93: 378-381.

Reed, G. and T.W. Nagodawithana. 1991. Yeast Technology 2nd Edition. AVI. Van Nostrand Reinhold, New York.

Rose, A.H. and G. Vijayalakshmi. 1993. Baker's yeasts. Chapter 10. In: The Yeasts. Volume 5. Yeast Technology. 2nd Edition. Edited by: A.H. Rose and J.S. Harrison. Academic Press, London. pp. 357-398.

Schultz, A.S. and S. Pomper. 1948. Amino acids as a nitrogen source for growth of yeasts. Archives of Biochemistry, 19: 184-192.

Simpson, W.J. and J.R.M. Hammond. 1990. A practical guide to the acid washing of brewers' yeast. Ferment 3(6): 363-365.

Sobolov, M. and K.L. Smiley. 1960. Metabolism of glycerol by an acrolein-forming *Lactobacillus*. Journal of Bacteriology, 79: 261-266.

Sols, A., C. Gancedo, and G. Delafuente. 1971. Energy-yielding metabolism of yeasts. In: The Yeasts. Volume 2. Physiology and Biochemistry of Yeasts. Edited by: A.H. Rose and J.S. Harrison. Academic Press. New York.

Soumalainen, H. and E. Oura. 1971. Yeast nutrition and solute uptake. In: The Yeasts Vol 2. Physiology and Biochemistry of Yeasts. Edited by: A.H. Rose and J.S. Harrison. Academic Press. New York.

Stanbury, P.F., A. Whitaker, and S.J. Hall. 1995. Media for industrial fermentations. In: Principles of Fermentation Technology. Pergamon. Elsevier Science Ltd., Oxford, UK. pp. 93-122.

Thomas, K.C., S.H. Hynes and W.M. Ingledew. 1994. Effects of particulate materials and osmoprotectants in very-high-gravity ethanolic fermentations. Applied and Environmental Microbiology, 60: 1519-1524.

Thomas, K.C., S.J. Hynes and W.M. Ingledew, 1996a. Practical and theoretical considerations in the production of high concentrations of alcohol by fermentation. Process Biochemistry, 31: 321-331.

Thomas, K.C., S.H. Hynes and W.M. Ingledew. 1996b. Effect of nitrogen limitation on synthesis of enzymes in *Saccharomyces cerevisiae* during fermentation of high concentrations of carbohydrates. Biotech. Lett. 18:1165-1168.

Thomas, K.C., S.H. Hynes and W.M. Ingledew, 1998. Initiation of anaerobic growth of *Saccharomyces cerevisiae* by amino acids or nucleic acid bases: ergosterol and unsaturated fatty acids cannot replace oxygen in minimal media. Journal of Industrial Microbiology, 21: 247-253.

Thomas, K.C., S.H. Hynes, A.M. Jones, and W.M. Ingledew. 1993. Production of fuel alcohol from wheat by VHG technology. Effect of sugar concentration and fermentation temperature. Applied Biochemistry and Biotechnology, 43: 211-226.

Thomas, K.C. and W.M. Ingledew. 1990. Fuel alcohol production: effects of free amino nitrogen on fermentation of very high gravity wheat mashes. Applied and Environmental Microbiology, 56: 2046-2050.

Thomas, K.C. and W.M. Ingledew. 1992a. Production of 21% (v/v) ethanol by fermentation of very high gravity (VHG) wheat mashes. Journal of Industrial Microbiology, 10: 61-68.

Thomas, K.C. and W.M. Ingledew. 1992b. The relationship of low lysine and high arginine concentrations to efficient ethanolic fermentation of wheat mash. Canadian Journal of Microbiology, 38: 626-634.

Verduyn, C., E. Postma, W.A. Scheffers and J.P. van Dijken. 1990. Energetics of *Saccharomyces cerevisiae* in anaerobic glucose-limited chemostat cultures. Journal of General Microbiology, 36: 405-412.

Vos, P.J.A., and R.S. Gray. 1979. The origin and control of hydrogen sulfide during fermentation of grape must. American Journal of Enology and Viticulture, 30: 187-197.

Walker, G.M. 1998. Yeast Physiology and Biotechnology. John Wiley & Sons. Toronto, Canada.

Walker, G.M., R.M. Birch, G. Chandrasena and A.I. Maynard. 1996. Magnesium, calcium and fermentative metabolism in industrial yeasts. Journal of the American Society of Brewing Chemists, 54: 13-18.

Wang, S., K.C. Thomas, K. Sosulski, W.M. Ingledew and F.W. Sosulski. 1999b. Grain pearling and very high gravity (VHG) fermentation technologies for fuel alcohol production from rye and triticale. Process Biochemistry (in press).

Wang, S., W.M. Ingledew, K.C. Thomas, K. Sosulski, and F.W. Sosulski. 1999a. Optimization of fermentation temperature and mash specific gravity for fuel alcohol production. Cereal Chemistry, 76: 82-86.

Wang, S., K. Sosulski, F. Sosulski and W.M. Ingledew. 1997. Effect of sequential abrasion on starch composition of five cereals for ethanol fermentation. Food Research International 30: 603-609.

Yang, R.D., D.A. Grow and W.E. Goldstein. 1982. Pilot plant studies of ethanol production from whole ground corn, corn flour, and starch. Fuel Alcohol U.S.A., February, 13-16.

Yokoyama, A. and W.M. Ingledew. 1997. The effect of filling procedures on multi-fill fermentations. Master Brewers Association of the Americas Technical Quarterly, 34:320-327.

Chapter 6

Molasses as a feedstock for alcohol production

J.E. Murtagh
Murtagh & Associates, Winchester, Virginia

Introduction

Molasses differs from other feedstocks for alcohol production such as corn, milo and potatoes in that these plant products contain carbohydrate stored as starch. As a result, these feedstocks must be pre-treated by cooking and enzymatic action to hydrolyze starch into fermentable sugars. In contrast, the carbohydrates in molasses are already in the form of sugars and need no pretreatment.

Basic sugar chemistry

The simplest form of sugar is glucose, which is fermented by normal distillery and brewery yeasts. It has the formula $C_6H_{12}O_6$ and is made up of molecules with a single ring structure (Figure 1). Very slight rearrangements of the atoms in the molecule can give other sugars with distinctly different properties. For instance, if the end groups are rotated it becomes galactose, which is not readily fermentable by normal yeasts (Figure 2). Alternatively, by rearranging the ring structure, it becomes the fermentable fructose (Figure 3). There are many other simple sugars, but only glucose and fructose will be considered here.

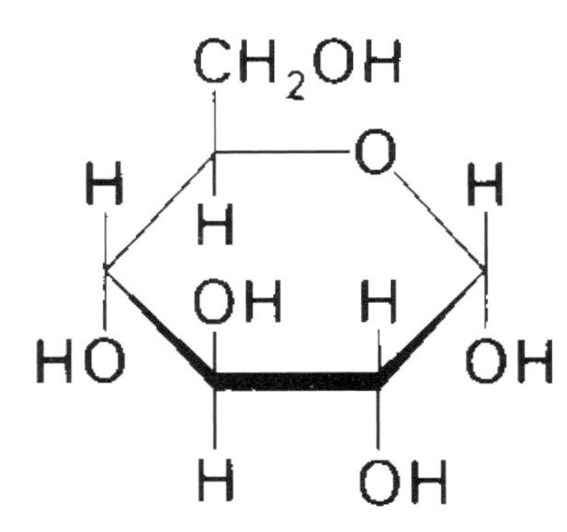

Figure 1. Structure of glucose.

Glucose does not exist extensively in the free state in nature. It is mostly polymerized as starch or cellulose, in which long chains of glucose units are formed. Glucose also exists in combination with fructose to form the disaccharide (two sugar molecules) or common table sugar (Figure 4). Sucrose is the principal sugar contained in molasses and is readily fermentable either directly, or as its glucose and fructose components.

It should be noted that the formula of sucrose is not exactly double that of glucose (or fructose), in that a molecule of water has been displaced in the synthetic process. This means that water is added when sucrose is broken down:

$$\underset{\text{Sucrose}}{C_{12}H_{22}O_{11}} + \underset{\text{Water}}{H_2O} \rightarrow \underset{\text{Glucose}}{2C_6H_{12}O_6}$$

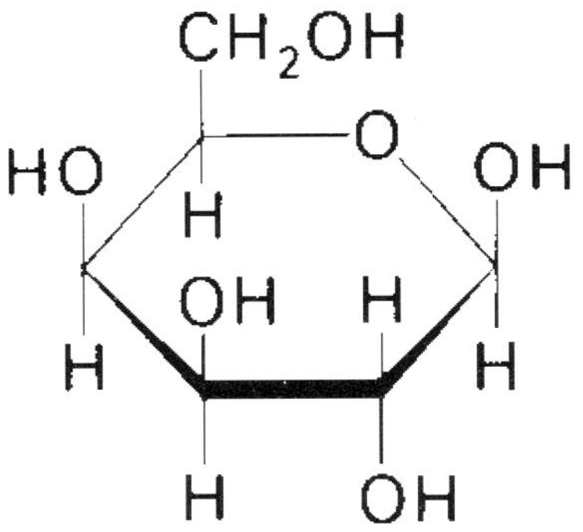

Figure 2. Structure of galactose.

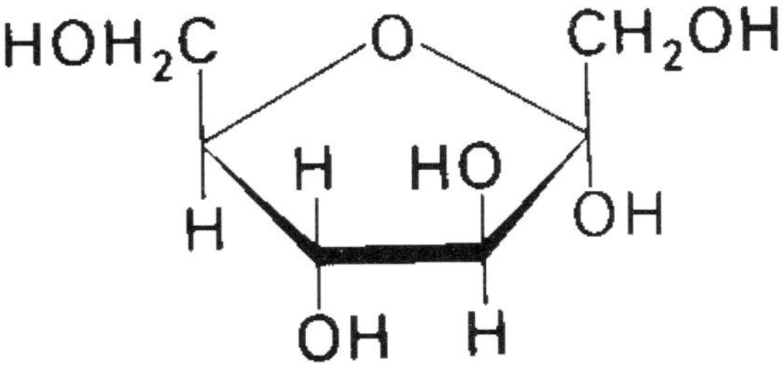

Figure 3. Structure of fructose.

Degradation with added water is referred to as hydrolysis. (This word comes from the Greek roots 'hydro', meaning water, and 'lysis', meaning cutting or breakage.) This addition of water should be taken into consideration when making any calculations of the amount of simple sugars (monosaccharides) produced from disaccharides when determining potential alcohol production by fermentation. For example, hydrolysis of 100 lbs of sucrose uses 5.26 lbs of water and produces 105.26 lbs of glucose and fructose (Figure 4).

Molasses

There are at least six basic types of molasses that may be encountered as fermentation feedstocks (Table 1). For the purposes of this section, consideration will generally be confined to blackstrap molasses, although reference will be made to some of the peculiarities of the other types.

Table 1. Six types of molasses.

Cane (blackstrap) molasses
High test (cane) molasses
Refiners cane molasses
Beet molasses
Refiners beet molasses
Citrus molasses

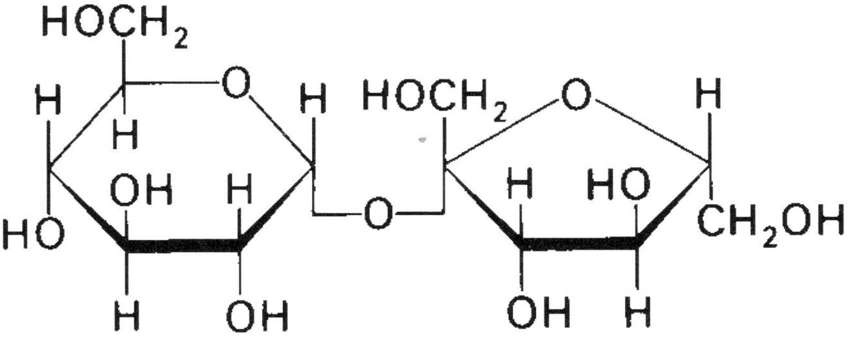

100 lbs sucrose (MW 342) + 5.26 lbs water (MW 18) → 105.26 lbs glucose and fructose (MW 360)

Figure 4. Structure of sucrose and its hydrolysis to glucose plus fructose.

Blackstrap molasses production

In the production of cane sugar, the cane is crushed in a mill to squeeze out the juice. The juice is heated, clarified by filtration and the addition of lime (to remove cane fibers and sludge) and then evaporated to concentrate the sugar and cause it to crystallize. The syrup containing the crystals is then centrifuged to separate the crystals and the syrup residue (which still has a high content of sugar). The residue is referred to as 'A molasses'. It is evaporated and centrifuged again to recover more crystalline sugar; and the syrup residue is now referred to as 'B molasses'. The process may be repeated to yield more sugar and a 'C molasses' as residue.

Sugar mills normally evaporate and centrifuge a maximum of three times, but the number of treatments will depend on marketplace economics. When sugar prices are high, a fourth processing may be practiced with production of a 'D molasses', which has lost most of its available crystallizable sugar. On the other hand, when sugar prices are low, the 'A molasses' may be sold directly.

Repeated evaporation and centrifugation decreases the sugar content of molasses and increases the viscosity and concentration of salts and other impurities. The result is thick, viscous, brown liquid which is very heavy. The concentration of molasses is normally measured in degrees Brix. The Brix scale is the same as the Balling scale, which is a measure of what the sugar content of a liquid would be if all the dissolved and suspended solids were sugar. Expressed another way, it is the sugar content of a sugar solution with the same specific gravity as the sample. Thus, an 80^0 Brix molasses has a specific gravity of 1.416, which is the same as a sugar solution containing 80% sugar by weight. Brix is measured using a hydrometer originally intended only for sugar solutions.

The world molasses trade tends to center on New Orleans, Louisiana, even though the US domestic molasses production is relatively low. Ships bring molasses into New Orleans for transfer to river barges and distribution to animal feed manufacturers and other users throughout the US midwest. Molasses prices are generally quoted f.o.b. New Orleans on the basis of 79.5^0 Brix and about 46% w/w 'Total Sugars as Invert' (TSAI).

Blackstrap molasses is really sugar mill garbage; and the mills generally cannot guarantee the sugar content in advance any more than a city could guarantee the percentage of paper or cellulose in its garbage if a company wished to use it for ethanol production. In years when sugar prices are high, mills like to get the sugar content of molasses as low as economically feasible. In some cases, additional chemical methods such as the Quentin and Steffen's processes may be used to improve sugar recovery. Unfortunately, these processes yield a molasses with very low apparent sugar fermentability and should be avoided.

The normal test used to measure TSAI is very crude in that it measures sample reducing power and assumes that all the reducing compounds are sugars. Thus, if sodium sulfite or another reducing compound were added to molasses, it would appear as sugar. The term 'invert sugars' refers to the monosaccharides glucose and fructose, which are produced by hydrolysis of the sucrose. Therefore there is no need to make allowance for the 5.26% water taken up in the hydrolysis.

The advent of high performance liquid chromatography has made it possible to measure the constituent sugar content of molasses. New Orleans molasses merchants now accept contracts to sell molasses based on actual sugars rather than TSAI although they still tend to buy on a TSAI basis.

Economics of using blackstrap molasses for alcohol production

Based on the Gay-Lussac equation for ethanol production from glucose by fermentation (Figure 5) we should obtain 48.89 lbs of carbon dioxide (CO_2) and 51.11 lbs of ethanol if we ferment 100 lbs of glucose. This is the maximum theoretical output; however Pasteur concluded that it was virtually impossible to obtain more than 95% of the maximum theoretical yield as the reaction is not as simple as the Gay-Lussac equation implies. In addition to ethanol, yeast also produce other substances such as glycerol and succinic acid. Many distillers, therefore, use 95% of the 51.11% w/w figure as a realistic basis for calculating expected ethanol yields. (When comparing yields from different plants, one should know exactly how each plant calculates theoretical yields.) For costing purposes, one can expect to obtain about 58 gallons of ethanol from a ton of molasses at 46% sugars (TSAI). This is calculated as follows:

(a) 1 ton molasses at 46% sugars contains 920 lbs sugar.

(b) 920 x 51.11% (Gay-Lussac yield) = 470.21 lbs ethanol.

(c) 470.21 ÷ 6.58 lbs ethanol/gallon = 71.46 gallons ethanol.

(d) 71.46 gallons x 95% (Pasteur yield) = 67.89 gallons.

(e) 67.89 gallons x 85% (low-average plant efficiency) = 57.71 gallons/ton of molasses.

If the price of molasses after freight is added is around $65 per ton, the feedstock will cost well over $1.10 per gallon of ethanol less any return on the concentrated molasses stillage residue (called condensed molasses solubles or CMS). CMS is primarily used as a molasses substitute, mainly for dust suppression in animal feeds. In addition, CMS is increasingly being used as a fuel to replace part of the oil used in the boilers.

CMS generally sells in the US at an f.o.b. price of around $25/ton at 50° Brix. Assuming a cost of $10.00/ton for evaporation and $5.00/ton for freight to the point of sale, the net return on CMS production is around $10.00/ton. Based on over 3 tons of residue (stillage) at 10° Brix produced from a ton of molasses after fermentation and distillation and a requirement of 5 tons of stillage to make 1 ton of CMS at 50° Brix, the CMS return is around $6.00 per ton of original molasses. Therefore, the feedstock cost would be about $1.00 per gallon of alcohol produced. At this cost, using molasses for alcohol production in the US is not economically feasible when compared with other feedstocks. However, in sugar-producing countries where molasses is valued at the New Orleans price less freight costs of $30-35 per ton, use of molasses for alcohol production may be economically attractive.

Fermentation of molasses for ethanol production

Pre-treatments

Dilution

Blackstrap molasses at 80° Brix will not ferment without dilution as the sugars and salts exert a very high osmotic pressure. It is therefore necessary to dilute the molasses to below 25° Brix. Yeast will not start fermenting rapidly above this point; and contamination may develop before the yeast become established since molasses is laden with contaminating bacteria. Some distillers choose to pasteurize molasses before fermentation; but it is difficult to justify the costs unless pasteurization is incorporated into the clarification process.

When diluting molasses, it must be remembered that the Brix scale measures on a weight % basis and all calculations must be based on weight and not volume. 80° Brix molasses has a specific gravity of 1.416, therefore a gallon weighs about 11.8 lbs and a ton contains about 169.5 gallons. Example calculations for diluting 80° Brix molasses to 40° and 25° Brix are given in Table 2.

$$\underset{\text{Glucose, 100 lbs}}{C_6H_{12}O_6} + \text{YEAST} \rightarrow \underset{\text{Carbon dioxide, 48.89 lbs}}{2CO_2} + \underset{\text{Ethanol, 51.11 lbs}}{2C_2H_5OH}$$

Figure 5. The Gay-Lussac equation for ethanol production from glucose by fermentation.

Typical sugar content of molasses in the US is relatively low (46%). Therefore, when the molasses is diluted to 25° Brix the sugar content is only about 14.3% (Calculated: 25 x 46/80 = 14.3). This is only sufficient to yield 7-8% v/v of ethanol in the fermented beer. Distilleries generally need a higher final ethanol content to economize on energy for distillation; but the fermentation cannot begin at a much higher Brix without running into problems of slow starts and bacterial contamination. Some distilleries overcome this problem by diluting the first portion of molasses going into the fermenter to about 18° Brix, which allows the yeast to get established very rapidly. When the Brix reading in the fermenter is down to about 12° Brix, molasses diluted to around 35° Brix is added. This allows beer ethanol levels of around 10% to be attained. This procedure is the first step toward what is known as 'incremental feeding', which in itself is a step toward continuous fermentation. In the full incremental feeding system, one may start fermentation near 14° Brix and keep feeding more molasses at 35° Brix to keep the Brix reading in the fermenter between 12 and 14 until the vessel is filled. Then, to go semi-continuous or continuous, it is merely necessary to allow the fermenter to overflow into one or more other vessels.

A multistage approach to prevent scale

There are various possible pretreatments of molasses prior to fermentation, all mostly aimed at reducing the suspended and dissolved compounds that might cause scaling and blocking of distillation columns. The main problem is the presence of calcium compounds, which are mostly derived from the lime added to the hot cane juice in the clarification process in the sugar mill. Thus, one of the usual important specifications for molasses as an alcohol feedstock is that it should not contain more than 10% ash. Ash content above 10% can lead to serious scaling problems.

The scale formed in the distillation columns when running molasses beer is mainly calcium sulfate, otherwise known as gypsum. If sulfuric acid is used to acidify the dilute molasses in the fermenter to control bacterial contamination, some of the calcium salts will be converted to calcium sulfate, which is fairly insoluble. Some distillers acidify partially diluted molasses with sulfuric acid in a pretreatment in an attempt to remove the calcium salts and thereby reduce the scaling. Unfortunately, this does not work very well. Calcium sulfate is not as insoluble as many people believe; and it is peculiar in that its solubility decreases at higher temperatures

Table 2. Calculations for diluting molasses.

Example 1: Dilute 1 ton of 80° Brix molasses* to 40° Brix	
a. Total tons: 80 ÷ 40 = 2 tons at 40° Brix	
b. Gallons produced at 40° Brix	
1 ton molasses at 80° Brix	169.5 gallons
1 ton of water	240.0 gallons
Gallons produced at 40° Brix	409.5 gallons
Example 2: Dilute 1 ton of 80° Brix molasses* to 25° Brix	
a. Total tons: 80 ÷ 25 = 3.2 tons at 25° Brix	
b. Gallons produced at 25° Brix	
1 ton of molasses at 80° Brix	169.5 gallons
2.2 tons of water	528 gallons
Gallons produced at 25° Brix	697.5 gallons

*80° Brix molasses has a specific gravity of 1.416.

instead of increasing (as occurs with table salt, sugar, etc.). This means that when the molasses beer enters the preheaters or the steam-heated stripping column, the calcium sulfate precipitates as scale. The solubility also decreases with increased alcohol concentrations; so if too much alcohol is refluxed onto the stripping column beer feed plate, scaling will increase.

Scale tends to block the holes in the stripping column sieve trays and reduce distillation capacity. If the sieve holes are 3/8 inch in diameter (about 10 mm) and if there is just 1 mm of scale around the holes (barely noticeable on visual inspection), then hole diameter is reduced to 8 mm and the column capacity is reduced to 64% of the original (open area is proportional to the square of the diameter of the hole). This serves to illustrate why scaling can be a major problem with molasses feedstocks.

Rather than resorting to the purchase of expensive centrifuges and other molasses pretreatment systems, one should use a multi-stage approach to settle out the solids in the various stages of operation:

(1) *If molasses is first diluted to about 45° Brix with hot water and held above 70° C for a few hours, then much of the suspended solids will settle out.* This method has an advantage in that after the initial dilution, it is no longer necessary to use gear pumps or other positive displacement pumps to move the molasses as it can be handled by centrifugal pumps.

(2) *If possible, use fermenters with a steep bottom slope to improve separation of the solids.* After fermentation the solids, which may amount to about 1% of the total volume, may be discarded before pumping the remainder of the fermenter to the beer well. Alternatively, draw off the beer from above the bottom of the fermenter without agitation and then discard the solids sludge from a lower outlet.

(3) *There should be a second decanting in the beer well.* It is preferable to draw from above the bottom of the beer well, and to have a number of sample valves between the bottom and the draw-off point to monitor solids sludge accumulation in the bottom. The sludge can develop the consistency of toothpaste, and may be drawn off from a bottom outlet periodically.

(4) *There should be a good proof control system in the rectifying section.* This ensures that high proof ethanol does not periodically (or continuously) go down to the beer feed plate of the stripper and cause scale precipitation. In plants using molecular sieves for ethanol dehydration, particular attention should be given to the recycle return to the lower part of the rectifier to ensure that it does not upset the proof control system.

(5) *Select a stripper column design less susceptible to scaling problems such as the baffle tray or disc-and-donut system.*

(6) *Use hydrochloric acid instead of sulfuric acid* to adjust the pH in the fermenters or in yeast propagation.

(7) *Decant the stillage in the stillage holding tank before pumping to the evaporator to avoid sending sludge to the evaporator.*

If scaling occurs in spite of all provisions, one may add a specially developed chelating agent to the beer to minimize the scaling. There are also simple chemical means of descaling columns. It should not be necessary to resort to manual removal of scale.

It should be mentioned that decanting sludge from molasses beer may also help improve the quality of the distillate when producing beverage spirit for rum or vodka. Leaving the sludge in the beer sent to the still may tend to give it an odor like cooked turnips, which is difficult to remove by further rectification.

Nutrient use in molasses fermentation

In blackstrap fermentations it may be necessary to add some nitrogen and phosphorus to obtain optimum results. Nitrogen should not be in the form of ammonium sulfate as it will add to the scaling problem by forming calcium sulfate. Likewise, liquid ammonia is undesirable as it tends to raise the pH and encourage bacterial contamination unless it is counteracted with acid additions. (Some plants using liquid ammonia either use sulfuric acid to balance the pH, which introduces the undesirable sulfate anion, or use phosphoric acid, which generally introduces more phosphate than necessary.

Urea may be used to supply nitrogen in molasses fermentations for fuel ethanol production, but its use for beverage alcohol production should be approached with caution. Urea usage may lead to the production of carcinogenic ethyl carbamate, which is unacceptable in alcoholic beverages. If phosphorus is deficient in the molasses, diammonium phosphate may be added with a corresponding reduction in urea or other nitrogenous nutrient. Generally, blackstrap molasses requires no other added nutrients for fermentation.

In the case of high test molasses (HTM) the situation is somewhat different. HTM is a primary product of the sugar mill rather than a by-product. In years when the sugar market is very depressed, the mills produce high test molasses by simply concentrating the cane juice into a thick syrup and adding acid to partially hydrolyze the sucrose into glucose and fructose (to prevent crystallization). HTM may have about 70% sugars in 80° Brix, indicating a high percentage of sugar in proportion to salts and other impurities. It may need relatively more nitrogen and phosphate and may require some trace elements such as zinc.

The three B vitamins most likely to be essential for the satisfactory growth of any particular strain of *Saccharomyces cerevisiae* yeast are biotin, pantothenic acid (pantothenate) and inositol. Biotin is normally present in great excess in cane molasses, but is normally deficient in beet molasses. If a yeast strain that does not require biotin cannot be obtained (they are rare), it may be advantageous to mix in 20% cane molasses when trying to ferment beet molasses, or at least to use a 50:50 cane:beet molasses mix for initial propagation of the yeast.

Pantothenate is normally borderline-to-adequate in cane molasses, but is generally in excess in beet molasses. Mixing some beet molasses with blackstrap cane molasses may be advantageous if it is available. Pantothenate concentrations may also be low in HTM and in refiners cane molasses (the by-product of refining raw brown sugar to produce white sugar). In these instances, it may help to add some beet molasses to the mash although one may obtain yeasts that do not require external sources of pantothenate for growth.

Inositol may be deficient in HTM, but this deficiency is less critical as it is relatively easy to find yeasts that do not require external sources of inositol.

Bacterial contamination of molasses

If blackstrap molasses is old or has been produced under unsanitary conditions, it may contain various forms of bacterial contamination. The principal contaminant to be aware of is the bacterium *Leuconostoc mesenteroides*. This organism can cause the sucrose molecules to polymerize into long chains of dextran that are not fermentable but will appear as sugar on the TSAI test. If the contamination is very extensive, the molasses may appear to be 'ropey' when stirred on a stick. Dextran may also be found in refiners molasses produced from raw sugar that has been stockpiled for a number of years. Apart from the loss of yield, the presence of dextran may result in far greater foaming of the fermenters.

Another important occasional bacterial contaminant is *Zymomonas mobilis*. This bacteria can ferment sugars to ethanol, but has the side effect of reducing sulfur compounds to produce a hydrogen sulfide smell. This can be disastrous if one is trying to produce a good quality rum.

Fermentation conditions

Molasses fermentations may be conducted at 90-95°F, but if one is aiming for high final alcohol levels in the beer it is advisable to keep the maximum temperature down. This is because alcohol inhibition of yeast growth is intensified by higher temperatures.

Molasses fermentations are generally much more rapid than grain fermentations and can be completed in half the time required for grain. This means that much more heat is produced per hour, which necessitates better cooling facilities than for grain fermentations. Some grain alcohol plants share one external heat exchanger between two fermenters, but that practice is not feasible with molasses if throughput and efficiency are to be maximized.

Distillation of molasses beers

The main problem in distillation of molasses beer is scale formation in the stripping column. Methods for minimizing scaling have been discussed in the earlier section on pre treatments of molasses for fermentation.

A second problem, foaming in the distillation columns, is more commonly encountered with molasses than with grain beers. Foaming frequently limits the rate of distillation; and if uncontrolled the foam may pass from the stripping column into the rectifier and cause problems of proof, color, etc. Foaming is best controlled by continuous injection of a suitable antifoam into the beer feed line as this ensures that all the beer reaching the column is adequately dosed. A brief interruption of the antifoam supply can cause havoc in the columns when operating at high feed rates; so it is desirable to have duplicate antifoam pumping systems to ensure continuity.

By-product recovery: condensed molasses solubles (CMS)

Molasses stillage may be evaporated to produce CMS using a multiple effect evaporator. CMS from blackstrap molasses tends to become very viscous at about 50° Brix, the standard degree of concentration for sale in the animal feed trade. Beet molasses CMS is less viscous and may be concentrated to about 60° Brix. The viscosity may be reduced if the stillage is neutralized; so it is helpful to save any waste caustic soda solution from washing fermenters, etc., for addition to the stillage tank.

Chapter 7

Whey alcohol - a viable outlet for whey?

J. O'Shea
Alltech Inc., Dunboyne, County Meath, Ireland

Introduction

Whey is the main by-product from casein and cheese manufacture. In the past, cheese making was a traditional cottage industry using the milk from various animals such as goats, camels and reindeer. Today cheese is usually made in large production facilities from cow's milk by the application of microbiology and chemistry. Casein, the chief milk protein, is coagulated by the enzymatic action of rennet or pepsin, by lactic acid producing bacteria, or by a combination of the two. The next step involves separating the curds from the whey, a process facilitated by heating, cutting, and pressing. With the separation of the curds most of the butterfat and a number of the other constituents are removed. The butterfat usually comprises only 3-4% by volume of whole milk. Whey makes up to 80-90% of the total volume of milk that enters the cheese making process and contains about 50% of the nutrients. These nutrients are soluble whey protein, lactose, vitamins and minerals. Whey from cheese and rennet casein is known as sweet whey; and the manufacture of acid casein yields acid whey. Acid whey is produced by removal of the fat by centrifugation to make butter. The milk is then acidified to coagulate the protein. This is then centrifuged off as casein, leaving acid whey.

Table 1. Approximate composition of whey.*

Component	%
Water	94.50
Dry matter	5.75
Protein	0.80
Lactose	4.40
Ash	0.50
Fat	0.05

*This is an average analysis, the type of cheese produced will influence composition of the whey.

Approximately 100 kg of milk are required to produce 10 kg of cheese. Quantities of milk produced in the top 40 milk producing countries in the world are outlined in Table 2. The top 40 cheese producing countries are listed in Table 3. The whey production from all this milk is considerable and can present a disposal problem. The challenge to dairy production plants is to dispose of whey in an environmentally friendly and economic manner. The problems facing whey producers and the uses for whey are explored below.

Table 2. International production of milk (thousand tonnes).[1]

	Country	*1995*	*1996*	*1997*
1	USA	70500	70003	71072
2	India	32000	35500	34500
3	Russian Federation	39098	35590	34000
4	Germany	28621	28799	28700
5	France	25413	25083	24957
6	Brazil	16985	18300	19100
7	Ukraine	17060	15592	16300
8	United Kingdom	14683	14680	14849
9	Poland	11642	11696	12099
10	New Zealand	9285	9999	11131
11	Netherlands	11294	11013	10922
12	Italy	10497	10799	10531
13	Argentina	8792	8947	9795
14	Turkey	9275	9466	9466
15	Australia	8460	8986	9303
16	Japan	8382	8657	8642
17	Mexico	7628	7822	8091
18	Canada	7920	7890	7800
19	China	6082	6610	6946
20	Spain	6150	6084	5997
21	Columbia	5078	5000	5408
22	Irish Republic	5415	5420	5380
23	Romania	4646	4676	5126
24	Belarus	5070	4908	4850
25	Denmark	4676	4695	4632
26	Pakistan	4223	4379	4540
27	Switzerland	3913	3862	3913
28	Iran	3450	3809	3897
29	Kazakhstan	4576	3584	3584
30	Belgium/Luxembourg	3644	3682	3477
31	Sweden	3304	3316	3334
32	Uzbekistan	3665	3183	3200
33	Czech Republic	3155	3164	3164
34	Austria	3148	3034	3090
35	Sudan	2760	2880	2880
36	South Africa	2794	2592	2720
37	Finland	2468	2431	2463
38	Kenya	2170	2210	2210
39	Korea	1998	2034	2072
40	Chile	1890	1924	2040
Total world milk production		**421,810**	**422,299**	**426,181**

[1]National statistics, Eurostat, 1999.

Table 3. International cheese production (thousand tonnes).[1]

	Country	1995	1996	1997
1	USA	3493	3627	3627
2	France	1617	1635	1660
3	Germany	1453	1531	1591
4	Italy	982	985	940
5	Netherlands	700	709	713
6	Russian Federation	477	477	477
7	Argentina	369	376	405
8	Poland	354	397	397
9	United Kingdom	362	377	378
10	Egypt	344	346	349
11	Canada	301	312	323
12	Denmark	311	299	291
13	New Zealand	197	239	274
14	Australia	234	261	270
15	Spain	240	247	246
16	China	202	202	206
17	Iran	196	196	197
18	Greece	202	192	192
19	Czech Republic	137	139	139
20	Turkey	136	139	139
21	Switzerland	132	133	133
22	Mexico	126	123	123
23	Sweden	129	127	118
24	Japan	105	109	114
25	Austria	101	100	107
26	Irish Republic	84	98	97
27	Israel	92	92	92
28	Norway	85	84	89
29	Finland	96	95	88
30	Hungary	83	88	88
31	Syria	66	75	86
32	Sudan	75	76	76
33	Belgium/Luxembourg	72	73	76
34	Bulgaria	82	71	72
35	Ukraine	88	71	71
36	Venezuela	103	70	70
37	Portugal	67	67	65
38	Brazil	60	60	60
39	Columbia	51	51	51
40	Chile	48	49	51
Total world		**14,052**	**14,398**	**14,541**

[1]National statistics, Eurostat, 1999.

Problems with whey

Whey produced by dairies has historically been and remains today a potential pollution problem. As a consequence, the majority of cheese producing countries have strict laws defining the permitted biological and chemical oxygen demand (BOD and COD) limits for the effluent from dairy plants. A cheese manufacturer's ability to dispose of the whey produced may be the limiting step in the volume of cheese that can be produced.

A number of problems are common to all whey users. The first complication is that the whey is extremely perishable. This is especially true for cheese whey, which contains bacteria added during cheese production. The whey must therefore be protected from biological degradation; and the most effective method of protection is pasteurization. Alternatively, the whey can be cooled to 4°C or protected by the use of chemical additives, e.g., hydrogen peroxide. A second problem with whey is the very high water content, over 93%. This means that transport and concentration costs are high, and form a large proportion of the total processing cost. Another factor influencing the economics of whey processing in many countries is one of seasonality. Most cheese production occurs between March and September. This means that the processing plant is used for only half of the year. Finally, cheese whey contains casein fines and fat that must be removed prior to most processing methods. This is not particularly a disadvantage, since the recovered fines can be added back to the cheese or used to make processed cheese while the separated fat can be used to make whey butter (Figure 1). The fines are usually separated on vibrating screens and the fat by centrifugation.

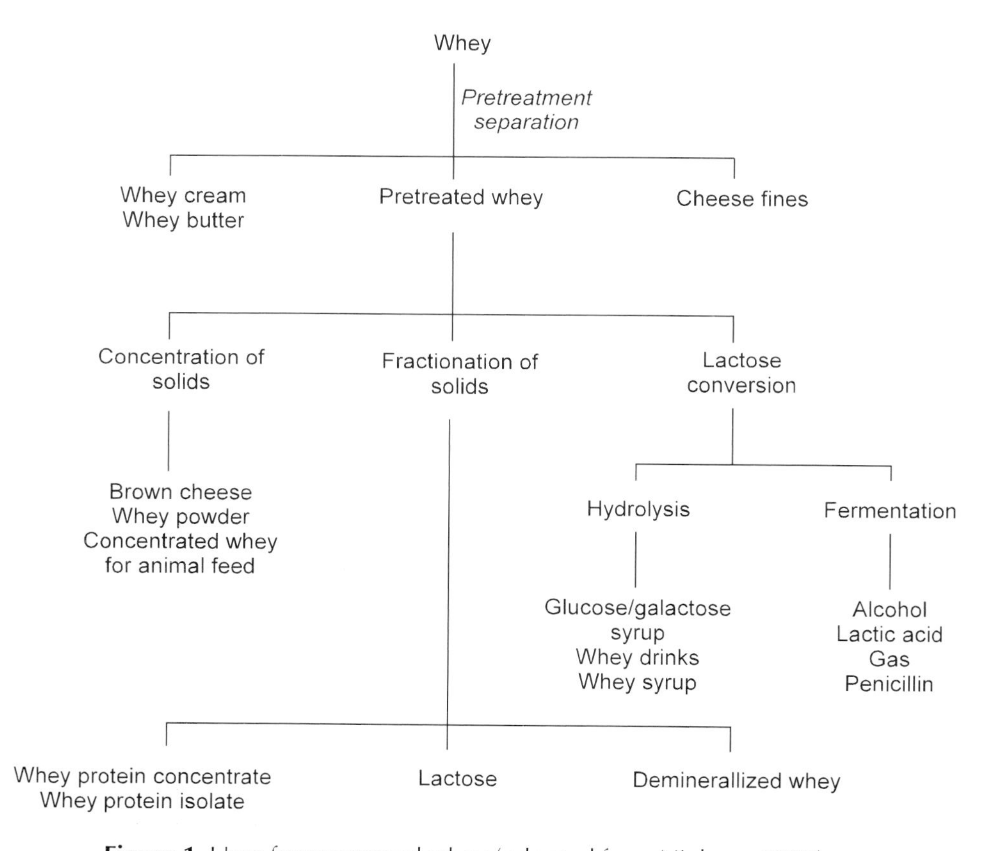

Figure 1. Uses for processed whey (adapted from Nielson, 1997).

Uses for whey

Whey as an animal food

One solution to the whey problem is to sell it to pig farmers, however there are many limitations to the use of whey as a pig feed. Due to its low dry matter content, whey-fed pigs typically produce very large volumes of urine. Many piggeries are not in a position to deal with the high humidity in barns housing whey-fed pigs, and respiratory problems are a very real deterrent to maximizing profits from intensive piggeries. In addition, the whey supply in many regions is seasonal; it dries up between October and April. The pig farmer is therefore vulnerable to these fluctuations in feed supply, and is more likely to opt for feed sources available all year such as barley plus soya. At the same time, the summer surplus of whey cannot be utilized by the pig farmer with a fixed number of pig spaces. In addition, whey is deficient in protein and fat-soluble vitamins and requires balancing with appropriate additives. The difficulties recently experienced by the European pig industry indicate how volatile this industry is and that it is not a guaranteed 'sink' for whey. Many whey producers would prefer to have several relatively secure outlets for their whey.

Whey powder

Whey is processed by several methods. One of the most common methods is the production of whey powder by drying. Table 4 outlines whey powder production in Europe from 1995 to 1998. Drying sounds like a simple process, but a number of stages are necessary to produce a high quality powder. The free-flowing whey powder resulting from drying has a number of applications in the food industry, including

Table 4. Whey powder production in Europe (thousand tonnes).[1]

Europe (15 countries)	*1995*	*1996*	*1997*	*1998 (estimated)*
Production	1280	1310	1280	1310
Consumption	1235	1235	1172	
Import	539	534	485	
of which, from outside EU	9	2	2	
Export	654	703	703	
of which, from outside EU	54	77	110	100

[1]Eurostat, 1999.

Table 5. Production of lactose in Europe (1995-1998)(thousand tonnes).[1]

Europe (15 countries)	*1995*	*1996*	*1997*	*1998 (estimated)*
Production	280	310	330	330
Consumption	195	225	230	240
Export to outside EU	85	85	100	90

[1]Eurostat, 1999.

incorporation into dry mixes, baked foods, confections, ice cream and processed cheese. Whey may also be made into cheese such as the Scandinavian primost and mysost (Edelman and Grodnick, 1986).

Reverse osmosis may also be used to concentrate whey. It works on the same principle as ultrafiltration except that membranes are such that only water is allowed to pass through them. Reverse osmosis is cheaper than thermal evaporation for removing up to 70% of the water from whey. Savings are therefore possible by using reverse osmosis for pre-concentration before evaporation or membrane processing (Burgess, 1980).

Lactose manufacture

One of the oldest methods of whey utilization is the manufacture of lactose. The first stage is concentration of the whey by low temperature evaporation, which ensures that the whey protein is not damaged. The concentrate, containing 60-65% solids, is fed to a crystallization tank and seeded with small crystals of lactose. Edible lactose is able to absorb flavor; and it is therefore used in certain applications in the food industry. Lactose may also be used as a wine sweetener. For pharmaceutical applications, e.g., tablet formulation, the lactose crystals must be further refined to remove traces of protein and minerals. Table 5 outlines lactose manufacture in Europe from 1995 to 1998.

Lactose hydrolysis

Lactose is not very sweet compared to cane sugar or even its constituent monosaccharides glucose and galactose. The sweetness of lactose can be considerably increased by hydrolysis. A number of technologies are available for lactose hydrolysis. These can be classified as either enzymatic methods using ß-galactosidase or ion exchange methods (Burgess, 1980).

Demineralised whey

One of the main disadvantages of whey as a food product is its high mineral content. Whey demineralisation has opened up the baby food market to whey producers. Two different technologies may be used for the demineralisation of whey, electrodialysis and ion exchange. Of these techniques most manufacturers use electrodialysis. In the ion exchange process whey is passed through a column containing a strong cation resin and then through a column containing a weak anion resin. The former absorbs the cations in the whey and releases hydrogen ions, while the latter neutralizes the hydrogen ions and absorbs the anions. The advantage of ion exchange lies in its relative simplicity and low capital cost. However, the regeneration costs are high and a large volume of waste is generated. Ion exchange is probably cheaper than electrodialysis for high levels of demineralization. In contrast, the capital costs of electrodialysis are high while running costs are relatively low.

Protein recovery from whey

Whey protein is a relatively high value product. Ultrafiltration is one of the processes used to extract whey proteins. Whey proteins with a molecular weight of 12,000-13,000 are extracted, leaving the permeate suitable for fermentation. The main feature of ultrafiltration is the semi-permeable membrane, usually made of cellulose acetate, polyamide or other copolymers. The membrane acts as a filter that is permeable to the low molecular weight salts and lactose, but impermeable to the higher molecular weight proteins. Pressure is applied to the whey side of the membrane and water, salts and lactose are forced through it while protein is retained. The protein concentration in the whey therefore increases in proportion to its reduction in volume and a whey protein concentrate containing up to 60% protein on a dry weight basis can be produced. After ultrafiltration, the whey protein concentrate is evaporated to 40-50% total solids by low temperature vacuum evaporation and spray-dried. The whey proteins have a high nutritional value.

Alcohol production

After ultrafiltration there remains a large volume of permeate that needs to be processed. The protein-free permeate is a yellow liquid containing 90% of the whey solids, including most of the lactose. The composition of whey permeate (Table 6) and that of skim milk are essentially similar.

Table 6. Average composition of whey permeate.

Component	%
Lactose	4.8
Protein	0.6
Non-protein nitrogen	0.1
Ash	0.5
Fat	0.5
Lactic acid	0.15
Total solids	**5.7**

It has been known for some time that whey can be fermented to produce alcohol; but in the past synthetic alcohol produced from crude oil has been cheap. Following the energy crisis in the 1970s, the price of synthetic alcohol increased, making fermentation alcohol production processes more economically attractive. Whey previously regarded as a waste product came to be viewed as a potential substrate for alcohol production. The technology was readily available; and today a number of industrial plants successfully produce alcohol from whey permeate.

Lactose is the major constituent of whey permeate and accounts for approximately 70% of the total solids. It must be stressed that this is an average analysis for sweet whey. The type of cheese produced will obviously influence the whey composition. Small amounts of other nutritionally desirable materials including riboflavin, amino acids and relatively high levels of calcium and phosphorus are also present in whey. To obtain the maximum reduction in BOD, it is necessary to develop a process whereby practically all the lactose is converted to alcohol.

In practice, a lactose content of 4.3-5.0% is obtained in the permeate. Using the yeast strain *Kluyveromyces marxianus* a yield of 86% is attainable from an initial lactose concentration of 4.3%. This yeast has the ability to ferment lactose, a disaccharide composed of glucose and galactose (Figure 2).

The alcohol concentration following fermentation of permeate is low thus requiring a very high energy input in the subsequent distillation. On the other hand, the pollution problem is greatly decreased with the total solids (lactose and protein) being reduced to 10% of that initially present.

Biomass produced during the fermentation will add to the effluent load. Whey pH ranges from 4.6 to 5.6. The temperature range for fermentation can be 25-35°C. Using a batch process the fermentation time will be between 18 and 24 hrs. At the end of fermentation the yeast cells may be separated for re-use. If a yield of 80% is attained from a lactose concentration of 4.4%, then production of 1 liter of 100% ethanol requires 42 liters of whey. The quality of the alcohol produced is very good with only very low levels of acetaldehyde and fusel oils. The permeate can be further concentrated; and the higher lactose content will increase alcohol content and thus lower distillation costs. This is very useful where permeate must be transported to alcohol plants.

One of the greatest difficulties facing a whey alcohol plant manager is infection. In certain instances there will be a carryover of lactic acid bacteria from the cheese plant. These microorganisms grow very easily on whey permeate thus reducing the amount of substrate available for alcohol production. Infection in such fermentations is usually controlled by using antibiotics.

Not withstanding such difficulties, utilization of whey as a substrate for alcohol production on a worldwide basis has become increasingly popular. This is evidenced by the diverse geographical locations of whey alcohol plants, examples of which are given in Table 7.

Considerations relevant to construction of a whey alcohol plant

Effluent treatment

Although ultrafiltration removes the protein from whey and the fermentation process converts the main ingredient (lactose) into alcohol, biomass is produced. The other ingredients of whey will contribute to the BOD/COD of the final effluent from the alcohol plant; and adequate effluent treatment facilities must be provided. One option is to use an anaerobic digestion system. This will convert a portion of the BOD into gas that can be used to fuel the boilers in the distillation plant.

Markets for alcohol

The alcohol markets in the US and Europe are very different. Because of government intervention in the US, ethanol production has grown from an insignificant amount in 1978 to a record 6.4 billion liters in 1998. In Europe the situation is very different. There has been no such government intervention and an oversupply of alcohol has existed for some time. Until recently a large portion of this alcohol was sold to Russia, but this market has diminished. The low price of oil has helped the majority of alcohol producers to keep production costs down; however as oil prices rise these production costs will also increase.

In the US, ethanol was promoted as a solution for a variety of complex problems. Among them: US dependence on foreign oil supplies, which was made apparent by two oil crises of the 1970s, and the low gasoline octane ratings caused by reduced use of lead after the approval of the Clean Air Act in 1977. Ethanol production from corn was seen as a way to boost farm incomes caused by the grain surplus in the wake of the Soviet embargo. Also, addition of ethanol to gasoline was seen as a means to reduce air pollution. These provide a ready market for any alcohol produced. Within the EU there has been much discussion about bio-ethanol, but only recently has there been any commitment to bio-fuels. Tax concessions for pilot plants producing bio-fuels have been allowed; and as a consequence new bio-fuel projects have been

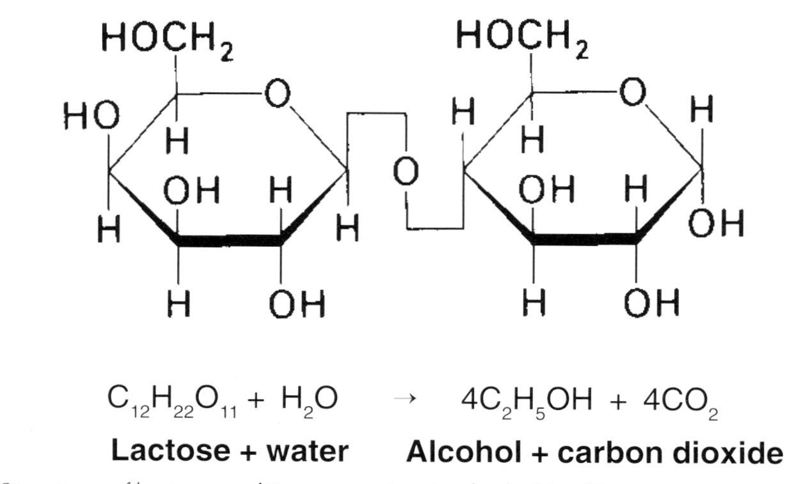

$$C_{12}H_{22}O_{11} + H_2O \rightarrow 4C_2H_5OH + 4CO_2$$

Lactose + water **Alcohol + carbon dioxide**

Figure 2. Structure of lactose and its conversion to alcohol by *Kluyveromyces marxianus.*

Table 7. Alcohol plants using whey as a substrate.

Ireland	Carbery Milk products, of Ballineen, County Cork, operate an alcohol plant using cheese whey as substrate. This alcohol is used for vodka and cream liqueurs (Murtagh, 1995).
United States	A plant to produce industrial alcohol from whey has been installed in California in one of the largest cheddar cheese plants in the world. This plant processes 1,000,000 kg of milk per day into cheese. The whey from the process is converted into permeate and protein by ultrafiltration. The protein is dried and the lactose in the permeate is converted into alcohol. The alcohol is then distilled in a distillation plant capable of processing 200,000 liters of whey per day, producing 1.5 million liters of pure alcohol per year.
New Zealand	Whey is the main by-product of the New Zealand dairy industry and is produced in the manufacture of cheese and casein. Anchor Ethanol was the first company in the southern hemisphere to manufacture alcohol from casein whey on a commercial basis. Anchor Ethanol began production at Anchor Products Repora in 1980 and a second plant was commissioned at Anchor Products Tirau the following year. These plants produce a combined total of 10 to 11 million liters of ethanol per year, over half of which is exported. This alcohol is the basis for a range of alcohol beverages. It is also used in the manufacture of solvents, methylated spirits, white vinegar, surgical spirit, food colouring, deodorants, perfumes, aerosols and pharmaceutical products (New Zealand Dairy,1999).
Russia	In Russia the range of feedstocks used for alcohol production is very diverse, including grains, fruit, whey, wine and molasses (Berg, 1999).

announced in the Netherlands, Sweden and Spain. France has made the most progress. In 1996 the French approved a draft law that made the use of oxygenated components in fuel mandatory by 2000.

Conclusions

The attachment of a whey alcohol plant to a dairy plant producing the appropriate substrate solves a number of major problems confronting whey producers. Alcohol production provides a guaranteed outlet for the whey that is not subject to the fortunes of pig markets or other elements outside their control. Whey production (or disposal) does not form a limiting step in cheese or casein production. The whey protein recovered by ultrafiltration is a valuable item, the sale of which contributes to process economics. There is also a universal move towards bio-fuels, therefore the markets for alcohol should be increasingly secure in the future.

References

Berg, C. 1999. World Ethanol Production and Trade to 2000 and Beyond. http://www.distill.com/berg.

Edelman, E. and S. Grodnick. 1986. The Ideal Cheese Book. http://cbs.infoplease.com/ce5/CE010398.html.

Eurostat. 1999. Institut National de Statistique, Brussells, Belgium, http://europe.eu.int/en/comm/eurostat/eurostat.html.

Burgess, K.J. 1980. Uses of waste dairy products. Technology Ireland, June, pp 43-44.

Murtagh, J.E. 1995. Molasses as a feedstock for alcohol production. *In:* The Alcohol Textbook (T.P. Lyons, D.R. Kelsall and J.E. Murtagh, eds). Nottingham University Press, Nottingham, UK. pp. 27-34.

New Zealand Dairy.1999. http://www.nzdairy.co.nz/public/educational/whey/ethanol/Ethanol~t.htm

Nielson, W.K. 1997. Whey processing. Technical Bulletin. APV Ireland Ltd, Dublin.

Chapter 8

Lignocellulosic feedstocks for ethanol production: the ultimate renewable energy source

R. Katzen[1], P.W. Madson[1], D.A. Monceaux[1] and K. Bevernitz[2]
[1]*KATZEN International, Inc., Cincinnati, Ohio, USA*
[2]*Neenah, Wisconsin, USA*

Production of ethanol from lignocellulosic feedstocks is of growing interest worldwide. Potential utilization of large volumes of primary and secondary lignocellulosic wastes, as well as renewable biomass (Table 1), to produce fuel and chemical feedstocks presents significant technical and economic challenges (Cowling *et al.*, 1976; Ferchak *et al.*, 1980; Klass and Emert, 1981; Tsao, 1978; Zermetz, 1979; GAO, 1981). The solutions to these problems can aid present and future generations. Production of fuel ethanol from biomass is the primary target for a commercial product. However, it is obvious that sugars produced for fermentation, or synthesis gas derived from biomass, can be utilized to produce other useful fuel and chemical products.

Chemistry of lignocellulosic biomass

The term 'holocellulose' is often used to describe the total carbohydrate contained in a plant or microbial cell. Holocellulose is comprised of cellulose and hemicellulose. Cellulose is further categorized into long-chain α-cellulose and the short-chain forms of ß-cellulose with a degree of polymerization (DP) of 15-90 units, and γ-cellulose with a DP of less than 15 units. Following is a description of terms used in lignocellulosic chemistry.

Cellulose is a structural material formed by a plant or microbial cell from glucose, a common six-carbon sugar. Glucose is a carbohydrate and therefore has the composition of carbon plus hydrogen and oxygen (CH_2O). Although glucose ($C_6H_{12}O_6$) is the smallest unit (monomer) that can be isolated from cellulose degradation, the basic building block of cellulose is actually cellobiose, a two-glucose anhydride unit (dimer). As glucose units are linked together into polymer chains, a molecule of water is lost, which makes the chemical formula for each monomer unit (glucan) $C_6H_{10}O_5$.

Table 1. Lignocellulosic feedstocks

Wood
Primary (native)
Plantation
Primary forest waste
Secondary processing waste
Agricultural residues
Straws (wheat, barley, rice)
Bagasse (sugarcane, sweet sorghum)
Stover (corn, milo)
Municipal waste
Waste paper

The cellulose molecular structure is an unbranched linear polymer. The length of a polymeric cellulose molecule is determined by the number of glucan units in the polymer, usually expressed as $(C_6H_{10}O_5)_n$, where n is the degree of polymerization. The DP of cellulose depends on the type of plant or microorganism from which it is isolated, as well as the method of isolation. Typical numbers for DP of native (non-disturbed) cellulose are estimated to be from 2000-14,000 glucan units, with commercial wood pulp DP in the range of 650-1500 units per glucan chain.

Due to the potential for each monomer unit of cellulose to form three hydrogen bonds with a monomer in a neighboring chain, the chains fit tightly together to form larger units known as microfibrils. The result is a very stable configuration - essentially free of interstitial spaces, making it anhydrous and quite recalcitrant to hydrolysis by acid, base or enzyme action. Chain regions with no special disturbances form crystalline areas known as micelles. Regions without extensive inter-chain hydrogen bonding are consequently less structured (amorphous) and are relatively more susceptible to hydrolysis. Native cellulose consists of cellulose fibrils bound together by an amorphous matrix comprised of pectin, hemicellulose and lignin.

Pectin and hemicellulose are macromolecular polysaccharides and are the most soluble fraction of native plant biomass. Pectin consists of mostly D-galacturonic acid with some residues of rhamnose, galactose and arabinose. Hemicellulose is not a form of cellulose, but is a mixture of straight-chain and branched forms of both 5- and 6-carbon sugars with varying types of linkages. Hemicellulose can be hydrolyzed to the 6-carbon sugars of glucose, mannose and galactose, and the 5-carbon sugars of xylose and arabinose.

Lignin is a very complex molecule constructed of phenyl propane units linked in a three-dimensional structure (Smook, 1982). Lignin is considered to be the glue holding the cellulose fibrils together, and as such is difficult to completely remove. Lignin confers stiffness to the cell and may also provide protection against microbial attack. Because lignin molecules have varying types of chemical bonds, the particular extraction method will determine what types of chemicals result.

Research

Although research on the production of ethanol and related products from cellulosic and lignocellulosic feedstocks is developing rapidly, there is a substantial record of prior research, development and commercialization that can be utilized in planning for the introduction of improved and more economic technology. The following will relate the prior history and long-established art of biomass conversion, particularly of lignocellulose to ethanol, with ongoing research and development in this area.

Concentrated acid hydrolysis

Hydrolysis with concentrated acid is relatively old art, harking back to the early days of lignin and cellulose chemistry (Table 2). The Klason lignin determination, developed in 1890 and still a standard analysis in the wood conversion industry, utilizes cold, concentrated (72%) sulfuric acid (H_2SO_4) to dissolve the cellulosic fraction of the raw material, leaving the lignin as a residue. The dissolved cellulose, separated from the lignin, is then hydrolyzed by dilution of the sulfuric acid and heating to yield sugars (Locke *et al.*, 1961). This basic technology is part of the two stage acid hydrolysis process developed at Purdue University's Laboratory of Renewable Resources Engineering (Ladisch, 1979).

A second method of concentrated acid hydrolysis is the Bergius process, which uses cold, concentrated (40%) hydrochloric acid (HCl) to dissolve the cellulosic fraction. Water dilution and heating then permits hydrolysis of the cellulose to sugars. Commercially, the acidified lignocellulosic material from this process was neutralized with caustic soda and used as animal feed in Germany during World War II.

More recent research on cellulose hydrolysis with concentrated HCl has been carried out in Japan (Locke *et al.*, 1961).

Table 2. Acid hydrolysis technologies.

Sulfuric
Concentrated: one stage
Dilute: one stage
Dilute/concentrated: two stage
Dilute/dilute: two stage
Hydrochloric
Concentrated: liquid phase
Concentrated: vapor phase
Dilute
Organic (autohydrolysis)
Steam pressure
Steam and mechanical pressure

The concentrated acid processes generally give higher sugar and ethanol yields than dilute acid processes, but the relatively mild corrosiveness of the concentrated acid at essentially ambient temperature is superseded by more corrosive conditions as the acids are diluted and heated to hydrolyze the dissolved cellulose to sugars. The intermediate acid concentrations are extremely corrosive, requiring either expensive alloys or specialized non-metallic construction such as ceramic or carbon brick lining, resistant both to corrosion and temperature. The resulting high investment and maintenance costs have greatly reduced the commercial potential for these processes.

Dilute acid hydrolysis

The technology for dilute sulfuric acid hydrolysis and extraction of sugars at elevated pressures and temperatures was developed by Scholler before World War II and employed in Germany for production of ethanol. Plants were also built in Russia, during and after the war, to produce ethanol or yeast from wood sugars. Another plant was built and operated at Ems, Switzerland. As part of the war effort, a plant was also designed and built in the US at Springfield, Oregon utilizing the US Forest Products Laboratory (Madison) variation of the Scholler process. The plant operated for only a short time because construction was not completed until after World War II ended.

Test runs in Oregon indicated serious corrosion, erosion, tar and pitch formation and mineral scale formation, problems that have bedeviled all of these installations. These problems create high investment and high cleaning and maintenance costs and have limited the commercialization of these processes (Hajny, 1981; Harris *et al.*, 1946; Saeman, 1979; Underkofler and Hickey, 1954; Hokanson *et al.*, 1978).

A Brazilian government agency contracted with the USSR for the design and construction of a demonstration plant utilizing dilute sulfuric acid technology with 30,000 liter/day capacity for conversion of Brazilian wood feedstocks to ethanol. This plant was put into operation in the 1980s, but results were apparently disappointing.

New acid hydrolysis technology

Newer approaches to acid hydrolysis technology that may lead to increased yields and reduced investment are under development in many countries (Gilbert *et al.*, 1952; Goldstein, 1981; Kamiyama *et al.*, 1979; Porteous, 1972; Anonymous, 1981). Some of these, such as the Swiss Inventa technology (Mendelsohn *et al.*, 1981), and the New Zealand Forest Service technology (Whitworth *et al.*, 1980), appear to be improvements on the US Madison-Scholler technology mentioned above.

Other newer technologies, such as the Purdue two stage process (Ladisch, 1979), utilize combinations of dilute and concentrated acid. The dilute acid is used under mild conditions for pre-hydrolysis of the hemicellulose while the concentrated acid is used to extract the cellulose from the lignocellulose residue after hemicellulose removal. Another version is the technology being developed at Dartmouth

University (Grethlein *et al.*, 1980; Thompson *et al.*, 1979; Lynd *et al.*, 1991), which provides for a two stage dilute acid hydrolysis at high temperatures for short times. The first stage has temperature and time adjusted to hydrolyze hemicellulose; while the second stage is set at higher temperatures and longer retention times to hydrolyze the residual cellulose. Pressurized dilute acid extrusion technology was under development at New York University in the 1970s (Brenner *et al.*, 1978). At Georgia Tech, various combinations of dilute and concentrated acid hydrolysis have been tested in a multi-purpose pilot plant.

None of these two stage processes have advanced beyond a relatively modest pilot plant scale. Longer term operation in demonstration plants will be required before firm investment and operating costs, as well as technical and economic feasibility, can be established.

Enzymatic processes

No enzymatic system for cellulose hydrolysis has yet been proven commercially without first removing (pretreating) the lignin present in lignocellulosic feedstocks (Table 3).

Failure of cotton clothing during military campaigns in the Pacific during World War II sparked research by the US military into cellulose degradation. Early work at the US Army Natick Laboratories (Katz *et al.*, 1968; Mandels *et al.*, 1974; Nystrom *et al.*, 1976; Reese, 1976) identified the Trichoderma fungus and its multiple enzyme system as responsible for the cotton (cellulose) failure (hydrolysis). Further research showed it essential to use delignified cellulose to enable the enzymes to produce reasonable yields of sugar within a moderate time frame. The long time element and low yield of the early work at Natick indicated relatively high investment and production costs. New research worldwide has advanced this basic technology (Cysewski *et al.*, 1976; Emert *et al.*, 1980a; Lyness *et al.*, 1981; Nisizawam, 1973; Puls *et al.*, 1977; Shoemaker *et al.*, 1981; Rosenberg *et al.*, 1980).

Table 3. Enzymatic hydrolysis technology pretreatment options.

Autohydrolysis
Steam pressure (Dietrichs)
Steam and mechanical pressure (Stake)
Steam explosion (Iogen)
Acid pre-hydrolysis
Dilute sulfuric acid
Dilute hydrochloric acid
Acetic acid
Concentrated sulfuric acid (cold)
Concentrated hydrochloric acid (cold)
Alkali
Sodium hydroxide
Ammonia
Organosolv
Methanol
Ethanol
Butanol
Hexamethylenediamine
Phenol
Mechanical
Attrition mill
Roller mill
Vibratory rod mill
Extruder

Research is under way at several laboratories to develop microorganisms that will secrete enzymes to break down lignin (Kirk *et al.*, 1977) that act as a protective coating for, and are apparently chemically bonded to, the hemicellulose. To date, the organisms under development have not demonstrated appreciable lignin removal without substantial degradation of the cellulose and with little improvement in sugar yield. Dilute acids, both sulfuric and sulfurous, have been tested in pretreatment methods to break the lignin-hemicellulose bond (Mandels, 1976; Millet *et al.*, 1974; DOE, 1978; Wilke *et al.*, 1979) and make the cellulose more accessible to enzyme attack. Initial results indicate potential feasibility and improvement in yields. However, the hemicellulose hydrolyzate produced by the dilute acid pretreatment will be more useful with suitable microorganisms developed for fermentation of the C_5 sugars to ethanol. Organisms such as Thermonospora and genetically-modified bacteria show potential in this regard (Ingram, 1991).

Alkali pretreatment has also been utilized on certain materials, particularly straws (Detroy *et al.*, 1981; Millet *et al.*, 1974; Toyama, 1976). This treatment cleaves the lignin-hemicellulose bond and at least partially solubilizes the lignin and decrystallizes the cellulose. Sodium hydroxide and ammonia have been used in such systems. Such pretreatment has indicated improved susceptibility of cellulose to enzyme hydrolysis. However, the C_5 sugars are modified by the alkaline extraction and may not be accessible for fermentation by the new organisms. The Ammonia Freeze Explosion pretreatment pro-cess for woody feedstocks has been developed by Dale and his associates at Texas A&M University (Dale *et al.*, 1989).

A non-chemical reaction route (organosolv) using organic solvents to remove lignin to provide a clean cellulose suitable for enzyme hydrolysis has been under development. Solvents such as methanol, ethanol, butanol, phenol and hexamethylenediamine, with water, have reportedly been used (Holzapple *et al.*, 1981). In each system, the solvent action is accompanied by partial pre-hydrolysis effected by the presence of water and organic acid catalysts (acetyl groups), producing hemicellulose sugars. Such solvent-delignified celluloses have proven susceptible to enzyme hydrolysis. The main advantage of using solvents over chemical pretreatment is that relatively pure, low molecular weight lignin can be recovered as a by-product. The C_5 sugars solubilized during organic solvent delignification are relatively clean and could be fermented to ethanol with appropriate microorganisms, such as the recombinant bacteria developed by Ingram and co-workers at the University of Florida (Ingram, 1991).

Another group of pretreatment methods can be described as thermomechanical, with the thermal energy in these systems generated by the mechanical action. This covers a wide variety of pretreatments such as vibratory rod milling (Millet *et al.*, 1979), the Natick roller mill system (Tassinari *et al.*, 1980), and the Gulf Oil/ University of Arkansas thermomechanical treatment in attrition mills (Emert *et al.*, 1980a). The Stake and Dietrichs technologies involve steaming under moderate pressures, plus mechanical attrition in a screw-propelled device (Dietrichs *et al.*, 1978; Taylor *et al.*, 1980). The Iogen technology (Foody, personal communication) involves steaming of wood chips at pressures up to 500 psig followed by explosion through a slotted die to provide both cleavage of the lignocellulosic linkage and shortening of the cellulose fibers. The Rugg technology developed at New York University (Gilbert *et al.*, 1952) involves a combination of acid, heat and pressurized operations in an extruder (Werner-Pfleiderer), which gives moderate yields of sugars. However, this device apparently combines the corrosion and polymerization problems of acid treatment with the erosion and corrosion problems of an extruder discharge (pressure reduction) mechanism. These latter operations indicate some promise, but involve substantial investment in mechanical pretreatment plus ongoing maintenance due to corrosion and erosion caused by processing the wood under pressure and discharging through a pressure-reducing device.

In some of these developments, experimental work has only been conducted through the hydrolysis stage. There is some indication that high temperature and acid combinations can lead to serious toxicity problems in subsequent fermentation due to the non-specific hydrolysis that can produce a variety of inhibitory materials such as furfural and its derivatives. Solutions to this problem are claimed; but actual methods of operation have not been defined, and the economics of such treatments are not identified. Energy cost is a critical factor in determining feasibility of these technologies (Datta, 1981).

Ethanol from biomass synthesis gas

Production of methanol from synthesis gas (carbon monoxide plus hydrogen) derived from wood is under development, and has been carried through the pilot plant stage (Feldman, 1980). This technology could be economic in areas where natural gas is too expensive and coal is not readily available as a basic feedstock for synthesis gas. Current research and recently issued patents have indicated development of catalytic systems for homologation reactions, which yield alcohols higher than methanol (Bartish, 1979; Koermer *et al.*, 1978). Thus, methanol produced in a primary synthesis can be converted to ethanol by reaction with additional carbon monoxide and hydrogen.

At this stage of development the catalytic systems are not highly selective, but yield a mixture of ethanol and higher alcohols as well as some unreacted methanol. This is not a problem where the ethanol is to be used as a fuel or octane improver in gasoline. However, this can lead to complications in production of industrial ethanol, which must be of high purity. It should be recognized that technical and economic evaluations of ethanol derived from lignocellulosic substrates should be compared with the ongoing developments in production of ethanol from biomass synthesis gas (Figure 1).

Development of cellulose-to-ethanol technology

Although modern technology for the acid hydrolysis of cellulose (particularly with two-stage operation) can lead to improved yields, there are not sufficient data available to define firm investment requirements. The optimistic estimates of low investment and operating costs indicated by some research groups have not been verified by engineers and contractors experienced in design and construction.

Enzymatic hydrolysis appears more promising, particularly in that the operating conditions are near ambient and corrosion, polymerization and fouling problems are relatively minor. These problems can be solved with conventional moderate alloy construction in certain parts of the process. The time element in the production of the enzyme mix and in the enzymatic hydrolysis have been overcome to a substantial extent by development of mutations of *Trichoderma reesei*, use of continuous process technology and development of the Simultaneous Saccharification and Fermentation (SSF) process (Emert *et al.*, 1980a).

Ongoing work in developing either improved mutations or genetically modified micro-organisms promises to yield more economic hydrolysis of cellulose, possibly even in the presence of substantial amounts of lignin. Combining hydrolytic enzyme production with fermentation in a single organism could result in improvements in this technology. Furthermore, a combination of dilute acid (or alkali) pre-treatment, as an option to mild thermomechanical pretreatment, and secondary enzymatic hydrolysis of the cellulose could result in an overall process with improved yields and better operating economics.

Among the groups conducting cellulolytic enzyme research, the partnership of Gulf Oil (now Chevron) and Nippon Mining (Bio-Research Corporation) of Japan led not only to advances in technology for production of the cellulolytic enzymes, but also the critical and major advance of developing the SSF process (Katzen, 1990; Mandels *et al.*, 1976; Gauss *et al.*, 1976).

A one-ton-per-day pilot plant built by Gulf Oil Chemicals at Pittsburg, Kansas operated with both SSF and continuous cascade fermentation starting in the late 1970s. Feedstocks included wood waste, pulp and paper mill waste, municipal solid waste and various agricultural residues. The key results of SSF application were a major decrease in retention time required for both the hydrolysis and fermentation operations and a major increase in yield relative to other known technologies. This was due to the fact that as fast as the cellulolytic enzymes produced glucose, the yeast converted the glucose to ethanol. Thus, the feedback inhibitory action of glucose on enzyme activity was essentially eliminated. Also, the Gulf researchers developed a continuous process for producing

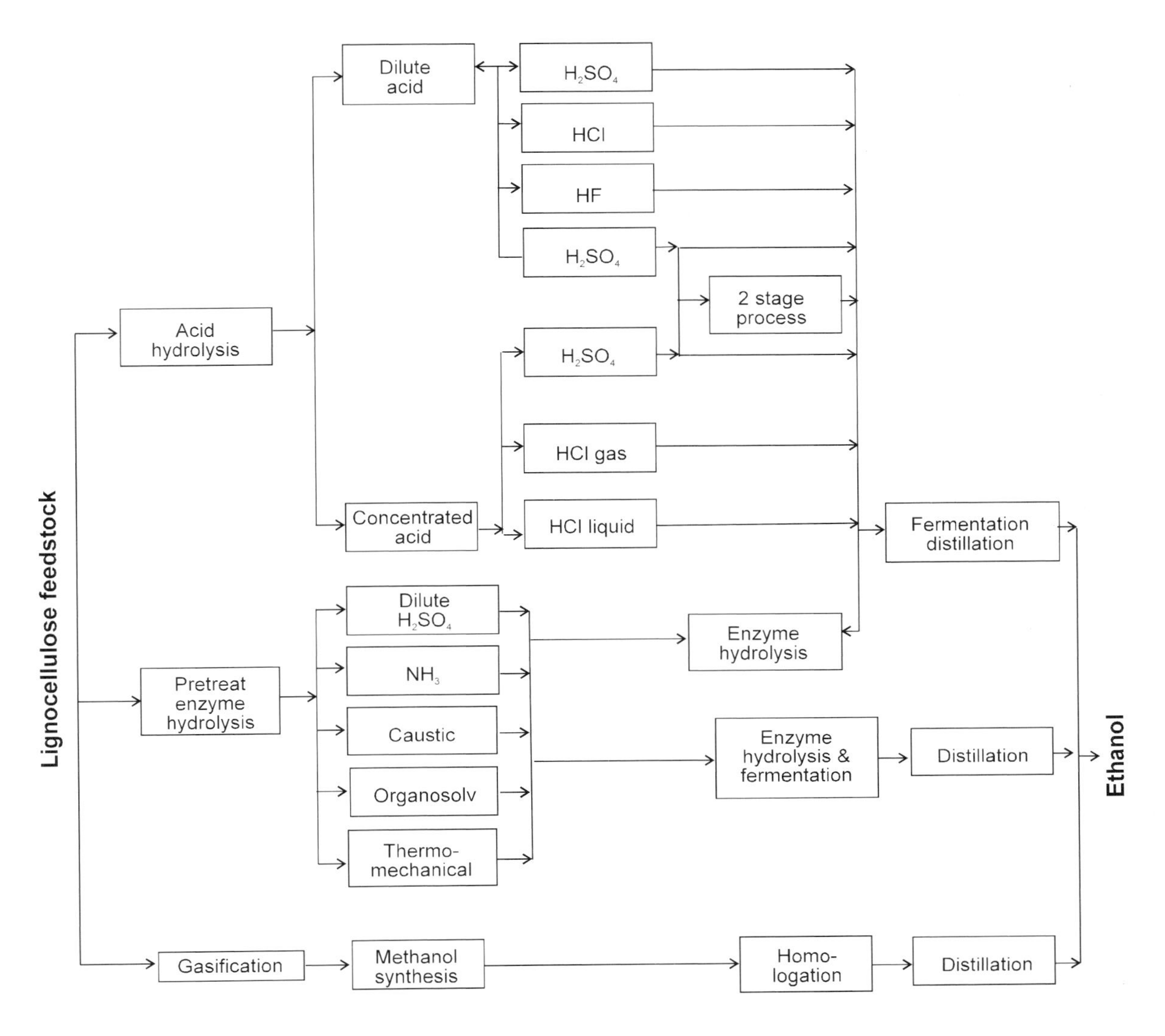

Figure 1. Ethanol from lignocellulose process matrix.

the Trichoderma fungus and the cellulolytic enzymes from the base feedstocks for the fermentation operation (Huff *et al.*, 1976). Coupled with a novel enzyme recycle process, enzyme cost was dramatically reduced.

Although the Gulf pilot plant operated for several years, there were shortcomings limiting its application to continuous SSF operation. In 1981, Gulf turned over this program to the University of Arkansas for further research and development. This included the improved enzyme production process, recovery and recycling of cellulolytic enzymes and continued evaluation of lignocellulosic feedstocks. Considerable emphasis was placed on pretreatment methods for breaking the lignin-hemicellulose-cellulose linkages to permit access of the enzymes to the cellulose. Such pretreatments included organosolv delignification, caustic, acid and thermomechanical processing.

Another opportunity arose to advance this technology when Procter & Gamble decided to apply this process to their pulp and paper waste, which had become a waste disposal problem. A one-ton-per-day pilot plant was erected at their mill complex in Pennsylvania. Six months of intensive experimentation in this pilot plant with both primary pulping waste and paper-making waste resulted in substantial improvement in the efficiency of conversion of cellulose to glucose and ethanol (Easley *et al.*, 1989).

Key improvements resulting from the operation of this pilot plant related to application of a mild pretreatment (since most of the lignin had been removed by the pulping process) and use of fed-batch techniques in the SSF process. As indicated in Table 4, runs of more than 180 hrs were made before a significant reduction in enzyme activity was indicated. Specific yields obtained during a typical extended fed-batch run of Table 4 are indicated in Figure 2. A key item to be noted is the feasibility of operating the SSF system at temperatures as low as 37°C, previously considered too low for economic enzymatic hydrolysis. Also, this work showed it was not necessary to use expensive vacuum recovery of ethanol to achieve good yields of ethanol per unit of enzyme activity.

An interesting development during this pilot plant operation involved the addition of a surfactant. The surfactant acts to facilitate access of the cellulolytic enzymes to the cellulose fiber (Katzen *et al.*, 1994). Yield of ethanol per unit of feedstock increased by more than 10%.

Work conducted for organizations interested in conversion of bagasse led to substantial pilot plant testing of a mild acid pre-hydrolysis pretreatment. Such pretreatment, carefully controlled with respect to acid use, temperature and retention time, led to essentially complete hydrolysis of the hemicellulose and recovery of the hemicellulose sugars in a near quantitative manner. The residue from this mild acid hydrolysis was processed in the SSF system, with high yields of ethanol achieved based on cellulose content.

The accumulated data and knowledge from the Gulf and Procter & Gamble pilot plants (Easley *et al.*, 1989; Emert *et al.*, 1980a; b), along with recent advances in improved cellulolytic enzyme systems have made it feasible to extend this technology by construction of commercial demonstration facilities using recombinant

Table 4. Fed-batch SSF conversion of cellulose to ethanol.

Conditions	*Organism*		
	Candida brassicae	*Saccharomyces cerevisiae*	*Saccharomyces cerevisiae*
Bench scale tests			
Total fiber, g (dry basis)	6804	1600	1000
Enzyme dose, g	183.7	43.2	27
Enzyme:fiber ratio	0.027	0.027	0.027
Vacuum-recovered ethanol, g	1427	NONE	NONE
Final ethanol concentraton, g/L	17.3	36.3	32.7
Final volume, L	24	11.42	8.12
Total ethanol produced, g	1569	415	265
Ethanol yield, gal/ton	69.8	78.5	80.2
Enzyme: ethanol ratio	0.117	0.104	0.102
Fed-batch pilot operation			
Fermentation time, hr	192	168	188
Volumetric production rate, g/L/hr	0.34	0.22	0.17
Average glucose concentration, g/L	2.4	1.05	0.82
Aeration rate, $ft^3/min/ft^3$	0.02	NONE	0.002
Temperature, °C	40	37	37

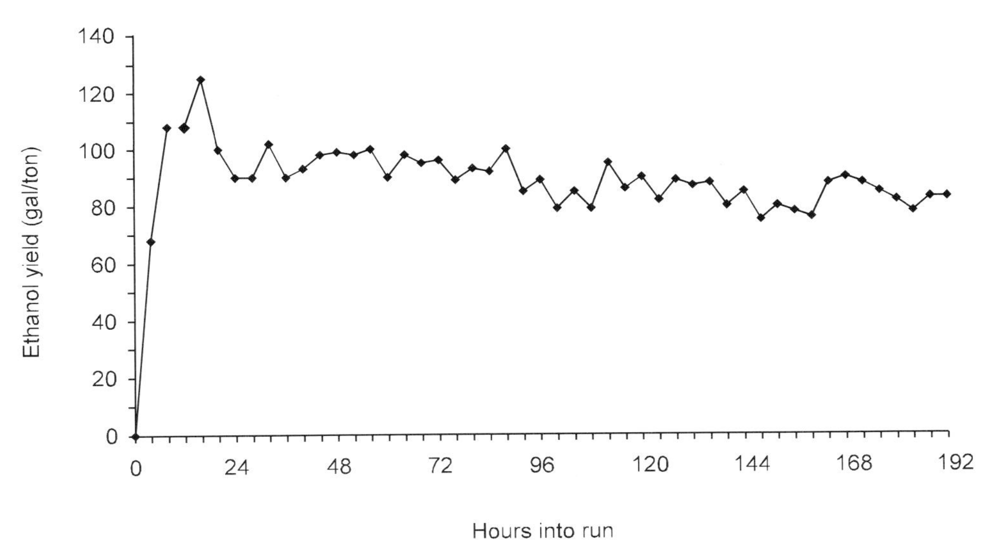

Figure 2. Procter and Gamble's SSF fed-batch performance.

bacteria or other select organisms to ferment both C_5 and C_6 sugars (Ingram, 1991) for production of ethanol from lignocellulosic feedstocks.

References

Anonymous. 1981. TVA Technology, Chemical Week, September 9, p. 54,

Bartish, C.M. 1979. US Patent 4,171,461, October 16.

Brenner, W. *et al.* 1978. National Technical Information Service, US EPA NYU/DAS 78-24.

Cowling, E.B. *et al.* 1976. Biotechnology and Bioengineering Symposium No. 6, p. 95.

Cysewski, G.R. *et al.* 1976. Biotechnology and Bioengineering, 18:1297.

Dale, B. *et al.* 1989. Technical summary of ammonia freeze explosion. Texas A&M University, October.

Datta, R. 1981. Process Biochemistry, June/July. p. 6,

Department of Energy. 1978. National Technical Information Service HCP/T 3891-1, January US Dept of Commerce, Springfield, USA.

Detroy, R.W. *et al.* 1981. Biotechnology and Bioengineering 23:1527.

Dietrichs, H.H. *et al.* 1978. Holzforschung 32:193.

Easley, C.E. *et al.* 1989. Cellulosic waste con-version to ethanol using fed-batch simul-taneous saccharification and fermentation, 13th Annual Conference Institute of Gas Technology, Paper 53, February 13-17.

Emert, G.H. *et al.* 1980a. Chemical Engineering Progress 76:47

Emert, G.H. *et al.* 1980b. US Patent 4,220,721, September 2.

Feldman, H.F. 1980. Tappi Journal 63:83.

Ferchak, J.D. *et al.* 1980. Solar Energy 26, p. 9 and 17.

Gauss, *et al.* 1976. US Patent 3,990,944, Nov. 9.

General Accounting Office (GAO). 1981. Nation's unused wood offers vast potential energy and product benefit, National Technical Information Service EMD-81-6, US Government Printing Office, Washington, DC.

Gilbert, N. *et al.* 1952. Industrial & Engineering Chemistry 44:1712.

Goldstein, I.S. 1981. Organic Chemicals from Biomass. CRC Press, Boca Raton, FL.

Grethlein, H.E. *et al.* 1980. National Technical Information Service USDOE/SERI, EG-77-S-01-4061.

Hajny, G. 1981. US. Forest Products Laboratory, FPL 385.

Harris, E.E. *et al.* 1946. Industrial & Engineering Chemistry 38:896.

Hokanson, A.E. *et al.* 1978. Chemicals from wood waste. Chemical Engineering Progress, January, p. 667.

Holzapple, M. *et al.* 1981. ACS-NYC Meeting, August.

Ingram, L. 1991. US Patent 5,000,000, March 19.

Huff G. *et al.* 1976. US Patent 3,990,945, Nov. 9.

Kamiyama, Y. *et al.* 1979. Carbohydrate Research, 73:151.

Katz, M. *et al.* 1968. Applied Microbiology, February. p. 419.

Katzen, R. *et al.* 1994. Development of bio-conversion of cellulosic wastes. 16th Symposium on Biotechnology for Fuels and Chemicals, Colorado Springs, CO, May 9-13.

Katzen, R. 1990. Ethanol from lignocellulose, agro-industrial revolution conference, Washington, D.C., June.

Kirk, T.K. *et al.* 1977. Developing Industrial Microbiology 19:52.

Klass, D. and G. Emert. 1981. Fuels From Biomass and Wastes. Ann Arbor Science Publishers, Inc. Ann Arbor, Michigan.

Koermer, G.S. *et al.* 1978. Chemical Product Research and Development, Industrial & Engineering Chemistry 17:231.

Ladisch, M.R. 1979. Process Biochemistry, January.

Locke, E.G. *et al.* 1961. Forest Products Journal, August, p. 380.

Lynd, L.R. *et al.* 1991. Science 251:1318.

Lyness, E. *et al.* 1981. Biotechnology and Bioengineering 23:1449.

Mandels, M. 1976. Biotechnology and Bio-engineering, Symposium No. 6, p. 221.

Mandels, M. *et al.* 1976. Biotechnology and Bioengineering, 16:1471.

Mandels, M. *et al.* 1974. Biotechnology and Bioengineering 16:1471.

Mendelsohn, H.R. *et al.* 1981. Chemical Engineering, June 15, p. 62.

Millet, M.A. *et al.* 1974. U.S. Forest Products Laboratory Report, June.

Millet, M.A. *et al.* 1979. Hydrolysis of cellulose, ACS Advances in Chemistry, Series No. 181, p. 71.

Nisizawam, K. 1973. Journal of Fermentation Technology 51:267.

Nystrom, J.M. *et al.* 1976. Biotechnology and Bioengineering, Symposium No. 6, p. 55.

Porteous, A. 1972. Paper Trade Journal, Feb. 7.

Puls, J. *et al.* 1977. Transactions of the Am. Inst. of Chem. Eng. September, p. 64.

Reese, E.T. 1976. Biotechnology and Bio-engineering, Symposium No. 6, p. 9.

Rosenberg S.L. *et al.* 1980. Hemicellulose Utilization for Ethanol Production. National Technical Information Service LBL-10800, April.

Saeman, J.F. 1979. American Chemical Society, JCS Symposium, p. 472.

Shoemaker, S.P. *et al.* 1981. Trends in the biology of fermentations for fuels and chemicals. Plenum Publishers, New York.

Smook, G.A. 1982. Handbook for Pulp & Paper. TAPPI, Atlanta, GA and CPPA, Montreal, Quebec.

Tassinari, T. *et al.* 1980. Biotechnology and Bioengineering. 22:689.

Taylor, J.D. *et al.* 1980. Bioenergy Conference Proceedings. p. 285.

Thompson, D.R. *et al.* 1979. Industrial & Engineering Chemistry, 18:166.

Toyama, N. 1976. Biotechnology and Bioengineering Symposium No. 6, p. 207.

Tsao, G.T. 1978. Process Biochemistry, October p. 12.

Underkofler L.A. and R.J. Hickey. 1954. Industrial Fermentation. Chemical Publishing Co.

Whitworth, D.A. *et al.* 1980. Ethanol from wood, New Zealand Forest Service Report.

Wilke, C.R. *et al.* 1979. Process development studies - bioconversion of cellulose and production of ethanol. National Technical Information Service LBL-10373.

Zermetz, C.A. 1979. Growing Energy - Land for Biomass Farm' National Technical Information Services PB-296-650, June.

Chapter 9

Alcohol production from cellulosic biomass: the Iogen process, a model system in operation

J.S. Tolan
Iogen Corporation, Ottawa, Canada

The production of fuel alcohol from cellulosic biomass is of growing interest around the world. Cellulosic biomass can be used to produce transportation fuel, with the overall process having little net production of greenhouse gases. Biomass is available as a by-product of many industrial processes and agricultural materials, or potentially can be produced from dedicated energy crops. The technology for biomass conversion has many significant technical and economic hurdles that have prevented commercialization to this point. However, significant progress in recent research has motivated Iogen Corporation to build a 300 t/wk plant in Ottawa, Canada to demonstrate the technology.

Process overview

The basic process steps are shown in Figure 1. A cellulosic feedstock material such as wood chips or straw is subjected to pretreatment, ie., cooked in the presence of acid to break down the fibrous structure. After pretreatment, the material has a muddy texture. Cellulase enzymes are added to the pretreated material to hydrolyze the cellulose to the simple sugar glucose, a process known as cellulose hydrolysis. The cellulase enzymes are made at the plant site by using a wood-rooting fungus in large fermentation vessels. After enzymatic hydrolysis, the sugars are separated from the unhydrolyzed solids, which include lignin and residual cellulose. These solids are burned to provide energy for the entire process (lignin processing). The sugars are fermented (sugar fermentation) to ethanol using simple brewer's yeast (to ferment the glucose) and more recently developed microbes for the more difficult sugars to ferment, primarily xylose. In ethanol recovery, the ethanol is recovered by conventional distillation and the purified ethanol is suitable for automotive fuel by itself ('neat ethanol' or E100) or as a blend with gasoline (formerly called gasahol, now called E30 for a 30% ethanol blend, etc.).

The remainder of this chapter describes the process steps and related technologies in more detail.

Feedstock selection

Feedstock composition

The term 'cellulosic biomass' refers to potential feedstocks that have cellulose as a primary constituent. Other major constituents of these

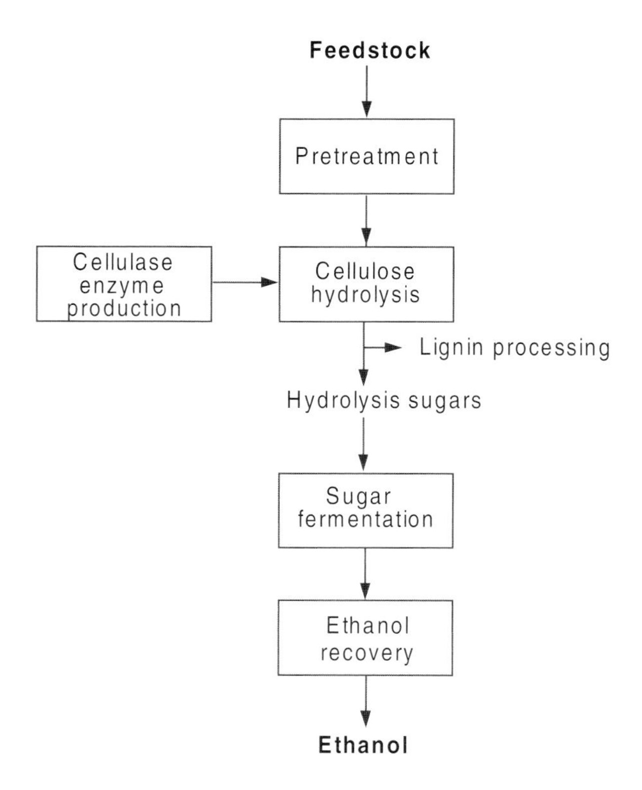

Figure 1. Iogen's ethanol process.

materials include hemicellulose and lignin. Minor constituents include proteins, ash, various organics and starch.

Cellulose comprises 40-60% of most plant material. Cellulose is a polymer of glucose, of DP (degree of polymerization, or chain length) of 1,000 to 10,000. Cellulose is a linear, unbranched polymer with glucose joined together by ß-1,4 linkages. Individual polymer chains run parallel to each other and form hydrogen bonds with up to three per monomeric glucose units. Several such chains form a microfibril. A microfibril region that has the full degree of hydrogen bonding forms a roughly cubic, 3-dimensional lattice. Such a region is crystalline cellulose and is very stable against attack by enzymes or acid. Other regions are not hydrogen-bonded to nearly this extent, and in the extreme are simply random configurations of glucose polymers. This is amorphous cellulose. Most natural cellulose is primarily crystalline cellulose.

The main source of ethanol is from the glucose, originating from the cellulose. However, a second source of ethanol is from the simple sugars that comprise the hemicellulose. Hemicellulose is a mixture of linear and branched polymers of the 5-carbon sugars xylose and arabinose and (less importantly) the 6-carbon sugars glucose, mannose, and galactose. Hemicellulose is readily dissolved and hydrolyzed to its simple sugars in dilute acid at moderate temperatures, around 120^{0}C. Hemicellulose comprises 15-25% of most plant material.

Lignin comprises 15-30% of most plant materials. It differs from cellulose and hemicellulose in that lignin is not comprised of carbohydrates, but rather consists of a complex three dimensional matrix of phenolic-propane units. Lignin confers water resistance and stiffness to the fiber and protection against microbial attack. Lignin does not participate in the pretreatment or hydrolysis processes except with a decrease in the degree of polymerization. The burning of lignin is the mode of energy generation for the process.

In normal operation, the minor constituents exert only a minor impact on the process. The protein is degraded by pretreatment to the point where it cannot be recovered economically. The ash content must not be too high in order to prevent abrasion to the process equipment. Such is the case if large amounts of silica (sand) are present due to the harvest practices. The starch is easily hydrolyzed to glucose and increases the overall ethanol yield. Table 1 shows the composition of several typical lignocellulosic materials.

Feedstock selection

The scope of feedstocks typically considered for ethanol production is listed in Table 2. These potential feedstocks are evaluated on the basis of desired feedstock properties. The desired qualities of a feedstock are:

1. *Low cost.* Naturally, the cost of the feedstock is an important part of the overall ethanol production cost. A desirable feedstock must be obtained and transported to the plant at low cost, at a maximum below $100/ton. This rules out waste paper.
2. *Availability.* A feedstock must be in sufficient quantity to supply a commercial plant. This requires perhaps 800,000 tons/year, which is not available from bagasse, for example.
3. *Uniformity.* For a high-speed production process, foreign matter such as hot dogs and tin cans present in municipal waste is unacceptable.
4. *Cleanliness.* High levels of silica can abrade equipment, as stated above. Heavy microbial contamination is unacceptable. High levels of toxic or inhibitory materials are not acceptable.
5. *High yield.* The main constituents, cellulose and hemicellulose, must be present in sufficient quantity to produce ethanol. This is a disadvantage of forestry waste, which is high in tree bark (mostly lignin and phenolic acids).
6. *High efficiency of conversion.* Not all feedstocks demand equal amounts of enzyme for conversion to cellulose. Softwood is notoriously difficult to hydrolyze.

Matching the feedstock list with the desired properties results in the leading feedstock candidates, which are agricultural residues such as straws and stover and energy crops such as aspen and grasses.

Table 1. Feedstock composition (mg/g total solids).*

Feedstock	*Cellulose*	*Starch*	*Xylan*	*Arabinan*	*Lignin*	*Ash*	*Protein*
Barley straw	406	20	161	28	168	82	64
Wheat straw	455	9	165	25	204	83	64
Wheat chaff	391	14	200	36	160	121	33
Switch grass	399	3	184	38	183	48	54
Corn stover	408	3	128	35	127	60	81
Maple wood	500	4	150	5	276	6	6
Pine wood	648	1	33	14	320	0	2

* from Foody *et al.*, 1999.

Table 2. Potential feedstocks.

Material	*Subclass*	*Comments*
Wood	Native forest	Difficult to process, especially softwood
	Forest waste	Cellulose/hemicellulose content is too low.
	Mill waste	Cellulose/hemicellulose content is too low.
Agricultural residues	Straws (wheat, barley, rice)	Possibility
	Bagasse (cane)	Not in sufficient quantity
	Corn stover	Possibility
Energy crops	Grass	Possible if yields improve
	Aspen wood	Possible
Waste cellulose	Municipal waste	Not uniform enough
	Waste paper	Too expensive

What about ethanol from corn or sugar cane from existing commercial processes?

Starch-based or sucrose-based processes are already widely used to make ethanol. The leading starch-based material is corn, which is widely used to make ethanol in the US. Starch is converted to glucose by grinding (in a dry milling process) or by steeping in dilute sulfurous acid (in a wet milling process), then using starch-degrading amylase enzymes. The glucose is then fermented to ethanol. Sucrose-based feedstocks include sugarcane (Brazil) and sugar beets (Europe). These feedstocks are ground and washed with water to extract the sucrose, which is then fermented to ethanol by yeast. Other feedstocks used to make small amounts of fuel ethanol in some regions include potatoes and Jerusalem artichokes.

Why ethanol from cellulose?

The conversion of cellulosic biomass to ethanol is more difficult than starch or sucrose. However, cellulose is available in much greater quantity and offers the potential for much greater ethanol production than other substrates. In addition, ethanol from starch and sucrose faces competition for the feedstock from the human food and animal feed industries, which exerts pressure on the price of the ethanol. Most cellulosic biomass is free of competition from other uses. Cellulosic biomass can be grown in a wider variety of climates and soils than starch and sucrose and therefore represents a new agricultural opportunity in many areas. Finally, ethanol from cellulose is expected to be neutral relative to the production of greenhouse gases (Singh, 1997). Corn, sugarcane, and sugar beets require large amounts of energy-intensive fertilizers and do not have the energy generation from the lignin by-product present in cellulosic biomass.

Corn, sugarcane and sugar beets all contain a small amount of cellulose and hemicellulose. Cellulose conversion technology represents an opportunity to improve the yields and decrease the wastes from these processes.

Many cellulosic materials, including straw and grass, contain up to 10% starch. This is converted to glucose during pretreatment and carried through to glucose fermentation where it is converted to ethanol.

Pretreatment

Process

Pretreatment is the process by which surface area of the feedstock is opened up for the subsequent enzymatic attack. In the absence of pretreatment, one might as well try to make sugar from a wooden table!

In the Iogen process, feedstock in the form of bales or chips is chopped to make small particles, which are conveyed to the pretreatment reactor. High pressure steam and sulfuric acid are added to the feedstock to reach 200-250 °C with 0.5-1% sulfuric acid. The material is maintained at this condition for less than 1 minute. The pressure is then released rapidly.

As a result of pretreatment, the fibrous structure of the feedstock is destroyed. The pretreated material has a muddy texture and a slightly sweet smell and is dark brown in color. Among different feedstocks, the cooking time varies while most of the other conditions are maintained relatively constant. The appearance of pretreated material is similar among feedstocks.

Chemical reactions

Of the major components, the first to react is the hemicellulose. The xylan portion is depolymerized and solubilized, and then hydrolyzed to xylose (Reaction 1). The presence of the exogenous sulfuric acid is particularly important in the formation of monomeric xylose. In the absence of exogenous acid, xylose oligomers are formed. Further, the added acid improves the uniformity of the process, because natural acid levels vary considerably. If the pretreatment reaction proceeds further, the xylose is dewatered to produce furfural (Reaction 2), which is undesirable.

Reaction 1. Depolymerization of xylan.

$$(C_5H_8O_4)_n + H_2O \rightarrow (C_5H_8O_4)_{n-1} + C_5H_{10}O_5$$

Reaction 2. Xylose is dewatered to form furfural.

$$C_5H_{10}O_5 \rightarrow C_5H_4O_2 + 3\ H_2O$$

Arabinose undergoes analogous reactions, but more slowly than xylose. Removing hemicellulose from the feedstock accomplishes two objectives. First, it makes the simple sugars that can potentially be fermented to ethanol. Second and more important, it exposes additional surface area on the feedstock, thereby allowing more cellulase enzyme to bind to the cellulose.

A small amount of cellulose reacts to form glucose, which is degraded to hydroxymethylfurfural (Reactions 3 and 4). Only a small amount of cellulose hydrolysis (<10%) is desired. More than this results in a decrease in yield and accessibility. The older acid hydrolysis process was much more harsh and will be described below.

Reaction 3. Cellulose degradation to glucose.

$$(C_6H_{10}O_5)_n + H_2O \rightarrow (C_6H_{10}O_5)_{n-1} + C_6H_{12}O_6$$

Reaction 4. Glucose degradation to hydroxymethylfurfural.

$$C_6H_{12}O_6 \rightarrow C_6H_6O_3 + 3\ H_2O$$

The lignin undergoes a depolymerization during pretreatment, but remains insoluble in water or acid. Protein is destroyed and starch is hydrolyzed to glucose.

One can draw an analogy to cooking a turkey to describe the trade-off between time and temperature: the longer the time, the lower the temperature needed. Acid also acts to decrease the temperature or time required for pretreatment.

The choice of reactor for this pretreatment is only important insofar as the desired chemistry must be delivered to the system. Numerous guns, vessels and tubes have been proposed and built to carry this out.

What other types of pretreatment might one consider?

Katzen *et al.* (1990) reviewed pretreatment in detail; and it is only summarized here. Pretreatment processes may be divided into (1) those that produce a stream directly for fermentation to ethanol, and (2) those that are followed by enzymatic hydrolysis. The former processes are necessarily more harsh and have a longer history.

Direct sugar production in pretreatment has been carried out using concentrated chemicals and dilute acid. Concentrated chemical pretreatment represents the cellulose conversion technology with the longest history, dating back to 1890 or earlier in Germany. The concentrated chemicals include acids, bases and salts. The principle behind the use of concentrated chemicals is to disrupt the crystalline cellulose structure, thereby dissolving and depolymerizing the cellulose. Among the chemicals used are 72% sulfuric acid, 40% hydrochloric acid, 40% sodium hydroxide, 65% zinc chloride and 40% calcium chloride. These methods have very high yields and low operating temperatures. Unfortunately, the economics of the process dictate that recovery and re-use of the pretreatment chemicals are at an absolute premium. The inability to obtain recoveries approximating 99.9% has hindered the commercialization of these processes. A second problem is the exotic materials required to handle these compounds.

Dilute acid hydrolysis represents the cellulose conversion technology with the most commercial experience. In dilute acid hydrolysis, the feedstock is treated at perhaps 180-200°C for 1-4 hr with 1-4% sulfuric acid. A glucose yield of 50% of the cellulose or higher is obtained. This process was used for ethanol production by Germany during World War II, Russia in the late 1940s, and pilot plants in the 1950s in Switzerland and Springfield, Oregon. All of these plants have been plagued by corrosion, low yields, high investments and overall poor returns. Perhaps one 100 ton/day plant built in Russia in the 1980s still operates with this process.

Pretreatment followed by enzymatic hydrolysis is newer than direct pretreatment, as cellulase enzymes were only discovered after World War II. These milder pretreatments include mechanical action, solvent-based pretreatments, alkali treatments and acid prehydrolysis.

Mechanical action was first tried at the Natick laboratories in the 1960s. The principle behind mechanical action is to increase the surface area of the feedstock particles. However, beyond producing small particles for uniform distribution of acid, there is no real advantage to further milling of the feedstock.

In solvent-based pretreatments, organic solvents such as ethanol and methanol are used to dissolve a portion of the lignin, thereby freeing up the cellulose for enzymatic attack. However, in the feedstocks used by Iogen lignin does not provide a significant barrier to cellulose. Therefore, Iogen does not use solvent-based pretreatment.

In alkali pretreatments, such as with sodium hydroxide or ammonia, the crystalline cellulose is converted to a different form, cellulose II or III, respectively. These forms of cellulose can be more easily hydrolyzed than native cellulose. However, destruction of the hemicellulose is reported in these systems.

Acid prehydrolysis is preferred by Iogen because it has fewer of these problems than the other methods. The low levels of acid preclude the need for recovery and corrosion is not a problem. The process is mild to hemicellulose and produces a material with a high surface area suitable for enzymatic hydrolysis.

Cellulase enzyme production

Production of cellulase enzymes

Cellulase enzymes convert cellulose to glucose, which can then be fermented to ethanol. Cellulase enzymes are made by a wide variety of microbes, but those best suited to cellulose hydrolysis are made by the wood-rotting fungus Trichoderma. This fungus was isolated during World War II in rotted US Army cotton tents in the South Pacific. Researchers led by Elwyn Reese and Mary Mandels at the US Army laboratories in Natick, Massachusetts determined that the microbe responsible for the destruction of the cotton was secreting a mixture of enzymes that hydrolyzed the cotton. Reese and Mandels determined cultivation conditions for production of cellulase in liquid culture. Selection of Trichoderma strains with higher productivities of cellulase was successfully carried out by Montenecourt at Lehigh University in the 1970s. Despite research and development of cellulase production from other microbes, Trichoderma remains the organism of choice to produce cellulase for ethanol production.

Cellulase is made in submerged liquid culture, in fermentation vessels similar to those used for producing antibiotics. The fermentors have volumes of 10,000-50,000 gallons and are maintained free of contaminating microbes. The liquid broth contains carbon source, salts, complex nutrients such as corn steep liquor and other nutrients in water. The most important nutrient is the carbon source. Glucose promotes growth of the organism but not cellulase production. The carbon source must include an inducing sugar to promote cellulase production. Well-known inducers of cellulase include the sugars cellobiose, lactose, sophorose, and other low molecular weight oligomers of glucose. The cellulase enzyme manufacturers devote a lot of research to developing better inducers.

The nutrient broth is sterilized before the start of the fermentation, typically by heating with steam. The fermenter is inoculated with the enzyme production strain once the liquid broth has cooled. Typical operating conditions are 30°C at pH 5; and these conditions are maintained by the addition of cooling water in external coils and by alkali, respectively. Trichoderma is highly aerobic and a constant stream of air or oxygen is used to maintain aerobic conditions. A cellulase enzyme production run lasts about one week. At the end of the run, the broth is heated to kill the microbes, then filtered across a cloth to remove cells. The spent cell mass is disposed of in a landfill. The resulting enzyme broth is a clear, light brown liquid, similar in appearance to weak tea.

Why have enzyme production on the ethanol plant site?

When used directly in an ethanol plant on-site, the crude broth is simply added to the cellulose hydrolysis tanks. If the cellulase is to be stored for long periods before use, it must be stabilized against (1) microbial contamination, which requires preservatives such as sodium benzoate, and (2) protein denaturation, which uses compounds such as glycerol.

The production and use of cellulase on the ethanol plant site therefore avoids the cost of preservatives and stabilizers, a big cost advantage. Another cost advantage for on-site enzyme production is that a small portion of the hydrolysis sugar can be used as feedstock for the enzyme production.

Is cellulase used for anything besides making ethanol?

Cellulase sales are $100 million annually to the textile, detergent, animal feed, beverage, and pulp and paper industries (Godfrey, 1996). Commercial cellulase enzymes for these industries are made by several microbes including Humicola, Aspergillus and Penicillium fungi, and Bacillus bacteria in addition to Trichoderma. Ethanol production requires aggressive action of cellulase to destroy cellulose, for which Trichoderma cellulase is superior. The other industries often require milder action and/or specific conditions better suited to other cellulases. For example, the textile industry uses cellulase to soften denim blue jeans. Humicola cellulases can be less aggressive in this application than Trichoderma cellulases, though Trichoderma cellulases can be modified to improve their performance. Detergents require alkaline conditions that are not easily accessible to Trichoderma enzymes.

Cellulose hydrolysis

Process description

In cellulose hydrolysis, cellulase enzymes convert the cellulose to glucose. The pretreated feedstock is conveyed to the hydrolysis tanks in a slurry that is 15-20% total solids (as high as can be handled). The slurry is adjusted to pH 5 with alkali and maintained at 50 °C. A single hydrolysis tank has a volume of 50,000-200,000 gallons. Crude cellulase broth is added as a liquid at a dosage of 100 liters per tonne of cellulose. The contents of the tank are gently agitated to move the material and keep it dispersed, but not as rigorously agitated as in a fermentation vessel.

The hydrolysis proceeds for 5-7 days. As it proceeds, the viscosity of the slurry drops; and the remaining insoluble particles, which are lignin in increasing proportion, diminish in size. At the end of the hydrolysis, 80-95% of the cellulose is converted to glucose. The remainder is insoluble and contained within the particles, which are mostly lignin.

Kinetics of cellulose hydrolysis

Trichoderma cellulase is a mixture of three types of enzymes: cellobiohydrolase (CBH), endo-glucanase (EG), and ß-glucosidase (BG). CBH enzymes act sequentially along the cellulose. Trichoderma cellulase includes two CBH enzymes, CBHI and CBHII, which together account for 80% of the total cellulase protein. Endoglucanase enzymes cleave at random locations on the fiber. Trichoderma makes at least four different EG enzymes: EGI, EGII, EGIII, and EGV. The EG enzymes account for about 20% of the cellulase protein. The third type of enzyme, ß-glucosidase, hydrolyzes the glucose dimer cellobiose to glucose. The properties of the Trichoderma cellulase enzymes are summarized in Table 3.

The enzymatic hydrolysis of cellulose proceeds as two consecutive reactions. The hydrolysis of cellulose to its soluble dimer, cellobiose (Reaction 5) is catalyzed by the CBH and EG enzymes. The CBH and EG enzymes work synergistically to hydrolyze cellulose. The hydrolysis of cellobiose to glucose (Reaction 6) is catalyzed by the soluble enzyme ß-glucosidase and proceeds according to standard Michaelis-Menten kinetics. ß-glucosidase accounts for less than 1% of the total cellulase protein. Hydrolysis of the cellobiose is important because glucose is readily fermented to ethanol while cellobiose is not. In addition, cellobiose is a potent inhibitor of CBH and EG; so the accumulation of even 5 g/L of cellobiose slows the hydrolysis significantly. There are several inherent difficulties in cellulose hydrolysis that have been the focus of much research. The first is the inherent shortage of ß-glucosidase, both because of its low concentration and because it is inhibited by the reaction end product, glucose. Cellobiose accumulates in a shortage of ß-glucosidase, thereby inhibiting the action of CBH and EG in hydrolyzing cellulose.

Three approaches have been proposed to overcoming the shortage of ß-glucosidase. The first is to produce ß-glucosidase in a separate fermentation by *Aspergillus spp.* The disadvantage of this is the added process cost of a second fermentation. The second and most widely discussed approach is to carry out a simultaneous saccharification and fermentation (SSF) process. In SSF, the enzymatic hydrolysis and glucose fermentation run simultaneously, with the notion that yeast consume the glucose to prevent inhibition of ß-glucosidase and subsequently CBH and EG. In Iogen's process, SSF systems have suffered non-optimal operation (the enzyme optimum is 50°C and the yeast function best at 28°C, so a compromise temperature of 37°C is used) and microbial contamination. As a result we have put this

Reaction 5. Hydrolysis of cellulose to cellobiose catalyzed by the cellobiohydrolase and endoglucanase enzymes.

$$(C_5H_{10}O_5)_n + H_2O \rightarrow (C_5H_{10}O_5)_{n-2} + C_{12}H_{22}O_{11}$$

Reaction 6. Hydrolysis of cellobiose catalyzed by ß-glucosidase.

$$C_{12}H_{22}O_{11} + H_2O \rightarrow 2\ C_6H_{12}O_6$$

Table 3. Properties of the Trichoderma cellulase enzymes.

Enzyme	*Molecular weight*	*Isoelectric point*	*Family*	*Concentration (%)*	*Reference*
CBHI	63,000	4.3	7	50-60	Shoemaker *et al.* (1983)
CBHII	58,000	6.0	6	15-18	Chen *et al.* (1987)
EGI	53,000	4.6	7	12-15	Penttila *et al.* (1986)
EGII	50,000	5.3	5	9-11	Saloheimo *et al.* (1988)
EGIII	25,000	7.4	12	0-3	Ward *et al.* (1993)
EGV	23,000	3.7	45	0-3	Saloheimo *et al.* (1994)

approach aside. The third option is to develop Trichoderma strains with high ß-glucosidase production included in the CBH and EG. This is the approach used at Iogen and it has largely overcome the shortage of BG.

Other important difficulties in the hydrolysis are the decrease in reaction rate as hydrolysis proceeds and the diminishing returns with enzyme dosage, i.e. doubling the amount of enzyme does not double the extent of conversion. These issues, which are probably linked, are not well understood and differ substantially from Michaelis-Menten kinetics. These effects can, however, be characterized empirically.

The enzyme components initially must adsorb to the surface of the insoluble substrate, cellulose. An equilibrium corresponding roughly to a Langmuir adsorption isotherm is reached within a few minutes. The adsorbed enzyme then acts on the cellulose at a rapid initial rate. The rate declines significantly after the first few minutes of hydrolysis, and after 24 hrs is less than 2% of the initial rate. The hydrolysis continues over several days at ever decreasing rates. Depending on the dosage used, the final cellulose conversion is 75-95% in a commercial plant.

The reason the rate slows down is not fully known. Speculation in the literature has centered on endproduct inhibition, increasing difficulty in hydrolyzing the substrate (substrate recalcitrance) and denaturation of the protein over time. However, straightforward experiments demonstrate that none of these factors account for more than a small fraction of the drop in rate. Further research continues in this area.

Improvements in enzymatic hydrolysis

Despite many years of research, cellulose hydrolysis remains the least efficient part of the process. To illustrate the problem, hydrolysis of pretreated cellulose requires 100 times more enzyme than hydrolysis of starch.

Why not use more enzyme and a shorter time? The enzyme manufacturing cost is still sufficiently high that the trade-off between enzyme dosage and hydrolysis time favors longer times and lower dosages.

What about recycle of cellulase or cellulose? These are potentially good ideas that have not been fully explored. The cellulase enzyme adsorbs to the substrate, so recycle of unconsumed cellulose would be one way to re-use the enzyme. Re-use of the enzyme in solution is also a possibility. The demonstration plant will be suited to evaluating these options.

What about continuous or fed batch systems? These are also ideas with a sound basis that have not been explored. Fed batch systems are especially interesting because they might allow high cellulose concentrations to be maintained at all times.

How about finding better cellulase enzymes? This has been explored extensively at several research labs, with as yet no cellulases found that are superior to Trichoderma cellulase. However, the new tools of molecular biology such as protein engineering can be brought to bear on this problem and might lead to success.

Are there opportunities with novel reactors, i.e. trickling bed, high shear, etc? This is a widely explored area with some improvements reported, but processes are not easily generalized across feedstocks and enzymes. A better understanding of the nature of the enzyme's action will be helpful in evaluating these reactors.

How about better pretreatment? Better hydrolysis would be a direct benefit of better pretreatment. Pretreatment has been widely studied, but there are always new ideas on the horizon.

Lignin processing

Process description

The hydrolysis slurry contains glucose, xylose, arabinose and other compounds dissolved in the aqueous phase, and insoluble lignin and unconsumed cellulose. The insoluble particles are separated by a plate and frame filter with the cake washed 2-3 times with water to obtain a high sugar recovery. The sugar stream is pumped to the fermentation tanks. The lignin cake is spray-dried to <10% moisture, then burned in a solid-fuel boiler to generate power for the entire plant.

Alternative uses for lignin

Lignin is a lattice of phenolic-propane units. Lignin has been the object of much study, and a good review of applications of lignin is provided by Chum *et al.* (1985). The potential applications of lignin fall into the broad categories of insoluble and chemically modified. Insoluble lignin is limited to high-volume, low value applications such as an ingredient in roads or cement. In these applications, lignin is a filler and competes with corn cobs, gravel, and ground bark.

Chemically modified lignin has a much wider variety of potential markets. The major types of chemical modifications are solubilization by reaction with e.g. sulfurous acid, or crosslinking the lignin by reaction with phenol. Crosslinked lignin is suitable for resins, glues and other such materials, where it competes with phenol-formaldehyde resins. Solubilized lignin such as sodium lignosulfonate is used in surfactants, detergents and biocides.

Sugar fermentation and ethanol recovery

The hydrolysis sugars, which consist of glucose, xylose, arabinose and various organic impurities in aqueous solution, are pumped to the sugar fermenters for ethanol production. The fermenters are large tanks with gentle agitation, sufficient to keep the contents moving. The fermenters are inoculated with Saccharomyces yeast, which readily ferment the glucose to ethanol. In addition, other strains are under development to ferment the pentose sugars to ethanol, which is not naturally very efficient.

At the time of this writing (July 1999), several strains were under consideration for pentose fermentation in the Iogen process and the selection had not yet been made. The strains under consideration include:

1. *Saccharomyces yeast genetically modified for uptake and metabolism of xylose.* This strategy is carried out at Purdue University (Ho and Chen, 1997).
2. *Pichia yeast with a natural ability for xylose uptake, but genetically modified to ferment the xylose to ethanol.* This strategy is carried out at the University of Wisconsin (Cho and Jeffries, 1998).
3. *Zymomonas bacteria genetically modified for xylose uptake and metabolism.* This strategy is carried out by the National Renewable Energy Lab in Boulder, Colorado (Doelle *et al.*, 1993; Fein *et al.*, 1983).

Two other strategies for pentose fermentation include genetically modified enteric bacteria such as *Escherichia coli* and Klebsiella, carried out at the University of Florida (Ingram *et al.*, 1987; Ohta *et al.*, 1990; 1991), and thermostable Bacillus strains developed at Imperial College. Iogen is not pursuing the enteric bacteria because of the anticipated difficulties in obtaining regulatory approval to use these strains. The Bacillus strains are not, at this writing, commercially available.

After fermentation, the broth containing ethanol and unfermented sugar is pumped to a distillation column. The ethanol is distilled off the top and dehydrated, then blended into gasoline in a 10% ethanol mixture. The still bottoms are sold as animal feed.

References

Chen, C.M., M. Gritzali and D.W. Stafford. 1987. Nucleotide sequence and deduced primary structure of cellobiohydrolase II from *Trichoderma reesei*. Bio/technology 5:274-278.

Cho, J. and T.J. Jeffries, T.J. 1998. *Pichia stipitis* genes for alcohol dehydrogenase with fermentative and respirative functions. Appl. Env. Microbiol. 64(4):1350-1358.

Chum, H.L., S.K. Parker, D.A. Feinberg, J.D. Wright, P.A. Rice, S.A. Sinclair and W.G. Glasser. 1985. The economic contribution of lignins to ethanol production from biomass. National Renewable Energy Lab, Golden, Colo.

Doelle, H.W., L. Kirk, R. Crittenden, H. Toh and M.B. Doelle. 1993. *Zymomonas mobilis*-science and industrial applications. Critical Reviews in Biotechnology 13(1):57-98.

Fein, J.E., H.G. Lawford, G.R. Lawford, B.C. Zawadski and R.C. Charley. 1983. High productivity continuous ethanol fermentation with a flocculating mutant strain of *Zymomonas mobilis*. Biotechnol. letters 5(19).

Foody, B., J.S. Tolan and J. Bernstein. 1999. Pretreatment process for conversion of cellulose to fuel ethanol. US Patent 5,916,780 issued June 29, 1999.

Godfrey, T. 1996. Scope of industrial enzymes. In: Industrial Enzymology (T. Godfrey and S. West, eds) MacMillan and Co., London.

Ho, N.W.Y. and Z-D. Chen. 1997. Stable recombinant yeasts for fermenting xylose to ethanol. PCT patent application WO 97/42307.

Ingram, L.O., T. Conway, D.P. Clark, G.W. Sewall and J.F. Preston. 1987. Genetic engineering of ethanol production in *Escherichia coli*. Appl. Environ. Microbiol. 53(10):2420-2425.

Katzen, R., P.W. Madsen, D.A. Monceaux and K. Bevernitz. 1990. Use of cellulosic feedstocks for alcohol production. In: The Alcohol Textbook (T.P. Lyons, D.R. Kelsall and J.E. Murtagh, eds) Nottingham University Press, UK.

Ohta, K., F. Alterthum and L.O. Ingram. 1990. Effects of environmental conditions on xylose fermentation by recombinant *Escherichia coli*. Appl. Environ. Micro. 56(2):463-465.

Ohta, K., D.S. Beall, J.P. Mejia, K.T. Shanmugam and L.O. Ingram. 1991. Genetic improvement of *Escherichia coli* for ethanol production: chromosomal integration of *Zymomonas mobilis* genes encoding pyruvate decarboxylase and alcohol dehydrogenase II. Appl. Environ. Micro. 57(4):893-900.

Penttila, M., P. Lehtovaara, H. Nevalainen, R. Bhikhabhai and J. Knowles. 1986. Homology between cellulase genes of *Trichoderma reesei*: complete nucleotide sequence of the endoglucanase I gene. Gene 45:253-263.

Saloheimo, M., P. Lehtovaara, M. Penttilla, T.T. Teeri, J. Stahlberg, G. Johansson, G. Petterson, M. Claeyssens, P. Tomme and J. Knowles. 1988. EGIII, a new endoglucanase from *Trichoderma reesei*: the characterization of both gene and enzyme. Gene 63:11-21.

Saloheimo, A., B. Henrissat, A.M. Hoffren, O. Teleman and M. Penttila. 1994. A novel, small endoglucanase gene, egl5, from *Trichoderma reesei* isolated by expression in yeast. Molecular Microbiology 61:1090-1097.

Shoemaker, S. V. Schweikert, M. Ladner, D. Gelfand, S. Kwok, K. Myambo and M. Innis. 1983. Molecular cloning of exocellobiohydrolase I derived from *Trichoderma reesei* strain L27. Bio/technology 1:691-696.

Singh, L. 1997. Scenarios of US carbon reductions: potential impacts of energy technologies by 2010 and beyond. Interlaboratory Working Group on Energy Efficient and Low-Carbon Technologies (ORNL, LBNL, PNNL, NREL, ANL). US Government Printing Office.

Ward, M., K.A. Clarkson, E.A. Larenas, J.D. Lorch and G.L. Weiss. 1995. DNA sequence encoding endoglucanase III cellulase. US patent 5,475,101 issued Dec. 12, 1995.

Chapter 10

Alternative feedstocks: a case study of waste conversion to ethanol

D.A. Monceaux and P.W. Madson
KATZEN International, Inc., Cincinnati, Ohio, USA

Introduction

Since the beginning of the US fuel ethanol industry in 1978, production capacity has grown to a present level of more than 1.4 billion gallons per year (Figure 1). During this time span, the majority of the growth has been as wet milling capacity, with a rapid expansion occurring during the period from 1981 through 1987. The optimistic outlook for the fuel ethanol industry during the early 1980s was led by a strong ethanol market with relatively stable pricing. This allowed a number of dry milling fuel ethanol facilities to acquire funding. Sizes of these plants ranged from farm-based operations with a capacity of less than 200,000 gallons per year to regional plants producing in excess of 60 million gallons per year.

Unlike wet milling facilities, which are able to distribute the cost of operations and feedstock over a variety of products based on the starch, fiber, protein and fat components of the grain, dry milling plants are limited to ethanol and distillers dried grains with solubles (DDGS). They are, therefore, held hostage to market prices of these two commodities. Further complicating this issue is the fact that no economic correlation exists between fuel ethanol pricing (which is linked to the rack price of gasoline) and grain pricing. Thus, the financial stability of fuel ethanol producers fluctuates dramatically as grain and fuel ethanol prices rise and fall (Figure 2).

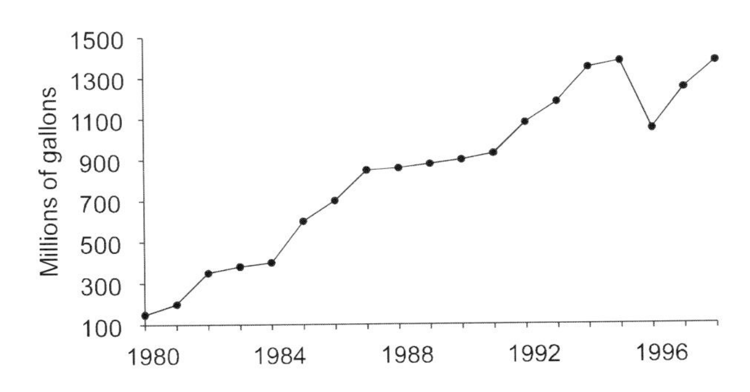

Figure 1. US fuel ethanol production.

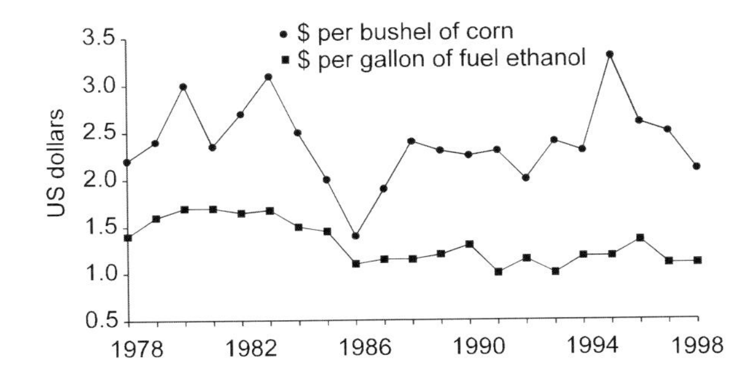

Figure 2. Price history of grain and fuel ethanol.

With small producers lacking economies of scale, and with grain comprising more than 60% of the cost of producing a gallon of ethanol in a dry milling plant, it is clear why a large number of small ethanol facilities ceased production in the mid to late 1980s (Figure 3). The driving force for this was a soft ethanol market combined with several years of poor growing conditions in the US corn belt. Many small plants driven into bankruptcy were able to re-start production later on due to the reduced debt service brought about by the low purchase price to the second owners. This was possible in plants located in the midwest, which typically could acquire corn at less than Chicago Board of Trade prices and at low freight costs. In the late 1980s, many plants with annual production capacities approaching 45 million gallons were unable to achieve a positive cash flow even with no debt service due to low ethanol prices combined with high delivered corn prices. It should, however, be noted that a number of these plants also had serious design deficiencies.

One might be led to conclude that all small ethanol producers outside the central states ceased production; but this, in fact, was not the case. Several small producers continued operating through what might be called 'creative acquisition of feedstocks'. Since feedstock represents such a large portion of the cost of ethanol production, small producers can maintain profitability even during periods of low ethanol pricing by significantly reducing feedstock cost.

Alternative feedstocks

A well-designed dry milling facility will contain most of the equipment required to handle many types of starch- or sugar-containing feedstocks. Often only minor modifications are required in feedstock handling operations. For example, a plant designed to process a dry feedstock such as corn may require only minor feedstock-receiving equipment changes to handle slurried or liquid feedstocks.

Table 1 presents a partial list of potential feedstocks. Typically, most of these feedstocks are available in small or unpredictable quantities, making it quite difficult to economically justify a dedicated fuel ethanol production facility. Exceptions exist in many large food processing plants, which generate significant quantities of sugar and starch-containing residues. In these cases, dedicated ethanol production facilities serve as a waste remediation process whereby high biochemical oxygen demand (BOD) effluent streams are converted to a liquid product,

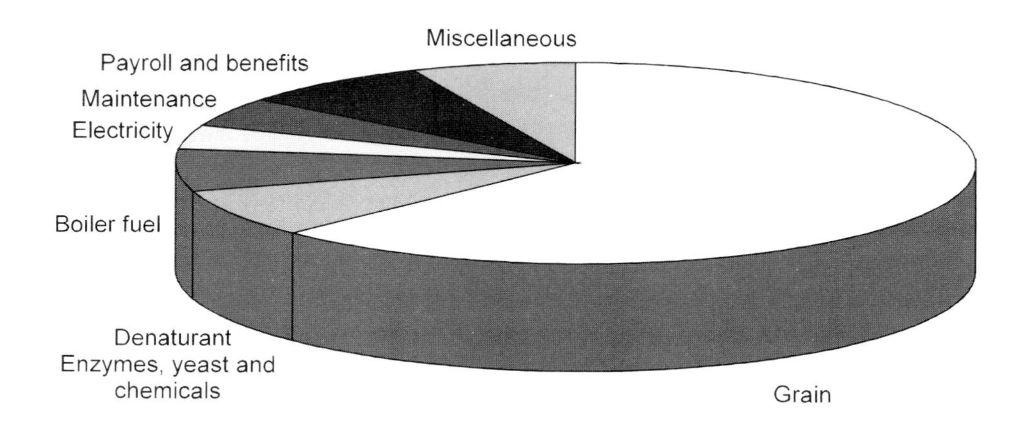

Figure 3. Distribution of operating costs in a corn dry milling ethanol plant.

ethanol, with a significant market value. This differs from typical waste treatment processes that convert soluble BOD to sludges, which then require expensive transportation and disposal. These low cost feedstocks are sufficiently attractive to entice many intermediate and large scale fuel ethanol producers to use them to supplement normal grain processing operations, thereby reducing the net cost of feedstock to the facilities.

Table 1. Potential feedstocks.

Offspec glucose and fructose syrups
Offspec dry starches and starch solutions
Low value starch, e.g. 'B' starch from wheat processing
Waste soft drink syrups
Brewer's spent grains
Damaged or spoiled grains
Expired seed grains
Food processing wastes, high in starches/free sugars
Brewery yeast slurries
Waste candies
Waste pet foods
Spoiled food products
Cheese whey
Spoiled fruit: apples, peaches, oranges, bananas
Citrus molasses
Honey
Raw sugar
Potatoes
Sweet potatoes and yams
Rice bran

A case study: potato waste

The following case study will discuss design and operational issues pertaining to a dedicated food processing waste-to-ethanol facility. Much of this discussion is based on two potato waste-to-ethanol plants built for the J.R. Simplot Company in the mid 1980s. These three million gallons/year fuel ethanol facilities were designed to receive potato processing waste, culled potatoes and plant washings high in starch as feedstock.

The general processes required to produce fuel ethanol from potato waste are similar to those for a corn dry milling operation. As is the case with many speciality feedstocks, processes must be designed with the properties of the specific feedstock in mind. In the case of potato waste, the properties requiring close attention include:

High water content
High sand and soil content
High fibrous tuber and vine content
Unique starch chemical and physical properties
Minimal storage life
Seasonal or erratic supply

Feed preparation

The feed preparation system, like a significant portion of the front end unit operations, must be designed for peak hydraulic and solids flows. These flows vary hourly, daily and seasonally.

The feedstock, which is a dilute starch slurry containing potato peelings as well as whole cull potatoes, vines and other residues, passes through a milling device. The purpose of the mill, like that in the grain plant, is to reduce the maximum particle size for cooking and subsequent processing. The incoming feedstock is heavily contaminated with soil microorganisms, and therefore cannot be stored for any appreciable time without a significant loss in ethanol yield.

Starch conversion

Due to the high water content in the cull potatoes and other process streams, the milled feed flows directly to the slurry tank without the need of additional dilution. This is critical, since the starch and sugar concentration of the incoming feed results in beer ethanol levels significantly lower than typically found in a grain dry milling plant. Minimizing water addition to the process reduces the necessary size of process equipment, as well as the energy consumption required for cooking, distillation and stillage processing.

The slurry tank is heated by recirculating the

contents through the slurry heater, which receives flash steam from downstream processes. Precautions are required since the gelatinization temperature of potato starch is much lower than that of corn starch (Whistler, 1984). Alkali is added to the slurry tank to control the pH at the α-amylase optimum. Mash from the slurry tank is continuously pumped through a jet cooker, which begins the starch conversion process along with providing a thorough sterilization of the incoming feed. Sterilization is critical due to the extremely high concentration of soil-borne bacteria in the feedstock. All pumps and process piping must be designed to withstand erosion associated with mash-entrained sand. The mash leaving the cooker is flash cooled to liquefaction temperatures prior to the addition of liquefying enzymes and entering the liquefaction tank. The flash cooling serves a secondary function of cost-effectively providing a small degree of mash dehydration (Figure 4).

The process of converting complex starches to dextrins is completed by holding the mash in the liquefaction vessel for a suitable time period. The mash is then acidified and cooled during transfer to the fermentation system. The pH adjustment, which in grain milling plants is normally accomplished with backset stillage, must be performed by the addition of acid. This is necessary to control water input into the fermentation system and thereby maximize the ethanol concentration in the beer stream to distillation.

Mash coolers must be designed to allow thorough cleaning. Plate-and-frame and spiral heat exchangers, often used in wet milling and dry milling facilities, are to be avoided in this instance. Due to the presence of fibers and vines that become entrapped in the exchangers, normal clean-in-place (CIP) systems are incapable of removing deposits and debris from these types of coolers. Thorough cleaning then requires labor-intensive dismantling of the equipment. For this reason shell-and-tube exchangers are preferred as they can be easily backflushed during the CIP. It is also advisable that spare mash coolers be installed so that the process can continue uninterrupted during cleaning periods.

Fermentation

The cooled mash enters the fermenter, which already contains the yeast inoculum and the saccharifying glucoamylase enzyme (Figure 5). The initial fermentation phase also includes yeast propagation *in situ*. The liquefied starch is then converted to glucose by the glucoamylase enzyme, which is then fermented to ethanol in a simultaneous saccharification and fermentation process (SSF). This minimizes process

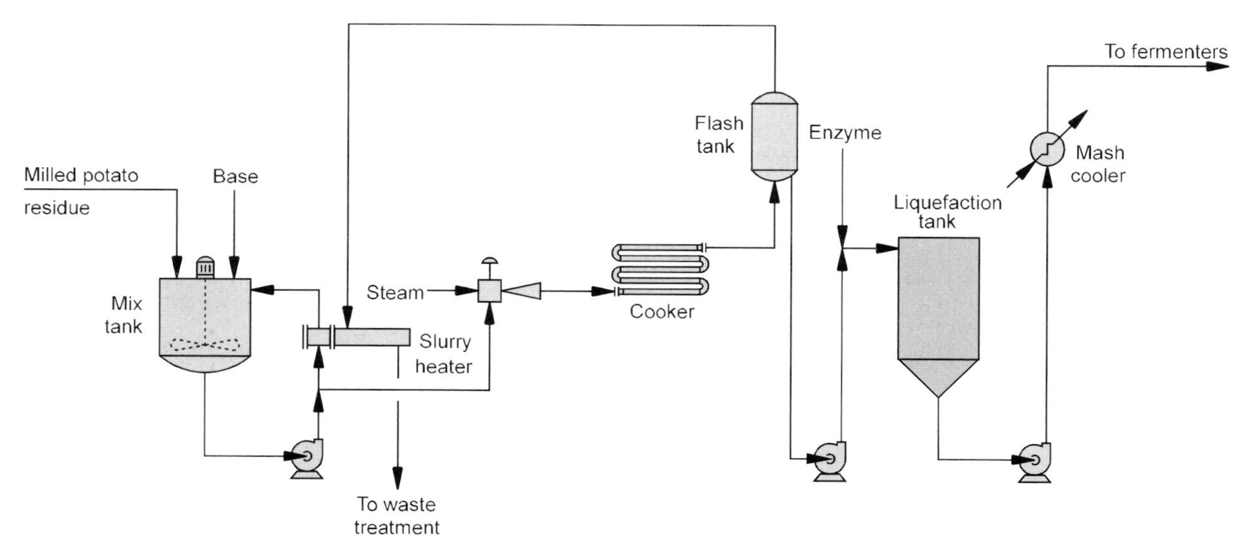

Figure 4. Mashing and cooking processes.

equipment and capital investment and reduces the potential for bacterial contamination while maximizing yields. The fermentation vessels are typically fabricated of carbon steel with sloped bottoms for ease of cleaning; and their contents are circulated through shell-and-tube fermenter coolers for temperature control. Steeply sloped bottoms are recommended for the tanks to assist in removing accumulated soil and sand upon emptying the fermenter.

Upon completion of the fermentation process, the contents are transferred to the beer well, which provides surge capacity for the process. Residues are washed to the whole stillage tank or to the sewer with the initial rinse. The fermenters and coolers are then chemically cleaned with a mild caustic solution in preparation for the next fill.

Distillation and dehydration

Fermented beer is preheated in shell-and-tube heat exchangers prior to entering distillation (Figure 6). Due to the low ethanol concentration in the beer, an energy-integrated distillation and dehydration system is necessary to minimize energy consumption. The beer enters the stripper section of the distillation tower, which removes the ethanol from the beer so that the residue (stillage) emerging from the base of the tower contains less than 200 ppm ethanol. Disc-and-donut baffle trays are recommended for this service due to their ability to withstand high concentrations of suspended solids. The beer stripper can be supplied with heat by thermocompression of the flash vapors from the stillage or with a steam-heated reboiler. The selected option is dependent primarily on downstream stillage-processing requirements, energy costs and regional environmental issues.

The stripped ethanol is concentrated to about 190° US proof (95° GL) in the rectifying section prior to entering the molecular sieve unit for dehydration to 199+° proof (>99.5° GL), to meet fuel-grade ethanol specifications.

Stillage processing

Evaporation

The stillage is pumped to the whole stillage surge tank, which provides surge capacity between the distillation system and the stillage processing operations. Whole stillage is processed in a decanter centrifuge, which separates the majority of the suspended solids (wet cake) from the liquid

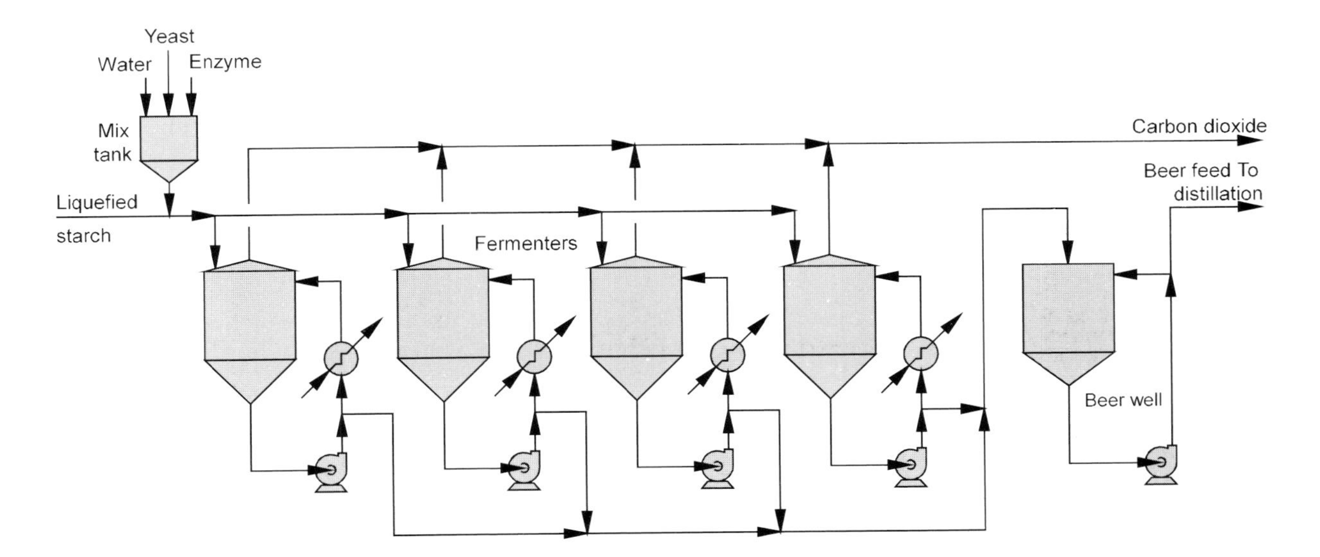

Figure 5. Simultaneous saccharification and fermentation.

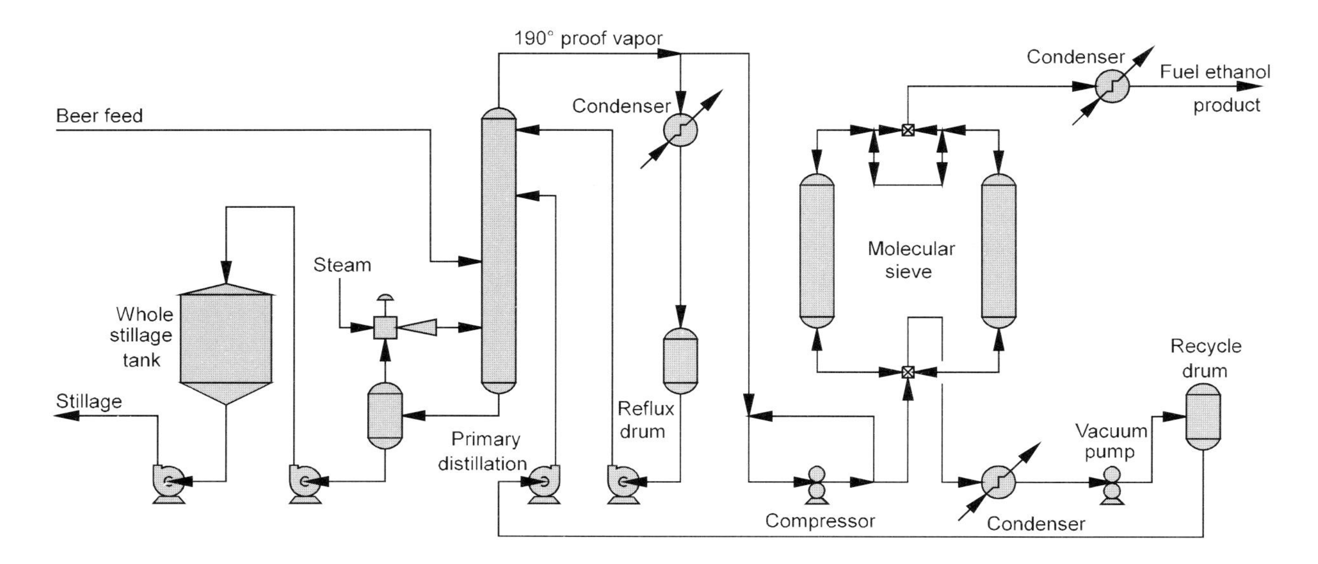

Figure 6. Distillation and dehydration.

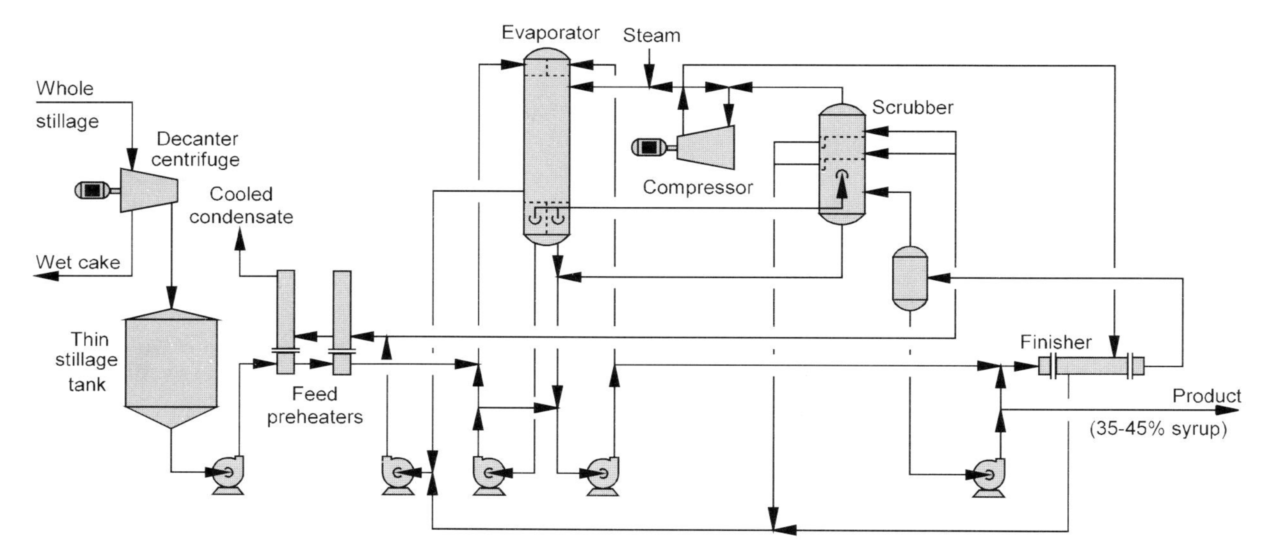

Figure 7. Mechanical vapor recompression evaporation.

(thin stillage) (Figure 7). The solids, with the consistency of wet sawdust, have sufficiently high protein and fat concentrations (Cullison *et al.*, 1987; Maynard *et al.*, 1979) to warrant sale as a cattle feed supplement.

The thin stillage, containing low levels of suspended solids, is concentrated in a multiple-effect, mechanical vapor recompression (MVR) evaporator. The resultant solubles concentrate (syrup), containing 25-30% dry matter, is also sold as a cattle feed supplement. In areas where the solubles concentrate must be transported long distances, the addition of a forced circulation finishing evaporator can increase the solids concentration to 35-45%, reducing freight costs and improving the value.

Due to the limited ethanol production capacity of plants of this type, the physical properties of the dissolved solids and the low fiber content of the stillage, it is generally neither cost effective nor technically attractive to install dryers for the wet cake and the syrup.

Biological treatment

An alternative to stillage evaporation is biological treatment. This is often warranted in areas that lack sufficient cattle to consume the recovered solids. Advances in anaerobic treatment processes have allowed this technology to achieve up to 95% reduction in BOD of high strength (10,000-50,000 mg/L) effluent streams. With this level of treatment, the processed effluent may be used for irrigation (Tchobanoglous *et al.*, 1979; Miorin *et al.*, 1977). If land application of primary anaerobic treatment effluent is not permissible, a second stage aerobic treatment process will typically meet environmental criteria for discharge into surface waters.

Summary

Plant designs such as that described provide cost-effective solutions to the ever-growing need to conform to environmental regulations. Converting waste carbohydrate streams to a renewable fuel provides a double environmental benefit. The choice of technology must provide energy efficiency with minimum labor and maintenance requirements while producing maximum yields if ethanol production is to be a cost-effective solution to a waste problem. With these goals in mind, the small fuel ethanol producer can continue to operate in this highly cyclical industry.

Acknowledgments

The authors would like to thank the management and staff of J. R. Simplot Company for assistance and support in the development, startup and operation of the facilities used as a basis for the case studies, with particular thanks to Paul Mann, David Kueneman, Bill Rutherford and John Totter.

References

Cullison, E. *et al.* 1987. Feeds and Feeding. Prentice Hall, New York.

Maynard, L.A., J.K Loosli, H.F. Hintz and R.G. Warner. 1979. Animal Nutrition, 7th ed. Mc-Graw-Hill

Miorin, A.F. *et al.* 1977. Wastewater Treatment Plant Design. WPCF Manual of Practice No 8, ASCE Manual on Engineering Practice No 36.

Tchobanoglous, G. *et al.*1979. Wastewater Engineering: Treatment, Disposal and Reuse, Metcalf & Eddy, McGraw-Hill.

Whistler, R.L. 1984. Starch. 2nd edition. Academic Press, New York.

Chapter 11

Production of Scotch and Irish whiskies: their history and evolution

T.P. Lyons
Alltech Inc., Nicholasville, KY, USA

Introduction

Whisky is the potable spirit obtained by distillation of an aqueous extract of an infusion of malted barley and other cereals that has been fermented with strains of *Saccharomyces cerevisiae* yeast. Various types of whisky are produced in a number of different countries in the world. They differ principally in the nature and proportion of the cereals used as raw materials along with malted barley, and also in the type of still used for distillation.

The principal types of whisky are also characteristic of particular geographical regions of the world. In Scotland, the characteristic product is manufactured using only malted barley as the raw material; and the fermented malt mash is distilled in relatively small pot stills.

The product, known as Scotch malt whisky, is produced in small distilleries, of which there are over 100 in Scotland. Scotch malt whisky is marketed both as a straight malt whisky, many brands of which have recently become extremely popular throughout the world, and also as a blend with another type of whisky produced in Scotland known as 'Scotch grain whisky' or (because it is distilled continuously in Coffey-type patent stills) as 'patent-still whisky'. Most Scotch whiskies available on the international market consist of blends with 20-30% malt whisky and 70-80% grain whisky. Within the blend, there may be as many as 20-30 individual malt whiskies and grain whiskies. These blends are, by law, matured for at least three years but in practice this period is much longer. Unblended Scotch malt whiskies are usually matured for a minimum of eight years.

The cereals used in the manufacture of Scotch grain whisky are malted barley, together with a high proportion (up to 90%) of wheat or corn (maize). Currently wheat is the main cereal, chosen on the basis of cost and the attraction of using a Scottish-grown cereal. All whiskies are legally protected and defined, mainly because of the huge revenues that governments obtain from their sale. The Scotch Whisky Order (1990) and the Scotch Whisky Act (1988), in defining Scotch Whisky, state that to be called Scotch Whisky spirits must be:

1. produced at a distillery in Scotland
2. from water and malted barley to which only whole grains or other cereals may be added
3. processed at that distillery into a mash

4. converted to fermentable carbohydrate only by endogenous enzymes
5. fermented only by the addition of yeast
6. distilled to an alcohol strength less than 94.8% so that the distillate has the aroma and taste derived from the raw materials
7. matured in Scotland in oak casks of less than 700 litres for a minimum of three years
8. be a product that retains the color, aroma and taste derived from the raw materials
9. produced with no substance other than water and spirit caramel added.

The word 'Scotch' in this definition is of geographical and not generic significance.

Irish whiskey (spelled with an e) is a distinctive product of either the Republic of Ireland or of Northern Ireland. In the Republic of Ireland, definitions were enacted by the Parliament in the Irish Whiskey Acts of 1950 and 1980. The 1950 Act distinguished pot still whiskey from blends and stated that the title 'Irish Pot-Still Whiskey' was reserved solely for spirits distilled in pot stills in the Republic from a mash of cereal grains normally grown in that country and saccharified by a diastase of malted barley. The 1980 legislation specified that the term 'Irish Whiskey' only applied to 'spirits distilled in the Republic or Northern Ireland from a mash of cereals saccharified by the diastase of malt contained therein, with or without other natural diastases'. This meant that unlike Scotch, Irish whiskey may be produced with the use of microbial enzyme preparations in addition to malt. The 1980 Act also specified that the whiskey must be aged for at least three years in Ireland in wooden barrels. While not possessing the 'smoky' taste and aroma of Scotch, Irish whiskey is usually more flavorful and has a heavier body than Scotch. Moreover, the whiskey is distilled not twice, as in Scotland, but three times to give a very strong spirit of 86° GL compared with the 71° GL whisky distilled in Scotland.

The tremendous popularity of whiskies manufactured in Scotland, Ireland, the US and Canada has prompted several other countries to try manufacturing whiskies, usually ones designed to resemble Scotch. Indeed, the number of countries with minor but nevertheless significant whisky-distilling industries must now be well over a dozen. Some countries, notably Australia and Japan, have whisky-distilling industries producing a sufficiently acceptable product for them to venture into the export industry. Other countries, including the Netherlands and Spain, have whisky-distilling industries that cater mainly, if not exclusively, for home consumption. The measures that some of these industries have taken to imitate Scotch become apparent when the spirits are sampled. One of the two whisky distilleries in Spain that located northwest of Madrid in the Guadarrama Hills near Segovia produces a very acceptable Scotch-type whisky. Its quality is attributed in part to the fact that the water used, which comes from the surrounding hills, closely resembles that used in highland Scotch whisky distilleries.

Recommended descriptions of Scotch whisky manufacture have come from Brander (1975), Daiches (1969), Dunnett (1953), Gunn (1935), Robb (1950), Ross (1970), Simpson (1968), Simpson *et al.* (1974), Ross Wilson (1959, 1970 and 1973) and Philip (1989). There are a few smaller books describing individual distilleries in Scotland. Recommended are those by McDowell (1975) and Wilson (1973; 1975). Recommended reading on Irish whiskey includes an account by McGuire (1973) and short review articles by Court and Bowers (1970) and Lyons (1974).

History of whisky production

The origins of the art of distilling potable spirits are obscure, and probably date back to ancient China. However, the first treatise on distilling was written by the French chemist Arnold de Villeneuve around 1310. The potency of distilled spirits caused many to be known as the 'water of life', a description that survives today in such names as *eau de vie* for French brandy and *akvavit* and *aquavit* for spirits in Northern Europe. The name 'whisky' is a corruption of 'uisgebaugh', the Gaelic word for water of life. Uisge was corrupted first into 'usky', which finally became whisky after several centuries. Dr. Johnson sang the praises of this potable spirit, although in his Dictionary of 1755 it is listed under 'u' and not 'w'.

Much to the chagrin of the Scotsman, it is likely that the first whisky was distilled not in Scotland but in Ireland. The spirit was known in Ireland when that land was invaded by the English in 1170. In all likelihood the art of distillation was imported into Scotland by missionary monks from Ireland. Two of today's main centers of Scotch distilling, namely the island of Islay and the Speyside town of Dufftown, were the sites of early monastic communities.

Whisky, principally Scotch whisky, has for many years been one of the most popular distilled beverages in the world; and it was in Scotland rather than in Ireland that its qualities came to be extensively appreciated. This has continued to the present day and in the intervening period many Scotsmen have felt compelled to record for posterity their thoughts and inspirations on the potable spirit. There are numerous histories of whisky distilling in Scotland, some more comprehensive than others. For good general accounts, the reader is referred to the texts by Brander (1974), Daiches (1969), Ross (1970) and Ross Wilson (1970).

Whisky distilling flourished in Scotland not least because consuming the spirit helped the inhabitants to withstand the climatic rigors of this northern region of Britain. The first recorded evidence of whisky production in Scotland is an entry in the Exchequeur Rolls for the year 1494. It reads "To Friar John Cor, by order of the King, to make aquavitae, eight bolls of malt". Production of whisky was therefore being controlled; and an Act of 1597 decreed that only earls, lords, barons and gentlemen could distill for their own use. To many Scots of this era whisky was a medicine, and in 1506 King James IV of Scotland had granted a monopoly for manufacture of 'aqua vitae' to the Guild of Barber Surgeons in the City of Edinburgh.

Taxation on whisky production first appeared in the 17th century. Breaches of the monopoly regulations and the need to raise money to send an army into England to help the English Parliament in its war against Charles I led to the Act of 1644, which fixed a duty of two shillings and eight pence Scot's on a pint of whisky (the Scot's pint was then about 1.5 litres). But the tax was short-lived and was replaced by a malt tax that later was also repealed.

At the time of the Treaty of Union between Scotland and England in 1707 there was a tax on malt in England, but not in Scotland. The English were irate; and in 1725 when Lord Walpole's administration decided to enforce the tax in Scotland, the first of a series of Malt Tax riots occurred. The English, meanwhile, had cultivated a taste for French brandy, there being very little whisky consumed at that time outside of Scotland. However around 1690 William III began to wage commercial war against the French, and imposed punitive taxes on imports of French brandy into England. The English reacted by acquiring a taste for gin, which was distilled locally. The scale of drunkenness that developed with the popularity of gin had to be controlled by law; and the Acts of 1736 and 1713 levied high taxes on gin manufacturers. Both of these acts contained clauses exempting Scotland, but not for long. The Parliament in London saw the prospect of a rich harvest of taxes in the distilleries of Scotland; and in a series of acts starting in 1751, production of whisky in Scotland was increasingly subjected to taxation.

The outcome of these punitive measures was not surprising. An extensive and thriving business in illicit distillation of whisky grew up in Scotland as described by Sillett (1965). Curiously, illicit production of Scotch hardly extended over the border into England, although there are a few records of the operation of illicit stills in the Cheviot Hills west of Newcastle-upon-Tyne. Following the Act of 1823, which introduced much stiffer penalties for illicit distillation, and to some extent because of the increased standards of living in northern Scotland, illicit manufacture of whisky declined. Indeed, many erstwhile illicit distillers emerged to become legal and registered distillers of Scotch whisky.

In 1826, Robert Stein of the Kilbagie distillery in Clackmannanshire, Scotland, patented a continuously operating still for whisky production. However, this invention was superseded in 1830 with the introduction by Aeneas Coffey of an improved version of this type of still. The appearance of continuous stills sparked off a period of turmoil in the Scotch whisky industry, it being claimed that the product from the continuous distillation of a mash that contained unmalted grain (described as neutral or 'silent spirit') could not be called whisky, since it had not been distilled in the traditional pot still. The battle was waged for about three quarters of a century; and in 1908 a Royal Commission decided that malt whisky and grain neutral spirit, when blended, could be labeled whisky.

The major factors which have affected the development of the whisky distilling industry in Scotland in this century have been economic. The industry has had to endure the privations of two world wars, the economic depression in Great Britain during the 1920s and prohibition in the United States from 1920 to 1933, which greatly affected export of Scotch to North America. Since 1945, however, the industry in Scotland has consolidated and expanded. The 20th century has also witnessed a considerable improvement in the quality of whisky distilled and blended in Scotland as a result of the acceptance of blending malt whiskies with grain whisky and the amalgamation of several smaller distilleries into combines.

Scotch malt whiskies can be divided into 'highland', 'lowland', 'Islay' and 'Campbeltown whiskies' (Simpson *et al.*, 1974). The 'highland line', which separates the areas in Scotland in which the first two types of spirit are distilled, is a straight line which runs from Dundee in the east to Greenock in the west (Figure 1). It then extends southwards, below the Isle of Arran. Any whisky produced north of this line, including those from Campbeltown and Islay, is entitled to be called a highland malt whisky, while whiskies distilled in areas south of the line are designated as lowland whiskies. Of the 104 malt whisky

Figure 1. Highland and lowland whisky-producing regions of Scotland.

distilleries in Scotland, 95 are highland malt whisky distilleries. Of these, no fewer than 49 are situated in an area measuring 50 miles east to west and 20 miles southwards from the Moray Firth. This area of Speyside has been called the 'Kingdom of Malt Whisky' (Cameron Taylor, 1970). Classification of the four whiskies distilled on the islands of Jura, Orkney and Skye is disputed. Some authorities list them along with the Islay whiskies as 'island' whiskies, others as highland whiskies, which is geographically correct. There are also eight grain whisky distilleries in Scotland.

Whiskey distilling in Ireland was, as has been noted, first recorded in the 12th century. By 1556, it had become sufficiently widespread to warrant legislation to control it. A statute proclaimed that year stated that a license was required to manufacture the spirit, but that peers, gentlemen owning property worth £10 or more and borough freemen were exempt (McGuire, 1973). Taxation of whisky distilling gradually became more excessive and collection of taxes became increasingly efficient. However, in 1779 there was an important change in the distillery laws. An attempt was made to limit the extent of evasion of spirit duty by prescribing a minimum revenue to be exacted from the owner of each still. The effect of this legislation was dramatic. In 1779, there was said to be 1,152 registered stills in Ireland. By 1790, this number had fallen to 216 and this inevitably fostered widespread illicit distilling (McGuire, 1973). This legislation lasted until 1823 when it was replaced by laws that taxed Irish whiskey on the volume of production, legislation that is essentially still in force today. Development of the Irish whiskey distilling industry in the present century has inevitably been influenced by economic circumstances and by the political division of Ireland into the Republic and Northern Ireland that occurred in 1922. Barnard (1887) described visits to 28 distilleries in Ireland, but closures and amalgamations followed such that when McGuire (1973) prepared his account, there were only two whiskey-distilling companies in Ireland, one with distilleries in Dublin and Cork in the Republic and the other with plants in Bushmills and Coleraine in Northern Ireland. These two companies have since amalgamated and have concentrated their distillery operations in Cork and Bushmills. There has also been a move towards production of a lighter Scotch-type whisky in Ireland to replace the heavier traditional Irish whiskey.

Outline of whisky production processes

Whiskies differ basically in the nature and proportion of the cereals used as raw materials and on the type of still used in the distillation process. These differences in the production process are illustrated in the flow diagram in Figure 2 for production of Scotch malt whisky (production of Irish whiskey is very similar). Detailed accounts of each of the unit processes in whisky production are given in subsequent sections of this chapter.

A characteristic of Scotch malt whisky is that the only cereal used in its manufacture is malted barley (Table 1). After milling, the meal is mashed in a mash tun (Figure 2) similar to that used in breweries for beer production. During mashing or conversion, enzymes in the malt catalyze the hydrolysis of starch into fermentable sugars. In the manufacture of Scotch grain whisky and Irish whiskey, other cereals are used along with malted barley to provide additional starch in the mash tun (Table 1). Owing to the high gelatinization temperature of their starches, unmalted cereals must be precooked before they are incorporated into the mash.

The wort, or clear mash, leaving the mash tun is cooled and fed into a vessel where it is mixed with yeast. In Scotland and Ireland these fermentation vessels have a relatively small capacity and are known as 'washbacks' (Figure 2).

Fermentation is conducted with strains of the yeast *Saccharomyces cerevisiae* that are usually specially propagated for the purpose, although Scotch malt whisky distillers may use some surplus brewers yeast (Table 1). The process is allowed to proceed to a point at which the

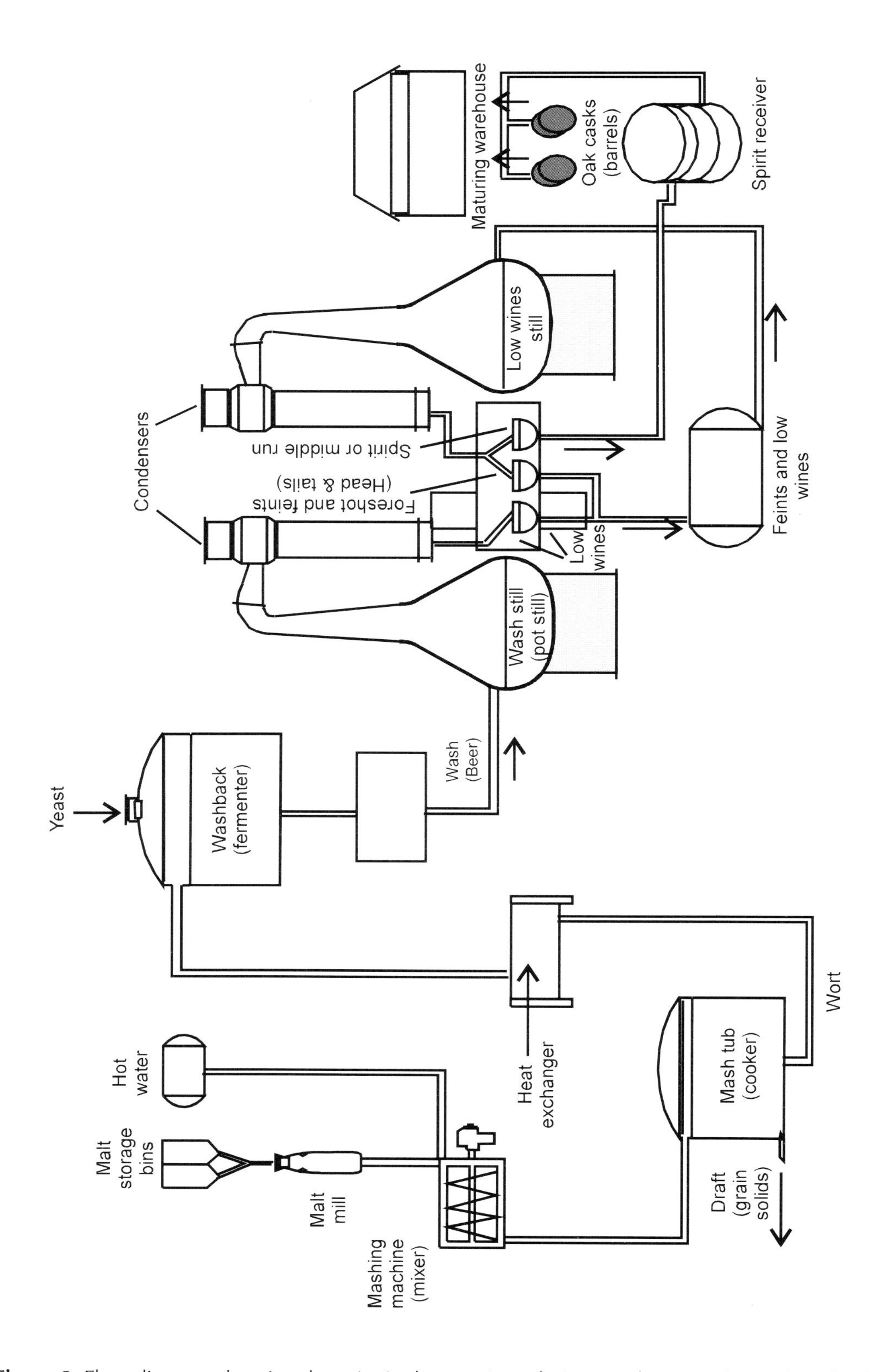

Figure 2. Flow diagram showing the principal operations during production of Scotch malt whisky.

Table 1. Raw materials and unit processes used in the production of scotch and Irish whiskies.

	Scotch malt	*Scotch grain*	*Irish*
Raw materials	Peated and unpeated malted barley	Wheat or corn and a small proportion of malted barley	Unmalted barley and unpeated, malted barley
Conversion	Infusion mash	Mash cook followed by conversion stand	Infusion mash
Fermentation	Distillers yeast and brewers yeast	Distillers yeast	Distillers yeast
Distillation	Two pot stills	Continuous still	Three pot stills
Maturation	At 62° GL in used charred oak bourbon whiskey barrels or sherry casks for at least three years	Up to 67° GL in used charred oak bourbon whisky barrels or sherry casks for at least three years	At 70° GL in used charred oak bourbon whisky barrels or sherry casks for at least three years

specific gravity of the fermented mash has usually dropped to below 1.000. In pot still distilleries, the fermented mash or 'wash' (beer) is fed directly to a still known as the wash still, from which the distillates are redistilled in the second or low wines still. In Ireland, and in one Scotch malt whisky distillery, a third distillation is carried out. Finally, the freshly distilled whisky is stored in charred oak barrels for minimum periods of time that depend on the legislation in the producing country (Table 1). Scotch malt and Irish whiskies are customarily matured for much longer than the legal minimum period.

Individual operations

Raw materials

Malted cereals

Malted barley is the principal malted cereal used in whisky production. Like the brewer, the whisky distiller uses barley cultivars of the species *Hordeum vulgare L.* and *Hordeum distichon* (Hough *et al.*, 1971). Malted barley is employed as a source of enzymes (principally amylolytic) that catalyze the hydrolysis of starches and in some instances serves as a source of starch that is converted ultimately into ethanol. These two demands must be finely balanced. In the manufacture of Scotch malt whisky, where only malted barley is used, care must be taken when the grain is sprouting during the malting process to ensure that only a limited amount of enzyme activity is produced. This is because enzyme is produced at the expense of the fermentable materials in the grain (referred to as 'extract'). However in the manufacture of other types of whisky, malted barley is often used as the only source of amylolytic enzyme in a mash bill that contains a high proportion of unmalted grain. In this type of whisky production the enzyme activity of the malted barley must be greater than that used in Scotch malt whisky manufacture.

Traditionally, barley used in the production of Scotch malt whisky was malted on the distillery premises using a floor malting system and dried over fires of coke and peat in the pagoda-shaped kilns which are still a feature of these distilleries. To a large extent, this system has been superseded by mechanical maltings, which produce malt for groups of distilleries. In order not to destroy the enzyme activity developed during malting, a balance must be achieved in the kiln between drying the green malt to a suitably low moisture level for storage, curing to give it the appropriate flavor and retaining sufficient enzyme activity (Simpson, 1968). In maltings attached to the distillery, the kiln temperature is increased slowly over a 48 hr period to achieve an even rate of drying and the desired flavor. The latter character is achieved by fuelling the furnace with peat during the early part of the kilning period when the green malt is moist and readily absorbs the peat smoke or 'reek'. In mechanical maltings the green malt is dried at a faster rate with a forced-air draught, but a supplementary peat-fired kiln is often used to produce flavored malts. The amount of peat used varies with different maltings. Some of the distilleries on Islay in Scotland specialize in producing a whisky with a very pronounced peat flavor, and they therefore use heavily-peated malts. Malted barleys used in the manufacture of Scotch grain and Irish whiskies are not dried over a peat fire. They generally have a greater enzyme activity, so the relatively small proportion of malted barley used in the mash contains sufficient enzyme activity to convert all of the starch in the mash (principally supplied by unmalted cereals) into fermentable sugars. The greater enzyme activity in these malted barleys is reflected in their nitrogen content. Malts used in production of Scotch grain whisky have a nitrogen content of 1.8% or higher (compared with a brewer's malted barley with a nitrogen content in the region of 1.5%) (Hough *et al.*, 1971). Malting aids, such as gibberellic acid and bromate, are not normally used in production of malted barley for whisky manufacture.

Because of the high cost of malted cereals, considerable effort has been expended by the whisky distiller in attempting to devise methods that allow prediction of the yield of alcohol expected using different proportions of malted barley in the mash bill. Unfortunately, the methods customarily used by the brewer, such as those recommended in Great Britain by the Institute of Brewing Analysis Committee (1975), have proved of limited value. The brewer has used measurements of diastatic power, expressed as the Lintner value, which is a measure of the extent of saccharification of soluble starch present in a cold water extract of the malt (Lloyd Hind, 1948) as an indicator of malt quality. Diastatic activity measured in this way includes contributions from both α- and ß-amylases. However, Preece and his colleague (Preece, 1947; 1948; Preece and Shadaksharaswamy, 1949 a,b,c) have shown that high ß-amylase activity, as determined by the Lintner value, is not always accompanied by high α-amylase activity. Pyke (1965) showed that the Lintner value of a malt is only useful for predicting the spirit yield in manufacture of Scotch grain whisky when the proportion of malted barley in the mashbill is low (Figure 3). Determination of α-amylase activities of the malt gave a less satisfactory correlation than the Lintner value, an observation which agrees with that made earlier by Thorne *et al.* (1945). Further evidence for the unsuitability of employing traditional malt specifications for predicting performance in whisky manufacture has come from Griffin (1972).

Although the degree to which a malted barley has been peated can to some extent be assessed by smell, such is the importance of this character in malt that it must be determined in a more rigorous fashion. Peat smoke or 'reek' contains a wide range of compounds, but it is generally held that the peaty character is imparted to the malt largely as a result of absorption of phenols. For some years, Scotch distillers used a method based on a reaction of phenols with diazotized sulphanilic acid. A lack of specificity in this method, coupled with the instability of the diazonium salt, prompted MacFarlane (1968) to recommend an alternative method involving extraction of phenols from malt with diethylether under acid conditions with absorptiometric measurement of the color developed when the phenols are reacted with 4-aminophenazone. MacFarlane applied the method to a wide variety of malts, both peated and unpeated, and reported values ranging from zero (for an unpeated grain) to as high as 9.4 ppm for a malt produced on Islay in Scotland. (It has been calculated that to obtain a malt with 10 ppm phenols, one tonne of peat must be used for drying each tonne of malted barley).

Scotch whisky distillers have been concerned by the possibility that colorimetric methods, such as those recommended by MacFarlane (1968) and Kleber and Hurns (1972), may not assay all of the organoleptically-important compounds that malt acquires as a result of peating. To examine this possibility, MacFarlane *et al.* (1973) produced peat smoke condensate on a laboratory scale and separated the oil and aqueous phases from the wax fraction, as only the two former would contain components that might appear in the distilled whisky. Six compounds, namely furfural, 5-methylfurfural, guiacol, phenol, ρ-cresol and 5-xylenol, were detected in the aqueous fraction by gas liquid chromatography (GLC). The peat smoke oil was more complex, and no fewer than 30 peaks, some of them created by more than one compound, were obtained by GLC. The compounds included hydrocarbons, furfural derivatives, benzene derivatives and phenols. The authors stressed that using their GLC techniques, 3,5-xylenol is masked by the peaks of m-ethylphenol and ρ-ethylphenol, two compounds thought to make an important contribution to peat aroma and taste. Figure 4 shows gas liquid chromatograms of extracts of a peated and a unpeated malt.

Unmalted cereals

Fewer problems are encountered in arriving at specifications for the unmalted cereals used in whisky manufacture, namely wheat, corn, rye and barley. The corn (varieties of *Zea mays*) used in mash bills for manufacturing Scotch grain whiskies is usually of the yellow dent type, generally obtained from France. Occasionally white corn is used, and it is reputed to give a higher alcohol yield. Corn is a popular grain because it has a high content of starch that is readily gelatinized and converted into fermentable sugars. The US has imposed controls on the quality of corn used for whisky manufacture. In the US there are three grades of corn, with only grades 1 and 2 being approved for spirit manufacture. In Great Britain, on the other hand, the corn used is normally grade 3 on the US scale.

Unmalted barley used in manufacture of Irish whiskey has a quality intermediate to that used for malting and that used for cattle feed. In this way, the maltster can select the best barley available on the market at the time of purchase. For many years, a small percentage (about 5% of the total) of unmalted oats (*Avena spp.*) was

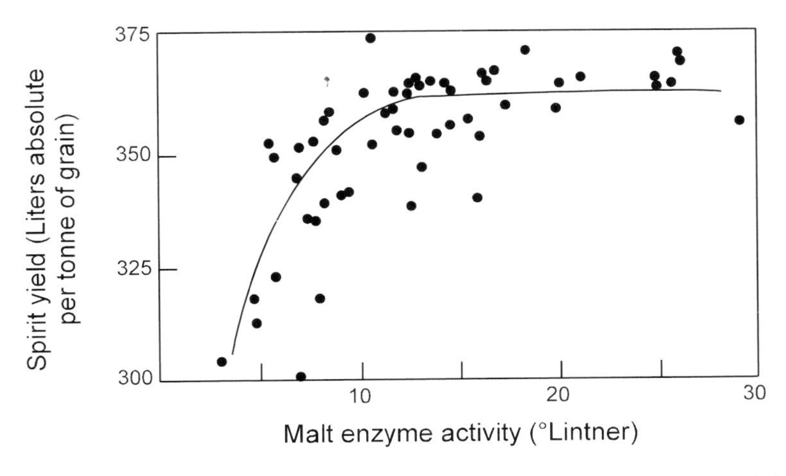

Figure 3. Relationship between the diastatic activity of a Scotch grain whisky mash bill and spirit yield. Laboratory fermentations were conducted using mash bills containing different proportions of malt mixed with corn, and therefore with different Lintner values. It can be seen that only when the proportion of malt is rate-limiting can spirit yield be correlated with the Lintner value (Pyke, 1965).

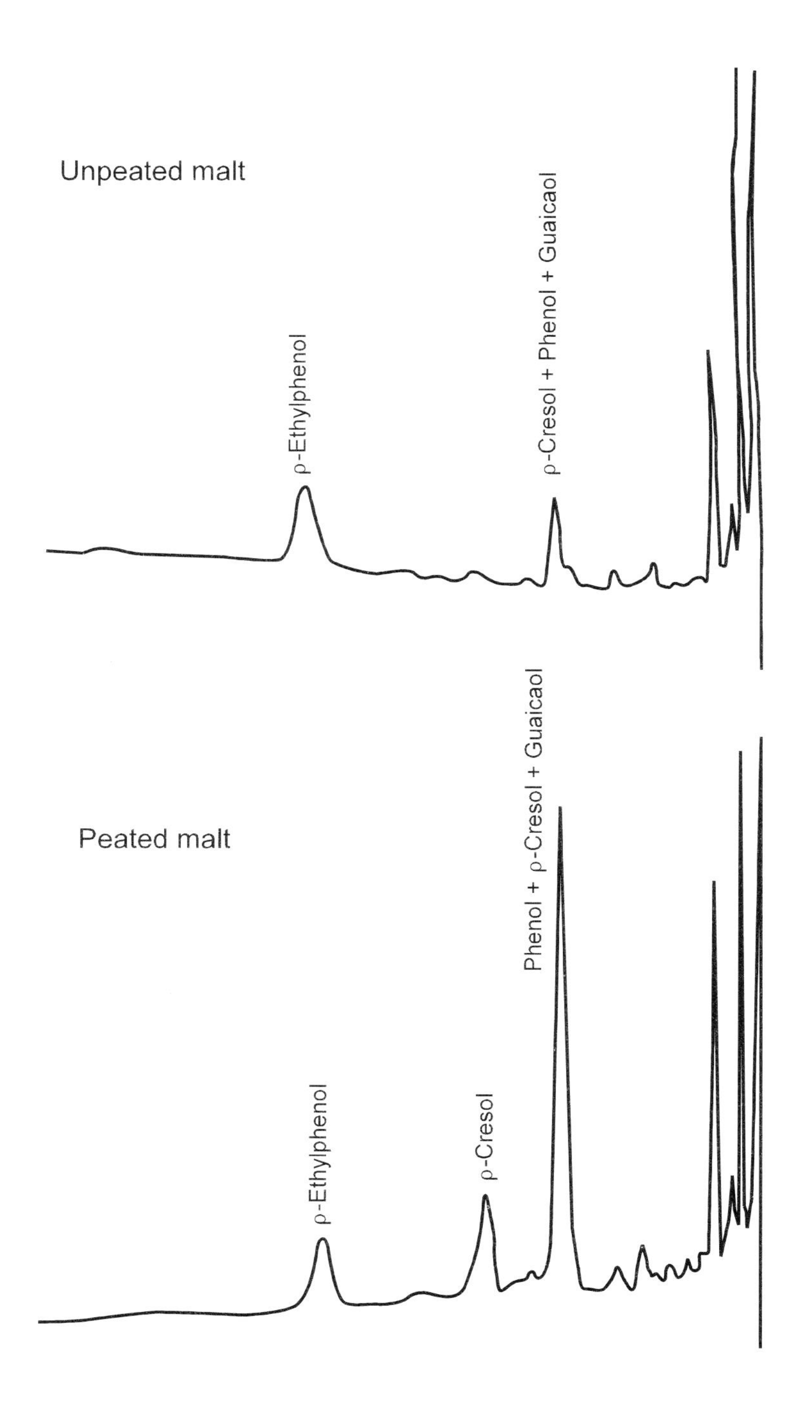

Figure 4. Gas-liquid chromatograms of extracts of unpeated and peated Scoth barley malts showing the contribution peating makes to the content of phenols in the malt. Peating leads to an increase in the size of the peak corresponding to phenol, ρ-cresol and guiacol, and of the peaks which lie to the right of the mixed phenol peak (attributable to furfurals and hydrocarbons). ρ-cresol is also detectable on the chromatogram of extracts of peated malt. The ratio of the area of the total phenol peaks to that of ρ-ethylphenol is used as an indication of the peatiness of the malt.

included in mash bills for manufacturing Irish whiskey. It was contended that these grains, with their large husks, improved the texture of the grain bed in the mash tun (to assist in straining off the clear wort from the mash solids) and that oats influenced the flavor of Irish whiskey. Whether either or both of these effects were important will probably never be known, for oats are no longer used in the production of Irish whiskey (Court and Bowers, 1970).

Mashing and cooking

Mash production in Scotch and Irish distilleries involves a process not unlike that used in breweries to prepare wort for beer manufacture. However, where cereals other than malted barley are used, the malt mashing is preceded by a high temperature cooking process.

Mashing

Regardless of whether the mash bill contains cereals other than malted barley, the main biochemical changes that take place during mashing are hydrolysis of starch, protein and other biopolymers in the meal to produce water soluble low molecular weight compounds that form a fermentable substrate (or wort). The major starch-liquefying and saccharifying enzymes are α- and ß-amylases, while the limit dextrins formed by action of amylases on amylopectin are further hydrolyzed by limit dextrinases.

Barley malt used in the manufacture of Scotch malt whisky is coarsely ground in a roller mill adjusted to give a grind no finer than the malt warrants. Too fine a grind can give rise to a 'set mash' that settles on the bottom of the mash tun to block and impede drainage of the liquid (Simpson, 1968). The mash tun is usually preheated with water and the perforated bottom plates are flooded before mashing commences. The meal from overhead bins is mixed with hot water at 60-65°C in the proportion of one part meal to four parts water, and the mash is homogenized by action of revolving rakes. In the traditional mashing process, the mash is usually loaded into the tun to a depth of about one meter and allowed to stand for about one hour, after which the wort is drained off from under the grain bed. This liquid extract, which has a specific gravity (SG) of 1.070-1.060 (Figure 5) is collected in an intermediate vessel known as an 'underback'. After being cooled to around 25°C in a heat exchanger, the wort is pumped into the fermentation vessel. The bed of grains in the tun is then re-suspended in water at 75°C and a second batch of wort is drawn off at a specific gravity of around 1.030 and passed into the underback. This process is known as the first aftermash and is repeated twice more; except that the dilute worts drawn off are not passed to the underback, but are returned to the hot water tank to be used in the next mash. The wort in the underback has a pH value of about 5.5, a specific gravity of 1.045-1.065, and an amino nitrogen content of about 150-180 mg/L. The spent grain residue or 'draff' is removed from the mash tun and sold as animal feed. Many distilleries have now reduced this process to just three water additions. Larger distilleries are now installing semi-lauter tuns in place of the traditional mash tuns. These have vertical knives, enabling the structure of the mash bed to be maintained whilst speeding wort draining, and sparging rings to allow simultaneous draining and sparging. The result is a faster more efficient extraction without loss of clarity or changing the composition of the wort.

Mash preparation in the manufacture of Irish whiskey is very similar to that used in Scotch production, but there are certain differences. Use of a high percentage of raw barley in the mash bill (up to 60%) has necessitated the use of stone mills or hammer mills to achieve the required grind. The unmalted barley is sprayed with water to give a moisture content of about 14% and then dried to around 4.5% moisture before grinding. The malted barley used is roller-milled. In recent years a plant has been installed in Ireland which uses a wet milling process that eliminates the need for watering and drying of

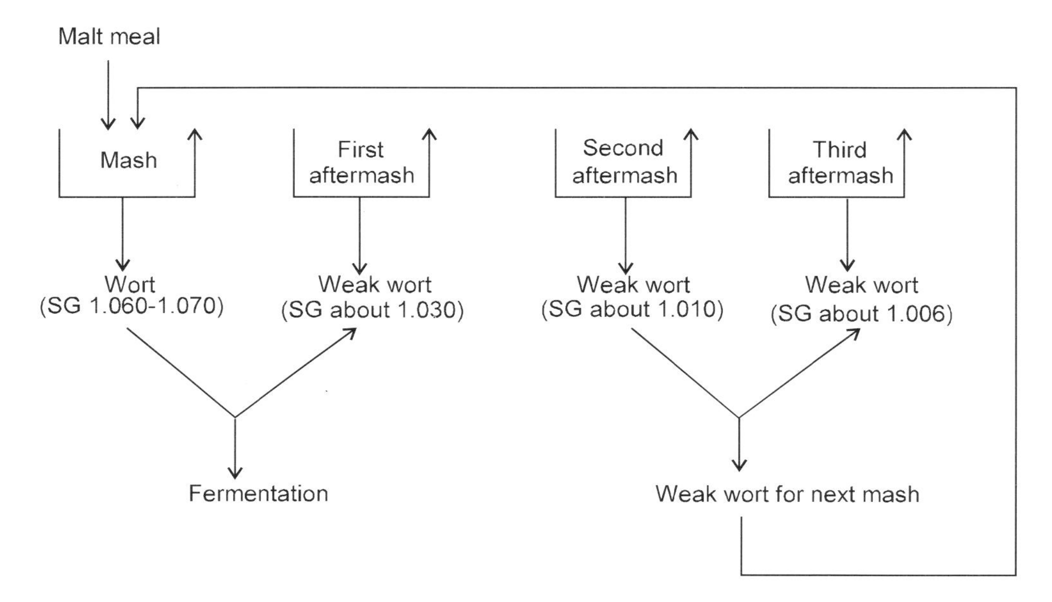

Figure 5. Flow diagram showing the mashing cycle in a Scotch malt whisky distillery.

the unmalted barley. Mash tuns used in Irish whiskey distilleries differ from those used to make Scotch in that they are larger, simply because these distilleries have stills with a greater capacity than those in Scotland. (This difference dates back to when there was a flat rate of tax per still in Ireland, rather than a tax per gallon of product.) Moreover, the mashing cycle differs in that the weak wort from the first aftermash is not passed to the underback, but is mixed with worts from the second and third aftermashes to be used in the subsequent mash (Lyons, 1974).

Cooking followed by malt mashing

The wheat, corn and rye used in production of Scotch grain whisky must be cooked before being added to the mash containing malted barley in order that the starch in these grains can become gelatinized and accessible to the malt enzymes. Traditionally, cooking was carried out as a batch process, but it has to some extent been superseded by continuous processes.

For batch cooking, the grain is freed from extraneous material by passage through screens and then ground either in a pin-type mill or a hammer mill. It is then conveyed to the cooker and subjected to a cooking cycle involving high temperatures and pressures designed to bring about complete gelatinization of the starch. Inadequate cooking will sometimes leave starch granules intact in the mash, while excessive heating can cause caramelization and therefore loss of sugar and decreased spirit yield. Pyke (1965) has provided an account of the cooking of corn in its production of Scotch grain whisky and a similar process is still used today. The conventional cooker is a horizontal cylindrical vessel capable of working at pressures up to 90 psi (63×10^4 Pa), and fitted with stirring gear. A typical cycle in cooking corn might consist of 1-1.5 hrs of heating with live steam injection to reach a temperature of 120°C and a pressure of 15×10^4 Pa. The mash is held at this temperature and pressure for a further 1.5 hrs. The liquid used for the mash is often that from the third aftermash mentioned earlier. At the end of the cooking time, the pressure is released and the hot cooked corn mash is blown directly into a saccharification vessel containing the malted barley suspended in hot water. Cold water is then added to bring the temperature in the mash tun to around 63°C. Good mixing is also essential at this stage; and failure to achieve it can lead to the entire mash solidifying. The need for rapid cooling of the corn mash was emphasized by Murtagh (1970), who showed that with slow cooling or holding at a temperature around 82°C lipid-carbohydrate complexes could be formed that are not fermentable and can lead to a loss of 1-2% in spirit yield.

After a suitable mashing period, the mash may be filtered as described previously. More commonly, the entire mash may be pumped via a heat exchanger into a fermenter. Pyke (1965) has published the composition of a typical Scotch grain whisky mash (Table 2).

Table 2. Chemical composition of a typical scotch grain mash*.

	%
Total soluble carbohydrate (as glucose)	9.00
Insoluble solids	2.20
Fructose	0.13
Glucose	0.29
Sucrose	0.28
Maltose	4.65
Maltotriose	0.96
Maltotetraose	0.45
Dextrin	2.54
Amino nitrogen (as leucine)	0.09
Ash	0.27
containing P_2O_5	0.09
K_2O	0.09
M_gO	0.02
	μg/ml
Thiamin	0.46
Pyridoxine	0.61
Biotin	0.01
Inositol	236
Niacin	11.1
Pantothenate	0.71

*Pyke, 1965.

In the manufacture of Scotch grain whisky, the proportion of malted barley in the mash bill is usually around 10-15%, a proportion far greater than that required merely to provide a source of amylolytic enzymes. It is a practice of long standing, and is done to obtain the required malt flavor in the grain whisky. This was noted as early as 1902 by Schidrowitz and prompted further comment from Valaer in 1940.

Continuous cooking has been adopted in recent years in Scotland. Stark (1954) listed the advantages of using continuous cooking. While practices vary to some extent in different distilleries, essentially the cereal is slurried at around 50°C and then pumped through a steam-jet heater into the cooking tubes where the residence time is normally 5-10 minutes. The continuous cooker has a narrow tube (6-16 cm in diameter) which reduces the incidence of charring and carbon deposition. The mash passes through the tube at about 20 meters/minute, with the temperature reaching near 65°C and the pressure 65 psi (40 x 10^4 Pa).

Fermentation

The objectives in fermenting a whisky distillery mash with strains of *Saccharomyces cerevisiae* are to convert the mash sugars to ethanol and carbon dioxide while producing quantitatively minor amounts of other organic compounds which contribute to the organoleptic qualities of the final distilled product. Fermenting vessels vary considerably in volume, depending on the distillery. The small Scotch whisky distillery at Edradour near Pitlochry, for example, has fermenters with a capacity of only 4,500 liters. In contrast, other Scotch and Irish pot whiskey distillers have fermenters with a capacity in the range 50,000 to 150,000 liters. Much larger fermenters are found in Scotch grain whisky distilleries. Traditionally, the smaller pot distillery fermenters were made of wood, usually of larch or Oregon pine, but in recent years they have been constructed of steel or aluminum. In some Scottish distilleries, timber is still used as a covering material.

When the fermenter is partly filled, the mash is inoculated with a suspension of *Saccharomyces cerevisiae*. The source of yeast varies with the location and size of the distillery. Distilleries in

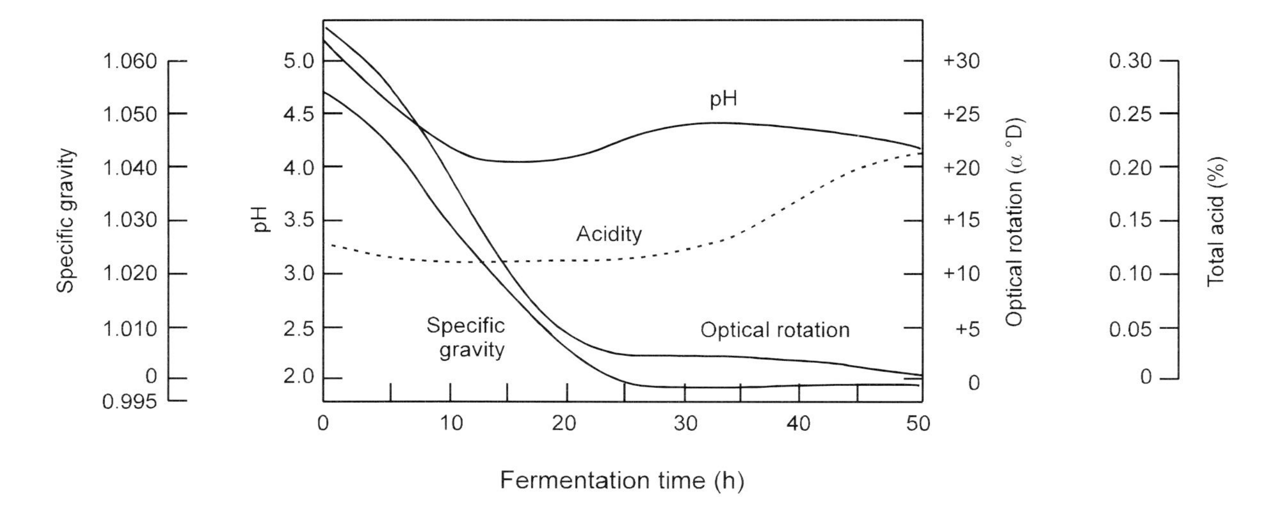

Figure 6. Changes in specific gravity, optical rotation, pH value and acidity during fermentation of a mash in the production of Scotch malt whisky (Dolan, 1976).

Scotland and Ireland, particularly the pot whisky distilleries, rarely have their own yeast propagation equipment. They rely on specially-propagated pressed yeast for use in fermentations. In Scotch malt whisky distilleries, this pressed yeast is augmented, usually on an equal weight basis, with surplus brewery yeast typically supplied in compressed form from breweries that may be as far as 500 km from the distillery.

The requirements for a strain of *Saccharomyces cerevisiae* yeast used in distillery practice have been described by Harrison and Graham (1970). Apart from the obvious need to select a strain that maintains very high viability in the pressed state (containing 25% dry weight), other very important properties of these strains are ability to tolerate concentrations of ethanol on the order of 12-15% (v/v) and the capacity to metabolize oligosaccharides such as maltotriose and maltotetraose in order to maximize the conversion of starch into ethanol and carbon dioxide.

Whisky worts usually have a specific gravity in the range 1.050-1.080, a pH value of around 5.0, a total acid content of 0.1% and an optical rotation of +30°. After inoculation, the yeast content is 5-20 million cells/ml. The bacterial count varies with the cleanliness of the plant and the extent to which the raw materials were endowed with a microbial flora. Scotch grain whisky fermentations create little if any foam because of their large content of suspended solids. However, in most Scotch malt whisky fermentations only a small proportion of the suspended solids in the mash is retained in the fermentation vessel. These fermentations tend to foam and the distillers have resorted to the use of various types of antifoams.

Changes over the time course of a typical fermentation in a Scotch malt whisky distillery are depicted in Figure 6. Fermentation proceeds vigorously for the first 30 hrs, during which time the specific gravity falls to 1.000 or below and the optical rotation to around zero. The sugars in the wort are utilized in a particular sequence with glucose and fructose being fermented first, followed by maltose and then maltotriose. The removal of sugars during fermentation of a Scotch grain whisky mash is shown in Figure 7 (Pyke, 1965). Over the first 30 hrs the pH value, after declining to around 4.2, rises to about 4.5. During the first 30 hrs the specific gravity drops at a rate of about 0.5° per hour accompanied by a massive evolution of heat. While many of the larger grain whisky distilleries have fermenters fitted with cooling coils, these are absent, or if fitted are relatively inefficient in most malt whisky distilleries where temperature can rise by the end of the fermentation to as high as 35-37°C. The distiller is concerned about the temperature rise during fermentation since this can cause the fermentation to stop or become 'stuck'. Temperature rise can be controlled by using a lower starting temperature or, because glycolysis of sugar is a heat-producing process, by using a lower initial concentration. Strains of *Saccharomyces cerevisiae* are well suited for malt whisky distillery fermentations since they can ferment efficiently over a wide temperature range. Fermentation is usually continued for at least 36 hrs and frequently longer, at which time the ethanol content of the wash is 7-11% (v/v). In larger distilleries, particularly those in the US, the carbon dioxide evolved is collected, liquefied and sold. Smaller distilleries, particularly the malt whisky distilleries in Scotland, usually do not have this facility.

It should be noted that mashes in malt whisky distilleries are not boiled, so any enzyme activity manifested at the temperature of the mash and any microorganisms that can survive at that temperature will continue to be active during the fermentation. The continued activity of limit dextrinases in unboiled distillery mashes increases the concentration of sugars available for fermentation by the yeast. Hopkins and Wiener (1955) calculated that with amylases alone the yeast cannot metabolize the equivalent of the final 12-16% of the starch.

Another important consequence of using non-sterile conditions in distillery fermentations is the activity of bacteria that pass through in the mash, which are encouraged to some extent by the relatively high temperatures to which the fermentations can rise. In addition to lactic acid bacteria, the flora can include other Gram-positive as well as Gram-negative strains. The concentration of the flora depends on a number of factors including the extent to which the lactic acid bacteria grew during yeast propagation, the extent of the flora on the cereal raw materials and on the standard of hygiene in the distillery. There is no doubt, however, that the controlled activity of this bacterial flora, and particularly of the lactic acid bacteria, is accompanied by

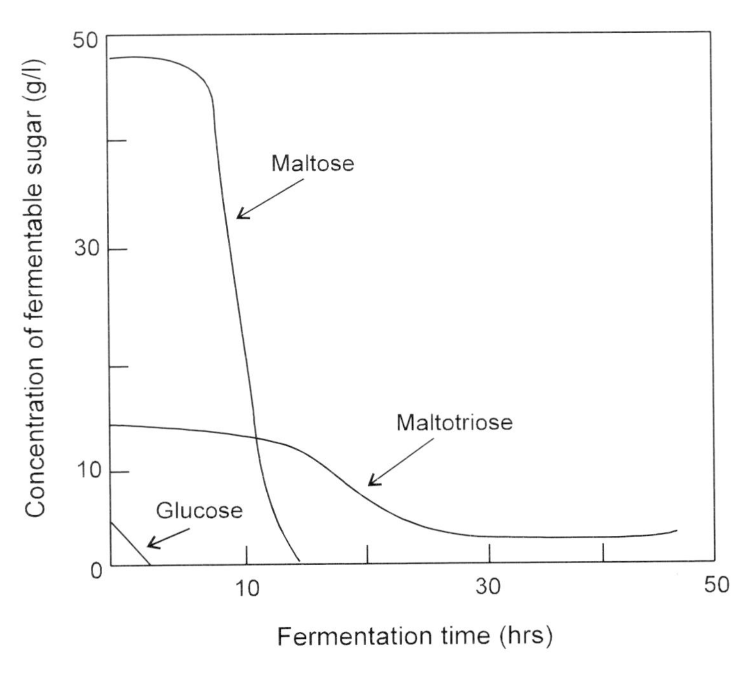

Figure 7. Time course removal of fermentable sugars from a Scotch grain whisky mash (Pyke, 1965).

excretion of compounds that contribute to the organoleptic qualities of the final whisky (Geddes and Riffkin, 1989).

During the first 30 hrs or so of malt whisky fermentation there is vigorous fermentation and the majority of the aerobic bacteria die. This, however, provides ideal conditions for growth of anaerobic or microaerophilic bacteria, principally lactic acid bacteria (mainly strains of *Lactobacillus brevis, L. fermenti* and *Streptococcus lactis*) with the result that the concentration of lactic acid in the fermented mash can be as high as 30 mg/L (MacKenzie and Kenny, 1965). A wide range of lactobacillus species have been identified in Scotch whisky fermentations including *L. fermentum, L. brevis, L. delbrueckii, L. plantarum, L. casei* and a bacterium resembling *L. collinoides*, in addition to *Leuconostoc spp., Streptococcus lactis* and *Pediococcus cerevisiae* (Bryan-Jones, 1976). More recently Barbour (1983) isolated many species that did not conform to recognized species of lactic acid bacteria, a point emphasized by Walker *et al.* (1990) who used DNA hybridization techniques to classify distillery bacteria. Growth of lactic acid bacteria is probably enhanced by yeast excretion of nitrogenous nutrients at the end of a vigorous fermentation. Kulka (1953) demonstrated the ideal nature of yeast autolysate for growth of lactobacilli. Bacterial activity in the fermenting wort also leads to removal of some acids. Actively growing yeasts secrete citric and malic acids, but MacKenzie and Kenny (1965) attribute the lower concentrations of these acids in malt distillery worts (as compared to brewery worts) to their partial removal by bacteria.

Occasionally, the extent of the bacterial flora in the fermenting wort can become too large. This causes problems due to sugar utilization by the bacteria that lead to an overall decrease in spirit yield. In addition, the bacteria may produce organoleptically-undesirable compounds and also release hydrogen ions causing the pH value of the wort to fall too low, thereby providing suboptimal conditions for action of certain enzymes. Examples of undesirable compounds that may be excreted by bacteria are hydrogen sulfide and other sulfur-containing compounds (Anderson *et al.*, 1972). Lactobacilli can also metabolize glycerol (excreted by the yeast during fermentation) to produce ß-hydroxypropionaldehyde, which subsequently breaks down on distillation to give acrolein (Harrison and Graham, 1970). Acrolein imparts a pungent, burnt and often peppery odor to the whisky (Lyons, 1974). In a later paper, Dolan (1976) concentrated on the problems arising in malt whisky distilleries when there is an unacceptably high concentration of bacteria in

Table 3. Changes in the concentration of bacteria during fermentation of a minimally infected and a heavily infected Scotch malt whisky wort*

	Minimally-infected wort		*Bacteria/ml*		*Heavily-infected wort*	
Age (hours)	*Gram-negative rods*	*Gram-negative cocci*	*Lactobacilli*	*Gram-positive rods*	*Gram-positive cocci*	*Lactobacilli*
At setting	2,000	3,700	n.d.	52,000	60	0.6×10^6
10	150	n.d.	n.d.	n.d.	n.d.	2.3×10^6
20	n.d.**	n.d.	1.53×10^6	n.d.	n.d.	18.8×10^6
30	n.d.	n.d.	10.2×10^6	n.d.	n.d.	96×10^6
40	n.d.	n.d.	10.2×10^6	n.d.	n.d.	502×10^6
50	n.d.	n.d.	50×10^6	n.d.	n.d.	1×10^9

*Dolan, 1976.
**None detected.

the mash. Table 3 shows changes in the concentrations of Gram-negative and Gram-positive bacteria and (separately) of lactobacilli during fermentation of a minimally-infected mash and of a heavily-infected mash. The time course of fermentation of an unacceptably-infected malt distillery mash (Figure 8) shows, in comparison with similar data for fermentation of an acceptable mash (Figure 6), a greater rise in the acid content of the mash after about 35 hrs and a lower optical rotation of the mash after about 40 hrs. In the fermentations there is often a difference of up to 4 hrs from the time a rise in the acid content is detected to the point when the pH value of the fermentation begins to fall. Dolan (1976) attributes this to the buffering capacity of the mash. The data in Table 4 show the effect of different levels of infection after 30 hrs fermentation of a malt distillery mash on spirit yield and the associated financial losses to the distiller. Dolan (1976) recommends upper limits of 1,500 bacteria, 50 Gram-positive and 10 lactic acid-producing bacteria per million yeast cells in the mash at the start of fermentation.

Much less has been published on the effect of retaining solid material in the fermenting mash. However, marine microbiologists have long known that the presence of solid particles in a liquid medium can affect bacterial growth, probably because of the concentration of nutrients at the solid-liquid interface (Heukelekian and Heller, 1940; Zobell, 1943). Moreover, Cromwell and Guymon (1963) found that formation of higher alcohols during fermentation of grape juice is stimulated by the presence of grape skins or inert solids. Beech (1972) made similar observations on cider fermentations. Merritt (1967), in the only detailed report on the role of solids in whisky distillery fermentations, states that a dry solid concentration of 50 mg/100 ml might typically be expected, although much will clearly depend on the design of the mash tuns used in individual distilleries. Merritt went on to report that a concentration of dry solids as low as 5 mg/100 ml causes an increase in yeast growth, and that solids also enhance the rate of production of ethanol and glycerol. There was also an effect on production of higher alcohols by the yeast (Table 5). With the possible exception of n-propanol, production of all of the major higher alcohols was increased in the presence of solids, the effect being particularly noticeable with isobutanol and 2-methylbutanol.

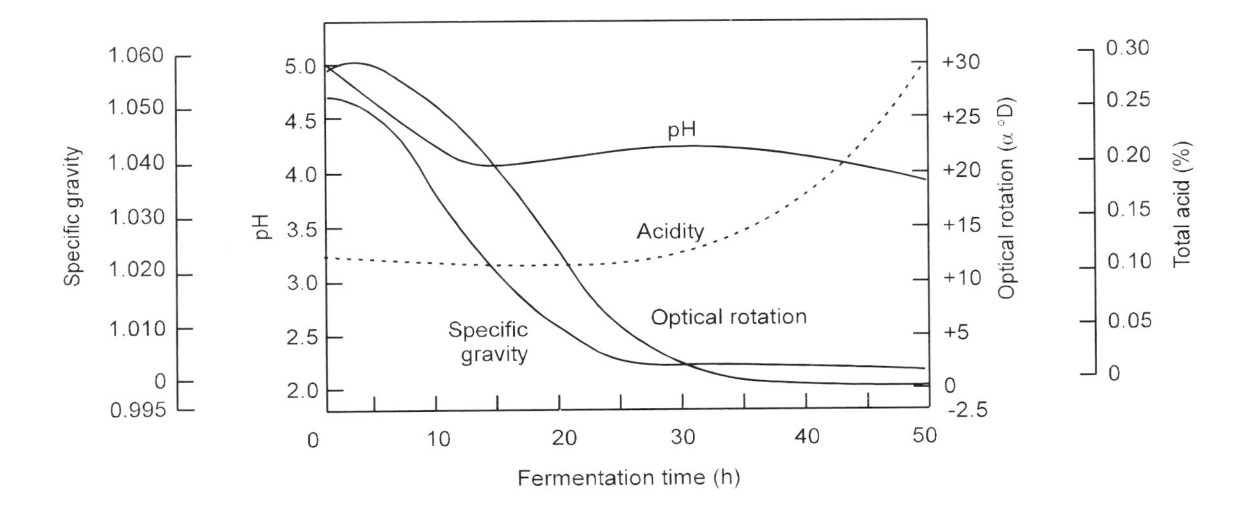

Figure 8. Changes in specific gravity, optical rotation, pH value and acidity during fermentation of a Scotch malt whisky mash containing an unacceptably high concentration of bacterial infection (Dolan, 1976).

Table 4. Effect of the level of bacterial infection on the loss of spirit incurred following fermentation of a scotch malt whisky mash.*

Infection rating	*Bacteria/ml in 30 hr old mash (million)*	*Approximate loss in spirit yield (%)*	*Approximate financial loss in US dollars (thousands)*	
			At filling	Duty paid
a	0 - 1	<1	<26	<369
b	1 - 10	1-3	26 - 79	369 - 1106
c	10 - 100	3-5	79 - 131	1106 - 1844
d	>100	>5	>131	>1844

*Dolan, 1976.

The effect of low insoluble solids content is a factor relevant to congener levels in malt whisky fermentations. In grain whisky production, where 'all-grains-in' fermentations are generally used, the degree of rectification during distillation is the principle determinant of higher alcohol levels in the spirit.

Distillation

Whether the fermented beer is distilled in a pot still, as in production of Scotch malt and Irish whiskies, or in a continuous still based on the Coffey design as in the manufacture of Scotch grain whiskies, the objectives are the same: selective removal of the volatile compounds (particularly the flavor-producing congeners) from the non-volatile compounds and to create additional flavor-producing compounds as a result of chemical reactions that take place in the still. Nevertheless, it is still most convenient to discuss whisky distillation under the separate headings of pot still and continuous distillation.

Pot still distillation

The copper pot still, which is the feature dominating any Scotch malt or Irish whiskey distillery, has changed hardly at all over the centuries, except of course in size. Traditionally, the onion-shaped stills were fired from beneath and had the vapor pipe or 'lyne arm' from the still projecting through the distillery wall to connect with a condenser in the form of a coil immersed in a water tank fed from a local stream (Figure 9). Internal steam-heated calandria are now preferred to direct firing because this decreases the extent of pyrolysis of the still contents and results, for example, in a lower concentration of furfural in the whisky. Variations

Table 5. Production of higher alcohols (mg/100 ml).

Mash	*Insoluble solids content (mg/100 ml)*	*n-Propanol*	*Isobutanol*	*2-methyl butanol (amyl alcohol)*	*3-Methyl butanol (isoamyl alcohol)*	*Total*
1	0	1.5	4.2	3.0	8.0	16.7
	25	1.5	6.5	4.4	8.8	21.2
2	0	2.8	4.5	2.7	9.5	19.5
	35	3.3	6.6	3.7	10.1	23.7
3	0	1.7	4.8	3.1	8.0	17.6
	50	1.7	6.8	4.6	9.5	22.6

in still design include expansion of the surface area of the column into a bulbous shape, water jacketing and return loops from the first stage of the condensation. (Nettleton (1913) provided a valuable account of early still design). In many distilleries, condenser coils have been replaced by tubular condensers that have an advantage in that they are designed to conserve the heat extracted from the distillates. Yet other pot stills are fitted with 'purifiers,' which consist of a circular vessel cooled by running water interposed between the neck of the still and the condenser. In Irish pot stills this purifier function is effected by a trough fitted around the lyne arm through which running water is circulated. Pot stills in Scotch and Irish distilleries are traditionally constructed of copper. The reason for this adherence to copper is more than tradition. It has been established that copper fixes some volatile sulfur-containing compounds that are produced during fermentation but undesirable in the distilled spirit.

Early whisky distillers realized that although the objective of distilling was to separate volatile constituents from the beer, collecting the distillate not as a whole but in several fractions and combining certain of these fractions gave a much more acceptable product. Pot still distillation in Scotland and Ireland differ not only in the size of the stills (25,000-50,000 liters in Scotland vs 100,000-150,000 liters in Ireland), but also in the different ways in which fractions are collected from the stills.

In Scotland, the beer is subjected to two distillations. In the first, carried out in the beer still, the beer is brought to a boil over a period of 5-6 hrs and the distillate is referred to as low wines. This distillation effects a three-fold concentration of the alcohol in the beer (Figure 10). The residue in the wash still, known as 'pot ale', is either discharged to waste or evaporated to produce an animal feed (Rae, 1967). Distillation of the low wines in the spirit is still more selective. The first fraction, which contains low boiling point compounds, is rejected as 'fore-shot heads'. At a stage determined by continued hydrometric monitoring, which usually occurs when the distillate has an alcohol content of approximately 70-73° GL, the distillate is switched from the fore shots tank to the whisky receiver tank. This switch has traditionally been made at the discretion of the distiller; and he has been

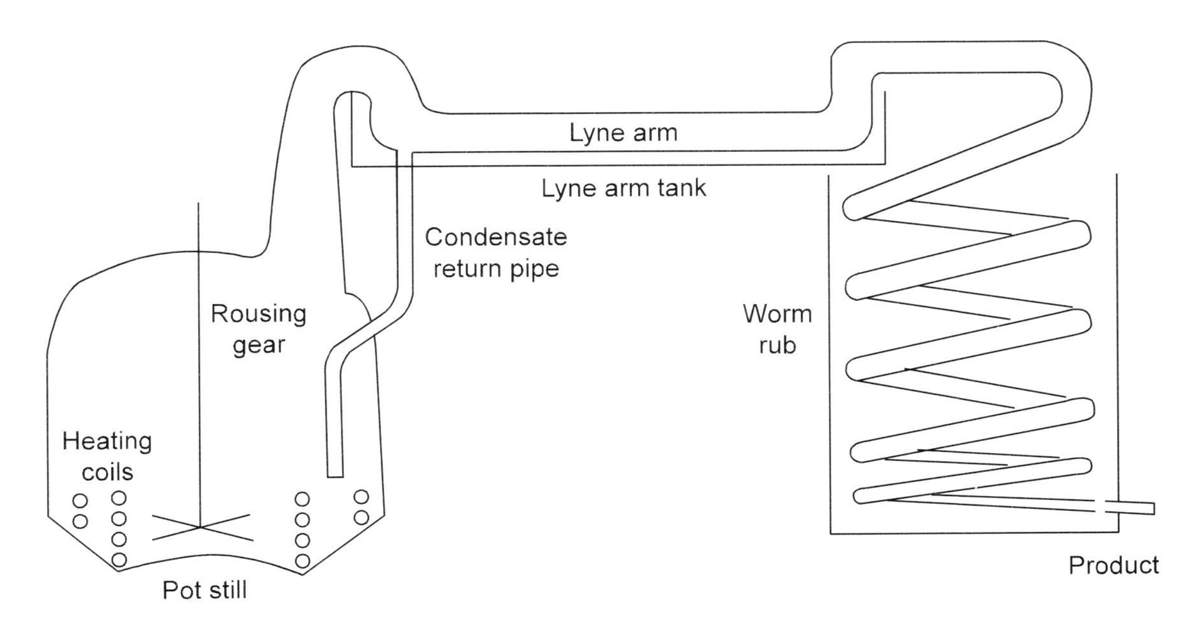

Figure 9. Diagram of an Irish distillery pot still. Designs used in Scotch malt whisky distilleries are similar except that the pot is onion-shaped and the still usually has a shorter lyne arm not surrounded by a lyne-arm tank (Lyons, 1974).

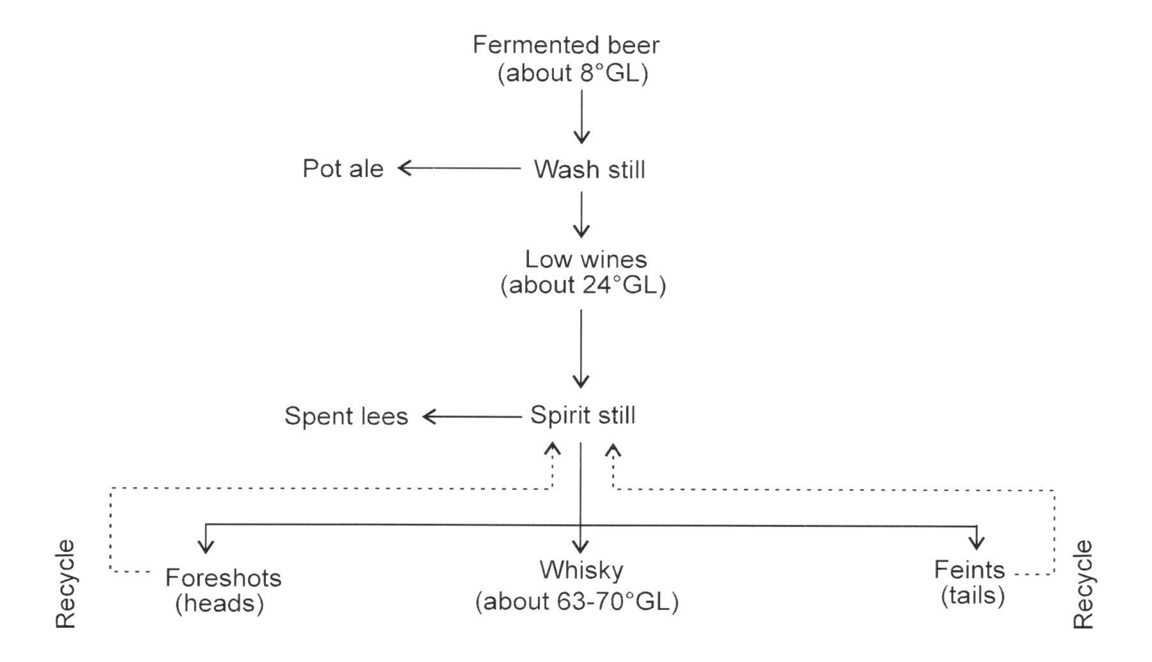

Figure 10. Flow diagram showing the stages in distillation of Scotch malt whisky.

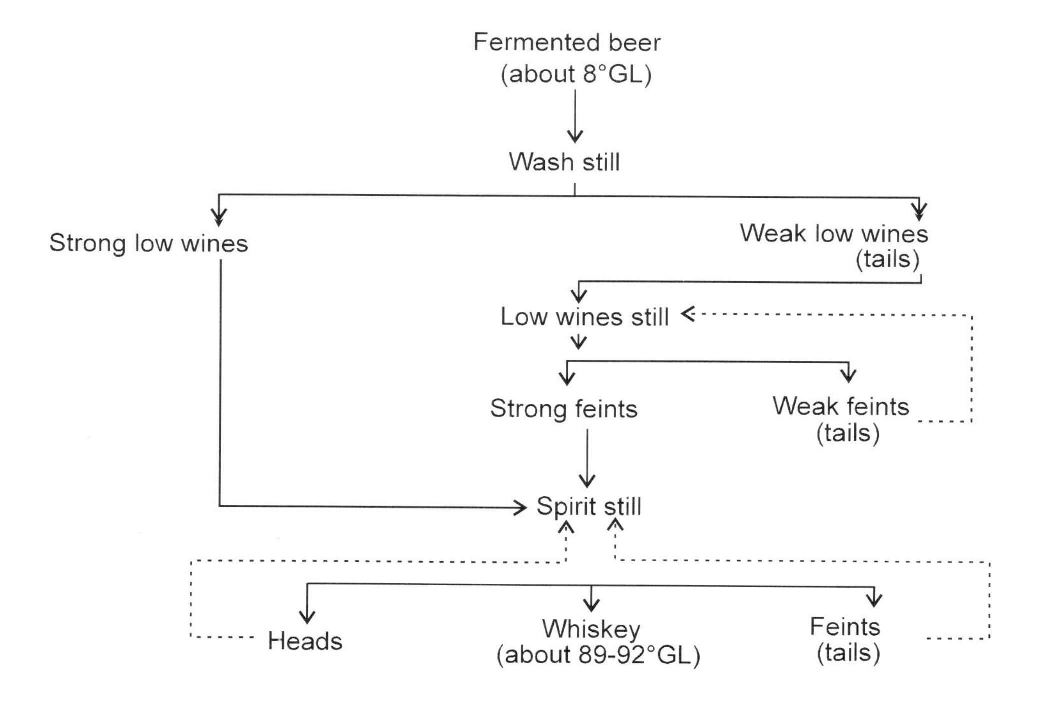

Figure 11. Flow diagram showing the stages in distillation of Irish whiskey.

aided by the disappearance of a bluish tinge when water is added to the distillate. The monitoring process takes place in a spirit safe secured with a Government Excise lock. Collection of the distillate is terminated when the alcohol content has fallen to a specified value, although distillation of the feints or tails is continued until all of the alcohol has been removed from the low wines. The residue remaining in the spirit still is known as 'spent lees', and like pot ale is either run to waste or evaporated to manufacture animal feed. The whisky distilled over in the middle fraction has an alcohol content of 63-70° GL.

The manufacture of Irish whiskey involves three rather than two distillations (Figure 11). The fermented beer is heated in the wash still, and the first distillation ('strong low wines') is collected until the distillate reaches a predetermined specific gravity. The distillate, then known as 'weak low wines', is switched into a separate vessel. The weak low wines are pumped into the low wines still and are re-distilled to produce two fractions termed 'strong feints' and 'weak feints'. Strong feints are mixed with the strong low wines in a spirit still. Distillates from this are collected in the same fashion as in production of Scotch. The whiskey collected usually is about 89-92° GL.

Continuous distillation

No fundamental changes have been introduced into the design of the patent or Coffey still over the past century. Automation, particularly of the beer feed, is now commonplace, as is continuous monitoring of other stages in the distillation process. Nevertheless, many Scotch and Irish producers of grain whiskies continue to use a still which, like the original Coffey still, has just two columns: a beer stripper (or analyzer) and a rectifier.

A description of the operation of two column continuous stills in the manufacture of Scotch grain whisky has come from Pyke (1965). In order to obtain whisky of high quality from these stills, they must be operated such that the alcohol concentration of the spirit at the spirit draw tray in the rectifier is not less than 94.17° GL. The manner in which the precise control of still operation can affect the composition of the whisky is shown in Figure 12. As illustrated, if conditions are changed in either direction on the abscissa, the concentration of congeners will alter with a possible adverse effect on final product quality.

Maturation and aging

Freshly distilled whisky of any type is very different from the spirit that is later bottled, either singly or blended. The transformation is brought about by storing the whisky in oak barrels for periods of time that depend on traditional practice and legal requirements. In general, whiskies are matured for far longer than the legally-required period of time. The raw spirit is taken by pipeline from the distillation plant to the tank house where it is diluted with water to the required strength and then transferred into barrels.

Maturation in barrels is accompanied by a loss of liquid by evaporation, and the relative rates of loss of water and of alcohol determine whether the aged whisky has a higher or lower alcoholic strength than that at filling. In Scotland, where the barrels of whisky are stored in cool, unheated, but humid warehouses, the alcoholic strength decreases (Valaer, 1940). In contrast Valaer and Frazier (1936) reported that in the US storage conditions cause an increase in alcoholic strength. Maturation in barrels is also accompanied by changes in the chemical composition of the whisky. These changes are attributable to extraction of wood constituents from the barrel, oxidation of components present in the original whisky as well as those extracted from the wood, reactions between components in the whisky and removal and oxidation of highly volatile sulfur components by the carbon char on the inner surface of the barrel.

Some of the earlier investigators reported on changes in the composition of the major classes

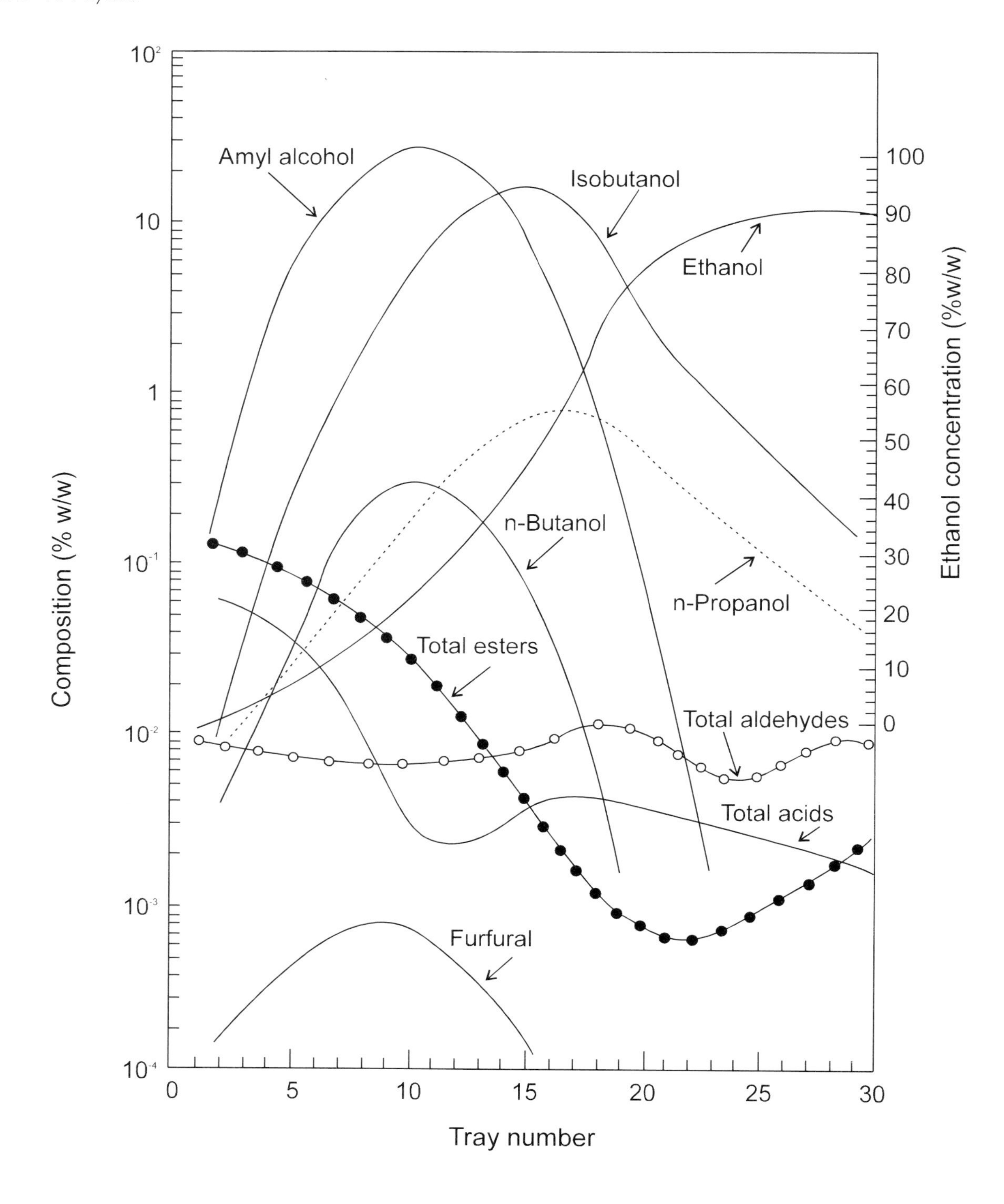

Figure 12. Changes in composition of the vapor at different trays in a Coffey still rectifier used in the manufacture of Scotch grain whisky (Pyke, 1965).

of organoleptically-important compounds during maturation in barrels. Thus, Schidrowitz and Kaye (1905) found increased concentrations of volatile acids in aged whiskies, a trend also described in the report of the Royal Commission on Whisky and Other Potable Spirits (1909). Several reports followed. Liebmann and Scherl (1949), for example, reported increased concentrations of acids, furfural, tannins and color with maturation in barrels. The arrival of the gas liquid chromatograph and HPLC greatly accelerated research on this topic; and more recent data on chemical changes that take place during maturation are described later in this chapter.

Maturation of whisky in oak barrels is an expensive process; and it is hardly surprising that consideration has been given to methods for acceleration. Jacobs (1947) described several of these methods including pretreatment of the beer with activated carbon, chemical treatment of the whisky to convert aldehydes into esters and use of oxidation treatments. Such techniques are not used in the industry, which adheres to the traditional nature of the whisky-producing process.

Blending and coloring

Whisky manufacturers take great pride in the quality of their straight or blended whiskies and in their ability to maintain the quality of a particular product over the years. Blending is conducted by experts who advise manufacturers on mixing or blending large volumes of whisky after appropriate sniffing and tasting sessions on experimental blends. Blended whiskies are filled into re-used barrels after dilution with water (or not) and stored for a further period of time referred to as 'marrying'. The water used for dilution is softened or demineralized, since water containing an appreciable concentration of salts can cause hazes in the whisky (Warricker, 1960). After marrying, and dilution if necessary, the color of the whisky may be adjusted to the desired value by adding caramel. Some brownish pigment is extracted from the casks, but this may not be sufficient to provide the desired color. Finally, the whisky is clarified for bottling by filtration through sheets of cellulose (Simpson, 1968). Chill filtration may also be practiced, since it removes tannin material from the whisky and prevents subsequent appearance of haze.

Effluent disposal and spent grains recovery

Traditionally effluents from whisky distilleries were disposed of in the most convenient manner. Spent grains were retailed, often quite cheaply, to local farmers as animal fodder while pot ale and spent lees were simply discharged into the local sewer, stream or river. This is no longer the case, mainly because of the distiller's awareness of the nutritional value (largely the protein content) of some of these effluents, and public awareness of environmental problems arising from uncontrolled disposal of effluents into waterways. Simpson has described the production of 'distillers dark grains'. These processes are widely used to dispose of effluents from whisky distilleries, especially where the traditional methods of disposal are forbidden or uneconomic.

Organoleptically important components of whisky

Modern analytical techniques have enabled major advances in the understanding of the compounds responsible for the organoleptic properties of whiskies. However, the first reports on the nature of flavor-producing compounds in whiskies antedate the era of gas liquid chromatography by nearly half a century. Two publications by Schidrowitz (1902) and Schidrowitz and Kaye (1905), dealing exclusively with Scotch whiskies, reported on the higher alcohol, acid and ester contents of some 50 different brands. They reported analyses of several Campbeltown Scotch malt whiskies. A report by Mann (1911) published a few years later also quoted values for acidity and levels of furfural, aldehydes, esters and alcohols in Scotch whiskies imported into Australia.

Concentrations of organoleptically important compounds

Since the introduction of GLC into distillery laboratories, several reviews have been published on the composition of the major flavor-producing compounds in whiskies (namely higher alcohols, esters, carbonyl compounds, organic acids, aromatic compounds) and on the identification of individual compounds that make up these fractions. Higher alcohols, which are still routinely determined with GLC using a polar stationary phase (Aylott *et al.*, 1994), are quantitatively the most important. Scotch malt whiskies are richest in higher alcohols, with contents often well over 2 g/L. Free fatty acids are relatively volatile and make a major contribution to the organoleptic qualities of whiskies. Concentrations of acids in some Scotch malt whiskies can be as high as 0.4-1.0 g/L absolute alcohol (Duncan and Philip, 1966). Roughly comparable concentrations of esters are found in whiskies, although those produced with pot stills generally have higher concentrations than those from continuous stills. Ester concentrations in Scotch and Irish pot still whiskies have been reported in the range of 0.27-0.87 g/L absolute alcohol (Valaer, 1940). Lighter whiskies contain lower concentrations of carbonyl compounds, although the concentration varies with the brand. Grain whisky may have as little as 20 mg/l of aldehydes, while in a mature Scotch malt whisky the concentration may be as high as 80 mg/L (Duncan and Philip, 1966).

Chemical nature of organoleptically-important compounds

Duncan and Philip (1966) reviewed chromatographic and other methods used to separate the various organoleptically important compounds from whiskies. Even though analytical methods such as capillary column gas chromatography linked to a mass selective detector (GC-MS) (Aylott *et al.*, 1994) have developed in terms of sensitivity and discrimination, a major problem in analyzing whisky by GLC is the overwhelming preponderance of ethanol and water. Only one volatile compound, namely isoamyl alcohol, is likely to be present in a concentration exceeding 0.01%; while most of the others are present in concentrations that rarely exceed 50 ppm. Indeed many compounds now understood to have an important impact on whisky flavor are present at ppb levels. Carter-Tijmstra (1986) described a technique for measuring dimethyl trisulfide, a compound with a threshold of only 0.1 ppb present in whisky at concentrations below 50 ppb. Analyses are most conveniently conducted on extracted fractions of the different classes of compounds. When direct analysis of whiskies has been employed, only a limited number of components have been determined (Morrison, 1962; Bober and Haddaway, 1963; Singer and Stiles, 1965).

Recent developments in headspace analysis using trapping and thermal desorption techniques (Pert and Woolfendon, 1986) have enabled analysis of the more volatile components of whisky flavor whilst avoiding interference from high concentrations of other congeners. Headspace analysis using trap and purge techniques would now be the method of choice for measuring highly volatile sulfur compounds such as dimethyl trisulfide.

Some idea of the variety of compounds detected in whiskies came from a compilation of both published and unpublished sources (Kahn 1969, Kahn *et al.*, 1969). Of the some 200 compounds listed there are 25 higher alcohols, 32 acids, 69 esters and 22 phenolic compounds. Undoubtedly, this list could now be extended quite considerably. Of the higher alcohols, isoamyl alcohol and optically-active amyl alcohol predominate, accompanied by lower concentrations of isobutanol and n-propanol. Characteristically, there are usually only low concentrations of n-butanol and secbutanol. The principal organic acid in whiskies is acetic acid, which can account for between 50 and 95% of the total content of volatile acids determined by titration. Of the remaining acids, caprylic, capric and lauric are quantitatively the most important (Suomalainen and Nykänen 1970a). Some of the characteristic flavor and aroma of Irish whiskies may be attributed to somewhat higher concentrations of the odoriferous butyric acid. Compared with other types of whisky, Scotch whisky characteristically contains more palmitoleic acid (and its ethyl ester) than palmitic acid. Suomalainen (1971) has suggested that the typical stearin-like smell of Scotch malt whisky may be attributed to long chain fatty acid ethyl esters. It is not surprising to find that ethyl acetate is the major whisky ester in view of the prevalence of acetic acid in the distillate. Concentrations of 95 mg/L have been detected in a blended Scotch whisky (de Becze, 1967). Other esters such as ethyl caprate are present in much lower concentrations, on the order of 2-10 mg/L. Of the carbonyl compounds, acetaldehyde is the principal component together with a range of other short chain aldehydes. Furfural, with an aroma resembling that of grain, also occurs with as much as 20-30 mg/L in Scotch malt whiskies (Valaer, 1940). Acrolein, a pungent and lachrymatory compound, is also present; and it has been suggested that it may contribute to the 'peppery' smell of whisky.

A variety of other organoleptically important compounds has also been detected in different whiskies, many of which result from maturing the whisky in charred oak barrels. Scopoletin and other aromatic aldehydes, including vanillin, were detected in bourbon by Baldwin and his colleagues (1967). These compounds were previously identified by Black *et al.* (1953) in ethanolic extracts of plain and charred American white oak from which the bourbon whiskey barrels subsequently used for the maturation of Scotch and Irish whiskies are constructed. (It should be noted that the US regulations for bourbon whiskey production require that new charred oak barrels be used for maturation; so there is a continuous supply of once-used bourbon barrels that are shipped to Ireland and Scotland for re-use.) A lactone dubbed 'whisky lactone', also appears in whisky following storage in oak barrels. This compound, ß-methyl-octalactone, was first isolated from Scotch whisky by Suomalainen and Nykänen (1970b). It has since been reported that both cis and trans diastereomers of the compound occur in whisky (Nishimura and Masuda, 1971). Other compounds detected in whisky include phenols (Salo *et al.*, 1976), glycerol and erthritol (Black and Andreasen, 1974), pyridine, α-picoline and various pyrazines (Wobben *et al.*, 1971).

Contribution of compounds to organoleptic properties

Published information on compounds responsible for the organoleptic qualities of whiskies is meagre and very largely confined to reports by Suomalainen and his colleagues from the State Alcohol Monopoly in Helsinki, Finland. In order to assess the contributions made by whisky components to the odor of these spirits, Salo *et al.* (1972) concocted a synthetic whisky with components that chromatographic analysis had revealed were present in a light-flavored Scotch whisky. The synthetic whisky was made using 576 g of a mixture of higher alcohols, 90 mg of acids, 129 mg esters and 17.4 mg carbonyl compounds in highly-rectified grain spirit diluted to 34° GL in water that had been ion-exchanged and treated with activated charcoal. This imitation whisky contained 13 alcohols in addition to ethanol, 21 acids, 24 esters and 9 carbonyl compounds. Caramel coloring was used to give it the color of a distilled and matured whisky. Odor thresholds of the individual compounds and groups of compounds were determined as described by Salo (1970). Experienced taste panel participants were easily able to distinguish the imitation whisky from a blended Scotch whisky; but when the concoction was mixed with an equal amount of the Scotch, only 6% correct judgments above chance were made. This suggested that the concentrations of, and interactions between, components of the synthetic whisky were not greatly dissimilar from that in the Scotch used for comparison.

Odor threshold determinations of individual components in the imitation whisky revealed that the contributions made by the mixture of alcohols and acids accounted for only about 10% of the total odor intensity, despite the fact that the alcohols themselves accounted for over 70% of the total concentration of organoleptically-important compounds in the concoction. Esters and carbonyl compounds had a much greater influence, particularly butyraldehyde, isobutyraldehyde, isovaleraldehyde, diacetyl, and the ethyl esters of acetic, caproic, caprylic, capric and lauric acids. Since just three of the most important carbonyl compounds could substitute for the whole carbonyl fraction, there would seem to be considerable homogeneity in the odor contributions made by these compounds. Interestingly, the relative contributions made by the different classes of compounds are not very different from the contributions Harrison (1970) reported that they make towards the taste of beers.

Threshold values can be assessed not only for individual components in whisky, but for the total aroma of the beverage. Salo (1975) examined this by diluting different whiskies with water until the characteristic whisky aroma could only just be recognized. Values for the threshold dilution of several commercial whiskies shown in Table 6 reflect the differences in aroma strength for several different commercial whiskies.

Origin of organoleptically-important compounds

The two main sources of the organoleptically-important compounds in whisky are the yeast used to ferment the wort and the charred oak barrels in which the whisky is matured. Suomalainen and Nykänen (1966) fermented a nitrogen-free medium containing sucrose with a strain of *Saccharomyces cerevisiae* and distilled the fermented medium either after removing the yeast by centrifugation or with the yeast remaining in the medium. Gas chromatographic analyses of these distillates are reproduced in Figure 13, which also shows a chromatogram of Scotch whisky for comparison. There is clearly a similarity among the three analyses; although differences, such as the higher proportion of isoamyl alcohol in the distillate from the fermented medium, can be detected. Also worth noting is the greater concentration of ethyl caprate in the distillate from the spent medium containing yeast as compared with that obtained by distilling the medium from which yeast had been removed. It would be interesting to learn of the importance of yeast strain in production of organoleptically-important compounds in whisky. Unfortunately, there is a lack of published data on this matter.

Table 6. Threshold dilution levels for nine different types of whisky*.

Whisky	*Threshold dilution (x 10^{-4})*	*One standard deviation range (x 10^{-4})*
Scotch malt	0.56	0.2 - 1.3
Scotch, old blend	0.87	0.5 - 1.5
Scotch, blend	1.20	0.3 - 4.2
Irish	1.30	0.4 - 2.0
Bourbon	2.40	0.6 - 11.5
Irish	4.50	2.7 - 7.5
Canadian	10.40	3.0 - 37.0

*Salo, 1975.

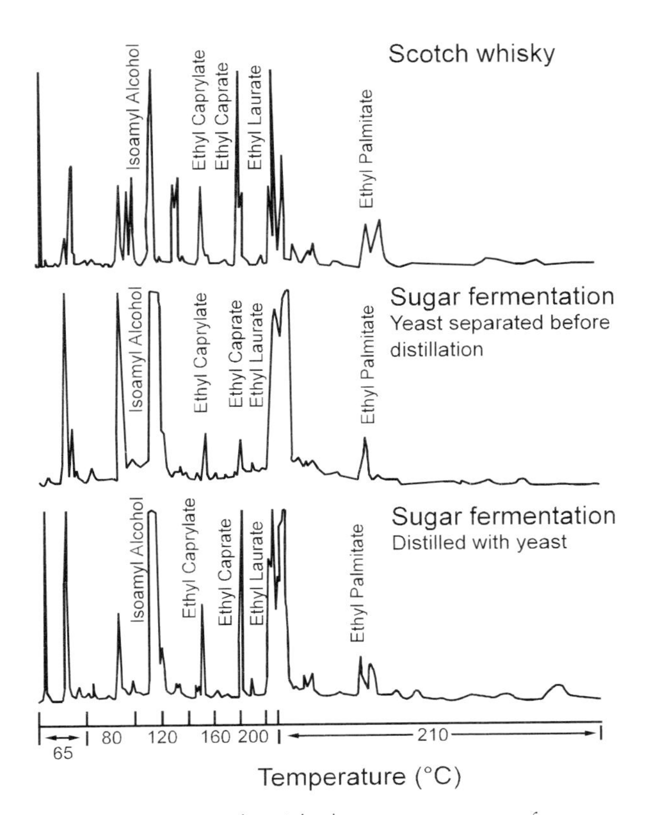

Figure 13. Gas liquid chromatograms of aroma compounds produced by yeast in a nitrogen-free sugar fermentation, with a trace for comparison of the aroma compounds detected in a sample of Scotch whisky (Suomalainen and Nykänen, 1966).

The complexity of the processes for determining the presence and concentration of compounds in whisky may be illustrated by looking at one compound for which this process has been elucidated in detail. In the 1980s ethyl carbamate, a naturally-occurring component of many alcoholic drinks, was identified as being undesirable. Canada specified by regulation a maximum limit for ethyl carbamate in whisky spirit of 150 ppb and the US set guidelines of 120 ppb. Control of this compound was only possible after the processes resulting in its presence in spirit were fully understood. Extensive research in Scotland revealed the mechanism of ethyl carbamate production and facilitated the introduction of very effective control measures. These measures maintain levels close to zero and always less than 30 ppb in distilled whisky spirit.

Cook (1990) reviewed the outcome of this research and the resulting control procedures. When barley sprouts during malting, a glycoside, epiheterodendrin (EPH) is present in the acrospire. This glycoside, which survives kilning and mashing, is extracted into the wort and converted by yeast enzymes to glucose and isobutyraldehyde cyanohydrin (IBAC). The IBAC is stable during fermentation, but when heated above 50°C at distillation it breaks down to form volatile nitriles. There are a multiplicity of potential routes nitriles can follow including reaction with other beer components or complex reactions often mediated indirectly by copper. Some of the volatile nitriles may escape these reactions and pass into the distillate. A number of reactions can remove the nitriles from the spirit, one being the reaction with ethanol to form ethyl carbamate.

Two control strategies may be used to prevent ethyl carbamate reaching the final spirit. Firstly, some varieties of barley produce low levels of the glycoside EPH (Cook, 1990), and plant breeders are now concentrating on incorporating this character in all varieties for the distilling industry. The second strategy relies on the low volatility of ethyl carbamate. Control of distillation conditions eliminates all the volatile precursor formed by copper mediated reactions. Non-volatile substances are formed prior to the final distillation in malt distillation or before rectification in grain distillation. Thus, ethyl carbamate can be minimized in the final spirit.

A considerable number of studies have been published on those organoleptically-important compounds arising either directly or indirectly from the oak barrels in which whisky is matured. The increase in coloring, tannin, dissolved solids and acid concentrations are not observed when whisky is stored in glass, which is proof of the importance of the oak barrels in the maturation process. An analysis of heartwood of the American oak (*Quercus albus*) gave cellulose (49-52%), lignin (31-33%), pentosans (or hemicelluloses, 22%) and compounds extracted with hot water and ether (7-11%; Ritter and Fleck, 1923). However, when charred oak sawdust was directly extracted with water or 96° GL ethanol, the extracts obtained differed markedly in odor from aged whisky. Moreover, none of the various fractions of ethanol-soluble oak extractives contained flavors that resemble mature whisky (Baldwin *et al.*, 1967). As a result, it is now generally held that the maturation process involves not only extraction of compounds from the oak but also chemical modifications of at least some of the compounds extracted from the wood.

For a long time, most of the work reported on this aspect of maturation of whisky came from the laboratories of Joseph E. Seagram and Sons in the US. More recently, accounts of the mechanisms of Scotch whisky maturation have been given by Philip (1989) and Perry (1986) and on the maturation of whisky generally, by Nishimura and Matsuyama (1989). A theme from all this work has been the identification of a number of mechanisms of maturation common to all whiskies. These divide into addition, subtraction and modification by reaction. There is addition of components from the oak wood, including those derived from lignin, tannins and oak lactones. There is the subtraction of volatile compounds from the maturing whisky by evaporation and adsorption on the charred surface of the barrel (Perry, 1986). Lastly there are reaction processes including establishment of equibria among acetaldehyde, ethanol and acetal (Perry, 1986), polymerization reactions (Nishimura and Matsuyama, 1989) and oxidation-reduction reactions (Perry, 1986; Connor *et al.*, 1990). Many of the reactions involve, and are indeed dependent on, the components extracted from the wood.

Looking at several of these reactions in more detail will serve to illustrate the complexity of the maturation process. The work at the Seagram laboratories, whilst focused on bourbon, is directly relevant to Scotch and Irish whiskies for which once-used bourbon barrels are extensively utilized for maturation. Changes in the concentrations of organoleptically-important compounds during a 12 year storage of a 109°

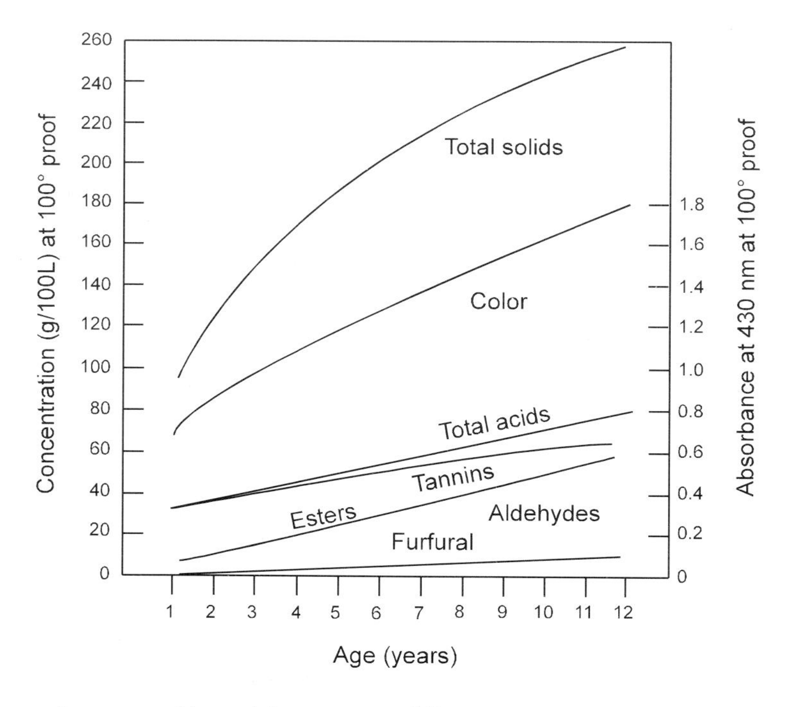

Figure 14. Changes in composition of the vapor at different trays in a Coffey-still rectifier used in the manufacture of Scotch grain whisky (from Pyke, 1965).

US proof (54.5° GL) bourbon, on a 100° proof (50° GL) basis, are shown in Figure 14. The nature and origin of some of these compounds have been examined in some detail. Among the aldehydes, scopoletin and the aromatic aldehydes syringaldehyde, sinapaldehyde, coniferaldehyde and vanillin are important. According to Baldwin *et al.* (1967), these compounds could be formed by ethanol reacting with lignin in the oak wood to produce coniferyl alcohol and sinapic alcohol. Under mildly oxidizing conditions, these alcohols could be converted into coniferaldehyde and sinapaldehyde, respectively. Vanillin could then arise from coniferaldehyde, and syringaldehyde from sinapaldehyde. The increase in aldehyde content during maturation is also attributable in part to formation of acetaldehyde by oxidation of ethanol. Formation of ethyl acetate probably accounts for the steady rise in the ester content of whisky during maturation.

Several other groups of compounds not described in Figure 14 are also important in the maturation process. Monosaccharide sugars are found in mature whisky, and probably arise from the pentosans and other polysaccharides in the oak wood. Otsuka *et al.* (1963) reported that a mature Japanese whisky contained xylose, arabinose, glucose and fructose, while Black and Andreasen (1974) added rhamnose to this list when they analyzed a mature bourbon. The latter workers found that the concentrations of arabinose and glucose increased at a faster rate than those of xylose and rhamnose over a 12 year maturation period. Salo *et al.* (1976) also detected low concentrations of mannose and galactose in a matured Scotch malt whisky in addition to the sugars already noted. The concentrations of sugars in mature whiskies (of the order of 100 mg/L) are too low to suggest any sweetening effect on the beverage. Phenols are also detectable in mature whisky, although some of these probably arise during mashing (Steinke and Paulson, 1964) or from malt produced using peat-fired kilns (MacFarlane, 1968). However, Salo *et al.* (1976) reported an increase during a one year maturation of a Scotch malt whisky in the concentration of eugenol, which is a major phenol extracted from oak chips by ethanol (Suomalainen and Lehtonen, 1976). Also present in mature whiskies are sterols, which may precipitate in bottled whisky stored at room temperature. Black and Andreasen (1973) found campesterol, stigmasterol and sitosterol in mature bourbon, in addition to sitosterol-D-glucoside, although the possibility that some of these were formed during mashing cannot be excluded. Finally, reference has already been made to the whisky lactone, ß-methyl-octalactone, and its origin.

Not surprisingly, the nature and amounts of compounds extracted from charred oak wood depend on the ethanol concentration of the whisky. It is to some extent an advantage to mature whisky at a high proof, since this requires fewer barrels and saves on storage space. Until 1962 the US Treasury Department limited the barrelling proof of whisky to a maximum of 110° US proof (55° GL). In anticipation of this limit being raised to 125° US proof (62.5° GL), Baldwin and Andreasen initiated a series of experiments in 1962 to establish the importance of barrelling proof on changes in color and concentrations of organoleptically-important compounds during maturation of bourbon whiskies. Their report in 1973 indicated that color intensity and congener concentration of whiskies matured for 12 years decreased as the barrelling proof was raised from 109° US proof (54.5° GL) to 155° US proof (77.5° GL). The one exception was the higher alcohol content, which remained approximately constant.

Acknowledgments

The author wishes to thank his friends and colleagues in whisky production in many countries of the world for their cooperation and willingness to supply information on the processes of whisky production. Particular thanks are due to Dr. Gareth Bryan-Jones of United Distillers, Menstrie, Scotland, for his assistance in providing information on recent research reports.

References

Anderson, R.J., G.A. Howard and J.S. Hough. 1972. 'Proceedings of the European Brewing Convention, Vienna. p 253.

Aylott, R.I., A.H. Clyne, A.P. Fox and D.A. Walker. 1994. Analyst 119:1741–1746.

Baldwin, S. and A.A. Andreasen. 1973. Journal of the Association of Official Analytical Chemists 57:940.

Baldwin, S., R.A. Black, A.A. Andreasen and S.L. Adams. 1967. Journal of Agriculture and Food Chemistry 15:381.

Barbour, E.A. 1983. Taxonomy of some lactic acid bacteria from scotch whisky distilleries. PhD Thesis, Heriot-Watt University.

Barnard, A. 1887. The whisky distilleries of the United Kingdom. Proprietors of Harpers Weekly Gazette, London, pp. 457.

de Becze, G.I. 1967. Encyclopedia of Industrial Analysis. Vol. 4:462. Interscience Publications, New York.

Beech, F.W. 1972. Progress in Industrial Microbiology I I:178.

Black, R.A. and A.A. Andreasen. 1973. Journal of the Association of Official Analytical Chemists 56:1357.

Black, R.A. and A.A. Andreasen. 1974. Journal of the Association of Official Analytical Chemists 57:111.

Black, R.A., A.A. Rosen. and S.L. Adams. 1953. Journal of the American Chemical Society 75:5344.

Bober, A. and S.W. Haddaway. 1963. Journal of Gas Chromatography 1 12: 8.

Brander, M. 1975. A Guide to Scotch Whisky. Johnson and Bacon, Edinburgh. p. 96.

Bryan-Jones, G. 1976. Lactic Acid Bacteria in Beverages and Foods. Proceedings of the Fourth Long Ashton Symposium. (J.G. Carr, C.V. Cutting and G.C. Whiting, eds). Academic Press, London. pp. 165-175.

Cameron Taylor, I.B. 1970. Whisky Export, 2nd issue, June-Sept. Aldiffe Publishing Co. Ltd., Wilmslow, Cheshire.

Carter-Tijmstra, J. 1986. Proceedings of Second Aviemore Conference on Malting, Brewing and Distilling. pp. 413–416.

Connor, J.M., A. Paterson and J.R. Piggot. 1990. Proceedings of Third Aveimore Conference on Malting, Brewng and Distilling. pp. 460–463.

Cook, R. 1990. Proceedings of Third Aveimore Conference on Malting, Brewing and Distilling. pp. 237–243.

Court, R.E. and V.H. Bowers. 1970. Process Biochemistry 5:17.

Cromwell, E.A. and J.F. Guymon. 1963. American Journal of Enology and Viniculture 12:214.

Daiches, D. 1969. Scotch whisky, its past and present. Andre Deutsch, London. p. 168.

Dolan, T.C.S. 1976. Journal of the Institute of Brewing. 82:177.

Duncan, R.E.B. and J.M. Philp. 1966. Journal of the Science of Food and Agriculture 17:208.

Dunnett, A. 1953. The land of scotch. Scotch Whisky Association, Edinburgh. p. 170.

Geddes, P.A. and H.L. Riffkin. 1989. Distilled Beverage Flavour (ed. Piggot, J.R. Ellis, Horwood, London. pp. 193-199.

Griffin, O.T. 1972. Process Biochemistry 7:17.

Gunn, N.M. 1935. Whisky and Scotland. Routledge, London. p 198.

Harrison, G.A.F. 1970. Journal of the Institute of Brewing 76:486.

Harrison J.S. and J.C.J. Graham. 1970. In: The yeasts. (A.H. Rose and J.S. Harrison, eds.). Vol. 3:283-348. Academic Press, London.

Heukelekian, H. and A. Heller. 1940. Journal of Bacteriology 40:547.

Hopkins, R.H. and S. Wiener. 1955. Journal of the Institute of Brewing 61:493.

Hough, J.S., D.F. Briggs and R. Stevens. 1971. Malting and brewing science. Chapman and Hall Ltd., London. p. 678.

Institute of Brewing Analysis Committee, Recommended methods of analysis. 1975. Journal of the Institute of Brewing 81:368.

Jacobs, M.B. 1947. American Perfumery and Essential Oil Review. p 157.

Kahn, J.H. 1969. Journal of the Association of Official Analytical Chemists 52: 1166.

Kahn, J.H., P.A. Shipley, E.G. Laroe and H.A. Conner. 1969. Journal of Food Science 34:587.

Kleber, von W. and N. Hurns. 1972. Brauwissenschaft 25: 98.
Kulka, D. 1953. Journal of the Institute of Brewing 59: 285.
Liebmann, A.J. and B. Scherl. 1949. Industrial and Engineering Chemist 41:534.
Lloyd Hind, H. 1948. Brewing Science and Practice, Vol. 1:505. Chapman and Hall, London.
Lyons, T.P. 1974. The Brewer. p 634.
McDowell, R.J.S. 1975. The Whiskies of Scotland. 3rd edit. John and Murray, London. p. 166.
MacFarlane, C. 1968. Journal of the Institute of Brewing 74: 272.
MacFarlane, C., J.B. Lec and M.B Evans. 1973. Journal of the Institute of Brewing 79:202.
McGuire, E.B. 1973. Irish Whiskey. Gill and MacMillan, Dublin. p. 462.
MacKenzie, K.G. and M.C. Kenny. 1965. Journal of the Institute of Brewing 71:160.
Mann, E.A. 1911. Government Analyst for Western Australia. Perth, W. Australia. pp. 1-12.
Merritt, N.R. 1967. Journal of the Institute of Brewing 73:484.
Morrison, R.L. 1962. American Journal of Enology and Viniculture 13:159.
Murtagh, J.E. 1970. M.Sc. Thesis: University College of North Wales, Bangor.
Nettleton, J.A. 1913. The manufacture of whisky and plain spirit. Cornwall and Sons, Aberdeen. p. 616.
Nishimura, K. and M. Masuda. 1971. Journal of Food Science 36:819.
Nishimura, K and R. Matsuyama. 1989. The Science and Technology of Whiskies. (J.R. Piggott, R. Sharp and R.E.B. Duncan, eds.). Longman Scientific and Technical, Harlow. pp. 235-263.
Otsuka, K. and K. Morinaga and S. Imai. 1963. Nippon Jozo Kyokai Zasshi 59:448.
Perry, D.R 1986 Proceedings of Second Aviemore Conference on Malting, Brewing and Distilling. pp. 409-412.
Pert, D. and E. Woolfendon. 1986. Proceedings of Second Aviemore Conference on Malting, Brewing and Distilling. pp. 417-424.
Philip, J.M. 1989. The Science and Technology of Whiskies. (J.R. Piggott, R. Sharp and R.E.B. Duncan, eds). Longman Scientific and Technical, Harlow. pp. 264-294.
Preece, I.A. 1947. Journal of the Institute of Brewing 53:154.
Preece, I.A. 1948. Journal of the Institute of Brewing 54:141.
Preece, I.A. and M. Shadaksharaswamy. 1949a. Journal of the Institute of Brewing 55:298.
Preece, I.A. and M. Shadaksharaswamy. 1949b. Biochemical Journal 44:270.
Preece, I.A. and M. Shadaksharaswamy. 1949c. Journal of the Institute of Brewing 55:373.
Pyke, M. 1965. Journal of the Institute of Brewing 71:209.
Rae, I.L. 1967. Process Biochemistry I, 8:407.
Ritter, G.J. and L.C. Fleck. 1923. Industrial and Engineering Chemistry 15:1055.
Robb, J.M. 1950. Scotch Whisky. W. and R. Chambers, London. p 197.
Ross, J. 1970. Whisky. Routledge and Kegan Paul, London. p. 158.
Royal Commission on Whisky and Other Potable Spirits. 1909. Appendix Q, p. 229.
Salo, P. 1970. Journal of Food Science 35:95.
Salo, P. 1975. Proceedings of the International Symposium on Aroma Research. Central Institute for Nutrition and Food Research, TNO, Zeist, Netherlands. p 121.
Salo, P., L. Nykänen and H. Suomalainen. 1972. Journal of Food Science 37:394.
Salo, P., M. Lehtonen and H. Suomalainen. 1976. Proceedings of the Fourth Symposium on Sensory Properties of Foods. Skövde, Sweden.
Schidrowitz, P. 1902. Journal of the Chemical Society. p 814.
Schidrowitz, P. and F. Kaye. 1905. Journal of the Chemical Society. p. 585.
Scotch Whisky Order. 1990. Stationery Office, London.
Sillett, S.W. 1965. Illicit Scotch. Beaver Books, Aberdeen. p. 121.
Simpson, A.C. 1968. Process Biochemistry 3, 1:9.
Simpson, B., A. Troon, S.T. Grant, H. MacDiarmid, D. MacKinlay, J. House and R. Fitzgibbon. 1974. Scotch Whisky. MacMillan, London. p. 120.

Singer, D.D. and J.W. Stiles. 1965. Analyst 90:290.

Stark, W.H. 1954. Industrial Fermentations (L.A. Underkofler and R.J. Hickey, eds). Chemical Publishing Company, New York. Vol. 1, pp. 17-72.

Steinke, R.D. and M.C. Paulson. 1964. Journal of Agricultural and Food Chemistry 12: 381.

Suomalainen, H. 1971. Journal of the Institute of Brewing 77:164.

Suomalainen, H. and M. Lehtonen 1976. Kemia-Kemi 3, 2:69.

Suomalainen, H. and L. Nykänen. 1966. Journal of the Institute of Brewing 72:469.

Suomalainen, H. and L. Nykänen. 1970a. Naeringsmidelidustrien 23:15.

Suomalainen, H. and L. Nykänen. 1970b. Process Biochemistry. July 1.

The Scotch Whisky Order. 1990. Statutory Instruments No 998. HM Stationery Office, London.

Thorne, C.B., R.L. Emerson, W.J. Olsen and W.H. Paterson. 1945. Industrial and Engineering Chemistry 37:1142.

Valaer, P. 1940. Industrial and Engineering Chemistry 32:935.

Valaer, P. and W.H. Frazier. 1936. Industrial and Engineering Chemistry 28:92.

Walker, J.W., D.M. Vaughan and H.L. Riffkin. 1990. Proceedings of Third Aveimore Conference on Malting, Brewing and Distilling. pp. 421-426.

Warricker, L.A. 1960. Journal of the Science of Food and Agriculture 11:709.

Wilson, J. 1973. Scotland's malt whiskies, a dram by dram guide. Famedram Ltd., Gartochan, Dunbartonshire, Scotland. p. 109.

Wilson, J. 1975. Scotland's distilleries. a visitor's guide. Famedram Ltd., Gartochan, Dunbartonshire, Scotland. p. 104.

Wilson, Ross. 1959. Scotch made easy. Hutchinson, Ltd., London. p 336.

Wilson, Ross. 1970. Scotch: the formative years. Constable, London. p 502.

Wilson, Ross. 1973. Scotch, its history and romance. David and Charles, Newton Abbot, Devon, England. p 184.

Wobben, H. J., R. Timmer, T. ter Heide and P. de Valois, P. 1971. Journal of Food Science 35:464.

Zobell, C.E. 1943. Journal of Bacteriology 46:39.

Chapter 12

Production of Canadian rye whisky: the whisky of the prairies

J.A. Morrison
Distillery Scientist, Calgary, Alberta, Canada

Introduction

The establishment of rye grain distillery operations east of the Rocky Mountain foothills in Canada in the late 1940s was an attempt to utilize a potentially abundant and relatively cheap source of starch in the form of rye grain for the production of beverage spirits. The fertile sandy soil existing in many parts of the prairies was perfect for the fall planting of rye varieties such as Cougar, Musketeer, and Kodiak. The September planting and subsequent July/August harvesting of rye meshed well with the farmer's timetable; and with the establishment of distilleries at certain main centers, rye became a welcomed cash crop for many Western Canadian farmers.

The decision by the distilleries to use rye grain as the main starch source at that early date flew in the face of a considerable body of experimental knowledge about rye grain that promised high-viscosity mashing, foaming fermenters and poor quality distillers dried grains. Most of this experimental knowledge emanated from American corn grain distillers using small amounts of rye grain mainly for the production of rye whisky flavoring used for blending. To these people, the problems of mashing rye grain were not worth solving and were handled by adjusting corn systems to inefficient operation levels.

This chapter presents the essential features of the process and quality control procedures developed over the years to understand, control and exploit the special attributes of rye grain in the manufacture of beverage spirits. This chapter includes flow charts and descriptions of the processes developed for rye grain spirits manufacture in the areas of grain handling, batch mashing and fermentation, dual-column continuous distillation and a multiple-effect-evaporator distillers dried-grain process.

Rye grain reception and quality control

The supply of grain to a distillery, be it rye or any other grain, will involve several stages of activity. They are 1) submission of a sample for quality evaluation, 2) grain reception and quality checking, 3) initial storage, 4) cleaning, 5) clean grain storage, 6) milling, and 7) meal storage.

Grain sample submission procedure

The rye grain used by distilleries in Western Canada is mainly of the Cougar or Musketeer variety and is supplied mainly from the 'rye crescent' of Alberta, located 50 to 70 miles (80 to 112 kilometres) south of Calgary, Alberta. In years of poor harvest, additional grain is usually available from the more eastern provinces of Saskatchewan and Southern Manitoba. Bushel weight, sprouts, appearance and dockage are combined with analytical data such as starch (reducing power removed by fermentation), protein (Kjeldahl), moisture (oven-dry) and sieve analysis (Tyler or US Standard); and the results are used by the grain department in their selection of the best grain inventories available.

It is important to note that the grain department's job is to select the best rye available and not just to make purchases according to a rigid preset specification similar to that used for corn purchases. While it is preferable to have minimum quality acceptance levels in mind, the conditions of cultivation and harvest in the region can vary greatly from year to year, causing wide swings in starch and protein content and in the degree of field sprouting.

Following selection of a particular supply of rye, a timetable for receiving it at the plant is established based on the current inventory and anticipated requirements. In a situation where grain is delivered by truck, and where weather can cause delays, an acceptable minimum inventory would be about a two week supply. Thus, in a distillery mashing 8,000 to 8,200 bushels per day (203-208 tonnes) on a seven-day-week basis, the inventory on hand at any time should be about 115,000 bushels (3,000 tonnes).

Grain reception quality checks

Upon arrival of a rye grain shipment, immediate on-site checks include dockage (trash), moisture (conductivity), ergot (caused by the organism *Claviceps purpurea*), sprouts and bushel weight. These features are graded against data received from the sample submitted earlier to determine whether the quality expected is present in the delivered grain. As the grain makes its way through the system and approaches the milling stage, samples are taken for sieve analysis, moisture, starch levels, protein and cleanliness. These data affect hammer-mill performance, expected alcohol yield and quality, dried grain production and plant microbiology.

The presence of significant levels of ergot (>0.5%) in rye can cause a secondary problem in distillery operations by making the stillage unfit for animal feed production. The Claviceps fungus causes the rye spikelets to develop hard, slightly bent, blue-violet crescents instead of normal grains. The fungus crescents contain alkaloids that are toxic to animals and humans alike, and must not be allowed to pass through the distillery process (Starzycki, 1976). The blue-violet color of the ergot bodies (sclerotia) make it easy to spot contaminated grain and can be used as a guide in the rejection of rye shipments.

A process flow sheet for receiving and quality monitoring of rye grain up to the grain meal storage stage is shown in Figure 1. Five quality stations are recommended with appropriate physical and chemical checks needed for good manufacturing practice.

Previous distillery experience has shown a correlation between the degree of whole grain cleanliness and development of the undesirable congener acrolein and its intermediates in the fermenters, which may result in a multitude of reactions with normal fermentation congeners during the distillation stage. Rye distillers have noted that the frequency of acrolein formation tends to increase toward the end of a production year when silos and grain-handling vessels are cleaned of dust and grain debris. The grain dust was allowed to join the whole grain entering the cleaning cycle and it is likely that extraordinary amounts of the dust by-passed the screening to some degree and entered the production cycle. This condition at times led to higher delta acid titers in the finished fermenters followed by the production of spirits with an exceptionally irritating odor.

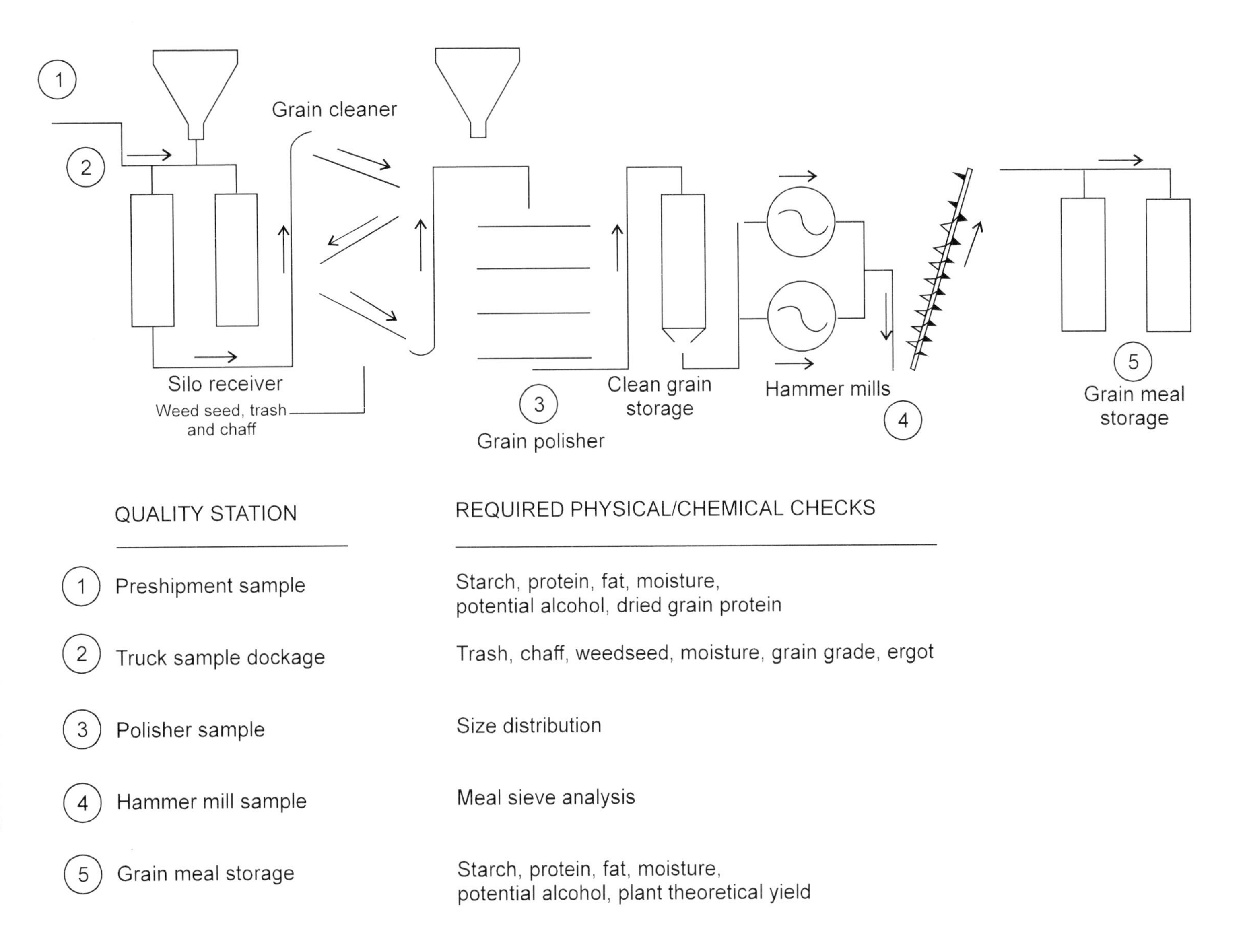

Figure 1. Receiving and monitoring rye grain (Morrison, 1992).

Acrolein is an acrid, lachrymatory compound formed when yeast glycerol is metabolized to ß-hydroxypropionaldehyde by bacteria. It is this compound that is degraded to form the acrolein molecule in the rectification process. Acrolein has been accused of causing the phenomenon of peppery spirits, which may give considerable discomfort to operators. However, acrolein's propensity to react with other congeners in the spirits to form stable complexes places some doubt on whether acrolein *per se* is the real culprit (Kahn *et al.*, 1969).

Physical and chemical laboratory data for good quality western Canadian rye are presented in Table 1.

Depending upon rate of delivery, the incoming grain may be bucket-elevated into uncleaned grain silos for later cleaning or it may go directly to standard trash and weed seed separators and then on to a Crippen-style polisher for final dust and chaff removal. The polished rye is then stored in silos to feed the Praterstyle 100 HP hammer mills with selected screens that convert the grain to meal. The screens have a variety of sizes of geometrically shaped holes and are capable of producing several spectra of meal. Meal samples are monitored for final moisture and sieve analysis and the milling is adjusted accordingly.

The laboratory procedure for meal sieve analysis involves placing a sample on a set of US

Table 1. Physical and chemical data for western Canadian rye grain*.

Physical analysis					
Grain size distribution (sieve analysis)					
Tyler screen No.	6	7	8	12	pan
Mesh opening, mm	3.35	2.80	2.38	1.68	
Grain found, %	0.10	15.70	40.30	41.80	2.1

Chemical analysis (dry basis)			
Amino acid content, g/100 g protein			
Glutamic acid	30.18	Glycine	3.98
Proline	11.40	Threonine	3.67
Aspartic acid	7.09	Ammonia	3.19
Leucine	6.56	Lysine	3.18
Valine	5.37	Cystine/Cysteine	2.51
Phenylalanine	4.91	Histidine	2.30
Serine	4.70	Tyrosine	2.07
Arginine	4.60	Tryptophan	1.33
Alanine	4.07	Methionine	1.28
Isoleucine	4.00		

Kjeldahl protein (N x 6.25), %	10.29
Starch (reducing power removal by yeast), %	64.05
Nitrogen-free extract (total carbohydrate), %	80.10
Lipids by ether extract, %	1.80
Ash content, %	2.10

Mineral content, mg/100 g			
Phosphorous	380	Potassium	520
Iron	9	Manganese	7.5
Calcium	70	Magnesium	130
Copper	0.9		

*Unpublished distillery data; Simmonds *et al.* (1976).

Standard or Tyler screens of varying mesh sizes that vibrate on a shaker for a determined time. Meal remaining on each screen is weighed and percentages calculated. The sample of meal tested should be a composite of samples taken over a period of time sufficient for all run conditions to be represented. On a set of Tyler screens with mesh ratings of 10, 20, 30, 40, 60, 80 and 100, it is recommended for rye that a total of 65% of the meal should rest on screens #30 to #60 and less than 10% should pass through screen #100 as flour. A typical sieve distribution analysis for rye meal is presented in Table 2.

Table 2. Typical meal sieve analysis for rye meal.

Tyler screen #	*Mesh (mm)*	*Meal (%)*
9	1.68	1
20	0.841	10
30	0.595	28
60	0.247	26
80	0.177	15
100	0.15	12
Through 100	-	8

Rye meal sieve analysis

Meal size distribution (sieve analysis) will affect cooker performance in two ways, each creating opposite effects. The fermentable extract obtained during the mashing process will be maximized when the grain is ground as fine as possible thus creating the maximum surface area for contact with the cooking liquid. However, excessively fine meal will create problems in equipment such as batch cookers when large quantities of the powdered meal are added rapidly to the liquid phase of backset stillage and/or water. In the manner of most dry meal materials, the rye meal can form dough balls when a thin coat of wetted mash forms a very impervious covering around a quantity of dry meal. The tendency to form dough balls is temperature-sensitive and will become significant at addition temperatures exceeding 55°C. This encapsulated meal will not be wetted and the starch will not be accessible to gelatinization processes and subsequent hydrolysis to lower molecular weight sugars.

Distillery experience has indicated that starch loss may exceed 5% when extensive dough ball formation occurs. Use of excessively fine meals in batch cookers can, however, be tolerated if mechanical distributors are employed in the meal addition system. The distributors rotate with the cooker agitator and add the meal in a spray pattern across the entire liquid surface.

In continuous mashing systems the extent of dough ball formation will depend on the degree of homogeneity of mixing in the slurry tanks and by the transfer pumps. The use of in-line mixers, which move the slurry through a high-sheer rotor pump, will minimize the presence of the yield-reducing dough balls.

Detection of dough ball formation in cooker systems becomes somewhat difficult in that the balls may not survive the pumping system and thus may escape visual detection. The application of dilute iodine solution to a small amount of mash may indicate the presence of unhydrolyzed starch by producing an unusually pronounced blue coloration. A positive test result will be useful only if extensive experience of normal expectations is at hand against which the new test can be measured. The unusual-blue iodine coloration test should provide the first reason to suspect dough ball formation, especially if there is also an unexplained alcohol yield loss.

The next step in the investigation would be to alter the meal sieve analysis to a lower flour and fines content by changing hammer mill screens. If there is any control on the meal addition rate this control should be adjusted to slow the addition, and thus reduce the ball formation potential. If the meal addition temperature exceeds 55°C, then a stepwise reduction to 50°C should be tried over a sufficient period of time for proper evaluation. Each step of the change should be monitored using the dilute iodine test and by recording the alcohol yield.

Finally, the daily sieve analysis of the meal can be used for determining maintenance needs on hammers or screens. A sudden change in the meal grind would suggest a need to inspect screens for punctured areas or to inspect the hammer points for sharpness and mechanical integrity. Screen-tear problems may show up as quantities of whole grain on the upper screens. The hammer maintenance problems usually manifest themselves in higher percentages of meal on the upper screens.

Rye mashing and fermentation

The hydration and gelatinization of rye starch

The mashing or cooking of rye meal, like that of most other grains, is for the primary purpose of solubilizing carbohydrate components so they can be hydrolyzed to fermentable materials. The solubilization is a two-fold mechanism of hydration and gelatinization. The temperature range in which the processes take place most efficiently is specific for each grain starch used in a distillery (Table 3). Rye grain starch hydrates/ gelatinizes over a range of 57 to 70°C (MacLeod, 1977). Carbohydrate-hydrolyzing agents such as amylolytic enzymes will act very slowly, if at all, on starch that has not undergone hydration and gelatinization.

The gelatinization temperature range for rye starch is determined microscopically by observing the temperatures at which birefringence or double refraction of the initially crystalline starch granules begins to degrade and is then totally lost. Birefringence refers to the pattern of striations, or dark areas on starch granules, that appear when unheated crystalline starch solutions are viewed through a microscope using plane-polarized light. The birefringence pattern, in combination with the granular shape, is specific for each starch and can be used most effectively for the purpose of identifying starch origins. As the starch granule begins the process of gelatinization the granule swells, loses its crystal form and the birefringence gradually disappear.

Rye meal begins the hydration process immediately on dispersion in water with the exposed starch granules in the flour and fines chemically binding with the water molecules as the first step in solubilization of the starch. This means that the hydration of starch will be facilitated by the proper sieve analysis, which in turn will determine the dispersion efficiency of the particles.

Simultaneous to the chemical hydration process taking place, the physical phenomenon of gelatinization of the starch will begin. As the temperature of a grain meal slurry is raised to a critical level, the starch granules start to lose their crystalline structure. As water makes its way between the tightly-packed, hydrogen-bonded layers, granules swell and eventually the starch moieties disperse into the liquid phase. This swelling or gelatinizing of the granule and the subsequent starch dissipation will result in a substantial increase in the viscosity of the mash. If suitable liquefaction enzymes are not present, the viscosity will overload most mechanical mixing systems.

Table 3. Gelatinization temperature ranges of grain starches.

Grain starches	°Celsius	°Fahrenheit
Barley (Hordeum)	52-59	126-138
Wheat (*Triciticum aestivum*)	58-64	136-147
Rye (*Secale cereale*)	57-70	135-158
Maize/corn (*Zea mays*)	62-72	144-162
Rice (*Oryza sativa*)	68-77	154-171
Sorghum (*Sorghum vulgare*)	68-77	154-171

Rye starch structure

Rye starch, like most starches, consists of two main carbohydrate polymers called amylose and amylopectin (Figure 2). The basic unit in both starch forms is the monosaccharide D-glucose; and it is the method by which the units are joined that results in the two distinct entities. Amylose consists of unbranched chains of up to a thousand units of D-glucose (molecular weights up to one million) in which each glucose molecule is joined to the next by an oxygen bond between the No. 1 carbon atom of one unit and the No. 4 carbon atom of the next to form an α-1,4 D-glucosidic link. Amylopectin starch contains the same linear chains of D-glucose units but the chains are much shorter, consisting of up to two dozen glucose units joined by α-1,6 D-glucosidic linkages to give it a highly-branched structure with molecular weights up to six million. The ratio of amylopectin to amylose in rye starch is usually reported to be in the order of 4:1(Headon, 1991).

Owing to the differences in physical form, the two types of starch have numerous distinctive physical properties that have been described extensively in the literature (Noller, 1958; Whistler *et al.*, 1953). Of importance to the scope of this chapter is the enzymatic degradation or hydrolysis of the two starch types by α- and ß-amylases.

Sanitization of rye mash

The gelatinization temperature of rye grain starch is always exceeded in distillery cooking practice because the second purpose of mashing is sanitization of the mash as a means of microbial control during the fermentation stage. The higher mashing temperatures allow shorter exposure times to be effective for pasteurization and hence a more efficient mashing process is possible. A balance must be struck between using temperatures high enough to effect an efficient sanitization without detroying sensitive carbohydrates. The indigenous enzyme activity in rye meal may under certain cooker conditions produce significant quantities of low molecular weight sugars early in the mashing process. These will be susceptible to some degree of caramelization when exposed to high temperatures and will be lost as fermentables.

Similarly, the starch granules in the flour and fines of the rye grain meal may undergo irreversible damage from high temperatures,

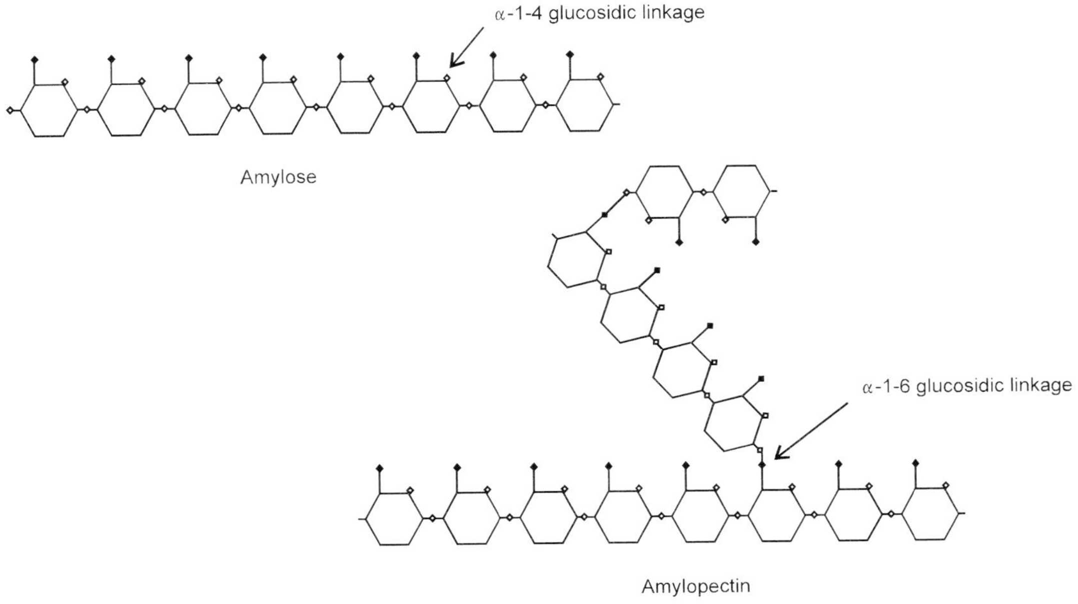

α-1-4 glucosidic linkage

Amylose

α-1-6 glucosidic linkage

Amylopectin

Figure 2. Structure of amylose and amylopectin starch polymers (Morrison, 1992).

especially if gelatinization of the granules has not taken place due to dough balling. The damage may be in the form of granular distortions that prevent gelatinization. Hence the starch cannot be hydrolyzed to fermentable sugar.

Hydrolysis of rye starch using malt enzymes

The third purpose of mashing is to hydrolyze large carbohydrate polymers to intermediate-sized dextrins and/or fermentable sugars through the presence or addition of physiologically active proteins called enzymes. This physiological activity is indispensable to fermentation by yeast, because yeast are incapable of converting the large carbohydrate polymers directly to ethanol. The extent of hydrolysis and the type of fermentable carbohydrate formed will depend on the enzyme source (e.g. malt (essentially sprouted barley), commercial microbial enzyme products or indigenous enzymes) and on the cooking conditions.

The use of malt as an enzyme source in the cooker supplies a mixture of liquefying and saccharifying enzymes and will yield a mash with a high content of fermentable sugars and a low content of dextrins, which is ready for fermentation. When malt is used with rye meal, the indigenous enzyme activity is sufficient to significantly alter the sugar:dextrin ratio and affect both the mashing and fermentation processes. Further dextrin breakdown would take place, albeit under less than ideal conditions, in the fermenter (called secondary conversion) as the yeast deplete the fermentable carbohydrate supply.

The main physiological activity of malt meal is caused by the presence of two amylases (starch-splitting enzymes), α-amylase and ß-amylase. In the presence of gelatinized (cooked) starch, each enzyme carries out the hydrolysis in ways both specific and collaborative. The α-amylase is called a liquefying (or dextrinizing) enzyme because of its ability to break down the large starch molecules into smaller, more soluble dextrins (short chains of glucose units) and thereby to reduce the viscosity of gelatinized-starch solutions. The ß-amylase is called a saccharifying (sugar-producing) enzyme because of its ability to hydrolyze starch solutions that have undergone gelatinization to yield the fermentable sugar maltose (a disaccharide, ie., composed of two glucose units). Both α- and ß-amylases in malt meal are able to cleave α-1,4 D-glucosidic linkages in both amylose and amylopectin starch, but do so in a different manner, to a different extent and to different sugar spectra. The α-amylase hydrolyzes the α-1,4 linkages in the amylose and amylopectin molecules in a random fashion to produce a carbohydrate spectrum consisting of glucose, maltose and various short-chain molecules known as dextrins (oligosaccharides), including some containing the α-1,6 D-glucosidic bonds which the α-enzyme cannot split. These amylopectin fragments have been given the name 'limit dextrin' as they cannot be reduced in size any further without the use of a α-1,6 linkage-breaking enzyme.

The size of the initial limit dextrin from amylopectin will determine how close to the α-1,6-branch linkages the enzyme can carry out the hydrolysis. The ultimate limit dextrin would be the trisaccharide panose with three units of glucose joined with α-1,4 and α-1,6 linkages. Of all the α-amylases available to the distiller, the α-amylase from malt is best able to approach the ultimate limit with the production of the panose (Whelan, 1964).

The ß-amylase, starting from the non-reducing end, hydrolyzes every second α-1,4 D-glucosidic linkage in amylose and in the straight chain portions of amylopectin up to the branch point containing the α-1,6 linkage. Hence, the carbohydrate spectrum resulting from this activity will consist of maltose and a relatively high molecular weight limit dextrin. This limit dextrin will succumb to attack by the accompanying α-amylase enzyme and will then be reduced to a true limit dextrin by the ß-amylase.

Hydrolysis of rye starch using microbial amylases

With the advent of thermostable bacterial α-amylases (as from *Bacillus amyloliquefaciens* and *Bacillus licheniformis*), the ability to 'custom hydrolyze' in the mash cooker has become a reality. Whereas the malt enzymes produce a carbohydrate spectrum with the disaccharide maltose as the major component, bacterial α-amylases tend preferentially to hydrolyze the starch molecule to maltohexose (with six glucose units) and other, larger dextrins. The higher molecular weight carbohydrates are less susceptible to heat damage (caramelization), less likely to retrograde to unusable forms and will not support bacterial growth to the same extent as maltose. Additionally, the bacterial amylases remain active for longer periods during the mashing process and may even survive into the fermenter. The conditions in the fermenter will be less than optimal for the bacterial amylases, but over a fermentation period of three days there is reason to believe that their contribution could be significant. In the fermenter the surviving amylases may be joined by the saccharifying fungal enzyme 'glucoamylase' (amyloglucosidase), which may be produced in the plant in mash form, or purchased as a commercial preparation. It will finish the conversion of the dextrins to fermentable sugars by hydrolyzing individual glucose units.

Rye distillery mashing procedure

At a typical distillery in western Canada rye grain meal is mashed in vertical batch cookers of 11,500 US gallons (43,600 liters) equipped with two-speed 60 HP agitators, nine steam-injection spargers and internal and external cooling coils (Figure 3). The steam spargers, internal coils and the agitators work together during the mashing cycle to provide the necessary degree of mixing for rye mashes. Modernized distilleries now initiate and control the process steps for mashing and transfer to fermenters from the main console of a microprocessor-style computer such as the Honeywell 2000.

Rye grain distilleries using batch cooking conduct the mashing in the following manner. Sufficient city water/recycled steam condensate is mixed with recycled backset (thin stillage) in a clean cooker to create a starting temperature of 30-35°C, a pH of 5.8-6.0 and a calcium concentration of 75 ppm. On full agitation the cooker receives 22,400 lbs (400 bushels = 10,160 tonnes) of rye meal from a microprocessor-controlled weighbelt or dropbin and the temperature is raised to 65°C by live steam. The microprocessor programs the addition of a thermostable α-amylase liquefying enzyme (such as Allcoholase I) at a rate of 0.05% of the grain weight and allows a hold of 10-30 minutes as required. The length of hold at this point will be determined by whether the viscosity of the mash is causing any impediment to mixing. The viscosity of the mash at this stage will depend on the level of intrinsic enzymes in the rye capable of extensive hydrolysis of the large starch, protein and gum molecules in the mash, if given sufficient time. If intrinsic enzymes are in short supply, quantities of a ß-glucanase enzyme (such as Allzyme) may be added at 65°C and a 30 minute hold put into effect. This is done to lower the molecular weight of the glucan gums (polymers of glucose with ß-type linkages) to at least the five glucose unit size.

Steam injection continues until the mash temperature is 85°C and then a final gelatinization hold of 10 minutes is allowed. After cooling to 78°C, a further addition of liquefying α-amylase enzyme is made at 0.05% of the grain weight. A 10-minute hold is taken and the mash is cooled to 70°C. It is then pumped to a holding vessel for transfer through a cooler to the fermenters.

If the production schedule permits a longer mashing period, a final gelatinization hold at 80°C for 20 minutes would suffice (instead of 85°C for 10 minutes) with a net saving of steam and possibly a decrease in caramelization of low molecular weight sugars. It is certainly worthwhile to stay flexible and to experiment with different mashing techniques now that energy is the major cost item.

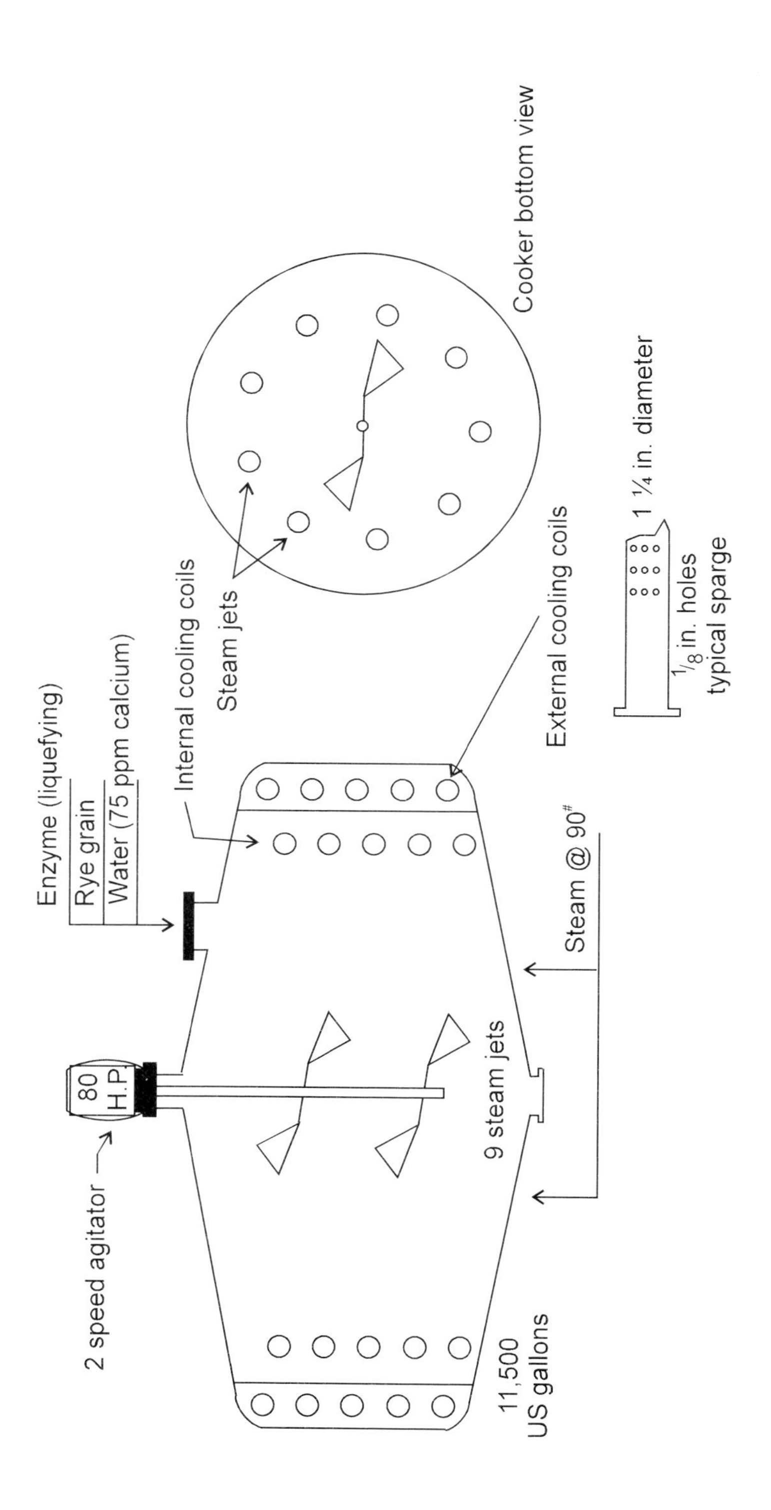

Figure 3. Vertical batch cooker for rye meal (Morrison, 1992).

It is important that the holding vessel be of sufficient capacity to be used for emergency storage of mash in case of maintenance problems or process delays that prevent the immediate transfer of mash to the fermenters. A surge volume in the holding vessel of at least twice the normal mash volume would be a good investment in plant flexibility. It is essential that the holding vessel be equipped with good agitation and a means of maintaining the mash temperature at 63-65°C.

The agitation and high temperature will maintain the mash in a pasteurized condition until delivery to the fermenter system takes place. Failure to maintain good microbial control at this stage may result in significant lactobacillus contamination with the loss of sugar to lactic acid production. It is also considered good manufacturing practice to have some steam sparged into the holding vessel when empty as a further retardant to bacterial growth. The final cooked mash analyzes at 21-21.5% solids, which calculates to about 264 US gallons liquid dilution per 56 lb distillers bushel (386 liters liquid dilution per 100 kg) of rye meal. Variations in the percent solids should be expected if cooker start temperatures, heat load in the steam and the proportions of hot backset and city water undergo variations from day to day.

Rye fermentation procedure

The fermentation of rye mash in western Canada is typically carried out in a series of 45,600 US gallon (172,000 liters) fermenters equipped with internal horizontal cooling coils and side-mounted Lightnin mixers. Most fermenters are insulated and use live steam injection as a means of raising the internal temperature to pasteurization levels as part of the cleaning cycle. Temperature probes and cooling water flow recorder controllers complete the equipment package on these vessels. In some older plants the fermenters are open-topped, do not have pasteurization capabilities, and are fitted with a platform of internal cooling coils horizontally positioned halfway up the wall.

Into a clean fermenter (and thoroughly clean and constantly-steamed lines) sufficient backset stillage (usually 25% of total fermenter dilution) and wash water are combined with three or four cooker mashes (depending on fermenter size) and a 1.88% by volume yeast mash. The mash and backset are cooled to 35-38°C by passing through banks of counter-flow pipe coolers. For open-top fermenters with less efficient cleaning, the backset may be acidified as a bacterial control measure. Additionally, the stillage-acid mixture may be used at high temperatures for continuous circulation through coolers and mash lines as a means of bacterial control when production delays occur. The combination of the heat, acid and grain solids in the mixture has an excellent scouring action on the cooler and mash lines.

The yeast mash addition can be the liquid yeast mash produced as described later in this chapter, or it can be dried distillers yeast either in dry powder form or after reconstitution in a dilute mash for several hours. The equivalent of 50 lbs (22.6 kg) of the powdered yeast should be added to each fermenter. A yeast cell count of 250 million cells/ml is recommended for fermenter treatment for efficient conversion of the sugar to alcohol. The liquid dilution of the prepared yeast mash slurry is on the order of 70:1, so the fermenter should have an initial viable yeast count of approximately 3 million cells/ml.

For fermenters 'set' (filled) with mash that has been liquefied with commercial thermolabile α-amylases only, a quantity of amyloglucosidase enzyme must be added. The enzyme can be added as either a prepared glucoamylase mash (described later) or an equivalent commercial amyloglucosidase (such as Allcoholase II). The dosage rate of 3,500 glucoamylase units (AMU) per 56 lb bushel of rye grain is needed for efficient saccharification. The target dilution in a rye fermenter is usually 36-39 US gallons liquid per 56 lb distillers bushel (526 liters liquid dilution per 100 kg) rye meal. This means that the total volume including grain meal, mashing water, backset stillage, yeast, enzyme mash and wash water will equal 36-39 US gallons for each 56 lbs of grain meal added. Set temperatures will vary according to cooling water temperatures, but generally are about 30°C with cooling water control settings at 33-34°C. Fermentation is continued for 72 hrs with slow, constant mixing. In cases where foaming becomes excessive, a cool water spray reinforced occasionally with antifoam is used as a means of control. This foam control is usually required when the plant receives large quantities of newly-harvested rye grain and/or when distillery-produced glucoamylase mash is not available.

Rye fermenter process control

Fermenter data are recorded at set times and then every 24 hrs throughout the 72 hr fermentation. Fermenter set data include mash sequence used, number of mashes, grain mash bill (percentages of grain types used if applicable), enzyme additions, yeast tub used and set temperature. Operator analyses include pH, titratable acidity, Balling hydrometer reading, temperature and time of fermenter activity. The rate at which pH and titratable acidity change in the fermenter can be an indication of cleaning operation efficiency and of the success in keeping the mash in a sanitized condition. The rate at which the Balling degrees change is an indication of the yeast cell population, propagation efficiency and the rate of consumption of sugars. The temperature recording will indicate whether the cooling water system is functioning properly in order to ensure that the yeast have an optimum working temperature.

At 72 hrs the fermenter is placed on full agitation for 1 hr and then sampled prior to transfer to the beer well to await distillation. A sample of each thoroughly mixed fermenter is sent to the laboratory for analyses, which include alcohol level and the resultant amount of absolute alcohol, the colloidal fraction, pH and final titratable acids. If the data indicate that the fermenter is still active, then it will be moved out of sequence for an additional half day and another fermenter may be transferred to the beer well a few hours early. The beer alcohol percentage and total quantity of absolute alcohol are used to calculate the fermentation alcohol yield entering the distillation system. This in turn will allow calculation of the distillation efficiency. The colloidal percentage is used to predict viscosity levels in the last stages of the dryer plant evaporation system and to indicate the adequacy of the enzyme additions in the cooker and fermenter.

Figure 4 is a flow sheet of the mashing and fermenter-filling procedures and includes the positioning of the yeast and enzyme systems. Note the quality stations and the process checks recommended for good manufacturing practice.

Yeasting, lactic cultures and glucoamylase

Yeast mash preparation - lactic souring

The lactic-yeasting operation at rye grain distilleries uses an all-rye grain mash slurry, which is pre-cooked at 65.5°C for 30 minutes and then cooled to 53°C. A small amount of thermolabile α-amylase enzyme is added for liquefication and the mash is then inoculated with a *Lactobicillus delbrueckii* culture and maintained at 53°C for 9-11 hrs. Initially the lactobacillus inoculum may be grown from an American Type Culture slant incubated at 53°C in sterilized rye. However, when the first culture has been grown satisfactorily on a plant scale, pails of the culture can be used as described below for future inoculation.

Rye meal can be used as a yeast mash slurry (unfortified by malt or other nutrients) as it meets the nutritional needs of the yeast. The cooking process promotes the release of starches and proteins from the meal, which in turn undergo conversion by the indigenous rye enzymes to smaller-sized carbohydrates and amino acids. This resultant mash is further converted and altered by the action of the lactobacillus incubation as described below to produce additional nutritients required by the yeast.

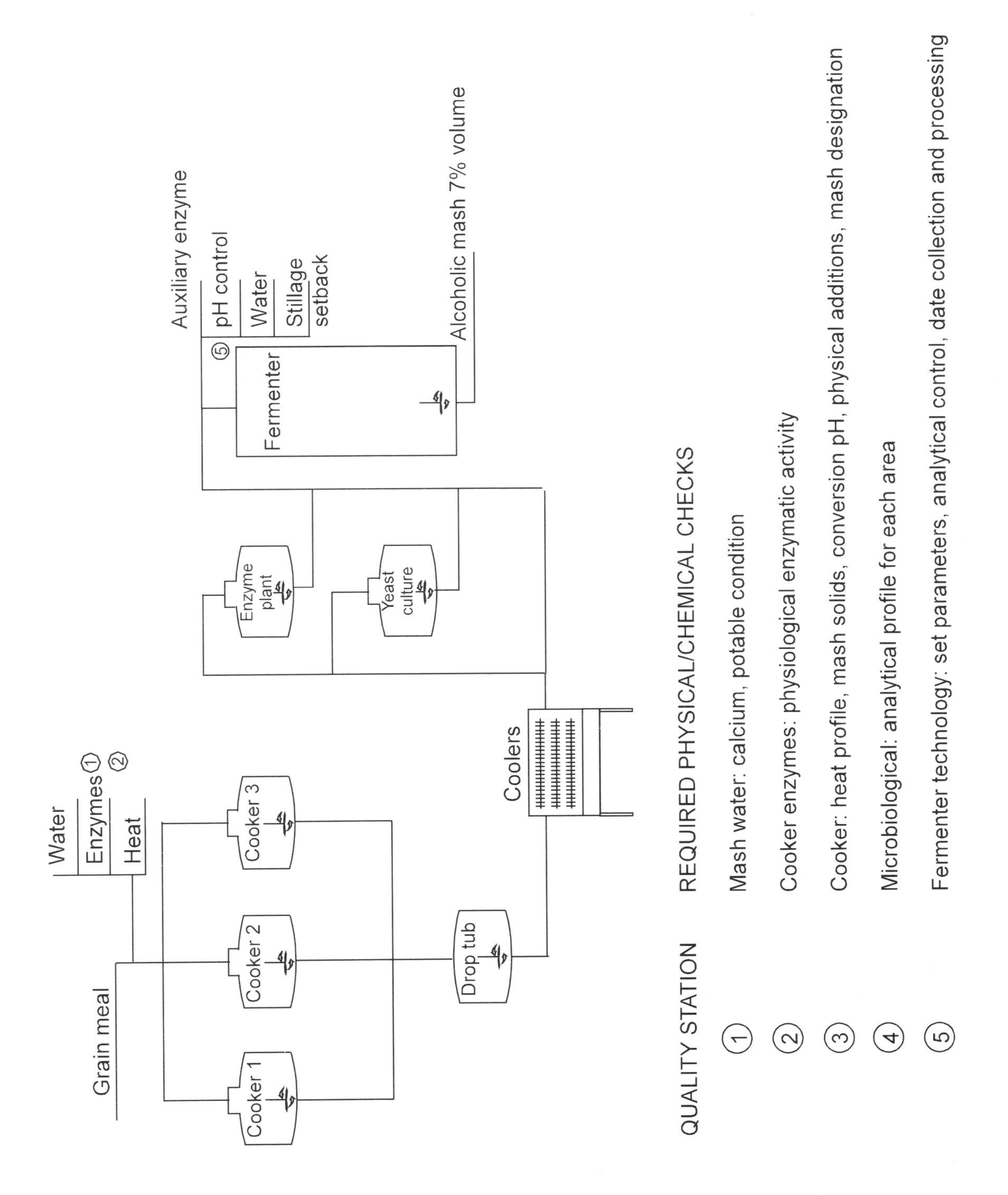

Figure 4. Mashing and fermentation flowsheet (Morrison, 1992).

During the lactic incubation period the lactobacilli multiply and produce lactic acid, which in turn lowers the pH to about 3.6. A healthy propagation of lactobacilli will generate up to 0.9% lactic acid in the 9-11 hrs and the mash viscosity will be lowered appreciably. Plant process quality control includes periodic pH checks and the titration of a 100 ml mash sample with standard sodium hydroxide to monitor acid development. As inoculum for future use, pails containing 6 US gallons (22 liters) of the completed lactic incubation (with a minimum of 0.9% lactic acid) are collected. Pails are cooled to room temperature, treated with powdered calcium carbonate to raise the pH to 5.5 and quick-frozen in upright freezers.

The deliberate cultivation of a lactobacillus culture in the preparation of yeast mash is good manufacturing practice with several benefits to the process. The selected temperature of incubation at 53°C encourages certain high acid-producing bacteria to predominate in the mash and to produce sufficient lactic acid in 11 hrs to drop the pH to 3.6. Deviations from the 53°C incubation temperature will decrease the purity of the population and the amount of lactic acid generated.

The time period and temperature of the lactobacillus incubation create an environment where bacterial spores are encouraged to metamorphose to a more easily-destroyed vegetative form prior to the final sterilization step. The lowered pH enhances sterility by creating an environment in which microbial proteins are made more susceptible to heat denaturation, which leads to the destruction of the organisms.

Finally, because of certain common biochemical pathways with yeast, the lactobacillus cultivation continues the conditioning of the rye mash by releasing certain growth factors that accelerate yeast growth. Presence of these growth factors, along with the nutrients in the rye mash, ensures healthy and prolific yeast cell growth.

Yeast mash preparation - sulfuric acidification

On occasion, timetable or maintenance problems make it difficult to allow the yeast mash sufficient time for a proper lactic acid hold. The mash is therefore prepared and precooked in the same manner as before; but the pH is adjusted to around 3.6 with sulfuric acid. This method allows no time for any bacterial spores to vegetate and subsequently be destroyed during sterilization; so an acid mash may undergo bacterial growth of considerable magnitude during the yeast incubation stage. Additionally, the intermediate nutrients mentioned above will not be produced and cannot contribute to healthy yeast growth. Thus, sulfuric acidification has the potential to produce inferior yeast inoculum for the fermenters. This method should be used only when all other options have been exhausted.

Yeast mash sterilization and inoculation

After acidification by lactic culture or sulfuric acid, the yeast mash is sterilized by heating to 121°C for 1 hr. It is then cooled to 30°C and treated with a small amount of thermolabile α-amylase at 79°C and a saccharifying enzyme such as glucoamylase mash at 57°C, and inoculated with large cans of *Saccharomyces cerevisiae* yeast culture. These large cans of yeast culture are prepared by standard propagation techniques from working yeast slants using a concentrated malt syrup as the medium. The inoculated yeast tub is incubated under controlled temperature conditions for 16-18 hrs with automatic agitation for 1 minute every 7.5 minutes. At the end of the 16-18 hr time period the yeast tub is cooled to 16°C and allowed to remain at that temperature until needed in the fermenters. The yeast propagations are usually out of phase by about 12 hrs so that each yeast vessel may inoculate a series of fermenters with a minimum hold time.

Yeast cell growth processes

Soon after the yeast culture is added to the sterile rye yeast mash, the yeast cells begin absorbing the available nutrients and will increase in size. Buds begin to appear at specific spots on the yeast cell, become larger and take on the appearance of the mother cell. If the budding is sufficiently rapid, the buds or daughter cells in turn form daughter cells and the yeast take on the appearance of chains of cells. Later in the incubation the chains will decrease in length and single or double budding cells will predominate. This budding process is the manner in which yeast cells replicate to produce a culture containing millions of cells in every milliliter of mash. Using phase contrast microscopy, the budding mechanism of yeast cells has been recorded on film and is presented in the literature (Stewart *et al.* 1981; Lyons, 1981).

The growth cycle described above is an aerobic process, meaning that oxygen (air) is essential. Oxygen becomes a limiting factor in cell growth because of its role in providing cell membrane lipid materials in the growing yeast cell (Ingledew, 1991). Agitating the yeast mash periodically throughout the incubation facilitates the air incorporation into the mash. The agitator churns the mash and brings all parts of the mash into contact with air at the surface. Generally, the yeast cell count obtained during the incubation will be a function of the amount of oxygen made available.

After the 16-18 hrs incubation, when the viable cell count should be approximately 200 million cells/ml, contents of the yeast vessel are cooled to 16°C. The cooling is done in such a way that the rapid cell growth potential is retained and may be recovered after the yeast are pumped to the fermenter as inoculum. The logic of slowing yeast cell development through cooling becomes clear by viewing the growth curve for yeast presented in Figure 5. Cooling is carried out at a point when the growth curve has maximum slope and the cells have the greatest potential for budding. The fermenter

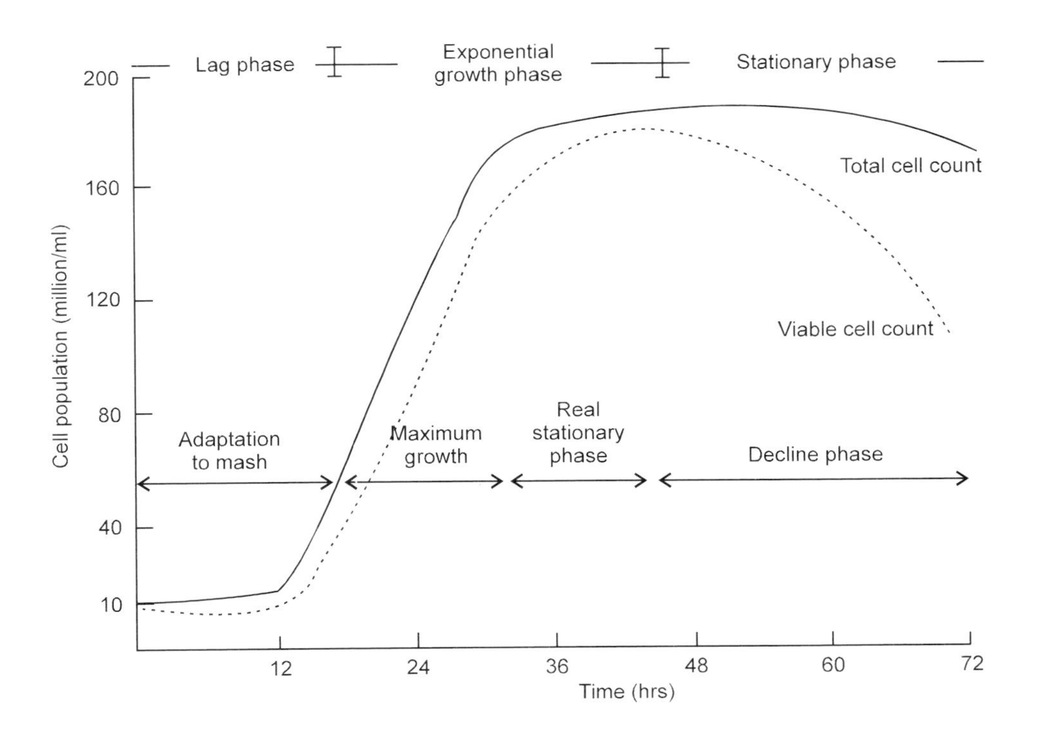

Figure 5. Typical yeast growth curve in rye grain mash (Lyons, 1981).

mash will in turn supply nutrients and oxygen to the yeast inoculum so that a new cell growth curve takes over and appropriate yeast counts are achieved in the fermenter.

Production of glucoamylase by submerged culture

Fungal amylases such as amyloglucosidase (glucoamylase) are produced by a submerged culture process in which a sterilized grain mash substrate (usually corn) is cooled to 37.7°C, subjected to a steady flow of sterile air (usually 0.5 volumes of air per volume of mash per minute), agitated rapidly (to assist air dispersion), inoculated with a strain of Aspergillus or Rhizopus and allowed to incubate for 4-6 days. Fungal inoculum may take the form of spores suspended in sterile water that is pressure-jetted into the reactor via a sterilizable transfer valve arrangement. Alternatively, a standard aseptic propagation procedure of fungal slant to broth flask to media flask to dona tank to reactor can be used. Both systems work equally well with the larger dona inoculum having the possible advantage of a shorter incubation period. The fungal spores will vegetate into individual cells and begin propagating new cells end to end to form filaments called hyphae. The hyphae branch and intertwine to form a mycelium (Pelczar, 1958). In about 40 hrs the mycelium will propagate throughout the grain medium mash and the mash will take on a thick, lumpy texture. A very characteristic mold odor will become apparent; and microscopic examination reveals extensive networks of elongated cells with very distinct walls dividing the cells.

Incubation is continued at full agitation and aeration for four days, and then a sample is removed for enzyme analysis, pH, citric acid level and microscopic examination. These data are compared to standard reactor data for a high enzyme-yielding incubation and adjustments in operating parameters may be made if necessary. Usually by the sixth day of incubation the fungus has developed its maximum level of enzyme, and the reactor contents are analyzed again for enzyme level and a calculation is made of the amount required for addition to fermenters.

Submerged culture glucoamylase mash is used as a source of saccharifying enzyme in the fermenter in conjunction with the use of thermostable bacterial α-amylases in the cooker. The enzyme combination is able to carry out the quantitative hydrolysis of starch to glucose owing to ability of the glucoamylase mash to hydrolyze the dextrins left in the cooker.

A typical glucoamylase production facility in a rye distillery will consist of 2-4 reactors of 3,400-3,900 US gallons (13,000 to 15,000 liters) and one or two scale-up reactors of 450 US gallons (1,700 liters). Operation is coordinated with the mashing and fermenting schedules in order to cultivate sufficient glucoamylase in mash form for use as the saccharifying enzyme in the fermenters. The mash additions to the fermenters are determined by the enzyme concentration of the mash and the need to supply 3,500 glucoamylase units for each 56 lb bushel of grain in the fermenter.

Distillation of rye whisky spirits

Beer still operation

At the end of the fermentation cycle, usually after 72 hrs, the contents of the fermenter are pumped to a holding tank known as the beer well. The beer well, essentially a surge tank, provides a continuous supply of beer to the first stage of distillation in the beer still. The beer still separates the volatile organic components of the beer from the grain solids and the carbon dioxide gas.

Rye distillers commonly use a 45 tray beer still to convert a fermented beer with 8% alcohol by volume into a solids-free high wine of about 120° US proof (60% alcohol by volume or 60° GL). The beer is fed through a preheater onto the top sieve tray of the stripping section of the beer still. The beer travels across the tray in a shallow layer and is prevented from flowing through the holes by an upward flow of steam. It then descends through the downpipes and across the trays repeatedly. The steam drives off the volatiles, which proceed up the column undergoing alcohol enrichment to become the high wine product at 120° US proof. At the same time, the beer moving down the stripping section becomes increasingly depleted of alcohol and arrives at the bottom tray as stillage. A stillage containing less than 50 ppm alcohol is normally considered acceptable for process control.

Figure 6 is an integrated flow sheet for the production of high wine from rye grain with nine stations for quality control analyses. Such flow sheets are invaluable in illustrating the total quality management necessary for good manufacturing practice. It is important to see the quality stations as loops in the process where the quality decisions at any stage affect the expected result at the following stage.

Rye whisky spirits manufacture

Rye whisky spirits are distilled from high wine using a two-column rectification system consisting of a 36-40 tray extractive distillation column followed by a 65 tray rectifying column. The system may also be equipped with two additional columns of smaller size, one for fusel alcohol separation and the other for concentration of the aldehydes and other heads congeners. The two column system is able to convert 34,500 US proof gallons of high wine at 120° US proof (108,000 liters at 60° GL) into rye whisky spirits at 192° US proof (96° GL).

Extractive distillation

An extractive distillation column is used to concentrate and remove the major portion of unwanted congeners before the alcohol stream enters the rectification column to be concentrated as whisky spirits. The whisky spirits will be essentially free of all objectionable impurities, but will contain controlled amounts of the congeners that create the proper balance of whisky character and flavor during the subsequent aging in oak barrels.

An extractive distillation column operates on the theory that higher (fusel) alcohols, aldehydes and esters are less miscible in water. Controlled water addition allows these compounds to go up the column and be drawn off as a heads stream while the diluted alcohol (containing controlled amounts of congeners) emerges from the bottom of the column and is fed into the rectifying column. Thus, an extractive distillation system can control the amounts of alcohol and heavy fusels that leave the top of the column and control the amount of congeners that exit out the bottom of the column with the diluted alcohol.

On a continuous basis, high wine at approximately 120° US proof (60° GL) is passed into a surge tank and diluted 1:1 with recycled rectifier bottoms. The 60° proof (30° GL) mixture is fed into the 36-40 tray extraction column about two-thirds of the way up the column. A 20° proof (10° GL) alcohol stream containing small levels of congeners emerges from the bottom of the column. This stream may be further diluted if live steam is injected at the base of the column. At the same time higher alcohols, aldehydes and some esters make their way to the top of the column and are taken off from the condensate-reflux loop and fed either to a fusel oil column or to a decanter. In the fusel oil column there is a bottom take-off containing appreciable quantities of recoverable alcohol. The decanter does a similar job by allowing a phase separation to take place, forming an oil layer with a higher concentration of the unwanted congeners and a water layer containing small amounts of congeners and a considerable amount of recoverable alcohol. The water layer is fed back into the extraction distillation column for alcohol recovery and the oil layer is removed for later processing and disposal.

Rectification

A rectifying column operates on the principle that an alcohol/water vapor leaving a boiling

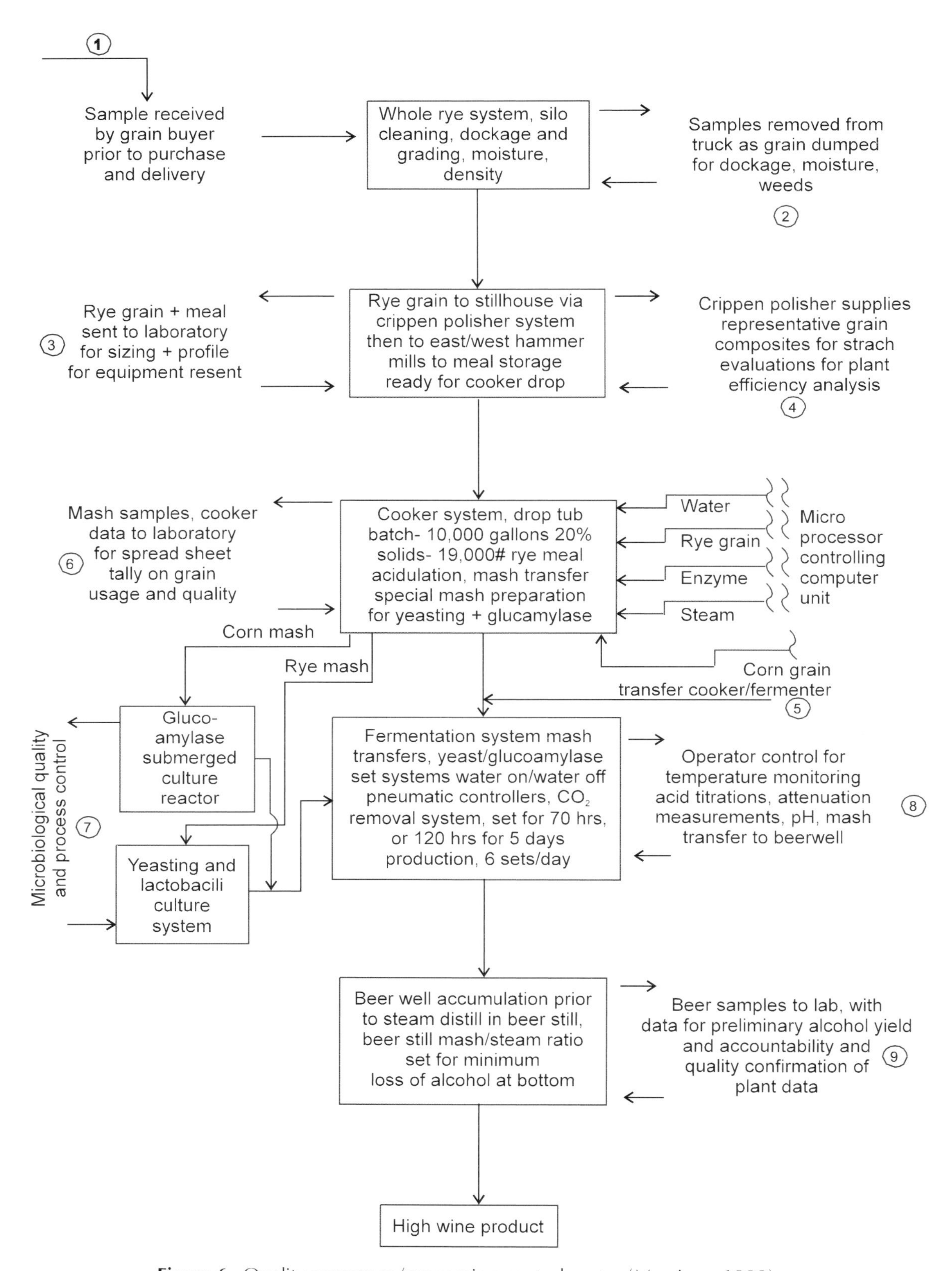

Figure 6. Quality assurance/processing control routes (Morrison, 1992).

layer of liquid on a tray passes up through a liquid layer on the tray above via the bubble caps, condenses and transfers its heat to the liquid. In turn, a portion of this enriched alcohol-water liquid is vaporized and rises to the next tray to repeat the process. Since the alcohol has a higher vapor pressure and evaporates more readily than water, the alcohol concentration of the rising vapor continually increases from plate to plate. Depending on the operation of the column, some portion of the condensed liquid on each tray will flow back down the column to replenish the liquid level on the lower trays. So, as the alcohol-enriched vapor passes up the column, the alcohol-depleted liquid passes down the column. At the top of the column total-vapor condensation occurs and the condensate is fed back into the column to assist liquid replenishment on the trays while the product is drawn off just below the top of the column. The amount of condensate flowing back compared to the product take-off from the column is known as the reflux ratio. The reflux ratio will determine the alcohol concentration on each tray and the alcohol concentration achieved at the product take-off point.

Thus, under proper operation, a rectifying column will move alcohol up the column increasing in strength from tray to tray until the required strength is achieved. Equally important, the tower will move water down the column decreasing the strength from tray to tray until the minimum-detectable alcohol level exists on the bottom tray.

The extractive distillation column bottoms flow at 20° proof (10° GL) passes into the rectifier about one-third way up the column (at tray 26) and proceeds to enrich to 192° proof (96° GL) on trays 58-60, which are the product take-off trays. Alcohol product strength, tray temperatures, base pressures and high wine feed rate are used by the distillation operators in their process control of the column. Gas-chromatographic monitoring of congener presence and magnitude on a number of control trays can follow this column operation. These data are fed back to the operator center on a daily basis to form an ongoing record of column conditions and are used for daily adjustment of column parameters.

Figure 7 is a flow sheet of the beer still and two-stage rectifier for the production of high wine and its processing into rye whisky spirits, respectively. Note the quality stations and the process control checks recommended for good manufacturing practice. The gas chromatographic monitoring of tray components and the use to which the data are put are explained below.

Samples for gas chromatographic analysis from the various trays in the rectification column are taken via sample valves. The sample is passed through a cooler (or simply a long take-off line) exposed to the lower room temperature and collected. For sample lines without proper coolers it will be necessary to take precautionary steps to avoid alcohol 'flashing' into the air when taking samples from the trays at high temperature. Safety equipment must be worn to protect against scalding with the boiling alcohol.

Various monitoring programs may be established. Samples can analyzed either from all of the selected trays several times daily or certain tray samples can be analyzed only once daily, with additional emphasis on product take-off and perhaps on the congener buildup on fusel oil take-off trays. The analysis of one or more congener substances present on various trays can be used in column process control. When tray analyses are available over a significant period of operation, it then becomes possible to make certain predictions about how the column will react if current run conditions are allowed to continue.

The congeners found and monitored on selected column trays are listed in Table 4. It is important to realize that the analytical monitoring is carried out to determine concentration trends from one production shift to another. For example, if the methanol concentration on the top tray of the column remains reasonably constant at 1,100 ppm; it probably means that the take-off rate of the heads is sufficient for good column operation. Removal of the methanol also means that other undesirable heads fractions are

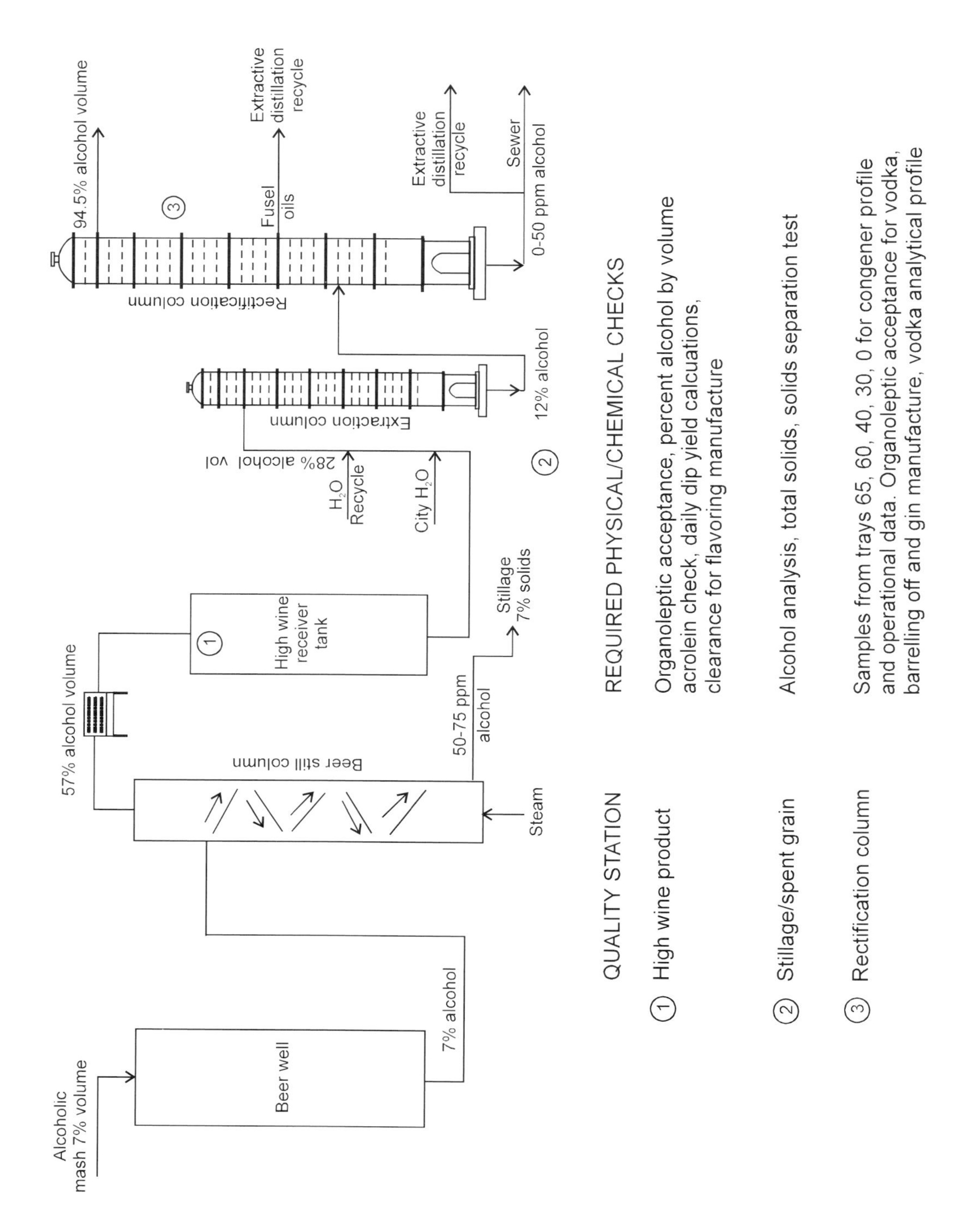

Figure 7. Distillation and rectification flowsheet (Morrison, 1992).

Table 4. Congeners found and monitored on rectifier trays.

Congener	*Tray number*				
	T65	*T60*	*T40*	*T30*	*T0*
Methanol (methyl alcohol)	X		X		
Ethanol (ethyl alcohol)			x		x
n-Propanol (propyl alcohol)		X	X		
Isobutanol (isobutyl alcohol)			X	X	X
n-Pentanol (n-Amyl alcohol)					X
Isopentanol (isoamyl alcohol)					X
Propenol (allyl alcohol)				X	

being removed and will not eventually spread down the column to appear in high concentrations on the product take-off tray. In this manner the methanol becomes an indicator of column conditions and with a little interpretation can lead to proper adjustments. The methanol analysis along with other chemical testing can also assist in column adjustment for the manufacture of high quality neutral spirit for vodka production.

The rectifier feed from the bottom of the extractive distillation column will contain small amounts of higher alcohols that tend to accumulate in the upper part of the rectifier. Propanol can be used as an indicator of this accumulation because of its ability to exist on certain trays without isobutanol or amyl alcohols being present. This becomes very important if fusel oil-free spirits or spirits with controlled amounts of congeners are desired. At a position in the column several trays down from the product tray, the n-propanol will accumulate significantly and decrease in magnitude as one moves up the column. If a vector is drawn to represent the n-propanol magnitude on each tray, then the bank of vectors form a triangle with the base on tray 40 (for example) and the apex on a tray somewhere above (Figure 8). As the n-propanol concentration increases in the rectifier, the apex of the triangle will rise higher and higher until n-propanol appears in the product tray. Using historical analysis it is then possible to predict when n-propanol will appear in the product by monitoring propanol

n-Propanol, ppm		Tray no.	Vector equivalent depiction
0	————————	62	
Trace	————————	60	+++
50	————————	58	++++++
200	————————	56	+++++++++
600	————————	52	++++++++++++
1800	————————	48	+++++++++++++++
5400	————————	44	++++++++++++++++++
15000	————————	40	++++++++++++++++++++++++

Figure 8. n-Propanol concentration as vectors on distillation trays (Unpublished distillery data).

concentration on tray 40. A level of n-propanol on tray 40 will become an operational specification or process control point below which propanol will never appear in the product spirits. Tray 40 then becomes an indicator tray as it indicates what will happen in the near future if the column conditions are unchanged. This information allows a decision about the take-off rate at the fusel oil tray (at tray 30 for example) to be made, thus reducing the total fusel alcohols in the column.

Rye distillation fusel oils

Fusel oil is the term applied to a group of aliphatic alcohols whose formation is related to the amino acid metabolism of yeast cells. The creation of these alcohols appears to be the result of transamination reactions in which amino groups are transferred from a supplied amino acid to an oxo acid molecule thus leading to the synthesis of a different essential amino acid. The original (amino acid) carbon skeleton, now an α-oxo acid, then undergoes decarboxylation and subsequent reduction to form the appropriate fusel alcohol. The creation of oxo acids also appears to be possible from carbohydrate metabolism by the yeast, so that fusel alcohol levels continue to rise even after amino acid requirements have been met (MacLeod, 1977; Ingledew, 1991). Total concentration and diversity of the fusel alcohols are significantly related to the amount and form of free amino acid nitrogen available in the grain mash.

The fusel oil alcohols consist of n-propanol, isobutanol and two pentanols (amyls), optically active amyl alcohol and isoamyl alcohol. The combined amyls may constitute the major portion of the fusel oil at a margin of 2:1 over the isobutanol and 10:1 over the n-propanol. Table 5 presents an analysis of fusel oil removed from a two-column distillation system for rye whisky spirits.

The rectifying column trays below the feed tray (about tray 26) are considered stripping trays; and their function is to decrease the alcohol concentration on each lower tray such that the column bottoms contain essentially zero alcohol. In reality, alcohol concentration will probably remain in the 30 to 50 ppm range with the column running efficiently.

The bottom take-off liquid is usually divided between a recycle stream to the extractive distillation column and excess liquid going to the sewer. Knowledge of the discharge rate and the alcohol content allows the calculation of total alcohol lost. Alcohol concentration in the rectifier bottoms is a function of the column temperature profile and the balance between the rate of alcohol fed into the column and the rate of alcohol removal. Increasing the amount of alcohol fed into the rectifier at any time will cause the alcohol to move to lower and lower trays. This is shown on the control panel as the increased alcohol concentration lowers boiling point temperature. If the imbalance persists, significant quantities of alcohol will appear in the bottoms and be detected by temperature reduction and gas chromatography. The alcohol content of the bottoms then becomes a means of monitoring the process and of suggesting alteration in run parameters. If the bottom effluent is a part of the waste stream from the plant, then the alcohol level will have additional significance, as it will be contributing to biological oxygen demand (BOD) of the waste stream.

Stillage and feed grain recovery systems

Stillage is removed from the bottom of the beer still at about 90°C after passing through a heat exchanger and is pumped to a 240,000 US gallon hold tank equipped with a steam sparger and side-mounted agitator. Running continuously at about 13,000 US gallons (49,000 liters) per hour, hot stillage containing around 8% solids is passed through three Sharples centrifuges to produce a liquid and wet cake containing about 2.5% and 25% total solids, respectively.

The centrifugate containing 2.5% solids is passed into a surge tank and subsequently fed to the first stage of a four-stage falling film evaporator that raises solids content to 30%. The falling film evaporator is followed by a two stage, rising-film, finisher evaporator that increases the solids to a final 38 to 40%. The evaporation system is basically atmospheric and operates in the 109°C range and uses thermal recompression to enhance the heat efficiency.

The wet centrifuge cake moves to a first stage mixer where it is mixed with low moisture recycle and fed to the drying ring of the first stage flash dryer with a cake product take-off for further processing at 90% solids. The 90% solids cake then moves to a second mixer and is homogenized with second stage low moisture recycle and the 40% solids evaporator syrup to produce an 80% solids material which is fed to the drying ring of the second stage flash dryer. At a 90% solids (10% moisture) the process yields about 65 tonnes of distillers dried grains (DDG) per production day.

Process control of the drying unit centers on monitoring flow rates, temperatures and pressures at the various stages. Of key importance is the moisture content, which is checked frequently by operators and laboratory staff. While each stage has a certain range of parameters for proper operation, it is essential to ensure that the product does not overdry and become a fire or explosion hazard. Equally, the product must not be stored at too high a moisture level or it may become susceptible to spontaneous combustion in the silos or during transportation.

Figure 9 is a flow sheet of the stillage-handling and drying procedures. Note the quality stations and the process checks needed for good manufacturing practice. Data collected on centrifuge solids ratios are used to determine operating conditions of the centrifuges and to

Table 5. Composition of a typical fusel oil from rye spirits distillation.*

Physical appearance	Clear, slightly yellow liquid	
Gross composition, %	Water	13.87
	Alcohols	86.13
		100.00
Alcohol composition, % of total alcohols		
Methanol (1 carbon atom)		0.14
Ethanol (2 carbon atoms)		11.92
Propyl (3 carbon atoms)		4.82
Isobutyl (4 carbon atoms)		27.21
Amyls (5 carbon atoms)		54.17
Others		1.49
		100.00
Composite analysis, %		
Water		13.87
Methanol		0.12
Ethanol		10.27
Propanol		4.15
Isobutanol		23.44
Amyl alcohols		46.66
Others		1.49
		100.00

*Unpublished distillery data.

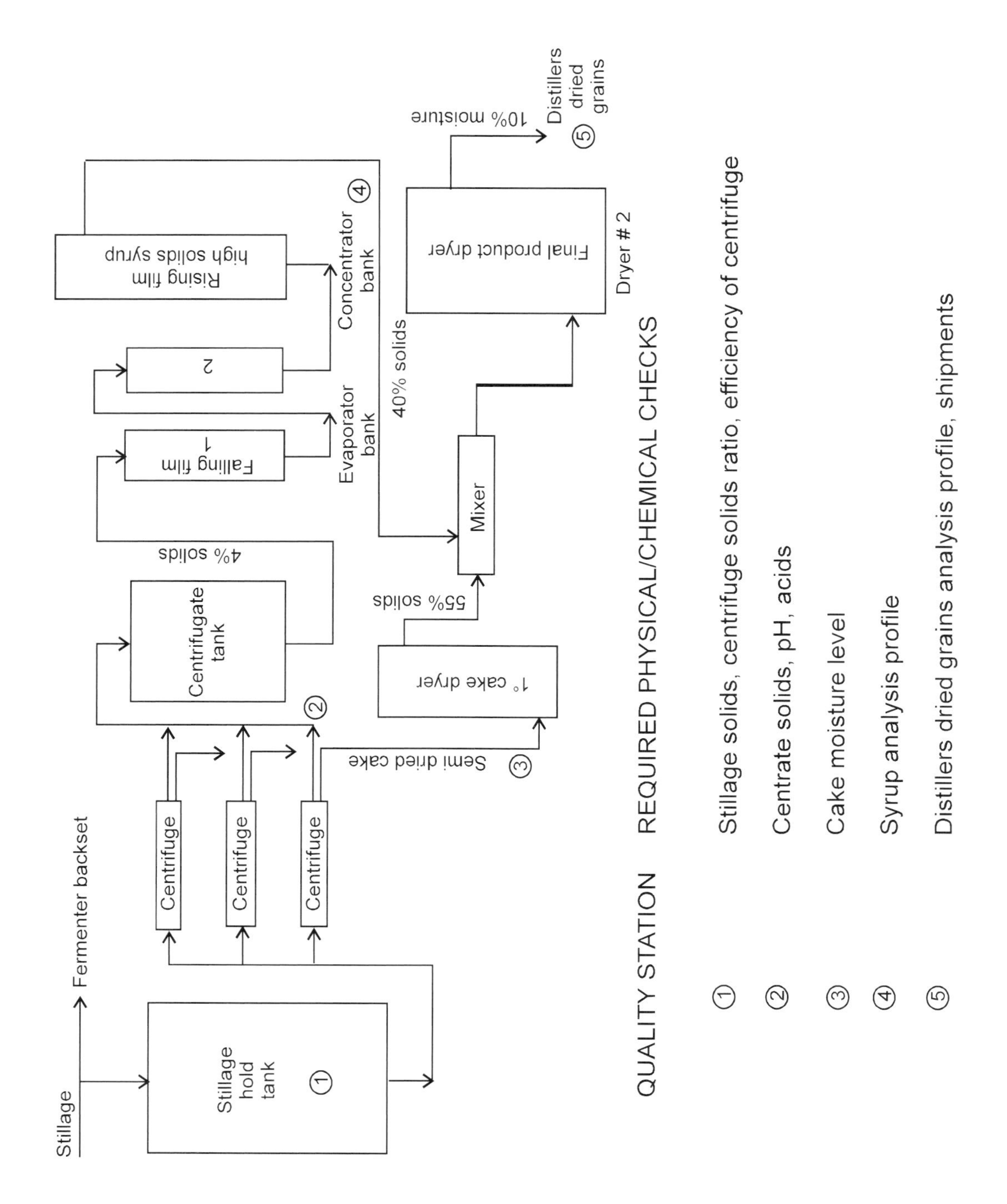

Figure 9. Stillage handling and rye distillers dried grain manufacture (Morrison, 1992).

determine syrup/cake ratios that will affect the recombining of cake with syrup in the second-stage mixer. The syrup analysis profile includes viscosity checks at the fourth and sixth stages of the evaporator system so that product take-off rates and recycling can be programmed.

The DDG analysis profile is a set of specifications for shipment. The profile includes crude protein (Kjeldahl), fat (ether extract), density (loose fall), sieve analysis, color, moisture (oven), ash content (calcium and phosphorus), total dissolved nutrients and total solids. Table 6 contains a typical analysis of DDG (with solubles) and a comparison with DDG produced from Canada grade #1 corn. The darker color usually associated with rye distillers dried grain is the result of high temperature effects on nonfermentable pentoses present in the rye. The pentoses will caramelize as well as combine with protein in the typical browning reaction, causing significant changes in the color of the DDG. Note also the difference in yield of DDG from each grain type. This difference in degree of concentration is the reason for the relatively high protein values present in the corn DDG. The protein content of the grain multiplied by the factor 56, divided by lbs of DDG produced from one distillers bushel of grain will give the protein content of the DDG.

References

Headon, D.R. 1991. Enzyme Biochemistry for Fermentation. 11th Alcohol School Proceedings, Alltech Inc., Lexington, Kentucky.

Ingledew, W.M. 1991. Biochemistry of Yeast. 11th Alcohol School Proceedings, Alltech Inc., Lexington, Kentucky.

Kahn, J.H., P.A. Shipley., E.G. Laroe, and H.A. Conner. 1969. J. Food Sci. 34:587.

Lyons, T.P. 1981. A Step to Energy Independence. Chapter 4, Alltech Publications, Lexington, Kentucky.

MacLeod, A.M. 1977. Economic Microbiology. Vol 1, Alcoholic Beverages (A.H. Rose, ed), Chapter 2, Academic Press.

Morrison, J.A. 1992. Aspects of distillery operation using Western Canadian rye grain. 12th Alcohol School Proceedings, Lexington, Kentucky.

Noller, C.R. 1957. Chemistry of Organic Compounds. 2nd Edition, Saunders Press.

Pelczar, M.J. 1958. Microbiology., McGraw-Hill, Toronto, Canada, Chapter 13.

Table 6. Comparative composition of distillers dried grains with solubles from western Canada rye and no. 1 Canadian corn.

	*Rye grain #2 Western Canada**	*Canada grade #1 Corn grain***
Ingredients	%	
Calcium	0.16	0.27
Phosphorus	0.70	0.89
Crude protein (Kjeldahl N x 6.25)	25.30	29.30
Crude fat (ether extract)	2.40	11.30
Ash content	3.80	5.80
Crude fiber	5.20	6.90
Total dissolved nutrients	65.0+	80.0+
Moisture (oven dry)	10.10	10.00
Yield, lb/bushel	23.00	17.00
Color of product	Dark tan	Light tan

*Unpublished distillery data.
**Preston, 1987.

Preston, R.L. 1987. Feedstuffs, October, 1987.

Simmonds, D.H. and W.P. Campbell. 1976. Rye: production, chemistry, and technology. (W. Bushuk, ed), American Association of Cereal Chemists, Inc., Monogram Series, Vol. 5, Chapter 4, St. Paul, Minnesota.

Starzycki, S. 1976. Rye: production, chemistry, and technology'. (W. Bushuk, ed.), American Association of Cereal Chemists, Inc., Monogram Series, Vol. 5, Chapter 3, St. Paul, Minnesota.

Stewart, G.G. and I. Russell. 1981. A Step to Energy Independence. Chapter 5, Alltech Publications, Lexington, Kentucky.

Whelan, W.J. 1964. Methods in Carbohydrate Chemistry. (R.L. Whistler, ed), Vol IV Starch, Academic Press, New York.

Whistler, R.L. and C.L. Smart. 1953. Polysaccharide Chemistry. Chapter X, Academic Press, New York.

Chapter 13

Production of neutral spirits and preparation of gin and vodka

J.E. Murtagh
Murtagh & Associates, Winchester, Virginia, USA

Introduction

Neutral spirit is basically purified, odorless, tasteless, colorless ethanol (or ethyl alcohol - C_2H_5OH). It may be produced from almost any fermentation feedstock if suitable distillation and rectification techniques are used to remove the other chemical compounds, or congeners, produced with the ethanol in the fermentation process.

Neutral spirit is used in the production of beverages such as vodka, gin, cordials and cream liqueurs. It is also used for a wide range of other industrial manufacturing applications, where it is usually referred to as 'high quality industrial alcohol'. Purified synthetic alcohol produced from petrochemicals may also be used in industrial applications, being virtually indistinguishable from neutral spirit produced from fermentation feedstocks, but it is outside the scope of this chapter.

Feedstocks

Almost any fermentation feedstock may be used to produce neutral spirit, but certain feedstocks may produce greater concentrations of particular congeners that require additional equipment for removal. For instance, potatoes and grapes tend to produce much higher levels of methanol than are obtained from grain fermentations; and it may be necessary to use an additional distillation column, a demethylizer, to obtain a satisfactory product.

The choice of feedstock is largely dictated by economics and availability. Thus for part of the year a plant may use a low cost, seasonally available, perishable feedstock such as sugarcane juice, grape juice, or cheese whey. The rest of the year the plant will operate using more expensive but storable feedstocks such as grain, raisins or molasses.

Feedstocks for neutral spirit production must be consistently of good quality. They must be free of major bacterial infections in order to maintain fairly clean fermentations and avoid excessive variation in the amount and type of congeners to be removed in subsequent processing.

Fermentation

The preparation and fermentation of the various feedstocks is essentially the same as described in the chapters on the production of whisky, rum and fuel ethanol. However, because the objective is to produce lower concentrations of congeners (since no distinctive flavor is required), more emphasis may be placed on use of bactericides and antibiotics (e.g. chlorine dioxide and penicillin) to control bacterial contaminants in the fermenters.

The fermentation process yields a beer containing varying concentrations of ethanol depending on the feedstock used. For example, the ethanol concentration from fermentation of sulfite waste liquor (from the paper/pulp industry) may be only about 1% by volume. Cheese whey fermentations normally yield 2-5% ethanol; sugarcane juice fermentations may yield 6-8% ethanol and grain fermentations may yield 10-12% ethanol.

Distillation and rectification

Beer distillation

Conventional US process

The first stage in the distillation process is separation of the ethanol and most of the congeners from the bulk of the liquid and solids in the fermented beer. This is normally carried out in a beer still similar to that shown in Figure 1.

The beer still usually consists of two columns. The first column is the beer stripper and the second is the concentrator. In some instances the concentrator may be stacked on top of the beer stripper to make a single, tall column. The beer stripper may contain about 20-25 sieve trays, or it may have other contact devices such as disc-and-donut baffle trays.

The beer is fed into the beer stripper near the top, while live steam is introduced at the bottom of the column. (A reboiler may be used at the bottom of the column for indirect heating, but requires some provision to avoid scaling from the beer solids.) The ethanol and most of the congeners are stripped from the beer by the steam (as described in the chapter on distillation) and they pass in vapor form to the concentrator. This column may have about 30-40 trays with sieve holes, bubble caps, or tunnel caps.

For convenience and clarity, the concentrator in Figure 1 is shown as having overhead condensers with gravity reflux of condensate back to the column. In many instances, particularly in recent installations, the condensers may be situated near ground level with the reflux being pumped back up to the top of the concentrator.

The ethanol rises through the concentrator and may be concentrated to about 95° GL. It is drawn off from a valve located a few trays below the top of the column and sent to a surge tank for temporary storage as an intermediate product.

If the conditions in the concentrator can be held in a steady state, various congeners will accumulate at different levels in the column. Isoamyl alcohol, which is usually the principle higher alcohol or fusel oil congener in the beer, tends to accumulate where ethanol concentration is about 65° GL (This may be about two or three trays above the fusel draw tray.) These compounds may be drawn off in a small purge flow and sent to storage for subsequent reprocessing to recover the ethanol in the flow.

The concentration peaks of the congeners in a typical beer still are shown in Figure 2. (For clarity and simplicity the beer stripper and concentrator are shown as a single column.) It will be seen that if the peak heights are not controlled below certain limits, the congeners will extend up the column and be present in significant amounts in the 95° GL intermediate product discharge line to maintain a steady proof of about 65° GL intermediate product. To avoid this, European beer stills usually include a heads removal section on top of the beer stripper as shown in Figure 3. In this system (developed by Emile Barbet (1922) of France, described by Mariller (1943) and copied extensively in Brazil and other countries) the beer passes through a pre-stripping section of four or more trays where the more volatile components, including the

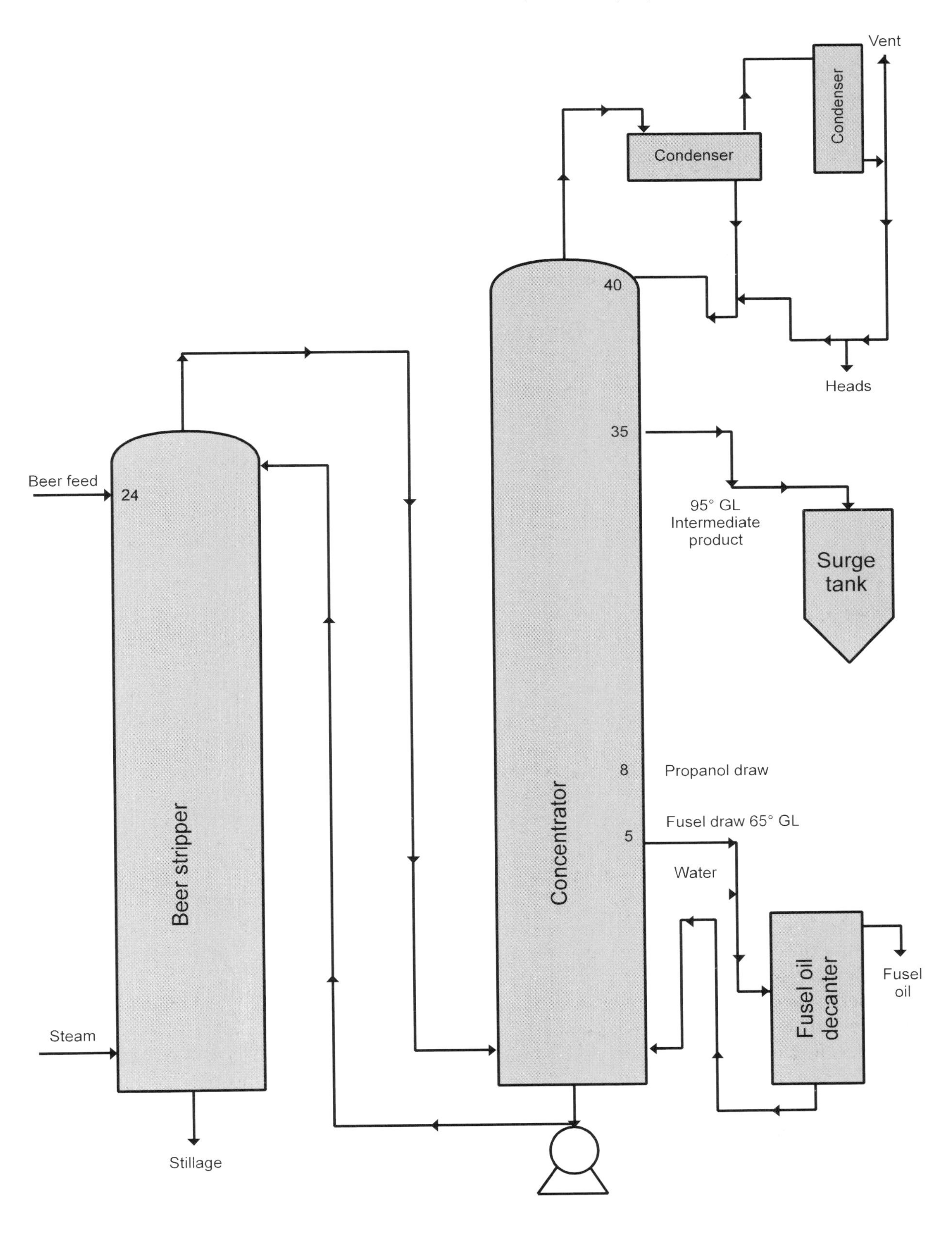

Figure 1. Conventional beer still used in neutral spirit production.

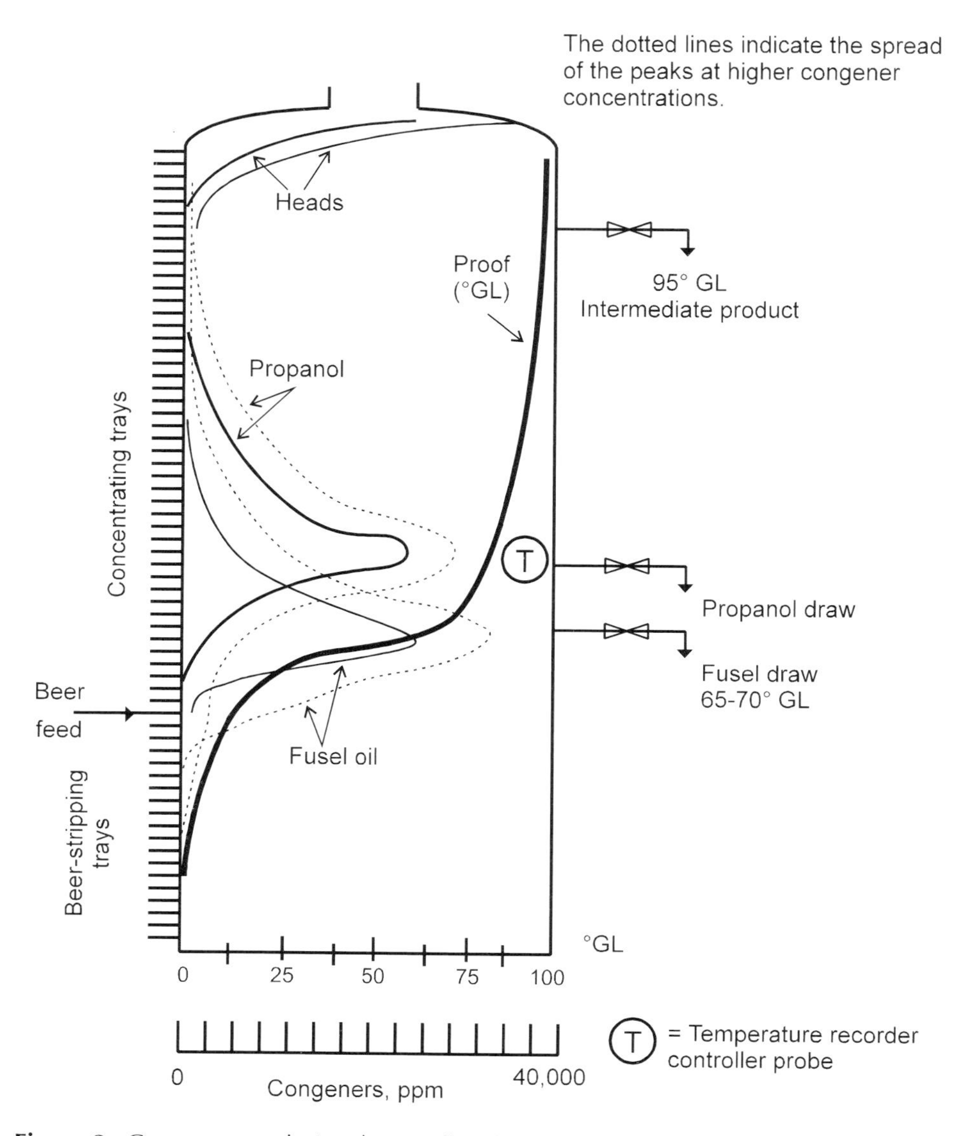

Figure 2. Congener peaks in a beer still with alcohol proof curve superimposed.

heads, are flashed off. The vapor rises into a heads concentrating section of six or more trays; and the heads may be drawn off as a purge stream from the reflux from the condenser. Meanwhile, the main bulk of ethanol and other congeners passes via the vapor line at the base of the pre-stripping section to the concentrator.

Splitting of the vapor flows to the heads concentrating section and the concentrator is regulated by a manually controlled valve located either on a vapor line between the pre-stripper and the heads concentrator or in the vapor line to the condenser.

The Barbet system not only removes much of the heads before the concentrator, it has an added advantage in that it permits removal of carbon dioxide in the beer feed to prevent it taking up space in the concentrator and reducing its capacity. The Barbet system also allows other volatile impurities such as sulfur compounds from molasses feedstock to be vented or removed in the heads.

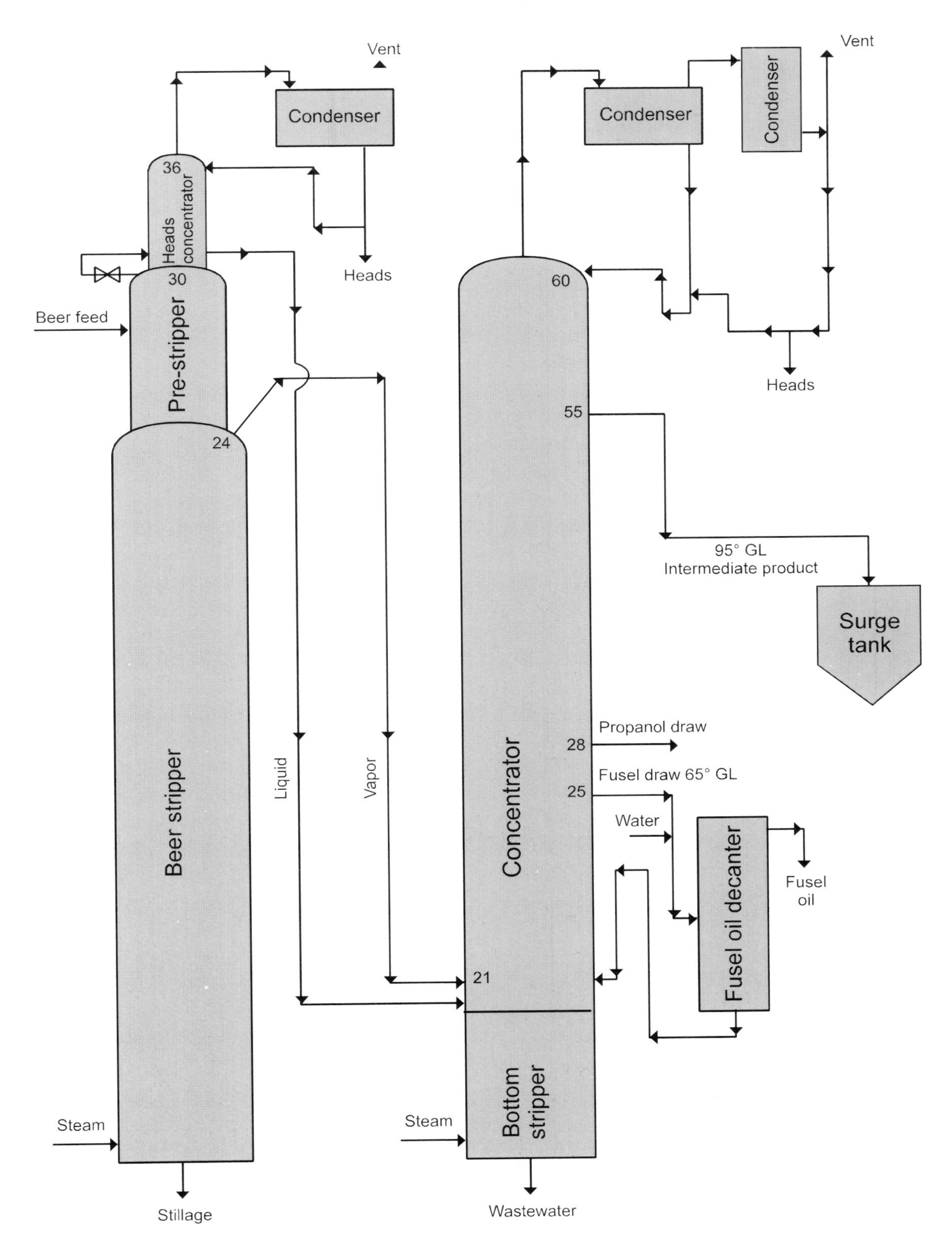

Figure 3. Modified Barbet still used in neutral spirit production.

Some Barbet systems also use the heads concentrating section on top of the beer stripper for concentrating heads piped back from the ethanol concentrator and from the rectification system. Conventional US systems normally have a separate heads concentrating column.

Another variation common on modified Barbet beer still systems is a bottoms stripping section below the concentrator (as shown in Figure 3). This might be just a downward continuation of the concentrator with about 20 stripping trays, or it may be a separate column, usually of a smaller diameter. Having a bottoms stripper on the base of the concentrator means there is no need for a reflux line from the concentrator to the beer stripper. This avoids any ethanol return to the beer stripper to cause scaling on the beer feed tray and reduce its capacity. This is of particular significance when using blackstrap molasses as the feedstock as it contains considerable quantities of calcium compounds that contribute to scaling. Also to avoid scale there may be a line to take the reflux from the bottom tray of the heads concentrating section down to the bottoms stripper below the concentrator.

Another variation in beer distillation is to have a greatly reduced number of trays in the ethanol concentrator. Some beer stills may have only six trays in a concentrator located on top of the beer stripper. This means that the intermediate product proof will be much lower, at about 70° GL. A beer still of this type is considerably less expensive in capital costs of fabrication. However, there is very little saving in operating costs, as there is not much reduction in steam usage because the energy input required to adequately strip the ethanol from the beer is also sufficient to operate the ethanol concentrator to near the 95° GL point. A reduced number of ethanol-concentrating trays also means there is no provision for removal of fusel oil and other higher alcohols, so the only way they can be removed is in the rectification process. Reducing the number of ethanol-concentrating trays also reduces the capacity for concentrating the heads at the top of the column. Thus more heads will tend to come out in the intermediate product.

Rectification

Extractive distillation

Depending on the type and efficiency of the beer distillation system, the intermediate product will have some noticeable odor and will contain varying amounts of higher alcohols and heads that must be removed in the rectification process to yield a truly neutral spirit. However the congeners do not separate very well from the ethanol in the 95° GL intermediate product by normal fractional distillation. For this reason it is necessary to employ an extractive distillation technique using water.

It should be noted that the efficiency of separation of the congeners can be increased greatly if the intermediate product is dehydrated to 100° GL. Young (1988) proposed in a US patent that the intermediate product should be dehydrated and then subjected to two fractional distillation steps to remove the congeners that are both more and less volatile than ethanol. With the advent of molecular sieve ethanol dehydrators, this process could possibly be more cost effective than extractive distillation of the 94° GL intermediate product.

The extractive distillation process relies on the fact that some of the congeners with higher boiling points that are normally less volatile become more volatile than ethanol in the presence of water. Thus, in a rectification system (as shown in Figure 4) the intermediate product is diluted in an extractive distillation column to remove the congeners as a heads stream. The diluted ethanol is then reconcentrated in the rectifying column.

The extractive distillation column may have about 40 trays. The 95° GL feed from the intermediate product surge tank enters the column at about tray 30; and hot water is added on the top tray while live steam is injected at the base. By regulating the feed rate, dilution water and steam flow, most of the congeners will tend to go up the column and accumulate in the reflux loop to be drawn off in a heads purge. The optimum conditions will vary for the different

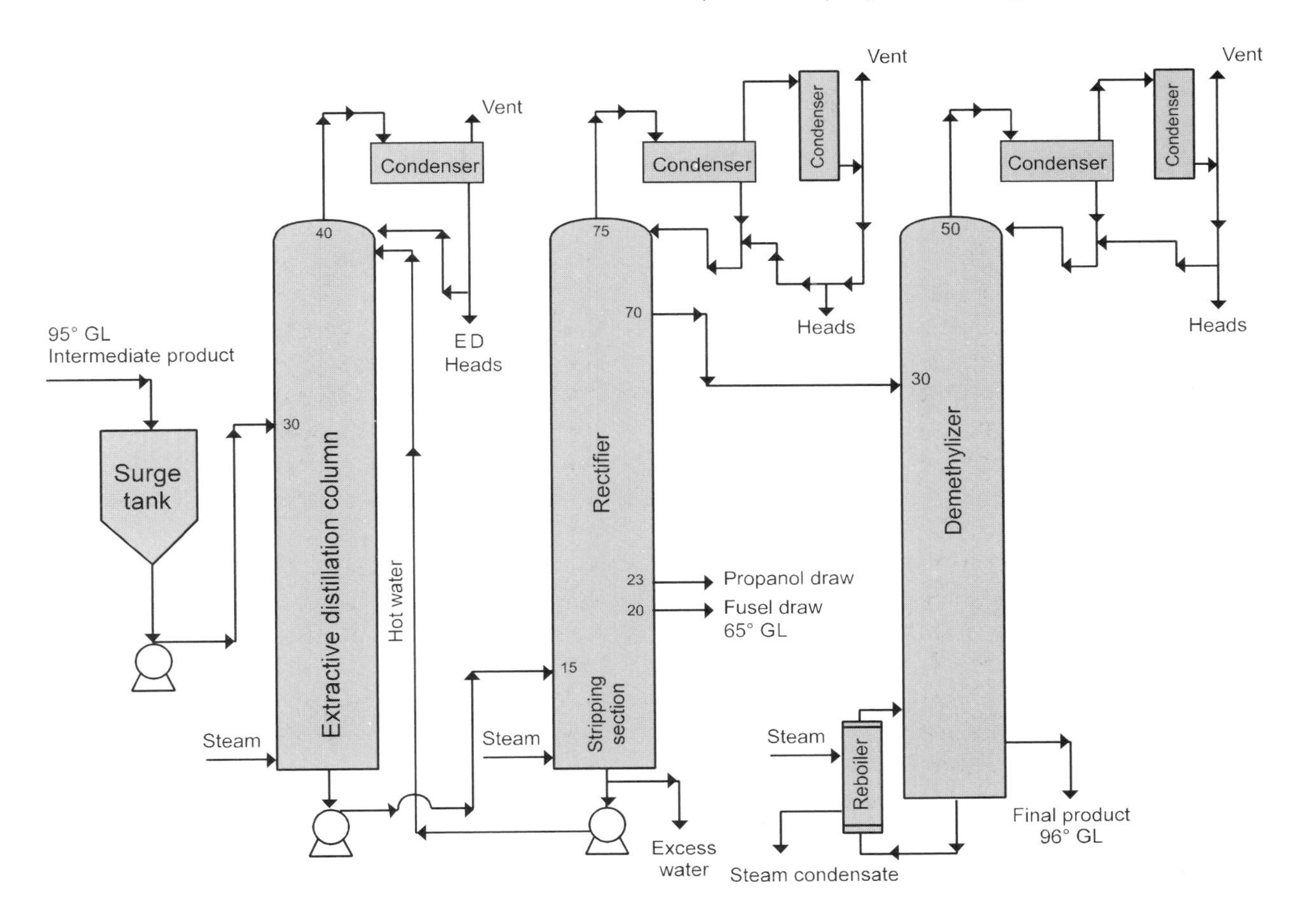

Figure 4. Three column rectification system used in neutral spirit production.

congeners; so some compromise will be necessary depending on the mix of congeners present.

The theory of extractive distillation and its practical implementation in ethanol rectification has been covered very extensively by Chambers (1951, 1953, 1959). Two graphs derived from the Chambers publications are presented in Figures 5 and 6. Figure 5 indicates how the optimum conditions for removal of isoamyl alcohol and n-propanol may be established by comparing the proof in the 'pinch zone' of the column (around tray 15) with the proof at the base of the extractive distillation column. The graph shows separate extraction limit lines for n-propanol and isoamyl alcohol; and it will be noted that less water and more steam are required to remove the latter than the former. The product obtained with conditions generally in the area between the n-propanol and isoamyl extraction limit lines is described by Chambers as 'bland quality', while the product obtained wholly within the n-propanol extraction limit line is described as 'vodka quality'. Figure 6 shows how the efficiency of fusel oil removal is affected by the steam usage in the extractive distillation column and the proof at the base of the column. It will be noted that the optimum working area is where the base proof is relatively low and the steam usage is relatively high.

Final rectification

The dilute ethanol stream emerging from the base of the extractive distillation column will contain

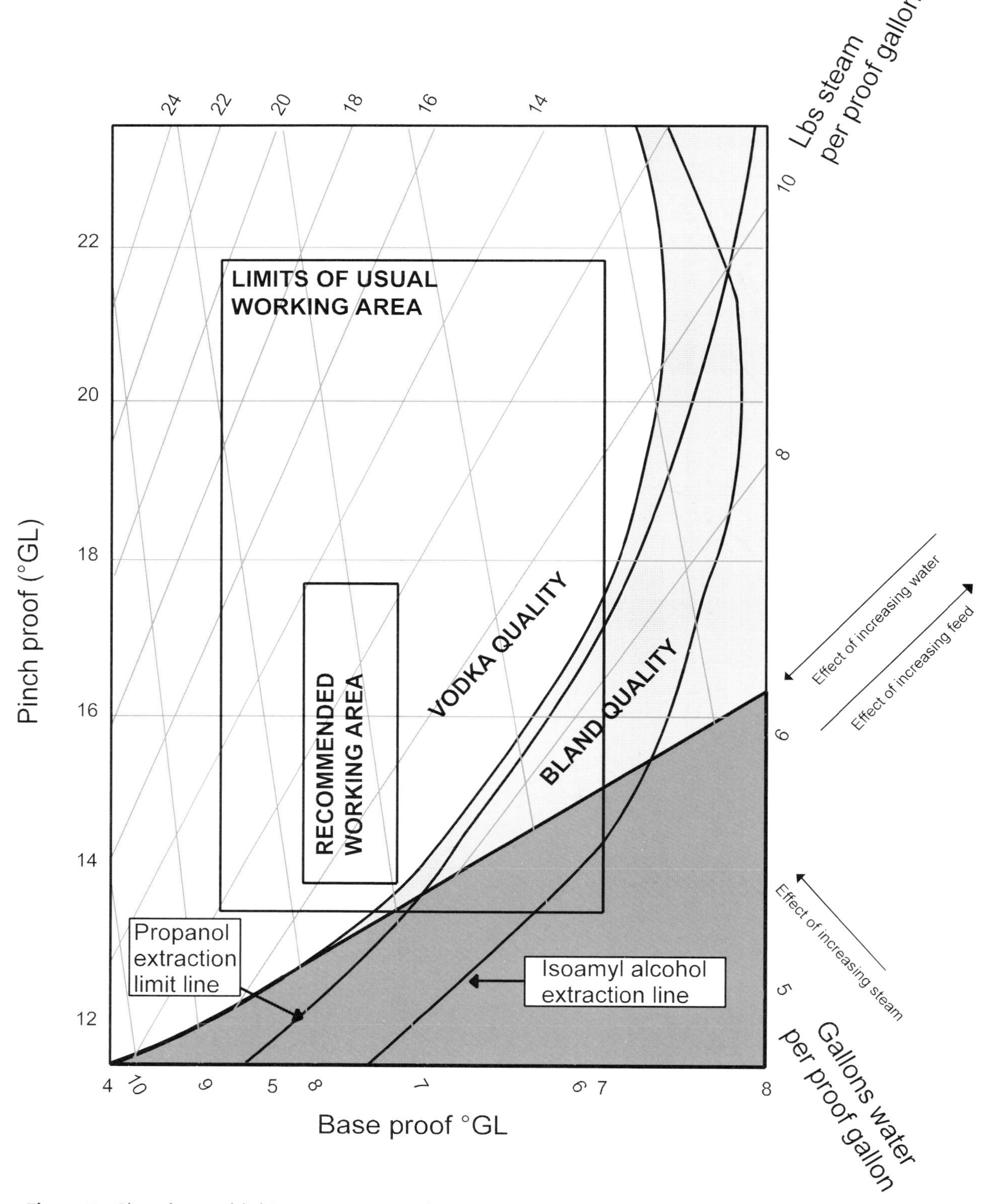

Figure 5. Chart for establishing optimum conditions for removal of propanol and isoamyl alcohol from an extractive distillation column.

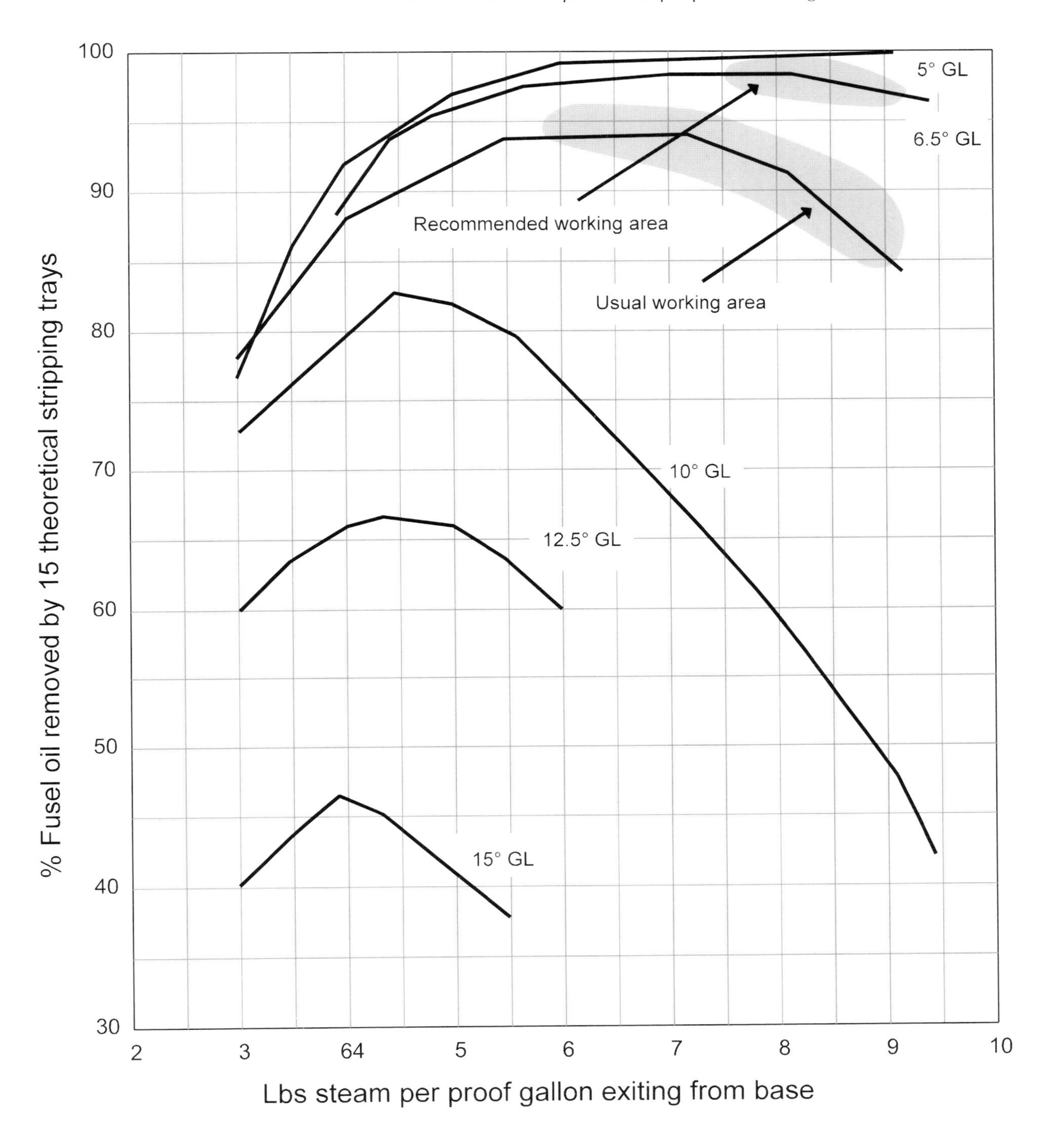

Figure 6. How the efficiency of extractive distillation varies with the base proof and steam usage.

the congeners not removed in the column heads. The congeners will include methanol, diacetyl, some aldehydes and relatively low concentrations of isoamyl alcohol, isobutanol and n-propanol. This stream is reconcentrated in a rectifying column (or rectifier) that usually has about 75 trays.

The feed stream from the extractive distillation column enters the rectifier at about tray 15. Live steam is injected at the bottom of the column, and the lower trays serve as a stripper to separate the water from the ethanol and congeners. The water emerges from the base of the rectifier; and part of the flow is recycled to supply the hot water requirements at the top of the column. The ethanol and congeners travel up the rectifier and concentrate in much the same fashion as in the beer still concentrator. Thus the residual isoamyl alcohol accumulates at about 65° GL, normally near tray 20, where it is drawn off as a purge stream to be sent back to the beer still. The traces of other high alcohols, mainly isobutanol and n-propanol, accumulate at about 80° GL around tray 23 and are similarly recycled to the beer still.

The rectifier product is drawn off at about 96.1° GL, usually about five trays from the top of the column (i.e. at plate 70), to leave some trays for heads concentration. The heads are drawn off as a purge stream from the final condenser reflux and are sent to the heads concentrating column (or the heads concentrating section in a modified Barbet beer still).

The control system used in the rectifier is usually the same as that used in the beer still concentrator, working on 'mid column' temperature.

Demethylization

When grain is used as the feedstock, the rectifier product is in fact the final product. However, where the feedstocks tend to produce more methanol as in the case of potatoes, grapes, and to some extent molasses, the rectifier product is sent to a demethylizing column. This column may have about 50 trays and will have indirect heating via a reboiler at the base.

The rectifier product is fed into the demethylizer at about tray 25 or 30. The methanol and any other heads components rise up through the column and concentrate at the top. They are drawn off as a heads purge stream and either sent to the heads concentrating column or sold as low grade industrial alcohol. The ethanol and any higher alcohol congeners in the rectifier product will descend the column and emerge from the base as the final product.

Variations in rectification systems

It is not possible to cover in this chapter all the variations in rectification systems used for neutral spirit production. However, one common variation is a fusel oil concentrating column which receives the fusel flows from the beer still concentrator and from the rectifier. The isoamyl alcohol concentrates in this column and is then sent to the fusel oil decanter for separation. However, if the proofs are controlled steadily in the other columns to concentrate the isoamyl alcohol sufficiently for a direct feed to the decanter, this column should not be necessary.

There are several variations in column arrangements that may save energy but do not materially affect overall neutral spirit production. For example, the extractive distillation column may be operated under pressure to heat the demethylizer. (In effect, the reboiler on the demethylizer becomes the first condenser for the extractive distillation column). Similarly, the rectifier may be operated under pressure to 'cascade' heat to a reboiler on the beer still, which in turn may be operated under vacuum to save energy and reduce the incidence of scaling.

Integration of beer still and rectification systems

Figure 7 shows an integrated beer still and rectifying system for the production of neutral spirit. In this example, a modified Barbet beer still is shown with the heads from the concentrator and the rectifier being

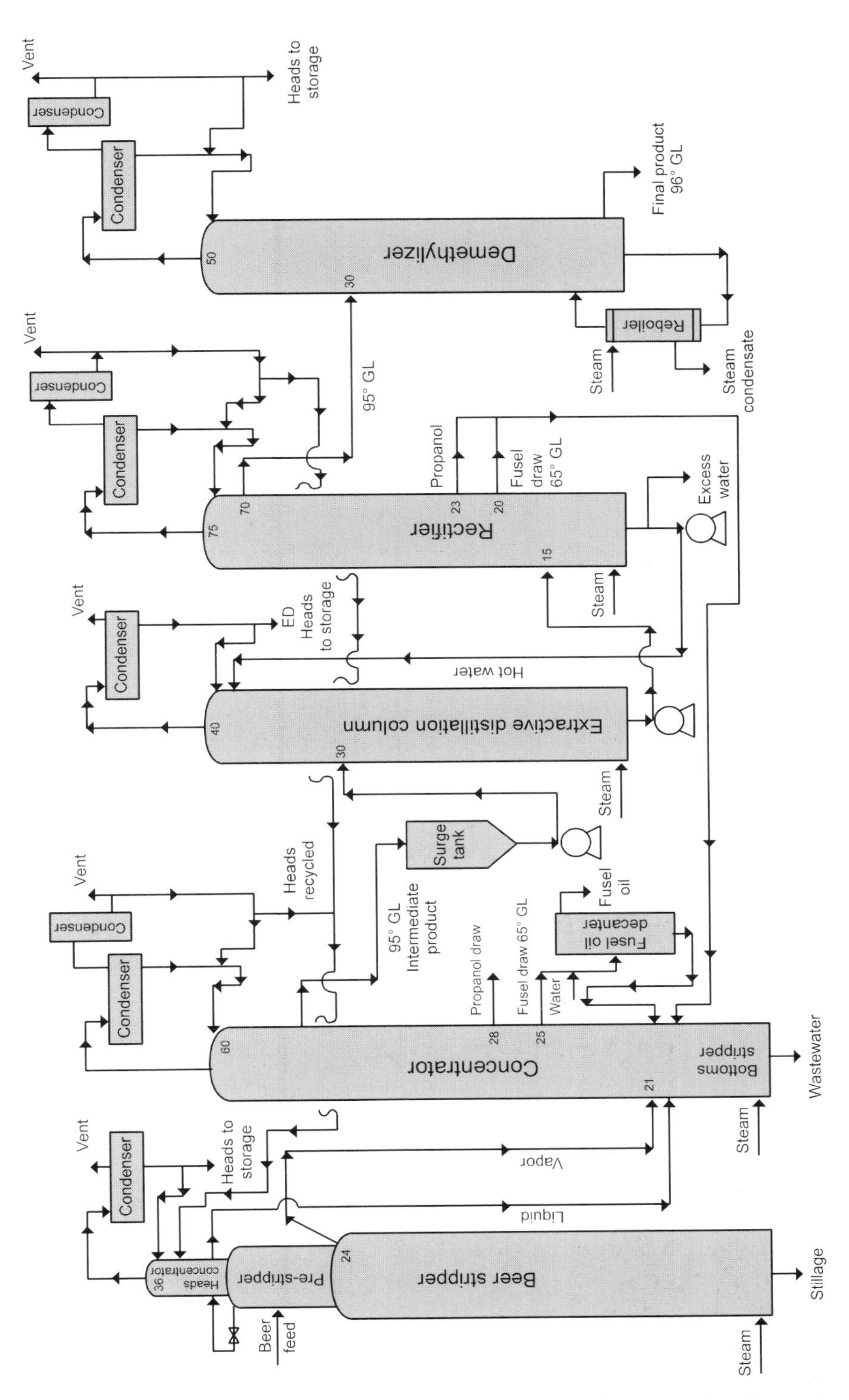

Figure 7. Complete beer still and rectification system used in neutral spirit production.

concentrated in the heads concentrator located on top of the beer stripper.

Handling and storage of neutral spirit

As neutral spirit should be odorless and tasteless, care must be taken in handling and storage to prevent it acquiring odors and tastes from other materials. Any piping and tanks used for neutral spirit should be checked carefully, drained and flushed with some neutral spirit (which is recycled to distillation) prior to use. Contrary to usual beliefs, storage tanks need not be fabricated from stainless steel. Carbon steel tanks are quite acceptable provided that the main discharge valve is located above the bottom of the tank to allow rust accumulating in the sump to be drained.

Rubber or plastic hoses should be avoided if possible, and flexible stainless steel hoses used where necessary. If rubber or plastic hoses or gaskets must be used, they should first be tested for any reaction with ethanol. This may be done by immersing shavings of the material in a jar of neutral spirit for at least 48 hrs. Compare a sample of the spirit diluted to about 20° GL visually and organoleptically with a fresh sample of the same neutral spirit. If there is some cloudiness (due to hydrocarbons), or if the sample has picked up an odor not present in the original, the hose or gasket is unacceptable.

Process control

The process control used in neutral spirit production is much the same as that used in the production of whisky, rum or fuel ethanol, but in general more use is made of gas chromatography and organoleptic testing. The final product should be run into daily production tanks (closed receivers). These tanks should be switched periodically (normally every 24 hrs) and isolated until the product has been approved. The normal tests conducted are proof, acidity, gas chromatography and organoleptic evaluation.

There are no official standards for neutral spirit in the US; so specifications are established between the supplier and purchaser. In 1983, the European Community established the specifications for neutral spirit (Table 1). These specifications included an extraordinarily high limit of 500 ppm for methanol. This political decision appeased Northern European ethanol producers who use potatoes and grapes as feedstocks. Their product consequently had high levels of methanol; and they did not want to be forced to invest in demethylizers. In practice, however, European ethanol purchasers generally set a maximum specification of 50 ppm for methanol. The standards for neutral spirit in Mexico, Bolivia, Brazil and Ecuador are also shown in Table 1. It will be noted that the standards vary greatly and that the Ecuadorian standard, which is one of the most stringent, also includes a limit of 45 ppm for all congeners.

Gas chromatographic analyses should not be relied upon exclusively in grading neutral spirit. Organoleptic testing should be considered at least equally important as there are many compounds such as diacetyl with odor thresholds well below levels detected by normal gas chromatography.

Very often plants have two quality grades for neutral spirit. The top quality is for use in the preparation of vodka, while the second is sold to less discriminating customers who can accept a trace of odor or taste or a slight excess of a congener in the gas chromatographic analysis. Any product failing to meet these grades is either sold as a mid-grade industrial alcohol or is reprocessed through the rectification system.

In controlling the process, it is advisable to take samples periodically from all of the draw points on the columns (the fusel oil, other higher alcohols, heads, and intermediate product valves). Additionally, use gas chromatography and sniffing to detect excessive accumulation of any congeners and adjust the purge draw rates accordingly. Some distillers recommend daily checking of the propanol concentration in the higher alcohol draws as high levels at these points could lead to 'spillovers' into the beer still or rectifier products.

Table 1. Specifications for neutral alcohol.

	European	*Mexico Community*	*Bolivia*	*Brazil*	*Ecuador*
Organoleptic characteristics (aspect)	No extraneous flavor detectable in the raw material.	Clear, colorless aspect, characteristic odor	Clean, clear, transparent with no suspension or coloration	Clear and without suspended matter	
Color APHA				5 max	
Specific gravity at 20°C, g/ml			0.805 max.	0.8068 max.	
Miscibility with water					
Minimum alcohol content					
Alcoholic strength by volume	96%				
Alcohol concentration at 15°C		95.5° GL	96.0° GL		96° GL
Alcoholic strength by weight				94.10%	
Maximum values of residue elements*					
Total acidity, mg acetic acid	1.5 (15 ppm)	3.0 (30 ppm)	0.5 (5 ppm)	0.8 (8 ppm)	1.5 (15 ppm)
Esters, mg ethyl acetate	1.3 (13 ppm)	8.0 (80 ppm)	8.0 (80 ppm)	1.8 (18 ppm)	2.0 (20 ppm)
Aldehydes, mg acetaldehyde	0.5 (5 ppm)	6.0 (60 ppm)	6.0 (60 ppm)	0.7 (7 ppm)	0.5 (5 ppm)
Higher alcohols, mg methyl-2-propanol	0.5 (5 ppm)	6.0 (60 ppm)	Traces	0.2 (2 ppm)	0.5 (5 ppm)
Methanol, mg	50 (500 ppm)	4.0 (40 ppm)		1.8 (18 ppm)	2.0 (20 ppm)
Dry matter, mg	1.5 (15 ppm)	5.0 (50 ppm)			
Volatile bases containing nitrogen, mg N	0.1 (1 ppm)				
Nitrogen compounds				Absent	
Furfural	Not detectable	Negative		Negative	Negative
Alkalinity		Negative	Negative	Negative	
Benzene		Absent			
Sulfur compounds, mg				0.05 (0.5 ppm)	
Fixed residues, mg			Traces	0.2 (2 ppm)	
Congeners, mg					4.5 (45 ppm)
Fusel oil content					
Permanganate test at 20°C, minutes		5 minutes			
Permanganate time at 15°C, minutes				30 minutes	

*expressed in mg per 100 mls alcohol at 100% volume.

Organoleptic testing is usually performed on samples of neutral spirit that have been diluted with odorless water to about 20° GL. There are various methods of comparison, but a preferred method is to have one sample glass marked as the standard and another standard sample identified only by a number or letter included with four or five other samples to be checked, which are similarly lettered or numbered. The persons performing the test should be unaware of the identities of the samples or the hidden standard until after completion. They should rate the samples on a 1-10 scale similar to that below:

10 Excellent - no discernible odor
9 Very good - only a very slight odor
8 Good - no unpleasant odor
7 Slightly good - slight unpleasant odor
6 Borderline - noticeable unpleasant odor
5 Unacceptable - yoo much off odor
4 Bad
3 Very bad
2 Terrible - unpleasant
1 Sickening

If a product quality problem cannot be corrected easily by increasing the draw rate of heads or other congeners, the best action is to slow the rectification operation and reduce throughput until the problem is corrected. In extreme cases, the rectifier may be put on 'total reflux'. This is a situation where the feed rate to the extractive distillation column is reduced to the point where there is just enough volume to meet the requirements of the congener draws, and none or almost no excess for the product draw. Under these conditions, quality problems can be corrected rapidly and the extractive distillation column feed rate can then be gradually increased to normal maximum rate.

It is obvious that obtaining a good 'clean' sample to use as an organoleptic standard is of great importance. Initially the standard may be obtained from another distillery, but a good opportunity to obtain a sample for use as a standard is at the end of an extended period of total reflux.

Production of vodka

The 1982 US Bureau of Alcohol, Tobacco and Firearms (BATF) regulations define vodka as 'neutral spirit so distilled or so treated after distillation, with charcoal or other materials, as to be without distinctive character, aroma, taste or color'. The use of charcoal filtration is now optional, whereas in earlier regulations it was mandatory and even the time and amounts of fresh charcoal to be used were specified. This change is in recognition of the relatively recent improvements in the quality of neutral spirits. Vodka is generally taken to be odorless, tasteless and colorless ethanol, but in the past in Eastern Europe vodka was lightly flavored with grasses or herbal extracts. It should be stressed that not only is the charcoal treatment nonessential, it is also not particularly effective and will not make a poor quality, improperly rectified neutral spirit into a good quality vodka.

Neutral spirit should be diluted to about 55° before charcoal filtration. The old BATF regulations (1961) specified dilution to between 55° and 40° GL at a minimum contact time of 8 hrs with 10% of the charcoal replaced every 40 hrs to give a minimum usage of 6 lbs of new charcoal per 100 gallons of spirit treated. This was usually achieved by passing the diluted spirit through a series of eight or nine cylindrical charcoal filtration beds in a slow, continuous flow with one of the beds changed every day. The fresh bed would be connected last in the series. This meant that the beds were constantly being rotated; so the preferred arrangement was to set the beds in a circle to facilitate the changing. A simple alternative treatment method is to add charcoal to diluted neutral spirit in a tank and agitate or circulate it through a pump for a suitable length of time.

The water used in the initial and final dilutions should be clean, odorless and preferably demineralized. The demineralization is generally for aesthetic purposes as consumers do not like to see a white film of salts around the side of a bottle or glass if the vodka has been allowed to evaporate.

In countries where laws require that all 'spirits' be aged in wooden barrels, it may be necessary to add a small amount of sugar and/or glycerine to be able to classify vodka as a liqueur or 'compound spirit' rather than as an 'immature spirit'. The amount of sugar or glycerine used is normally the minimum required to provide 'proof obscuration'. This occurs when there is sufficient dissolved material to cause the apparent proof obtained by direct testing to differ fractionally from the real proof obtained by distilling the ethanol from a sample in a laboratory still and retesting after redilution to the original sample volume.

Great care should be taken in the bottling of vodka to prevent contamination with residues of other odorous products. The tanks and bottling systems should be washed thoroughly if previously used for other products. However, some bottlers prefer to keep a set of tanks and a bottling line dedicated solely to handling vodka.

For further reading on vodka processing see the reviews by Simpson (1977) and Clutton (1979).

Production of gin

The BATF definition of gin is

> 'a product obtained by original distillation from mash, or by redistillation of distilled spirits, or by mixing neutral spirits with or over juniper berries and other aromatics, or with or over other extracts derived from infusions, percolations, or maceration of such materials, and includes mixtures of gin and neutral spirits. It shall derive its main characteristic flavor from juniper berries and be bottled at not less that 80° proof (40° GL). Gin produced exclusively by original distillation or redistillation may be further designated as 'distilled'.

The regulation also states that dry gin (London dry gin), Geneva gin (Hollands gin) and Old Tom gin (Tom gin) are types of gin known under such designations.

This regulation means that gin may be produced by 1) distilling spirit with juniper berries and other botanicals, or 2) mixing spirit with a distilled gin concentrate, or 3) mixing spirit with a blend of essences of juniper and other flavorings.

The spirit used in gin production is usually neutral, but in the production of Geneva gin, which is popular in the Netherlands and Quebec, it is a heavily flavored distillate referred to as 'malt wine'.

'Distilled gin' is normally produced in batch operations using pot stills. The pot still is usually filled with neutral spirit diluted to 45-60° GL, and then the juniper berries and other botanicals are added. The berries and botanicals may be added directly to the spirit either in loose form or contained in a cotton sack. Alternatively, the mixed botanicals may be suspended above the liquid surface either in a cotton sack or in a wire mesh rack.

In the gin distilling process the pot still is heated by steam indirectly through a calandria in the bottom of the pot. The distillate coming over in the first few minutes of flow is normally discarded as heads for reprocessing. The main bulk of the distillate is then taken as product, and the final portion distilling below a predetermined proof (of about 45°GL) is discarded as tails for reprocessing. The pot still product is then sent to the bottling department for dilution and bottling. There is usually no storage or blending of different gin batches.

In the preparation of 'gin concentrate' the distillation process is much the same as for distilled gin, but a much greater quantity of botanicals is added in the pot still. The gin concentrate is then simply blended with neutral spirit prior to bottling.

Gin essences are prepared by blending essential oils and other extracts derived from juniper berries and botanicals. With the introduction of highly concentrated gin essences, it is possible to use as little as 0.01% by volume of the essence in a blend with neutral spirit.

Some internationally known brands of gin are produced by all three methods (i.e. distilling, concentrate blending, and essence blending) in different countries without appreciable variance in taste and odor when normal quality control procedures are used.

The quality and type of juniper berries and the mix of other botanicals largely determines the nature of the end product. For example, the flavor of London dry gin is strongly influenced by large amounts of coriander seeds in the botanical mix. Simpson (1966; 1977) and Clutton (1979) have listed several botanicals commonly used in gin production (Table 2). Another frequently used botanical is the chamomile flower (*Chamaemelum nobile*).

Table 2. Botanicals used in production of gin.[1]

Common name	*Botanical name*
Juniper berries	*Juniperis communis*
Coriander seed	*Coriandrum sativum*
Liquorice root	*Glycyrrhiza spp.*
Fennel seed	*Foeniculum vulgare*
Cubeb berries	*Piper cubeb*
Cinnamon bark	*Cinnamonum zeylanicum*
Nutmeg	*Myristica fragrans*
Aniseed	*Pimpinella anisum*
Grains of paradise	*Afromomum melegueta*
Cassia bark	*Cinnamomum cassia*
Sweet orange peel	*Citrus sinensis*
Bitter orange peel	*Citrus aurantium*
Cardamom seeds	*Elettaria cardamomum*
Angelica root	*Archangelica officinalis*
Lemon peel	*Citrus limon*
Orris root	*Iris pallida*
Callamus root	*Acorus calamus*
Caraway seed	*Corum carvi*

[1]Adapted from Simpson (1966, 1977) and Clutton (1979).

As with vodka, great care should be taken in handling and bottling gin. Unlike vodka, however, the problem is not picking up flavors from other products. The risk is contamination of other products with gin. If it is not possible to use a dedicated set of tanks and bottling equipment, everything coming in contact with gin should be thoroughly washed before use on any other beverage.

References

Barbet, E. 1922. Distilling and rectifying column. US Patent 1,427,430.

Bureau of Alcohol Tobacco and Firearms. 1961. Regulations under the Federal Alcohol Administration Act. Washington, DC.

Bureau of Alcohol Tobacco and Firearms. 1982. Regulations under the Federal Alcohol Administration Act. Washington, DC.

Chambers, J.M. 1951. Extractive distillation. Chemical Engineering Progress 47 (11):555-565, 1951.

Chambers, J.M. 1953. Alcohol distillation process. US Patent 2,647,078.

Chambers, J.M. 1959. Distillation process. Canadian Patent 573,218.

Clutton, D. 1979. The production of gin and vodka. Brewers Guardian, October, pages 25, 27, 29 and 30.

Mariller, C. 1943. Distillation et rectification des liquides industriels. Dunod, Paris.

Simpson, A.C. 1966. Gin manufacture. Process Biochemistry, October, pages 355-358 and 365.

Simpson, A.C. 1977. Gin and vodka. In: Alcoholic Beverages (A.H. Rose, ed), Academic Press, London. pages 537-559.

Young, A.T. 1988. Potable spirit production. US Patent 4,784,868.

Chapter 14

Production of American whiskies: bourbon, corn, rye and Tennessee

R. Ralph

Ron Ralph & Associates Inc., Louisville, Kentucky, USA

Introduction: definitions of bourbon, corn, rye, wheat, and Tennessee whiskies

The US Bureau of Alcohol, Tobacco and Firearms (BATF) has set specific guidelines to define all types of alcoholic beverages produced in the United States. The general definition of whisky is 'a spirit aged in wood, obtained from the distillation of a fermented mash of grain.' This spirit can be produced from any grain or combination of grains; but corn, rye and malted barley are the principle grains used. Whisky is an alcohol distillate from a fermented mash produced at less than 190° proof in such a manner that the distillate possesses the taste, aroma and characteristics generally attributed to whisky; stored in an oak container, and bottled at not less than 80° proof. Also, whisky may contain mixtures of other distillates for which no specific standards of identity are noted.

Bourbon, rye, wheat whiskies are produced (distilled) at a proof no higher than 160° from a fermented mash of not less that 51% corn, rye or wheat and aged in new, charred oak barrels at a proof no greater than 125°. Also, these whiskies may include mixtures of whiskies of the same type. Corn whisky differs in that it may be aged in used or uncharred new oak barrels. Also, corn whisky may include a mixture of other whiskies. Tennessee whisky has the same definition of all four whisky types, but to be labeled 'Tennessee' it must be produced and aged in wood in the state of Tennessee.

All whiskies conforming to Section 5.22 of the BATF regulations must be aged a minimum of two years. To be designated a 'straight' whisky, it must conform to all regulations for its type and be aged not less than two years. 'Light' whisky is another type of whisky produced in the US. It is distilled at more than 160° but less than 190° proof and aged at least two years in used or uncharred new oak barrels.

In the US regulations, neutral spirits, vodka, Scotch whisky, Irish whiskey, and Canadian whisky are further defined. Neutral spirits are distilled spirits produced from any material distilled at or above 190° proof and bottled at not less than 80° proof. Vodka is a neutral spirit distilled and treated with charcoal or other materials to be without distinctive character, aroma, taste or color. Scotch, Irish and Canadian whiskies are defined as 'distinctive products of Scotland, Ireland, and Canada, respectively, and produced and distilled under the laws of those countries'.

History of North American whisky production

Whisky production began in the US in 1733 when the British government passed the Molasses Act. Until that time, the colonists produced distilled spirits from molasses. The Molasses Act imposed a duty on molasses of non-British origin. Since the American colonists imported most of their molasses from the French and Spanish islands, they were greatly concerned. Since non-British molasses was cheaper and more abundant, smuggling and ignoring the Molasses Act (and the later Sugar Act) was the basis for much of the 'Spirit of '76'.

Pre-revolution grain whisky production was small; although history notes that settlers in western Maryland and Pennsylvania produced rye whisky from their abundant rye grain crops and that rye whisky began to replace the popular molasses-based rums. After the Revolution, the Embargo Act cut off the supply of molasses; and with abolition of the slave trade by the new Congress, both molasses and slaves were smuggled into the US. These events increased the cost of molasses and accelerated the decline of rum.

The westward migration

Early settlers crossing the Allegheny Mountains included many Scots and Irish immigrants who were grain farmers and distillers with knowledge of pot still operation from their homelands. They produced the rye whisky that became the first 'American' whisky. When Alexander Hamilton needed money to pay the debts incurred during the American Revolution, he pushed an excise tax levied on distilled spirits through Congress. As news of the tax spread, the uproar and public outrage was so intense that President Washington sent 13,000 troops into western Pennsylvania to quell the 'Whisky Rebellion'. As the troops entered from the east, many farmer-distillers packed their stills and headed west to Kentucky to avoid both the tax and the army.

The farmers found Kentucky soils not as suitable for rye and wheat crops as soils in Pennsylvania and Maryland. They discovered that corn was much easier to cultivate. The first writing that expounded on corn growing in Kentucky comes from the Jesuit Hierosm Lalemont. He noted 'to mention the Indian Corn only, it puts forth a stalk of such extraordinary thickness and height that one could take it for a tree, while it bears ears two feet long with grains that resemble in size our large Muscatel grapes' (Carson, 1963).

Whisky production grew rapidly in the early frontier areas as the settlers found in whisky a means of moving grain to market. A pack horse could carry only four bushels of corn, rye or wheat; but that same horse could carry 24 bushels of grain that had been mashed and distilled into two kegs of whisky. Also, the price of whisky was more than double the price the farmer could get for grain.

Bourbon's 'accidental' history

As settlements moved westward, the demand for spirits increased. Riverboats had become a means for shipping barrelled whiskies to their destinations. A version of 'bourbon history' recounts how a Baptist minister, Elijah Craig, burned or 'charred' the inside of fish barrels to rid them of the fishy smell so he could fill the barrels with whisky to be shipped by raft down the Mississippi River to New Orleans. The whisky from the charred oak barrels 'aged' during shipping and storage. This aging improved the character of the whisky, gave it color and smoothed the taste (Carson, 1963).

Another version of this history tells of a careless cooper who accidentally let the staves catch fire (char) when heating them for pliability to make into barrels. Not wanting to lose money, he did not tell his distiller customer about the charred staves in the barrels. Months later, after the distiller filled the barrels and shipped the whisky downriver, the distiller heard pleasing compliments about his whisky. After discovering the cooper's 'mistake', the distiller asked him to repeat the charring process for all of his barrels.

Contrary to popular belief, none of this 'history' occurred in Bourbon County, Kentucky, and no one really knows how bourbon whisky was first made. The only historical evidence indicating Bourbon County as the source of bourbon whisky comes from a 1787 indictment of James Garrad (later a Kentucky governor) and two others by a Bourbon County grand jury for retailing liquor without a license. The only certainty in any of the lore is that Kentucky has a county named Bourbon and produces a whisky by the same name (Connelley and Coulter, 1922).

Essential traditions

Despite uncertainty about origins of the bourbon name, a tradition of good whisky making was handed down from fathers to sons for generations. Formulas, mash bills, yeasting methods and skills for operating the stills were passed along, even though many farmer-distillers could not read or write. They did not know acrolein from fusel oil, but they did have the special knack for making the 'cuts'. They knew good, clean yellow corn and plump rye. They faithfully guarded their yeast and yeast methods though many could not have said whether yeast belonged to the animal or vegetable kingdoms. Two exceptional bourbon whiskies of the 19th century were Old Taylor and Old Crow. Old Overholt was reportedly the best of the rye whiskies.

During the 19th century, the pot still evolved into the continuous column 'beer still' with a doubler or thumper. The continuous still operations allowed distillers to move to larger fermenters, more and larger cookers and automated grain handling. As the distillery operations grew and became increasingly automated, some of the smaller distilleries fell by the wayside or promoted their brands as better as a result of their 'old time, small distillery tradition and quality'. They touted this tradition and sold it as part of the product. The well known Maker's Mark bourbon is a prime example of tradition and sound practice bottled and successfully marketed to modern consumers.

As grain whisky and bourbon production grew in the 19th century, the US government increased the excise tax and the number of regulations. Costs passed on to the consumers dampened their enthusiasm for drink; but government regulation incensed leaders of the Temperance Movement because the tax and regulations drew attention to the production and sales of liquor. Most whisky at that time was sold 'from the barrel', and quality standards were almost nonexistent. It was not until the end of the century that consumers could purchase whisky sold in a corked and sealed bottle. Old Forester was the first product with 'guaranteed quality' put on the label. Old Heritage was the first bourbon with a strip stamp over the cork, thereby becoming the first 'bottled-in-bond' bourbon.

The prohibition era

All of the improvements in the 'character' of whisky production and consumption were to no avail when at 12:01 a.m. on Saturday, January 17, 1920 all beverages containing more than 0.5% alcohol were outlawed in the US (Tennessee, the state with the first registered distillery in the country, Jack Daniel Distillery, became the third state to vote to go dry in 1910, ten long years before the passing of the Volstead (Alcohol Prohibition) Act.) The 18th amendment to the Constitution, which prohibited the production and sale of alcoholic beverages, sounded a death knell for many distilleries. A drive through the countryside revealed closed distilleries choked with weeds with facilities in ruins. Brown-Forman Distillery in Louisville, Kentucky, was one of the few that survived because it produced its Old Forester Bourbon 'for medicinal purposes only'. Other distilleries lay in wait for 'the experiment' of prohibition to end.

Passage of the 21st amendment ended Prohibition after 13 years. At midnight on April 7, 1933 wines and beer were again legally sold; and on December 5 of that year at 5:32 p.m. bour-bon was again on the market. The American distilled spirits industry surged into production, building larger, more modern distilleries. The resumed legal production and sales were reinforced by the federal government when it passed the Federal Alcohol Control Act, which eliminated the sale of bulk whiskies to the wholesale and retail trades. This Act also formed the Alcohol and Tobacco Tax Division of the Internal Revenue Service. Though creating bur-eaucracy and taxation, the new Act also set regulations and definitions for producing spirits in the United States.

Distillation's new era

The new regulations defined the production of bourbon, rye, corn and blended whiskies as well as gin, brandies, rums, cordials and vodka according to their spirit type. The Act also noted the use of geographical designations with an origin defined for Scotch, Canadian, and Irish whiskies. This evolution of the industry led to the organization of companies that had goals of producing top quality spirits within the new regulations. New companies such as National Distillers and Schenley joined the American distilleries that survived prohibition, Brown-Forman and Jim Beam, along with the Canadian distillers, Seagrams and Hiram Walker. These distillers based their production methods on precisely defined procedures from yeasting to maturation.

During World War II, North American distilleries ceased whisky production and began manufacturing industrial alcohol for the war effort. Distillers gained the resources for further technical improvements; and at the War's end, the industry in the United States was technically ready to produce better bourbons, ryes, and Tennessee whiskies than ever before.

While the basis for modern American whisky production was developed during World War II, today's modern American distillery operates with recent innovations in controls (including computerization) and heat reclamation as the most significant advances. Even with all of the modern technology, the US distiller still carefully controls his yeasts, mash bills, distillation methods and maturation criteria, essential factors for good quality products initiated by the early pioneer distillers.

Production and maturation operations

In the production of American whiskies, six factors determine the character and flavor for each type of whisky:

1) Grain proportions in the mash bill
2) Mashing technique
3) Strain of yeast
4) Fermentation environment
5) Type and operation parameters of distillation equipment
6) Type of barrel used and the maturation process

These factors were recognized and carefully controlled in the distilling operations of the early settlers (Lyons, 1981). Families and companies that ensured consistency and control over these factors remain in business today. Those who failed to strictly adhere to a regimen controlling each factor have fallen by the wayside.

A matter of distinction

Popular bourbon producers use specific process differences to make their products distinct. The Jack Daniel Distillery continues to produce its whisky with the same processes used more than 100 years ago. Their strict adherence to tradition along with a down-home, folksy image have proven successful marketing tools for the whisky.

At Maker's Mark, grain selection and proportions are used to produce a superior whisky. They use wheat instead of rye in the mash bill. The wheat and good, consistent control of all factors, especially the barrels, make Maker's Mark the smoothest of the bourbons. Their claim of 'handmade' is justified because of their intense attention to production parameters.

All American whiskies maintain standards for the six production factors; and the variations among distilleries in adherence to standards for these factors determine flavor and cost differences. All American distillers start with a careful grain purchasing program. Though price is a criterion, they all use #1 or #2 yellow corn, #1 plump northern rye, and choice northern malted barley. Any off odors or below-grade grain are rejected at the distillery. Very stringent grain standards are a common feature of all American distilleries (Table 1).

Mash bill

The mash bill may vary, with a typical bourbon having a mash bill of 70% corn, 15% rye and 15% malt. A typical Tennessee whisky may have 80% corn, 10% rye and 10% malt while a typical rye whisky will have a mash bill of 51% rye, 39% corn and 10% malt. All grains are ground, with the hammer mill being the most common type of processing; however some roller and attrition mills are still in use. The milling (Figure 1) is checked for grind by a sieve analysis. A typical

Table 1. Specifications and analyses of corn, rye, wheat and malt.

	Specification	*Typical analysis*
Corn (#2 recleaned)		
Grain odor	No musty, sour or off odor	'Meets spec'
Moisture, %	14.0 (maximum)	12-14
Cracked grains and foreign material, %	2.0 (maximum)	1-2
Damaged kernels, %	3.0 (maximum)	0-1.5
Heat-damaged kernels, %	0.2 (maximum)	0-0.1
Bushel weight, lbs	55.0 (minimum)	55-60
Rye (#1 plump)		
Odor	None 'Meets spec'	
Moisture, %	14.0 (maximum)	10-14
Thins, %	2.0 (maximum)	1-2
Dockage, %	2.0 (maximum)	1-2
Bushel weight, %	56.0 (minimum)	56-60
Malt		
Bushel weight, lbs	35.0 (minimum)	35-38
Moisture, %	6.0 (maximum)	4-6
α-amylase	60.0 (minimum)	60-64
Diastatic power	22.0 (minimum)	22-26
Bacteria count, CFU/g	1,000,000.0 (maximum)	400-500,000
Wheat		
Odor	None 'Meets spec'	
Moisture, %	14.0 (maximum)	10-14
Thins, %	2.0 (maximum)	1-2
Dockage, %	2.0 (maximum)	1-2
Bushel weight, lbs	56.0 (minimum)	56-60

sieve analysis for grains in a bourbon mash bill is shown in Table 2.

Mashing

Mashing techniques vary considerably, but the major difference is whether pressure or atmospheric batch cooking is used. Bourbon, rye, wheat, Tennessee and corn whisky are mashed using batch cookers. Only the 'blend' or 'light' whisky producers use continuous cookers. Pressure cooking is usually done at 124°C while atmospheric cooks are done at 100°C. Cooking time varies from 15 minutes to 1 hr. Conversion time and temperature are very consistent among distilleries. Malt is never subjected to temperatures greater that 64°C; and conversion time is usually less that 25 minutes to minimize contamination. All distillers use backset (centrifuged or screened stillage from the base of the still), but the quantity of backset will vary based upon the beer gallonage (gallons of water per 56 lb distillers bushel of grain) to be used. American whiskies have beer gallonages in the 30-40 gallon range. High energy costs for by-product recovery have encouraged some distillers to use lower beer gallonage ratios for spirits. However, bourbon and Tennessee whisky producers continue to use 30-40 gallon beers. The cooling of cooked mash to fermentation temperature is achieved using vacuum (barometric condensers) or cooling coils.

Yeasting

All whisky producers use *Saccharomyces cerevisiae*, however the yeasting techniques vary tremendously between the 'modern' and 'traditional' distillers. The modern distillers have elaborate yeast laboratories and will propagate a new yeast from an agar slant every week. They are very aseptic and accurate, assuring continuity of the same flavor. The 'traditional' distillers use yeast stored in jugs; and though they backstock weekly, the potential for gradual yeast culture changes and contamination can lead to flavor variances. These distillers take extra effort and care to ensure that their yeasting does not cause ester, aldehyde or fusel oil variances in the distillate.

The most common grains used for yeasting are small grains, rye and malted barley. These grains are cooked in a separate cooker to about 63°C, and the pH is adjusted to 3.8 with lactic acid bacteria grown in the yeast mash. Lactic acid production is then stopped by increasing the temperature to 100°C for 30 minutes to kill the bacteria. This aseptic, sterile mash is then ready for the yeast from the dona tub grown in the laboratory (Figure 2). The yeast fermentation temperature is controlled at 27-30°C; and the yeast propagates until the Balling drops to half the original 22° Balling reading. This yeast mash will have a yeast concentration of 400 million cells/ml. Both modern and traditional distillers regularly have clean, sterile yeasts free of bacterial contamination that may cause side fermentations and unusual congeners in the

Table 2. Typical sieve analysis for grains in a bourbon mash bill.

US Sieve #	*Corn*	*Rye*	*Malt*	*Wheat*
16	15	22	2	20
20	21	25	8	26
30	17	13	14	12
40	13	9	16	8
50	10	7	13	6
60	3	2	8	2
Through 60	21	22	34	22

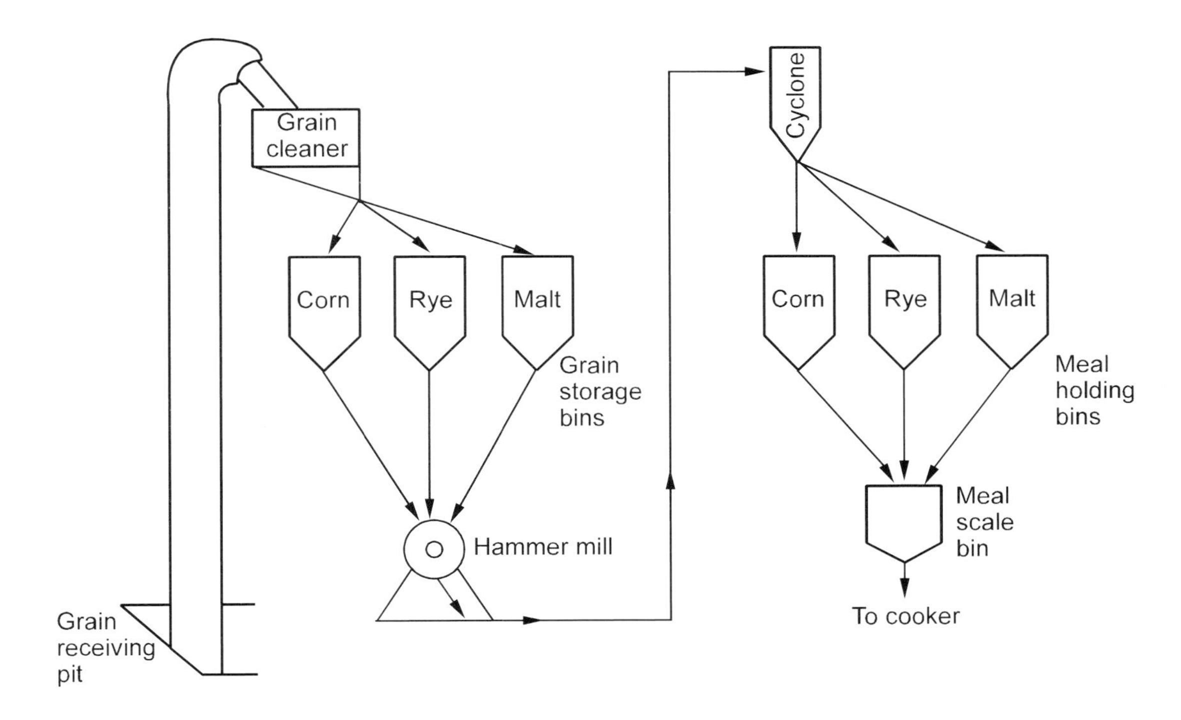

Figure 1. Typical grain-handling and milling facility.

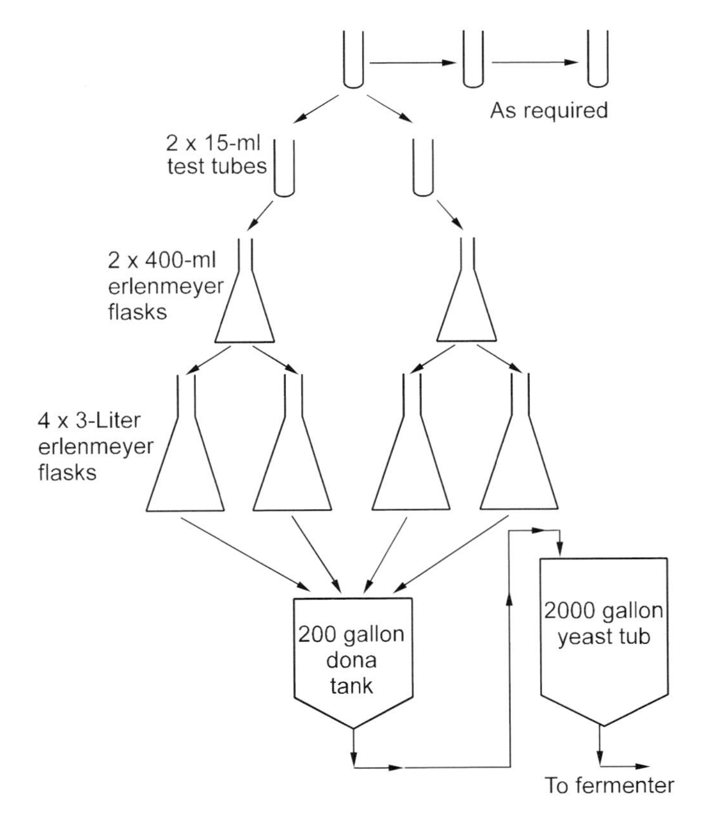

Figure 2. Stages in yeast propagation.

distillate. The 'lactic souring' and the alcohol content of the finished yeast mash (8%), along with sterile dona and yeast tank methods contribute to the excellent reputation American whiskies have for fermentation congener consistency. The advantage of using small grains are: preservation of enzymes for secondary conversion, low steam requirements and shorter processing time. Also, because of its nutrient value, barley malt is the most important constituent of yeast mashes. Corn is not used in a yeast mash because it does not contain the growth factors required for yeast and lactic bacteria growth.

Fermentation

Fermentation is the simplest part of the production process, but requires more control with efficient equipment in order to have stable, consistent results. After the two or three cooks required are completed, cooled, and transferred to the fermenter, the fermenter is 'set'. Setting the fermenter means filling the fermenter with cooked mash, inoculated yeast and backset. The yeast mash is pumped in as soon as the first cook is added to the fermenter. The addition of backset and/or water is done at the end of filling to bring the fermenter to the desired final beer gallonage. Most distillers use a 30-36 gallon mash, and the water:backset ratio determines the set pH. Set pH values of 4.8-5.2 are considered to be the best starting point. The modern distillers have closed-top fermenters with cooling coils or external heat exchangers to control fermentation temperature. They usually set fermenters at 27-29°C and control at 30-31°C. Traditional distillers will have metal or wood open-top fermenters without any device to control fermentation temperature. They usually set their fermenters as cool as they can (18-21°C) and let the fermenters work up to 31-32°C. All of this is controlled by mash cooker cooling, addition of cold water and weather factors. Contamination is controlled by cleaning fermenters, ensuring no mash pockets in pipes and regular steaming of fermenters.

Both traditional and modern distillers ferment their beers for at least 72 hrs, and some for as long as 96-120 hrs. Three and five day fermentations are the norm. During these periods the Balling will drop to 0.0 and the pH from 5.0 to 3.8 while the alcohol concentration rises to 8-10%. All the changes that happen during fermentation are checked daily by performing 'beer chemistry'. Balling, pH, acids, and fermenter temperature are monitored and recorded daily. The pH is the main indicator of contamination and potential fermentation problems and is regularly measured by traditional and modern distillers (Table 3).

Distillation

Upon completion of the fermentation process, the beer with 8.0-10.0% alcohol is transferred from the fermenter to the beer well. The beer well is a holding tank for the fermented beer, such that a continuous feed to the beer still can be maintained. Beer wells are usually 1-1.5 times the size of a fermenter. They also have continuous agitation to prevent solid grain particles from settling to the bottom of the vessel. All American whisky producers use a continuous still (Figure 3), though some have a second distillation 'doubler' or 'thumper' (Figure 4). The basic difference between a doubler and a thumper is whether the unit is operated with a liquid level (doubler) or essentially dry (thumper). Both the doubler and the thumper provide a second distillation.

The beer is pumped into the upper section of the first continuous column, the beer still, six to ten plates from the top. Live steam is introduced at the bottom. Beer stripping plates 1-18 have perforations, a downcomer from above and a dish on the plate below to hold the beer liquid at a set level so the plates are never dry, as the beer moves back and forth across each plate. The steam passing up and through the perforations, controlled by pressure or flow rate, strips the lighter, more volatile alcohol from the water/grain mixture on the plates. When

Table 3. Typical analysis of beers from bourbon, rye and corn whisky production.

	Bourbon	*Rye*	*Corn*
Set sample			
Balling	13.4	13.4	12.3
Titratable acidity	4.5	3.6	2.8
pH	4.5	5.0	5.2
Temperature, °C	27.0	24.0	27.0
24 hr sample			
Balling	2.6	4.0	3.6
Titratable acidity	5.1	4.6	4.2
pH	4.2	4.4	4.3
Temperature, °C	30.0	31.0	29.0
48 hr sample			
Balling	2.4	3.6	1.0
Titratable acidity	7.8	7.5	6.1
pH	3.8	3.9	3.8
Temperature, (°C)	30.0	30.0	30.0
Drop sample			
Balling	0.4	1.5	-0.4
Titratable acidity	8.2	7.9	7.1
pH	3.8	3.9	3.8
Temperature, °C	30.0	30.0	30.0
Alcohol, % by volume	6.73	5.8	6.8
Residual carbohydrates, %	8.0	8.4	4.2
Residual carbohydrates, % maltose	0.73	0.6	0.46

this alcohol gets to the top 4-8 bubble cap plates it is concentrated to about 100° proof (50° GL). As the vapors continue up into the beer preheater (a heat exchanger to heat beer going into the column), some alcohol is condensed and refluxed to the top of the beer column. The rest flows to the doubler or thumper as vapor. From either of these pot still-type chambers, the vapor goes to the condenser where the product is drawn off at 130-140° proof (65-70° GL). Bourbon cannot be distilled above 160° proof (80° GL). If it comes off at more than 160° proof, it must be called 'light' whisky. The stillage from the base of the beerstill is pumped to the dryer house to be processed. The 'high wine' from the stills is pumped to the cistern room where it is held, tested and reduced to barrelling proof (Table 4).

By-product recovery

The by-product recovery does not usually receive the same care that the other processes demand. The only essential for this process is ensuring that the backset stays hot, around 99°C, so that it is absolutely sterile when used in the mash tubs, yeast tubs and fermenters. (Most distillers use 20-30% backset in their total process.) Furthermore, the hotter the backset remains, the greater the savings in energy costs during the cooking process. The stillage from the base of the beer still has about 7-10% total solids. When used for backset it is screened or centrifuged to prevent solids accumulation in the cooker and fermenters. Nearly all distillers have a dryer house, though a couple of traditional distillers continue to sell their 'slop' to nearby farmers.

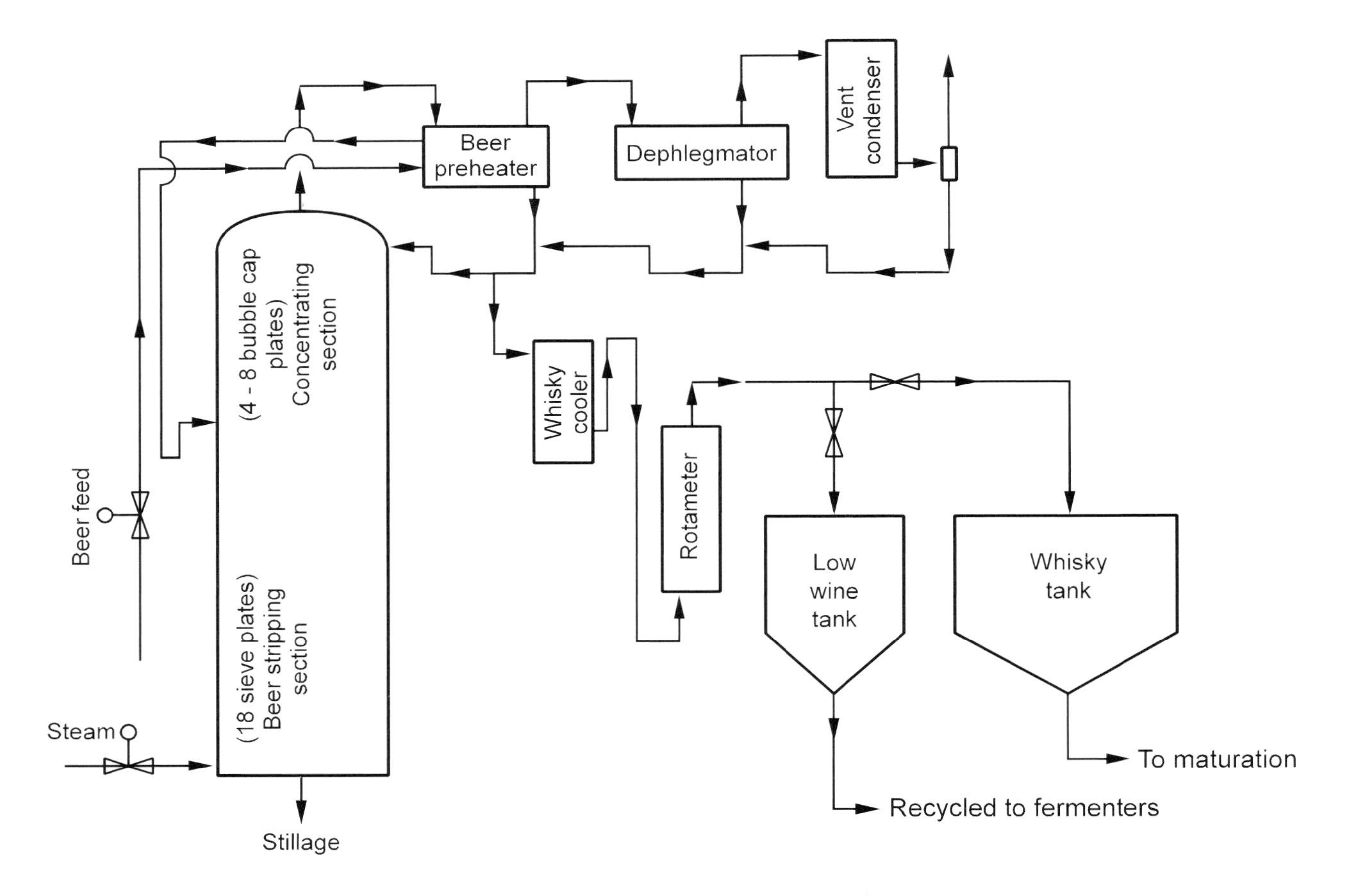

Figure 3. Bourbon whisky still.

Table 4. Typical operating data for whisky distillations.

	Bourbon	*Rye*
Product proof	130°	130°
Still pressure (inches of water)	48	42
Steam rate, lb/hr	12,000	12,000
Beer feed rate, gallons/minute	120	117
Reflux rate from beer preheater, gallons/hr	350	300
Reflux rate from dephlegmator, gallons/hr	700	750
Reflux rate from vent condenser, gallons/hr	30	65
Draw-off of product, gallons/minute	12.5	10.0
Still losses, %	0.0004	0.00025
Water temperature to vent condenser, °C	21	21
Water temperature from dephlegmator, °C	79	79
Beer solids, %	0.06	0.06

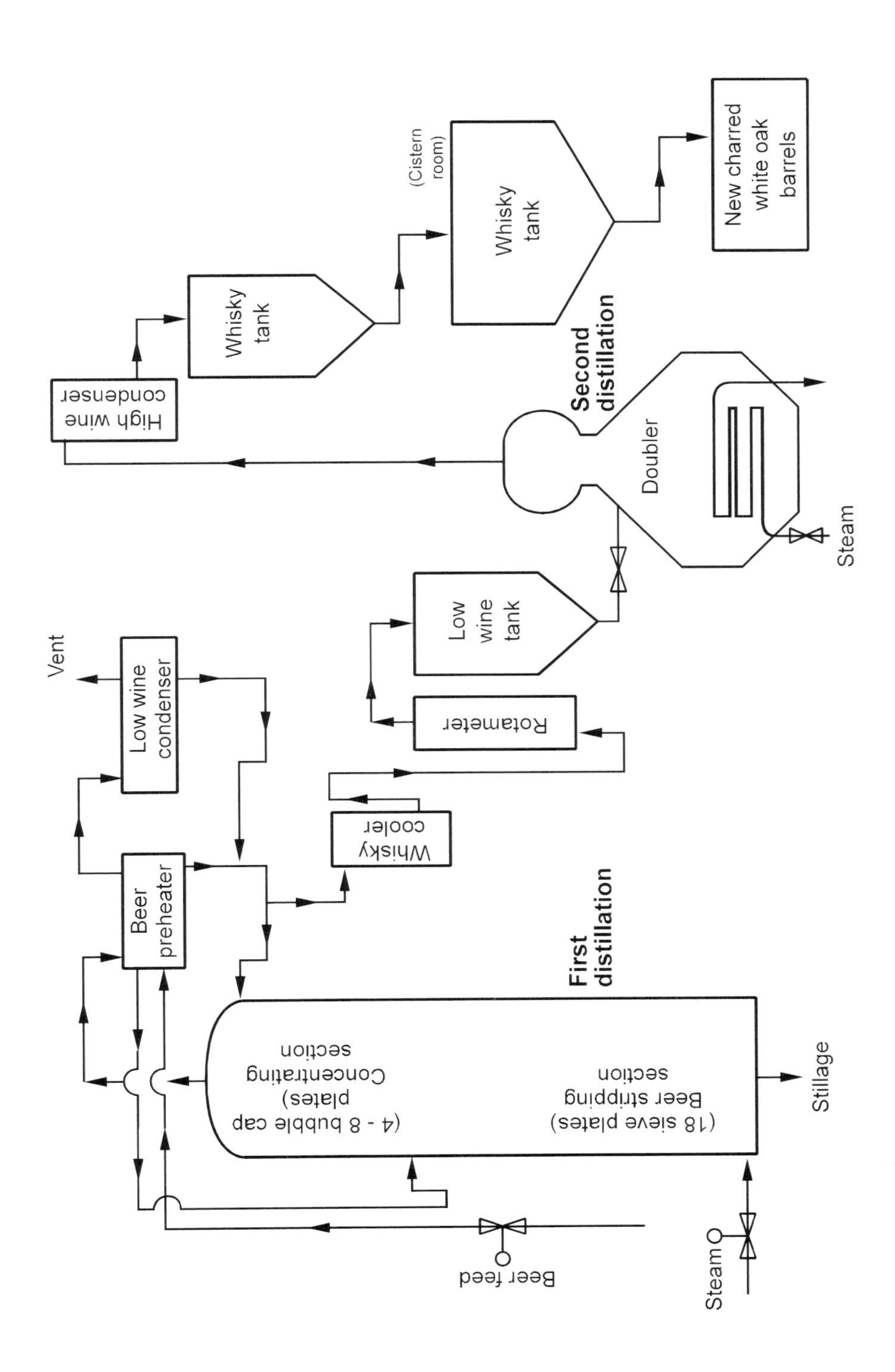

Figure 4. Bourbon whisky distillation system, including doubler.

For a while some, like the Jack Daniel Distillery, fed the stillage to cattle in wet-feeding operations; but environmental restrictions have generally eliminated such practices. Also, modern dryer house operations have become very profitable and the profit helps lower the cost of making whisky.

A modern dryer house will have a whole stillage tank, a centrifuge, press or screen device, thin stillage tank(s), fiber dryer, evaporator and dried grains storage bins. The centrifuge, press or screening device separates the heavy, fibrous grains, which are fed to a dryer. The thin stillage is fed into an evaporator and concentrated to a syrup with 40% solids. This syrup is mixed with dried fibrous material in a mixer and fed to the dryer(s), where it is dried to 8-10% moisture (Table 5).

Distillers dried grain is high in protein (24-30%), and is sold primarily in the cattle feed market. One distiller dries the fiber and solubles portions separately, selling the fiber portion based on its fiber content and the solubles as distillers solubles in the high protein (40%) markets. The distillers dried grains with 24-30% protein will sell for $180/ton, making it a very profitable by-product. Distillers solubles will bring a greater return owing to its higher protein content. In the modern distillery, the value of the by-product recovered will be credited against grain costs and is a major factor in maintaining low proof gallon production costs.

Cooperage and maturation

Cooperage and maturation are the processing factors that distinguish bourbon and American whisky from the other whiskies of the world. Only bourbon whisky regulations require that it be matured in a new charred, white oak barrel. Other whiskies of the world may require the use of small wooden barrels, but no other whisky but bourbon goes through the care and expense of requiring small white oak barrels to be freshly charred.

The making of the bourbon whisky barrel is a very traditional, but exact science and craft. Staves and heading are quarter-sawed from mature, white oak timber. Actually, some physical variance exists within a single tree, but no more than between trees. After being quarter-sawed with the medullary rays not less than 45 degrees to the stave surface, the staves and heading are air-dried in a stave yard. The more traditional distilleries require wood to be air-dried at least one year. The modern distillery usually has no air-drying specifications and allows its barrels to be produced from wood that has been air-dried for six months or less. All of the wood is kiln-dried at the cooperage, with staves and heading dried to 12 and 10% moisture, respectively. The kiln-drying is essential to prepare the wood for the planing, milling, edging and joining operations that cannot be done on wet wood. More important, however, is that proper drying (air-drying a year followed by kiln-drying),

Table 5. Specifications and analyses for wheat, rye and corn distillers dried grains (DDG).

	Wheat and rye DDG		*Bourbon and corn DDG*	
	Guaranteed analysis	*Typical analysis*	*Guaranteed analysis*	*Typical analysis*
Moisture, %	-	5.0	-	7.0
Protein, minimum %	20.0	22.0	25.0	27.7
Crude fat, minimum %	5.0	5.5	5.0	6.6
Crude fiber, maximum %	17.0	14.8	16.0	13.7
Ash, %	-	2.2	-	1.7

makes the wood chemistry satisfactory for flavoring the whisky during maturation.

As the whisky goes through maturation (3-4 years for the modern distillers and 4-8 years for the traditional distillers), two distinct types of reactions occur: reactions between the distillate components (regardless of the barrel); and reactions that occur when the distillate extracts chemical compounds from the wood. A major factor differing among the distillers is the warehouse environment. Most modern distillers heat their warehouses in winter. One particular distiller even controls the heat cycles to ensure constant aging. The traditional distillery seldom has heated warehouses and depends strictly on Mother Nature to determine the number and range of its heating and cooling cycles.

Whether the heat cycle is natural or forced, the greatest rate of change or formation of congeners occurs in the first 12-16 months. Only ester formation occurs at a fairly constant rate. Proof increases at a fairly constant rate of 4-5% each year of aging. Other specific changes occuring when the distillate reacts with the charred wood are: a) aldehyde formation, specifically acetaldehyde, which comes from the alcohol via oxidation, b) acetic acid formation with greatest activity in the first year of maturation, and c) ester formation (ethyl acetate) from the alcohol via oxidation. The components coming from the wood are tannins, sugars, glycerol and fructose. The hemicellulose in the wood appears to be the source of the sugars found in aged American whiskies.

The depth of charring and 'toast level' in the barrel determines the color of the whisky. Color formation is almost instantaneous when the distillate is put into the charred barrel, with 25-30% of the color formed in the first six months. Some color development occurs each year of maturation until the whisky is dumped from the barrel. The final product is then filtered, its proof reduced with demineralized water and bottled. Compared to the sweet rums, 'breathless vodka', fruity gins, light Canadians and Scotches, the American whiskies have flavor and bravado with a balance that is pleasant to the taste. American whisky produced and matured as described has a big, pungent aroma that leaves no doubt that it is bourbon or Tennessee whisky.

Though the modern and traditional distillers have different levels of technology, they both use the same basic processes for making bourbon or other American whiskies. Individual plant nuances produce different flavors, but they all have an American whisky bouquet and taste.

References

Carson, G. 1963. The social history of bourbon. Dodd, Mead & Company, New York.

Connelly, W.E., and E.M. Coulter. 1922. History of Kentucky, Vol. II. The American History Society, Chicago and New York.

Lyons, T.P. 1981. Gasohol, A Step to Energy Independence. Alltech Technical Publications, Nicholasville, Kentucky.

Chapter 15

Production of tequila from agave: historical influences and contemporary processes

M. Cedeño Cruz and J. Alvarez-Jacobs
Tequila Herradura, S.A. de C.V. Ex-Hda San Jose del Refugio, Amatitán, Jalisco, México

Introduction: Agave beverage alcohol products

Tequila is classically associated with Mexico, particularly with Jalisco, a state located in the west of the country. This beverage is obtained by distillation of fermented juice from only the agave plant *(Agave tequilana* Weber var. *Azul)* if 100% agave tequila is required. Fermentation is carried out by inoculated strains of *Saccharomyces cerevisiae* or in some cases by a spontaneous process. Up to 49% (w/v) sugars may come from a source other than agave, usually sugar cane or corn syrup, if 100% agave tequila is not required. This tequila could not be labeled as 100% agave tequila.

Different kinds of tequila are produced in 50 registered companies in Jalisco. The tequila products made by these companies differ mainly in proportions of agave used, production processes, microorganisms used in the fermentation, distillation equipment used and the maturation and aging times. The product known as 'silver' or white tequila must be distilled to a concentration not exceeding 55% alcohol in volume (v/v) and not less than 38% (v/v) from a fermented wort containing not less than 51% sugars from the agave plant. 'Gold' tequila is the white product to which caramel color (generally) has been added. Rested tequila (Reposado) and aged tequila (Añejo) are white tequila matured in wood containers, preferably oak casks, for at least 2 and 12 months, respectively, according to present regulations. In practice, the aging period is normally longer and depends on the characteristics each company wants to give to the final product. The products known as 100% agave, which are becoming more popular, could be white, rested or aged and are distilled from fermented wort with only agave as a source of sugars. The labels of these products must indicate that they were obtained using only agave and under Mexican government supervision.

Tequila is differentiated from the beverage known as 'mezcal' by the type of agave used in its elaboration. Mezcal is made from *Agave potatorum,* which grows in the state of Oaxaca. Most mezcal producers use a rudimentary fermentation and distillation process (Sánchez, 1991). There is no technical reason for, or any improvement in the organoleptic characteristics from, the worm inside the bottles of some mezcal brands. The worm is primarily a commercial ploy. Worms are grown in agave plants and introduced manually in the bottling line.

'Pulque' is another beverage obtained by fermentation of the juice obtained from several species of agave, *A. atrovirens* and *A. salmiana* among others, by a complex succession of yeast and bacteria that produce ethanol, a diversity of chemical compounds, and some polymers that give a sticky consistency to the final product (Rzedowski, 1978; Sánchez Marroquin and Hope, 1953). Pulque is sometimes mixed with fruits or vegetables, but has poor stability as it is neither distilled nor pasteurized.

Agave plants still serve as food in some states of México; and other fermented regional beverages are produced (e.g., 'Sotol' in the state of Chihuahua and 'Bacanora' in the state of Sonora), but only tequila and more recently mezcal have reached international recognition. Another difference between tequila and mezcal and all other regional drinks is that both are subject to an official standard that for tequila is NOM006-SCFI-1993 (Secofi, 1993), and production is supervised by the Mexican government.

Origin and history of tequila

The word 'tequila' is believed to originate from the tribe of *ticuilas* who long ago inhabited the hillside of a volcano bearing the same name located near the city of Tequila. Another possible origin is the Nahuatl word *tequitl,* which means *work* or *employment*, and the word *tlan,* which means *place*. Therefore 'tequila' would mean *place in which labor or work is done*.

The most ancient information revealing the existence of agave and its different uses is from the era before the Spaniards in several codices preserved to the present time (a codex, from the Latin *codex* meaning board or writing tablet, is a manuscript volume, especially of a classic work or scripture). The most important is the *Tonalmatlnahuatl* codex, which notes that certain tribes had learned to cook agave plants and used them as food and to compensate for the lack of water in desert lands. Also, these tribes discovered that cooked agave when soaked in water would ferment, producing a very appreciated beverage. In fact, this primitive and rudimentary method was used for centuries to produce beverages from agave, considered a sacred plant possessing divine properties. In other codices, such as *Nutall, Laud, Borgia* and *Florentine,* there are many references to uses of the agave plant for soap manufacture, a source of fiber, footwear, medicine and sewing needles as well as thread, paper and rope. In fact, Indians could distinguish the different species of agave by color, size, stem, leaf width and the different uses given each plant (Muria, 1990). The great religious importance of agave was apparent in those codices, as only warriors and priests used fermented drinks in ritual ceremonies.

In prehispanic México the general name for all species of agave (or mezcal as it is also known) was *Metl,* which is a representation of the goddess *Mayahuel.* The alcoholic drink produced was called *Iztac octli* (white wine). The first Spaniards to arrive in México referred to the plant as 'maguey', a name used for an identical plant they had seen in the Caribbean Islands, where they first encountered new world plants and animal life (Bottorff, 1971). It was not until arrival of the Spaniards, who brought knowledge of distillation techniques, that tequila took its present form (Goncalves, 1956).

There are only two main regions for tequila production in México. The oldest, the Tequila-Amatitán region that comprises the Amatitán village, developed at the end of the seventeenth century. The second region, the Jalisco Highlands, appeared in the last decade of the 19th century (Luna, 1991). The first tequila production process with a commercial purpose was established in the city of Tequila around the end of the 18th century. The main consumers were in the mining zones located in the state of Jalisco. The Spaniards tried to suppress consump-tion of tequila in order to reduce competition for brandy and other wines imported from Spain with a decree signed by Carlos III forbidding its sale and production under the pretext that its consumption was the cause of several illnesses.

The results were negative. The governor of the region later issued a decree imposing a tax on tequila in order to enrich the royal coffers, thus permitting its sale in all of New Spain. By the end of the 19th century, expansion of the tequila industry, helped by the railway, was evident. However it was not until the first casks were exported to the US that tequila was known beyond México's borders.

The Agave plant

According to Granados (1985), the genus *Agave*, which means 'noble' in Greek, was defined by Linneaus in 1753 when he described the plant *A. americana* as the first agave species known to science. Agave plants, which are often confused with cacti, belong to the family Agavaceae and are succulent plants with spirally-arranged leaves forming a rosette. Some have definite trunks, but more often they are nearly stemless. The leaves are bluish green in color, over 1 m long in mature plants, and end in a sharpened brown thorn. As Backman (1944) pointed out, the widespread distribution of some 300 species, combined with the fact that the plants require approximately 8-12 years to mature and hybridize very easily, make the taxonomic and phylogenetic study of the genus *Agave* extremely complicated and very difficult. The family Agavaceae includes 20 genera and nearly 300 species. Of these, around 200 are found in México.

Several agave species are important from an economic point of view. Fiber derived from *A. foucroydes,* grown in the state of Yucatán, is known worldwide for its use in producing ropes and carpet. More recently, the pulp remaining after removal of the fiber has found use in animal feed. *A. salmiana* and *A. atrovirens* are valued for pulque production and *A. potatorum* for mezcal production. Finally, *A. tequilana* Weber var. *Azul,* named around 1900 by the German botanist Weber (Diguet, 1902), is used to produce tequila.

Cultivation and harvest

The blue agave, as it is known, is the only species out of hundreds of Agavaceae with the appropriate characteristics for tequila production. These include a high inulin concentration, low fiber content and the chemical compounds present in the plant that contribute to the final taste and flavor of tequila to give the beverage its particular character. Some have attempted to produce tequila in other states with other agave species, but without success. The blue agave is cultivated in the state of Jalisco in the two regions with the right climate and soil composition for its growth, namely the Highlands and the Tequila-Amatitán region. The temperature conditions for good agave yields are a minimum of 3°C, an optimum of 26°C and maximum of 47°C. Soil should be fertile but not very deep, 30 to 40 cm. Good drainage is required to avoid effects of flooding, which are very harmful for development of the agave. The plants must be planted at 800 to 1700 m above sea level where the annual rainfall is about 800 to 900 mm. The correct planting time is immediately before the rainy season, from June to September, so that the plants do not suffer from water stress during the first year of growth.

Propagation is accomplished by the vegetative route in agave. Sexual reproduction via seeds is not usual. Asexual bulbils develop in the inflorescence at the base of the flowers, producing small plants that after some time detach themselves from the floral peduncle and fall to the soil where they root. Another mode of asexual propagation is by suckers, which are a characteristic type of lateral bud or branch developing at the base of the main stem. Plants developing near each mother plant are separated at the age of 3-4 years. These baby plants are called 'first-class seed' because they are better and healthier (Sánchez, 1991). In practice, people use the word 'seed' to refer to such young plants, but from a botanical point of view these are rhizome shoots or suckers (Valenzuela, 1992). Plant cell culture is used experimentally by some agave producers, but is not being used commercially due to unavailability of trained technicians and laboratory facilities to small agave producers. Some developments are being carried out to improve agave plants, or to obtain plants resistant to pests; and these new plants will be ready in the next two years (CIATE, 1993).

Land for agave cultivation must be cleared and deep-ploughed, sometimes twice. Agave is planted approximately 2-4 m apart in straight lines called ruts. Sowing is done by hand in holes 15 cm in depth. Plant density is around 2000-4000 plants per hectare, depending on the plantation system used; and yields can be between 30,000 and 200,000 kg/ha, assuming that the weight of a harvested plant varies from 15 to 50 kg. This variation is caused by differences in soil conditions, quality of plants sown, rainfall, pests and fertilization. Sometimes agave is sown intercalated with nitrogen-fixing crops such as peanuts, beans, chickpeas or soybeans. After the agave has been in the soil for a year, visual inspection is carried out to replace sick or dead plants with new ones. This operation is called re-seeding. The percentage of dead plants depends on soil and plant characteristics, but is generally from 8 to 15% (Pérez, 1990).

Agaves regularly host borer insects that live in the stems, leaves and fruits during the larval stage. These include butterflies of the family Megathymidae and moths of the family Prodoxidae. Also, the fungi *Diplodia theo-bromae* and *Colletotrichum agavae* may cause serious damage to agave leaves (Halffter, 1975; Agricultural Research Service, 1972).

Ehrler (1967) discovered that unlike stomata in most plants, agave stomata close during the day and open by night. This prevents loss of water through transpiration during hot daylight hours, although it results in a hotter leaf surface than most plants could tolerate. The thick cuticle overlying the epidermis, which is quite evident in tequila agave, apparently prevents damage to the leaf from high temperature. This waxy cuticle produces turbidity in tequila because it dissolves in the distillation step and produces a haze in the final product when it is diluted or cooled. One way to avoid this is to treat tequila with activated charcoal and to filter it through pure cellulose filter pads. This, unfortunately, results in the loss of some aromas.

Agave fertilization is based on soil composition, plant age, and the type of chemical compound used. The normal procedure is to use urea as a nitrogen source in amounts of 30-70 g per plant added directly into the soil. In some areas, phosphorus and potassium fertilization is also required. As some agave regions are also involved in cattle, swine, and chicken breeding, manure is sometimes employed for fertilization (GEA, 1992).

The average maturity time for agave is 10 years. Every year following planting, fields are cultivated to loosen the soil and weed and pest control is carried out. Each plant matures individually. Harvest begins at eight years. The leaves are cut from the base and left in the field to recycle nutrients. The harvested plants free of leaves look like large pineapples and weigh from 20 to 90 kg. They are transported to the distillery. Only the better plants, meaning those of good size and high inulin content (measured as reducing sugars) are harvested; but on the 12th year, all plants remaining (the weakest plants) are cut. These are called 'drag' and are generally discarded. Agave composition varies seasonally, but an average would be (wet basis) 27% (w/w) reducing sugars, a juice content of 0.572 ml/g and a pH of 5.2.

Tequila production

Tequilas differ greatly depending on agave quality and origin (Highlands or Tequila-Amatitán regions). The production process also strongly influences quality of the final product. Some distilleries still employ rudimentary production methods, just as they did several decades ago (Pla and Tapia, 1990). Most companies, however, employ technological advances that improve process efficiency and consistency; and some have implemented a ISO-9000 standard. Whatever processes are used, tequila manufacture comprises four main steps: cooking, milling, fermentation and distillation.

Processing harvesting agave

Agave is transported from the fields to the factories as soon as possible to avoid weight losses, because today most distilleries pay by weight and not by inulin content. The heads are unloaded from the truck in the receiving area of the factory and must be protected from the sun and rain in order to avoid withering and fungal growth. Although the agave has already been inspected during growth and at harvest, it is examined again to reject visually unacceptable plants or those damaged by pests. At this point, a representative sample of agave is taken for laboratory analysis. Modified AOAC (1990) procedures are used to determine reducing sugar content (after acid hydrolysis of inulin) along with pH, moisture, dry weight, juice and ash content.

Agave heads usually weigh between 20 and 60 kg, although some can reach 100 kg. They are cut to sizes that facilitate uniform cooking and handling. Different agave cutting systems exist, but the use of axes and a specialized tool called 'coa' are the most popular. The heads are cut in halves or quarters, depending on the weight, and the pieces are arranged manually in an oven or autoclave. Band saws can also be used to cut the agave heads. Band saws are faster and less labor-intensive than the manual procedure, but the belts break frequently because of the resinous consistency of agave. Some factories tear uncooked agave first with a knife and place the resulting pieces mixed with water into autoclaves to be cooked.

Other raw materials for tequila production

When producing 100% agave tequila, the only source of carbohydrate is the inulin hydrolyzed from agave in the cooking step. For other kinds of tequila the law permits the use of other sugars in amounts of up to 49% by weight in wort formulation. There are no legal specifications regarding the type of adjunct sugar sources to be used in tequila manufacture; and theoretically any kind of fermentable sugar can be used for the formulation of wort. In practice and from an economical point of view, only cane sugar, *Piloncillo*, cane molasses and acid- or enzyme-hydrolyzed corn syrup are employed. Cane sugar is received in 50 kg bags and stored in a dry, cool place for subsequent utilization. *Piloncillo* consists of brown cones of crystallized complete cane juice, sometimes individually wrapped in corn or cane leaves and packed in sacks. Cane molasses is also used but it is difficult to handle and there is risk of spoilage over a long storage period. Acid- or enzyme-hydrolyzed corn syrup may be used to formulate the wort. All sugars used in wort formulation are routinely analyzed by measuring solids content and reducing and fermentable sugar content.

The cooking step: hydrolysis of inulin

Cooking the agave serves three purposes. First, the low pH (4.5) together with the high temperature hydrolyze inulin and other components of the plant. The correct composition of these compounds is still unknown. In addition, cooked agave has a soft consistency that facilitates the milling operation.

In the pre-Hispanic era, agave cooking was carried out in holes filled with stones heated using wood for fuel. The stones retained the heat for the time needed to cook the agave. Nowadays some distillers have replaced the stuffed stone holes with brick ovens and heating is conducted by steam injection after the cut, raw agave has been introduced into the oven. Oven cooking is slow, and steam injection lasts around 36-48 hrs to obtain temperatures of 100 ^{0}C. After that period, the steam is shut off and the agave is left in the oven for a further two days to complete the cooking process. During this step, a sweet liquid called 'cooking honey' is collected and used later as a source of free sugars, mainly fructose. Also during this step some of the sugars are caramelized; and some of the compounds that contribute significantly to the aroma and flavor in wort formulation are due to its high content of fermentable sugars (>10% w/v). Finally, the oven door is opened to allow the cooked agave to cool. The agave is then ready for milling.

In most distilleries brick ovens have been replaced by steel autoclaves. Autoclaves have superior efficiency and allow good pressure and temperature control, enabling a homogeneous and economic cooking. In a typical autoclave cooking operation, steam is injected for 1 hr so that the condensed steam washes the agave. This condensed liquid is called 'bitter honey' and is discarded because it contains waxes from the agave cuticle and has a low sugar content (<1 % w/w). Steam is injected for an additional 6 hrs to obtain a pressure of 1.2 kg/cm^2 and a temperature of 121 °C. At the end of that time the agave remains in the autoclave for another 6 hrs without additional steam, cooking slowly in the remaining heat. This step produces a syrup with a high sugar concentration (>10% by weight) that is later used to formulate the initial wort. To calculate the yield and efficiency of this step, the amount of cooking honey and its reducing sugar content as well as the cooked agave are measured.

The main difference between autoclaved and oven-baked agave is that careful control of cooking time, temperature and steam pressure must be maintained in autoclaves to prevent overcooking or burning the agave. Overcooking gives a smoky taste to the tequila, increases the concentration of furfural in the final product and reduces ethanol yield due to the caramelization of some of the fermentable agave sugars. This is why some factories with both cooking systems reserve the ovens for their better-quality products. Although it is easier to obtain well-cooked agave in an oven than in an autoclave, there is no difference in terms of flavor and fermentability between agave cooked in autoclaves or in ovens if both are correctly controlled.

Extraction of agave juice: milling

Milling has gone through three historical stages. In ancient days cooked agave was crushed with wood or steel mallets to extract the juice. Later, a rudimentary mill consisting of a large circular stone 1.3 m in diameter and 50 cm thick was used. Driven by animals, the stone turned in a circular pit containing cooked agave and extracted the juice. The resulting juice was collected by hand in wood basins and carried to fermentation tanks. By the 1950s modern systems were implemented in which cooked agave was passed through a cutter to be shredded (except in factories that did this operation before cooking); and with a combination of milling and water extraction, sugars were extracted. The mills used for agave are similar to those used in the sugarcane industry but are smaller in size (normally 50 cm wide). This system is still employed in most distilleries.

Juice obtained in milling is mixed with the syrup obtained in the cooking step and with a solution of sugars, normally from sugarcane (if the tequila to be produced is not 100% agave), and finally pumped into a fermenter. Although the amount of sugar employed as an adjunct is regulated by law and must be less than 49% by weight at the beginning of the fermentation, each factory has its own formulation.

The milling step generates a by-product called bagasse, which represents about 40% of the total weight of the milled agave on a wet weight basis. Bagasse composition (dry weight basis) is 43% cellulose, 19% hemicellulose, 15% lignin, 3% total nitrogen, 1% pectin, 10% residual sugars and 9% other compounds. The bagasse, mixed with clay, is used to make bricks; but it is also the subject of research to find alternative uses. Examples are use of bagasse as an animal feed or as a substrate on which to grow edible fungi. Attempts are also underway to recover its components (cellulose, hemicellulose and pectin) using high-efficiency thermochemical reactors (Alonso *et al.*, 1993), to obtain furfurals, make particle board, or enzymes (cellulase and pectinase). Many of these projects are at the laboratory stage, and there is not enough information to evaluate feasibility.

In milling, as in all steps of the tequila process, a sugar balance is computed to determine the yield. If the yield decreases, the extraction pressure in the mill and the water/agave ratio are increased to improve efficiency.

Fermentation

Wort formulation

To produce 100% agave tequila, only agave may be used and the initial sugar concentration ranges from 4 to 10% w/v, depending on the amount of water added in milling. When other sugars are employed, they are previously dissolved and mixed with agave juice to obtain an initial sugar concentration of 8-16%, depending on sugar tolerance of the yeast strain. Wort formulation in most of the distilleries is based solely on previous experience. A few distilleries base wort formulation on composition of raw materials and nutritional requirements for yeast growth and fermentation. In these distilleries response surface methodology is the preferred method to optimize nutrient concentrations, using fermentation efficiencies and the taste of the resulting tequila as responses (Montgomery, 1984). To complement nutritional deficiencies of agave juice and sugars employed in the growth and fermentation steps, urea, ammonium sulfate, ammonium phosphate or magnesium sulfate could be added. Because the pH of the agave juice is around 4.5, there is no need for adjustment and the same wort composition is used for both inoculum growth and fermentation.

Yeasts

Some companies do not inoculate a specific strain of S. *cerevisiae* and instead allow natural fermentation to proceed. Others inoculate the wort with fresh packages of baker's yeast or a commercial dried yeast to obtain initial populations of 20-50 x 10^6 cells/ml. The dried yeasts were originally prepared for wine, beer, whisky or bread production; and sometimes the quality of the tequila obtained using these yeasts is not satisfactory, with large variations in flavor and aroma. To achieve high yields and maintain a constant quality in their tequila, some companies have been using yeast strains isolated from a natural fermentation of cooked agave juice. Nutrients are added; along with special conditions such as a high sugar concentration or temperature. These isolated and selected yeast strains have been deposited in national microbial culture collections, the most important being the Biotechnology and Bioengineering Department Culture Collection of CINVESTAV-IPN, located In México City.

Inoculum growth

When an inoculum is used, it is grown in the laboratory from a pure culture of a strain of *S. cerevisiae* maintained on agar slants in lyophilized form or frozen in liquid nitrogen. All laboratory propagation is carried out under aseptic conditions using a culture medium with the same ingredients used in the normal process but enriched to promote cellular growth. The inoculum is scaled-up with continuous aeration to produce enough volume to inoculate fermentation tanks at 10% of the final volume. Populations of 200-300 x 10^6 cells/ml are normally achieved. Strict cleanliness is maintained in this step as bacterial contamination is highly undesirable. When contamination is detected, antibiotics or ammonium bifluoride are used as antimicrobial agents. Once an inoculum is grown, it is maintained by mixing 10% of the volume of an active culture with fresh agave juice and nutrients. Although inoculation with commercial yeast greatly improves yield and turnover time, some companies prefer a more complex (in terms of the microbial diversity) fermentation. While yields might be lower and turnover time higher, the range of microorganism produces more compounds contributing to a more flavored tequila. It is also important to recognize that a change in taste and flavor could negatively affect the market for a particular brand of tequila.

Fermentation of agave wort

Once a wort is formulated with the required nutrients and temperature is around 30 °C, it may be inoculated with 5 to 10% (volume) of a previously grown *S. cerevisiae* culture with a population of 100-200 million cells/ml. Otherwise, microorganisms present in the wort carry out the fermentation. If an inoculum is not added, the fermentation could last as long as seven days. With an inoculum the fermentation time ranges from 20 hrs in the faster process to three days in the slower one.

Production of ethyl alcohol by yeast is associated with formation of many fermentation compounds that contribute to the final flavor of the tequila. These are organoleptic compounds or their precursors produced either in subsequent maturation of the wort before it goes to the distillation step, in the distillation process, or in the barrels if tequila is aged. The factors influencing formation of the organoleptic compounds in alcoholic beverages have been reviewed by many authors (Engan, 1981; Berry, 1984; McDonald *et al.*, 1984; Ramsay and Berry, 1984; Geiger and Piendl, 1976). Experience in the tequila industry is that the amount of organoleptic compounds produced is lower in fast fermentations than in slow fermentations. As a consequence, the flavor and general quality of tequila obtained from worts fermented slowly is best. The rate of fermentation depends mainly on the yeast strain used, medium composition and operating conditions. The wort sugar content decreases from an initial value of 4-11% to 0.4% (w/v) reducing sugars if an efficient yeast strain is employed. Otherwise, the residual sugar content could be higher, increasing the production costs.

Fermentation vessels vary considerably in volume, depending on the distillery. Their capacity ranges from 12,000 liters for small tanks to 150,000 liters for the largest ones; and they are constructed of stainless steel in order to resist the acidity of the wort. Ethanol production can be detected almost from the onset; and a pH drop from 4.5 to 3.9 is characteristic of the fermentation. The alcohol content at the end of fermentation lies between 4 and 9% v/v, depending on the initial sugar concentration. Alcohol losses may be significant because many fermentation tanks are open, allowing evaporation of alcohol with carbon dioxide. Some of the largest distilleries have a cooling system that keeps fermentation temperature within a tolerable range for yeast, but small producers do not have these systems. The fermentation temperature can exceed 40 °C, causing the fermentation to stop with an accompanying loss of ethanol and flavors that consequently decreases yields and affects the quality of the tequila. Fermentations carried out with pure agave juice tend to foam, sometimes requiring the addition of silicones. In worts with added sugars, foaming is usually not a problem.

Non-aseptic conditions are employed in fermentation, and in consequence bacterial activity may increase. The size of the bacterial flora depends on a number of factors including the extent to which bacteria grow during yeast propagation (if used), the abundance of bacteria on the raw materials and hygiene standards in the distillery. There is no doubt that the activity of these bacteria contributes to the organoleptic characteristics of the final product. Occasionally, the size of the bacterial population in fermenting wort may become too large (>20 x 10^6 cells/ml), in which case the bacteria use the sugars, decreasing ethanol yields and sometimes excreting undesirable compounds. The same compounds used in the propagation step may be used here to decrease common bacterial contaminants found in tequila worts. Lactobacillus, Streptococcus, Leuconostoc, and Pediococcus are the most common contaminants, but Acetobacter may be found in fermented worts that are left inactive for a long time prior to distillation.

In contrast to other distilled beverages, the organoleptic characteristics of tequila come from the raw material (cooked agave) as well as from the fermentation process. In most of the processes used in the tequila industry, fermentation is spontaneous with the participation of microorganisms from the environment, mostly yeasts and a few bacteria. This peculiarity brings about a wide variety of compositions and organoleptic properties. However, the special characteristics of the wort make it a selective medium for the growth of certain kind of yeasts such as *Saccharomyces* and to a lesser extent, acetic and lactic bacteria. In experiments isolating microbial flora from musts of various origins, different microorganisms were isolated. In most cases, this difference in flora is responsible for the wide variety of organoleptic characteristics of tequila brands (Pinal, 1999). Where a single purified strain is used, the final flavor and aroma are more neutral since the bouquet created by the contribution of several stains is richer than that obtained from only one type of yeast. Moreover, when yeast produced for bakery applications is used, the final product is also more neutral. It is also recognized that when non-100% agave tequila is made, a poorer bouquet is obtained because since a more defined medium yields a more defined product.

Organoleptic compounds generated during fermentation

Fusel oil

As in many other alcoholic fermentation processes, higher alcohols are the most abundant compounds produced along with ethanol. We have found (in decreasing order of abundance) isoamyl alcohol, isobutanol, active isoamyl alcohol and phenylethanol. It has been established that production of isoamyl and isobutyl alcohols begins after the sugar level is lowered substantially and continues for several hours after the alcoholic fermentation ends. In contrast, ethanol production begins in the first hours of the fermentation and ends with logarithmic yeast growth (Pinal *et al.*, 1997).

The most important factor influencing the amount of isoamyl alcohol and isobutanol is the yeast strain. It was found that a native strain isolated from tequila must produces a higher amount of such compounds when compared with a strain usually employed in bakeries. These results agree with those reported for Scotch whisky (Ramsay and Berry, 1984) and beer (García *et al.*, 1994).

The carbon:nitrogen ratio also has a significant influence on higher alcohol production. In tequila musts, which contain mainly fructose (≈95%) as a carbon source and an inorganic nitrogen source (ammonium sulfate), it was found for both native and bakery yeast strains that low carbon:nitrogen ratios result in low amounts of isoamyl alcohol: 19 mg/L in bakery strains and 30 mg/L for native strains *vs* 27 and 64 mg/L, respectively, for high carbon:nitrogen ratios. A similar relationship exists for isobutyl alcohol production (Figures 1 and 2).

Temperature is a third factor affecting isobutyl and isoamyl alcohol production with higher temperatures (e.g. 38 *vs* 32°C) yielding higher concentrations of those alcohols. On statistical analysis it was found that in addition to the direct effects of yeast strain, carbon:nitrogen ratio and temperature, the interaction of these three factors also had an impact on higher alcohol concentration in tequila (Pinal *et al.*, 1997). These results are consistent with the fact that with high carbon:nitrogen ratios there is a tendency to use amino acids as a nitrogen source, which implies the production of fusel oil as a by-product (by the Erlich pathway). On the other hand, variables such as the type of nitrogen source (urea or ammonium sulfate) or the amount of inoculum used for fermentation had little or no effect on the production of higher alcohols. Pareto diagrams involving all the variables tested are depicted in Figures 3 and 4.

Methanol

Another characteristic compound present in tequila is methanol. It is the general idea that

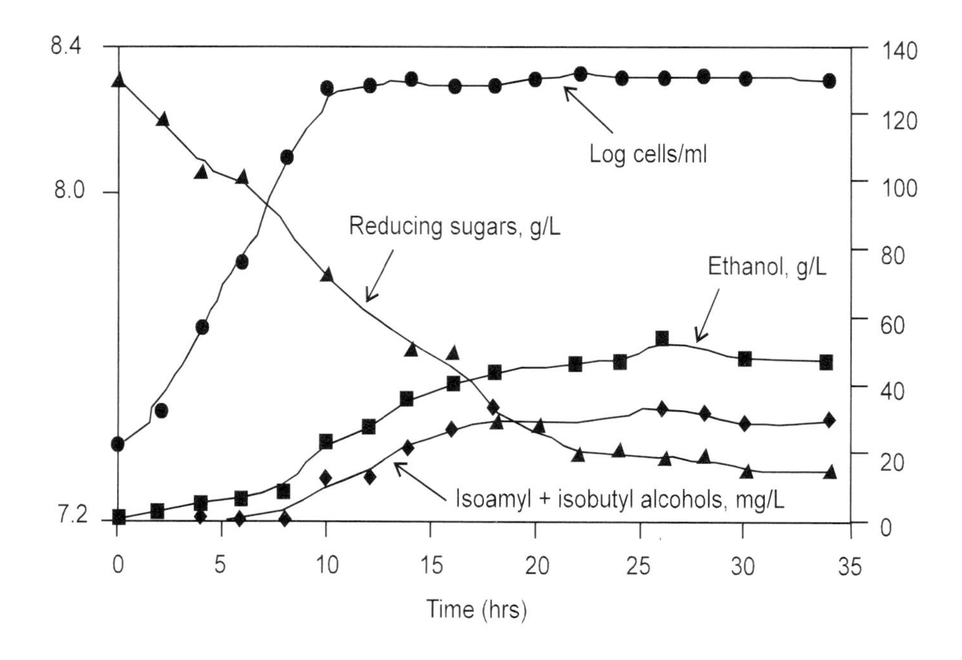

Figure 1. Higher alcohol production in tequila wort by a bakers yeast strain.

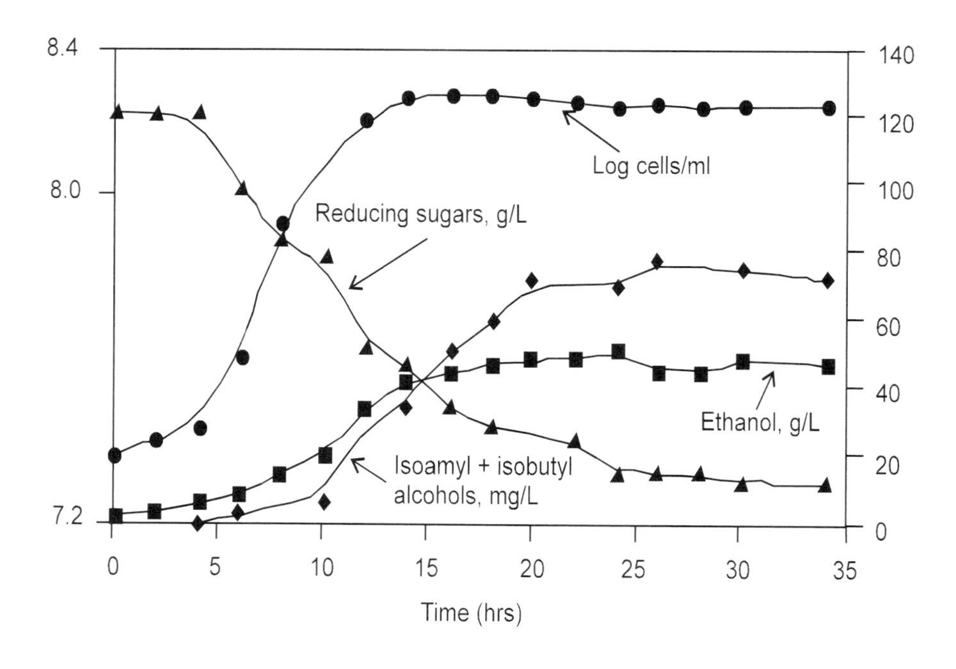

Figure 2. Higher alcohol production in tequila wort by a native yeast strain.

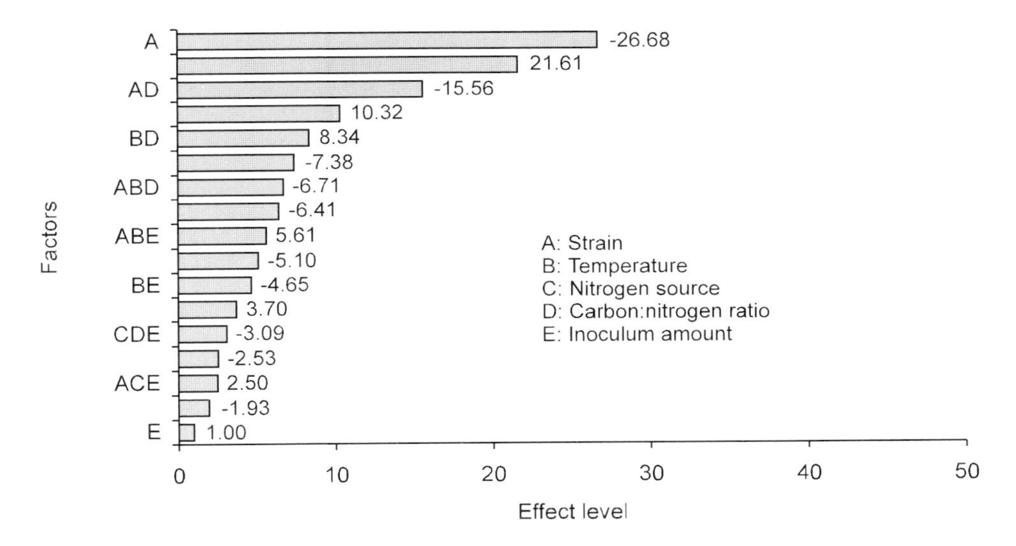

Figure 3. Pareto diagrams for isoamyl alcohol production in tequila.

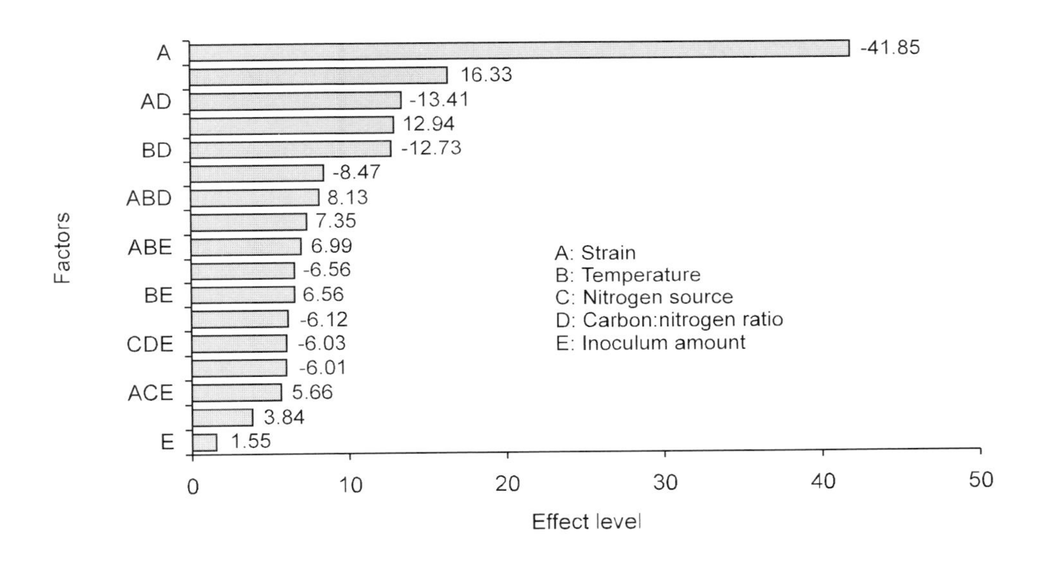

Figure 4. Pareto diagrams for isobutyl alcohol production in tequila.

methanol is generated through hydrolysis of methylated pectins present in the agave plant. Nevertheless, it is also believed that some yeast strains, natural or inoculated, have pectin methyl esterases. Some authors describe various amounts of methanol in musts with the same composition but fermented with different strains (Téllez, 1999).

Aldehydes

Along with the production of ethyl acetate, the oxidation of ethanol also generates acetaldehyde, an intermediate in the production of acetic acid. It is well known in commercial practice that an oxidation process instead of fermentation begins after the sugar concentration declines, thus provoking the increase in the acetaldehyde level. However, there is no formal report of such phenomena in tequila, as it is described in beer (Hammond, 1991).

Organic acids

Small organic acids (up to six carbons) and larger molecules (fatty acids) are produced during fermentation. The smaller molecules can be products of intermediate metabolism of the normal microbial flora; and their production depends on the presence of oxygen. The larger fatty acids are synthesized for membrane structures during cell growth and can also appear at the end of the fermentation when lysis takes place. The presence of octanoic and decanoic acids in the final product has been described particularly for tequila.

Esters

Esters are very important compounds in their particular contribution to flavor and aroma, since they have the lowest organoleptic threshold values (Ramsay and Berry, 1984). In particular, ethyl acetate is the most abundant and important compound of this family. Ethyl acetate has been reported to be the second most abundant compound in tequila after isoamyl alcohol. The quantity of this compound present in the final product can vary widely, since it is synthesized from acetic acid (in form of acetyl-CoA) and ethanol. Acetic acid can also be produced by the oxidation of ethanol when the fermentation has ceased and an oxidative process starts on the surface of the fermentation tank by *Saccharomyces* and many other yeasts such as *Brettanomyces*. Therefore, long fermentation periods (a current practice in the tequila industry) yield high ethanol oxidation. In addition, in open fermentation tanks with worts at low pH containing alcohol, ethanol is also transformed to acetic acid (itself a precursor of ethyl acetate) by bacteria of the genera Acetobacter. Besides ethyl acetate, the presence of several other esters has been described including ethyl and isoamyl esters. Some of the most important esters found in silver tequila are listed in Table 1.

Table 1. Most abundant esters found in silver tequila.[1]

Ester	%
Ethyl acetate	17.77
Ethyl decanoate	2.78
Ethyl lactate	2.74
Ethyl octanoate	1.92
Ethyl dodecanoate	0.95
Ethyl butanoate	0.63
Isoamyl acetate	0.58
Ethyl propanoate	0.57
Ethyl hexanoate	0.48
Ethyl hexadecanoate	0.48

[1]Estarrón *et al.*, 1999.

Distilling

Distillation involves the separation and concentration of the alcohol from the fermented wort. In addition to ethanol and other desirable secondary products, wort contains solid agave particles consisting mainly of cellulose and pectin, and yeast cells in addition to proteins, mineral salts and some organic acids. Although a great number of types and degrees of distillation are possible, the most common systems used in the tequila industry are pot stills and rectification columns. The pot still is considered the earliest form of distilling equipment. It is of the simplest design, consisting of a kettle to hold the fermented wort, a steam coil, and a condenser or a plate heat exchanger. Pot stills are often made of copper, which 'fixes', according to Thorne *et al.* (1971), malodorous volatile sulfur compounds produced during fermentation. Batch distillation using pot stills is carried out in two steps. First the fermented wort is distilled to increase the alcohol concentration to 20-30% by volume, separating out the first fraction called heads, and the last fraction, called tails.

Composition of these fractions varies depending on many factors including the yeast strain employed, wort nutrient composition, fermentation time and distillation technique; but in general, heads are rich in low boiling point compounds such as acetaldehyde, ethyl acetate, methanol, 1-propanol, 2-propanol, 1-butanol, and 2-methyl propanol, which give a very pleasant flavor and taste to tequila. Heads are normally mixed with the wort being distilled. The tails contain high boiling point components such as isoamyl alcohol, amyl alcohol, 2-furaldehyde, acetic acid and ethylactate, giving a strong taste and flavor to the tequila; and when the concentration is above 0.5 mg/ml, the final product becomes unpleasant. This fraction is not used.

In the second step, the liquid obtained from the first stage is re-distilled in a similar pot still in order to obtain a final product that is 110° proof if it is sold in bulk (reducing transport costs) or 80° proof if it is to be bottled. Some companies obtain high proof tequila and dilute it with demineralized water or water purified by reverse osmosis.

In continuous distillation systems, the fermented wort enters the feed plate of the column and flows downward, crossing a series of trays. Steam is injected from the bottom in a coil and strips the wort of its volatile components. Vapors condense higher in the column, depending on component volatility, allowing liquids to be drawn off or recycled at the various plates as appropriate. Sometimes tequila obtained in this way is mixed with tequila from pot stills to balance the amount of organoleptic compounds because in general, tequila obtained through continuous columns has less aroma and taste than tequila obtained from pot stills.

The presence of methanol in tequila is still a subject of discussion because whether methanol is produced only by a chemical reaction or in combination with a microbial hydrolysis has not been satisfactorily demonstrated. The chemical reaction is demethylation of agave pectin by the high pH during cooking and the first distillation steps. The microbial reaction could be the hydrolysis of agave pectin by the enzyme pectin methyl esterase produced by some microorganisms during the fermentation step, but this has not yet been demonstrated. Preliminary results favor the first theory, but some research is still needed on this matter because it is important to maintain the methanol concentration within the limits established by the official standard, which is 300 mg methanol per 100 ml of anhydrous ethyl alcohol.

Effluent disposal

The discharge from pot stills or distillation columns is known as stillage, slops or vinasse; and in a typical tequila distillery 7 to 10 liters of effluent are produced per liter of tequila at 100° proof. Tequila stillage has a biological oxygen demand (BOD) of 25 to 60 g/L. In addition to the dissolved salts (mainly potassium, calcium, and sulfate ions) and the low pH (<3.9) of the stillage, there are significant disposal or treatment problems. A general solution to the disposal problem does not exist because every factory has its own production process and is located either in a city or near agave fields. As a result of the difficulties of treating vinasses and due to their high concentrations of dissolved matter, a host of utilization schemes have been proposed. Some of the methods indicated below are under investigation and others are in use.

Recycling reduces the volume of waste to be treated. Stillage can be recirculated, mixing 5 to 10% of the total volume of the waste obtained with clean water to substitute for the dilution water used to prepare the initial wort. This can be carried out for a number of cycles, usually no more than five, because the concentration of dissolved salts increases and could affect the fermentation process. Also, great care must be taken with the final taste and flavor of the tequila because some components present in stillage could affect the organoleptic characteristics of the final product. Currently, only one tequila company uses this system.

Direct land application as irrigation water and fertilizer in agave fields is under careful evaluation to determine the optimum loading rates and the effects on the agave over the long time it takes to reach maturity. Evaporation or combustion of stilllage could provide fertilizer or potash, but the high cost of such a process is a serious limitation (Sheenan and Greenfield, 1980). The production of biomass and biochemicals including fodder yeast is a possibility, but the remaining liquor still has a high BOD (Quinn and Marchant, 1980). Stillage may be used as a food supplement for cattle, but it has an undesirable laxative effect on animals. Biological, aerobic, or anaerobic treatment offers a real means of disposal, but the cost is likely to be as high as the fermentation costs themselves (Speece, 1983; Maiorella, 1983). Ultimately, tequila vinasses should be viewed as a raw material rather than a waste, and a strategy should be devised that maximizes economic and social benefits and reduces recovery costs.

Maturation

Distillation is the final stage of tequila production if silver or white tequila is the desired product. For rested or aged tequila, maturation is carried out in 200 liter white oak casks or in larger wood tanks. The time legally required is two months for rested tequila and 12 months for aged tequila. Tequila is generally matured for longer periods, depending on the characteristics each company desires for its particular brand.

As tequila ages in barrels, it is subject to changes that will determine its final quality. Thickness and quality of the stave, depth of the char, temperature and humidity in the barrelling area, entry proof (40-110° proof), length of storage and number of cycles for the barrel (in México barrels may be re-used several times) all affect the final taste and aroma of the tequila. Fusel oil content decreases during maturation owing to the adsorbant nature of the char, smoothing the final product. Complex wood constituents are extracted by the tequila, providing color and the particular taste. Reactions among certain tequila compounds yield new components; and oxidation reactions change some of the original components in tequila and those extracted from the wood. As a result of all of these changes, the concentration of acids, esters and aldehydes is increased, while the concentration of fusel oils decreases as tequila reposes in barrels.

After aging and dilution with demineralized water (if necessary) the color of the tequila may be adjusted to the desired value by the addition of caramel. Alternatively, some companies blend different batches of tequila to obtain a standardized final product.

Government inspectors supervise the entire aging process. Prior to bottling, tequila is filtered through cellulose filter pads or polypropylene cartridges. Sometimes afterwards a pretreatment with charcoal is used to eliminate turbidity.

Future developments

Research and future developments in the production of tequila and agave cultivation can take many directions, but the implementation of any change must allow quality of the final product to be maintained and provide substantial improvement in the process. There are several key areas where important developments could take place. Development of new varieties of agave endowed with resistance to pests or extremely dry environments, higher inulin content, low wax content in the leaf cuticle and superior growth rates would be of benefit. Mechanization would improve aspects of cultivation and harvest of agave. In addtition, optimization of the cooking and fermentation steps would improve yields and reduce the amount of waste, while yeast strain selection could improve ability to ferment musts with high concentrations of sugars. Finally, low cost alternatives for waste (bagasse and stillage) treatment are needed.

Summary

Tequila is a beverage obtained from the distillation of fermented juice from the agave plant *(Agave tequilana.* Weber var. *Azul).* The use of agave to produce drinks in México dates from long ago when certain tribes used it for religious ceremonies. It was not until the arrival of the Spaniards, who brought distillation technique knowledge, that tequila acquired its present form. The agave plant, which is generally confused with cacti, is propagated through traditional methods or by means of plant cell culture. It is grown for 8 to 10 years before it can be harvested and sent to the distillery.

The tequila production process comprises four stages. In the first step, agave is cooked to hydrolyze the polymers present in the plant, mainly inulin, into fermentable sugars. In some factories this step is accomplished using stone ovens and in others it is carried out in autoclaves. The second stage is sugar extraction from cooked agave through milling; and the agave juice obtained through this step can be mixed with sugars from other sources, normally sugar cane, if 100% agave tequila is not desired. The third and most important stage is fermentation in which sugars are transformed into ethanol and other compounds such as esters and organic acids. These, along with other substances derived from the cooked agave, give the characteristic flavor and taste to tequila. It is of great importance to have a good yeast strain and nutritionally balanced wort for tequila production, as losses could be as high as 35% of the total production if inefficient yeast is used or nutrients do not appear in the right proportions. In the last stage, fermented wort is distilled, normally using pot stills or in some cases rectification columns, to obtain the final product. At the end of the distillation process white tequila is obtained. Maturation in white oak barrels is required for rested or aged tequila. The minimum maturation times are 2 and 12 months, respectively, for rested and aged tequila as required by government regulations. At every step of the production process, most companies employ several quality control analyses in order to ensure the quality of the product and the efficiency of the process. Some major producers of tequila are certified through an ISO-9000 standard. Tequila production is governed by the official norm NOM-006-SCFI-1993, which must be followed by all tequila producers to guarantee a good quality final product.

Acknowledgement

The authors would like to express their gratitude to Tequila Herradura, S.A. de C.V., for its support in writing this chapter.

References

Agricultural Research Service. 1972. The Agave family in Sonora. *Agriculture Handbook No.399,* US Department of Agriculture, p. 195.

Alonso, G.S., L. Rigal and A. Gaset. 1993. Valoración química de bagazo de agave de la industria tequilera. Rev. Soc. Quim. Mex. 3(6):19.

AOAC 1990. Distilled liquors. Official Methods of Analysis of the Association of Official Analytical Chemists (H. Kenneth, ed) Association of Official Analytical Chemists, Arlington, Virginia, p. 690.

Backman, G.E. 1944. A karyosystematic study of the genus Agave. Am. J. Bot. 35:283.

Berry, D.R. 1984. Physiology and microbiology of the malt whisky fermentation. In: Progress in Industrial Microbiology. (M.E. Bushell, ed.) Elsevier, Amsterdam, p. 189.

Bottorff de B., V. 1971. A guide to tequila, mezcal and pulque. Minutae Mexicana, Mexico, D.F.

CIATE, 1993. Informe de actividades al consejo, Agosto, Guadalajara, Jal, México.

Diguet. L. 1902. Estudio sobre el maguey de tequila. Generalidades e historia. El Prog. Mex. 9:424.

Ehrler, W.H. 1967. Agave plant, efficient water user. USDA Agr. Res. 16(4):11.

Engan, S. 1981. Beer composition: volatile substances. In: Brewing Sciences, Vol. 2 (J.R.A. Pollock, ed.) Academic Press, London, p. 98.

Estarrón, M., T. Martín del Campo and R. Cosío. 1999. Identificación de los componentes volátiles que caracterizan la huella cromatográfica distintiva de tequilas. Technical Report for Tequila Herradura S.A.

García, A.I., L.A. García and M. Díaz. 1994. Fusel alcohol production in beer fermentation processes. Process Biochem. 29:303-309.

GEA. 1992. La fertilización, La Jima. *Boletin Informativo,* Gerencia de extensión agrícola. Tequila Sauza, 7(50).

Geiger, E. and A. Piendl. 1976. Technological factors in the formation of acetaldehyde during fermentation, MBAA Tech. Q. 13:51.

Goncalves de L. O. 1956. El Maguey y el Pulque, Fondo dc Cultura Económica, México.

Grandos, S.D. 1985. Etnobotanica de los agaves dc las zonas áridas y semiáridas. In: Biología y Aprovechamiento Integral del Henequén y Otros Agaves, CICY, A.C., México, p. 127.

Halffter, G. 1975. Plagas que afectan a las distintas especies de agave cultivado en México. Secretaría de Agricultura y Ganadería, México, D.F.

Hammond, J.R.M. 1981. Brewer's Yeast. In: The Yeasts, Vol. 5. Yeast Technology (A.H. Rose and J.S. Harrison, eds.) 2nd. ed. Academic Press, Redding, U.K.

Luna, Z.R. 1991. La historia del tequila, de sus regiones y de sus hombres. Consejo Nacional para la Cultura y las Artes, México, D.F.

MacDonald, J., P.T,V. Reeve, J.D. Ruddlesden and F.H. White. 1984. Current approaches to brewery fermentations. In: Progress in Industrial Microbiology. Vol. 19, Modern Applications of Traditional Biotechnologies, (M.E. Bushell, ed.) Elsevier, Amsterdam.

Maiorella. B.L., H.W. Blanch and C.R. Wilke. 1983. Distillery effluent treatment and by-product recovery. Proc. Biochem. 8(4):5.

Martinez, M. 1979. Catalogo de nombres vulgares y científicos de plantas Mexicanas. Fondo de Cultura Económica. México. p. 543.

Montgomery, D.C. 1984. Design and Analysis of Experiments, 2nd Ed., John Wiley & Sons, New York. p. 445.

Muria, J.M. 1990. El tequila. Boceto histórico de una industria. Cuad. Difus. Cient. 18:13.

Pérez, L. 1990. Estudio sobre el maguey llamado Mezcal en el estado de Jalisco. Programa de Estudios Jalisciences, Instituto del Tequila, A.C., México.

Pinal, Z.L.M. 1999. Tesis de Maestría en Procesos Biotecnológicos. Universidad de Guadalajara, Guadalajara, México.

Pinal, L., M. Cedeño, H. Gutiérrez and J. Alvarez-Jacobs. 1997. Fermentation parameters influencing higher alcohol production in the tequila Process. Biotechnol. Lett. 19(1):45-47.

Pla, R. and J. Tapia. 1990. El agave azul, de las mieles al tequila. In: Centro de Estudios Mexicanos y Centroamericanos, Instituto Francés de America Latina y el Institut Francais dc Recherche Scientifique pour le developement en Cooperation, México, p. 61.

Quinn, J.P. and R. Marchant. 1980. The treatment of malt whisky distillery waste using the fungus *Geotrichum candidum*. Water Res. 14:545.

Ramsay, C.M. and D.R. Berry. 1984. The effect of temperature and pH on the formation of higher alcohols, fatty acids and esters in malt whisky fermentation. Food Microbiol. 1:117.

Ramsay, C.M. and D.R. Berry. 1984. Physiological control of higher alcohol formation in Scotch whisky fermentation. In: Current Developments in Yeast Research (G.G. Stewart and I. Russell, eds) Pergamon Press, London, Canada.

Rzedowski, J. 1978. *Vegetación de México,* Ed. Limusa. México.

Sánchez, A.F. 1991. Comparación de metodologías de micropropagación de *Agave tequilana* Weber. Tesis de Ing. Agrónomo, Facultad de Agronomía, Universidad de Guadalajara, Guadalajara, México.

Sánchez-Marroquín, A. and P.H. Hope. 1953. Fermentation and chemical composition studies of some species. Agric. Food Chem. 246:1.

Secofi, Norma Oficial Mexicana. 1993. NOM006-SCFI-1993, Bebidas alcohólicas, Tequila. Especificaciones, Diario oficial del 13 de Octubre, p. 48.

Sheenan, G.J. and P.F. Greenfield. 1980. Utilization, treatment and disposal of distillery wastewater. Water Res. 14:257.

Speece, R.E. 1983. Anaerobic biotechnology for industrial wastewater treatment. Environ. Sci. Technol. 17(9):416A.

Téllez, P. 1999. Tesis de Maestría en Procesos Biotecnológicos. Universidad de Guadalajara, Guadalajara, México.

Thorne, R.S.W., E. Helm and K. Svendsen. 1971. Control of sulfur impurities in beer aroma, J. Inst. Brew. 77(2):148.

Valenzuela, Z.A. 1992. Floración y madurez del agave, La Jima. Boletin Informativo, Gerencia de extensión agrícola. Tequila Sauza 5:33.

Chapter 16

Feedstocks, fermentation and distillation for production of heavy and light rums

J.E. Murtagh
Murtagh & Associates, Winchester, Virginia, USA

Introduction

The US Bureau of Alcohol, Tobacco and Firearms regulations (1982) define rum as: 'an alcoholic distillate from the fermented juice of sugar cane, sugarcane syrup, sugarcane molasses or other sugarcane by-products, produced at less than 190° proof (95° GL), in such a manner that the distillate possesses the taste, aroma and characteristics generally attributed to rum, and bottled at not less than 80° proof (40° GL), and also includes mixtures solely of such distillates'.

It should be noted that unlike the US regulations for whisky, the regulations for rum do not include any requirement that it be aged in oak barrels for a minimum period of time. The US regulations also do not specify any geographic region in which rum may be produced, however the British regulations define rum as: 'a spirit distilled directly from sugarcane products in sugarcane-growing countries'.

The regulations are rather vague of necessity owing to the very wide variety of rums ranging from the light, nearly neutral, continuous still products to the heavily flavored, dark, navy-type pot still products. The diversity of rum types and their production methods makes comprehensive coverage of the industry almost impossible for a publication such as this. Thus, readers should consult the references for additional information. What are now considered the classic studies on rum production were published by Rafael Arroyo of Puerto Rico in the 1940s (see reference list). That was at a time when the island's rum industry increased production very significantly in order to meet the demand arising from a wartime shortage of whisky and grain spirits. Other more recent studies and helpful reviews of the rum industry have been published by Clutton (1974), Ianson (1971), Kampen (1975), Lehtonen and Suomalainen, (1977) and Paturau (1969).

Feedstocks: fresh sugarcane juice or diluted molasses

Cane juice

Using cane juice for rum production has both advantages and disadvantages. The main advantage is that no additional processing is required. The cane is simply crushed and the juice fed directly into the fermenters. Another advantage is that cane juice does not have a high content of dissolved salts and therefore does not cause as much scaling and blocking of distillation columns as may occur using molasses.

One of the main disadvantages of using cane juice is that it normally contains only about 12-16% w/w sugar, so that the alcohol content of the fermented material is limited to about 6 -8% v/v as compared to 10-13% v/v obtainable from molasses. Another significant disadvantage is that cane juice cannot be stored in bulk for any considerable length of time. Heavy contamination with bacteria and yeasts causes it to start to ferment spontaneously and very rapidly. This means that the distillery must be located in close proximity to the cane mill. It also means that the cane juice is only available during the cane harvesting season, which generally does not exceed six months per year. Thus, some rum distilleries may use cane juice during the harvest season and molasses for the remainder of the year. Cane juice may also have cost disadvantages in that it is a primary, or at least an intermediate product in the production of sugar. In contrast, molasses is generally a by-product, considered to be of lower value.

Cane juice normally becomes heavily infected with bacteria and yeasts in the crushing process. Pasteurization prior to fermentation is usually infeasible due to the large amount of suspended fiber in the juice, which would tend to block most heat exchangers. Thus the contaminating organisms may significantly reduce alcohol yields from cane juice.

Molasses

The use of molasses as a feedstock for alcohol production has been covered in an earlier chapter in this volume; therefore only a few comments are necessary with specific reference to rum.

The source of the molasses can have a strong influence on the aromatic quality of rum. This has been demonstrated in trials on making rum from beet molasses where it was found that it was not possible to obtain the same characteristic aromas as obtained from cane molasses (Arroyo, 1948b). Arroyo (1941, 1942) reported that fresh blackstrap molasses with its low viscosity, high total sugars, nitrogen, phosphorus and a low ash and gum content was preferable in the production of rums with desirable odors and tastes.

Fermentation methods

Heavy rum

Just as there are many different types of rum, there are many different methods of fermentation. In the production of heavily flavored rums, cane juice or diluted molasses mash may be allowed to ferment spontaneously using yeasts present naturally in the feedstock. This is relatively inefficient in terms of alcohol production; and the results may vary. However, it is the high level of bacterial contamination in these fermentations that produces many of the desirable congeners such as acids and esters. In some processes, e.g. in Jamaica, the spontaneous fermentation may be assisted by the addition of 'dunder' at the start of fermentation. Dunder is old stillage that has been stored in open tanks to allow development of a strong bacterial flora.

To attempt a more controlled fermentation in production of heavy rums, a pure culture of yeast may be used together with a pure bacterial culture. A specially selected strain of yeast, usually *Saccharomyces cerevisiae*, which is a budding yeast, or possibly other species such as the fission yeast *Schizosaccharomyces pombe*, is propagated from an agar slant. This culture is transferred every 12-24 hrs through a series of Erlenmeyer flasks each holding increasing amounts of sterilized, diluted molasses mash supplemented with malt syrup or other sources of nutrients. When about 20 liters of yeast culture have been produced, it may be used to inoculate about 200 liters of diluted molasses mash in a small plant propagator or 'prefermenter'. The contents of the vessel are aerated, and after about 8-12 hrs of propagation are used to inoculate a larger prefermenter containing about 2,000 liters of mash. This vessel may also be aerated and fed incrementally with more diluted molasses medium until it contains the equivalent of about 10% of the capacity of the plant fermenters.

It is generally considered desirable to have a minimum of about 50 million yeast cells/ml of medium when the fermentation has been completed. Thus, after allowing for yeast propagation during the filling process, the 10% yeast inoculum should have a cell count of about 200 million cells/ml to achieve this objective.

A pure culture of bacteria such as *Clostridium saccharobutyricum* may be added after 6-12 hrs of the yeast fermentation. Usually the bacterial inoculum amounts to about 2% of the fermenter capacity; and the pH of the fermenting mash is adjusted upwards to about pH 5.5 before addition to give more suitable conditions for the bacterial propagation. The bacteria produce a mixture of acids, predominantly butyric, together with others such as acetic, propionic, and caproic acids. These acids in turn react with the alcohol to produce desirable esters.

Fermentation for light rums

In light rum production, the emphasis is generally on maintaining clean, rapid fermentations to minimize development of undesirable congeners and to maximize fermentation efficiency. Pure cultures of yeast may be propagated as previously described. These may be accompanied by the use of antibiotics such as penicillin or bactericides such as chlorine dioxide, ammonium bifluoride or quaternary ammonium compounds to control bacterial contamination.

Some plants use 'mother yeasting'. In this system a pure yeast culture is propagated in the prefermenter. The pH is lowered to about 3.7 to reduce bacterial growth; and up to 90% of the yeast volume is used to inoculate a fermenter. The prefermenter is refilled with mash to repeat the process several times.

In other plants, the yeast may be recycled from one fermenter to another. This is done by centrifuging the yeast out of the fermented beer prior to distillation and then subjecting it to an acid washing at pH 2.2-2.5 to kill most of the contaminating bacteria. As an alternative to acid washing, the yeast may be treated with about 15 ppm chlorine dioxide at a pH of about 3.5. The objective of yeast recycling is to save on sugar that would otherwise be needed for yeast growth and to ensure a very high cell count in the fermenter inoculum. This provides a rapid fermentation in which bacteria lack sufficient opportunity to get well established.

Commercial dry yeasts may also be used in light rum fermentations. A large quantity of dry yeast may be added to the molasses mash at the start of fermenter filling. More commonly, in order to save cost a smaller quantity of the dry yeast may be rehydrated and propagated for several hours in a prefermenter before transfer to the fermenter.

Distillation

The method of distillation used has a considerable effect on the nature of the rum product. Heavy rums are usually produced by batch distillation, while light rums are normally produced by continuous distillation.

Batch distillation: heavy rum

Batch distillation may be performed with various types of equipment. The simplest form is a single, simple pot still. Here, the fermented beer is transferred into a copper tank or 'pot', as shown in Figure 1. The pot is heated either by an internal steam coil (calandria) or by a fire of wood or bagasse underneath, usually in a brick enclosure. The pot is fitted with a vapor pipe which leads to a condenser coil immersed in a water tank. As the beer is heated, the alcohol and other volatile congeners are distilled off, condensed and run into a storage tank. Usually the first fraction distilled contains much of the more volatile, pungent 'heads' congeners and is discarded as the 'heads cut'. The process is continued until most of the alcohol has been distilled out of the beer.

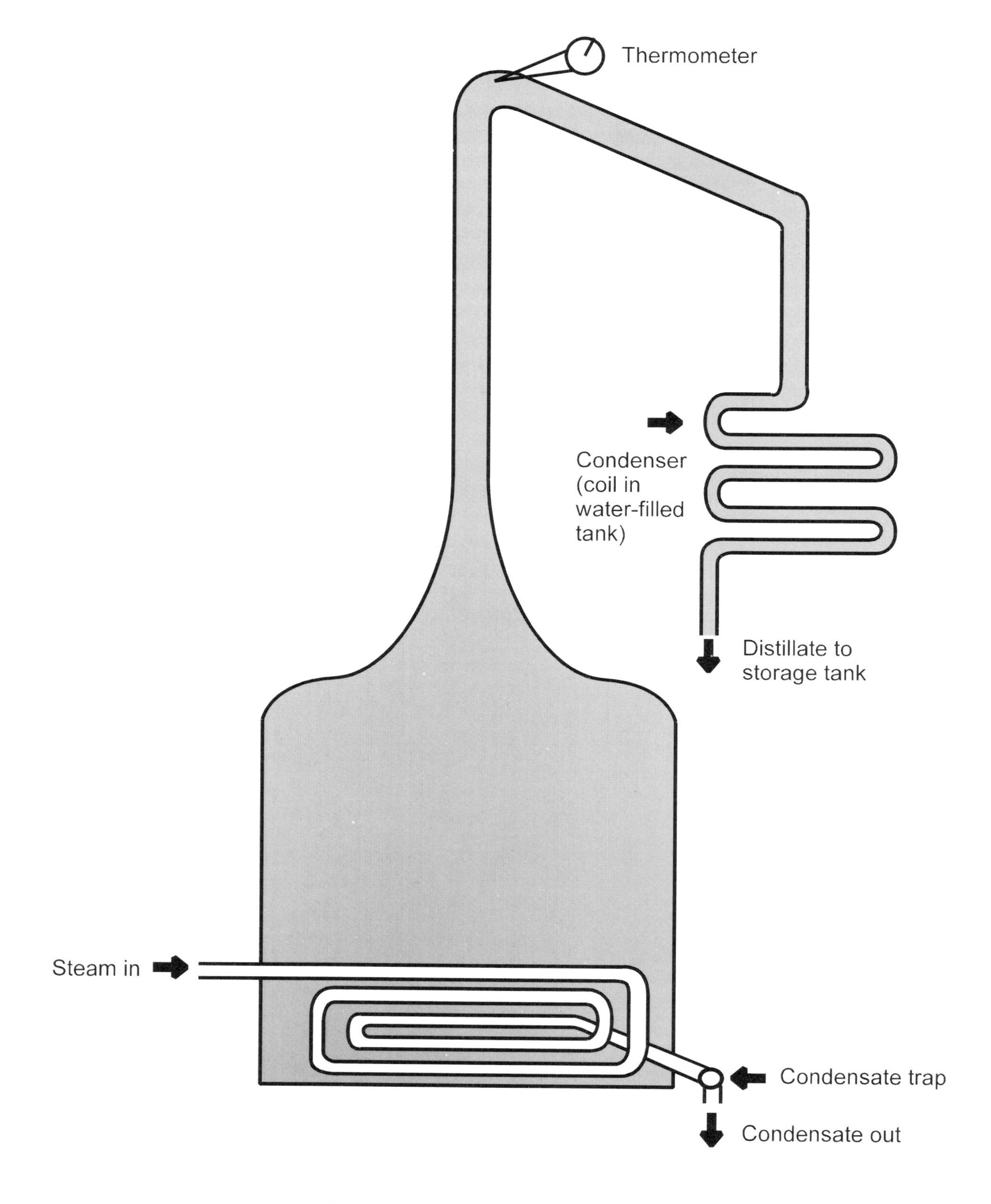

Figure 1. Simple pot still for heavy rum production.

The beer residue, or 'stillage', is emptied out of the pot and the distillate is returned from the storage tank to the pot to be redistilled to increase the proof. A heads cut may again be discarded together with a 'tails cut' taken toward the end of the distillation. The 'center cut', which is the bulk of the distillate, has the more pleasant aromas of fruity esters characteristic of rum. Normally, there are only two pot distillations, but there may be a third if desired.

Two or three pot stills may be interconnected, as shown in Figure 2, to eliminate the need to transfer the distillate back into the first pot for redistillation and to conserve energy. The efficiency of the pot still in separating congeners may be improved considerably by including perhaps five or six bubble cap trays between the pot and the vapor pipe, and a simple condenser above the trays to provide some reflux as shown in Figure 3. This is the design of pot still most frequently used by small, on-farm producers of common tafia and aguardiente rums. It is a simple progression from this pot still to the batch kettle and column still used by larger scale rum producers, as shown in Figure 4.

In some countries crude rum or aguardiente is produced on a small scale by numerous farmers using pot stills. Their product may then be sold to companies that redistill it in a batch system to standardize quality for sale as heavy rum or for blending with neutral or nearly neutral spirits to make light rums.

Continuous distillation: light rum

Continuous distillation is usually confined to the production of light rums. A two column beer still with a Barbet type prestripper, as shown in Figure 5, is normally preferred as it permits removal of heads, volatile sulfur compounds and some other undesirable fractions before the bulk of the alcohol enters the concentrating column. (Operation of this type of beer still is discussed in the chapter on neutral spirit production).

When a lighter or less flavorful rum than can be produced on a simple two column beer still is required, an extractive distillation column and a rectifier as employed in neutral spirit production can be used (Figure 6).

Combinations of batch and continuous distillation

There are many possible variations and combinations of distillation processes. For example, crude pot still distillates may be put through a simple continuous distillation unit to produce a more acceptable, standardized product. Arroyo has described two techniques for combinations of continuous and batch distillations to produce both heavy and light rums at the same time. In one system Arroyo (1948c) suggested that the beer first be distilled in a single continuous column with only about five concentrating plates. Provision for fusel oil removal allows some control over the fusel content in the distillate, which has a proof of about 70° GL. He then recommended that the distillate be diluted to about 40° GL and be submitted to a batch distillation.

The distillate from the batch column should be collected as five separate fractions based on monitored odor variations and rising distillation temperature. Arroyo found that the first fraction, which distilled between 69 and 72°C at a proof of about 91° GL, represented about 5% of the total distillate volume and consisted of unpleasant aldehydes, organic acids and esters. He recommended this fraction either be discarded or held for subsequent reprocessing to recover some of the ethanol. The second fraction, distilling over a temperature range of 72-77°C at a proof of 93-94° GL, represented about 10% of the total distillate volume and contained ethanol with appreciable amounts of aldehydes and esters. The third fraction, which distilled at a fixed temperature of 78°C and a proof of about 95.5° GL, was the largest in volume at 55-60% of the total and contained mostly ethanol with very small amounts of congeners such as aldehydes, esters and higher alcohols. It also contained the lowest

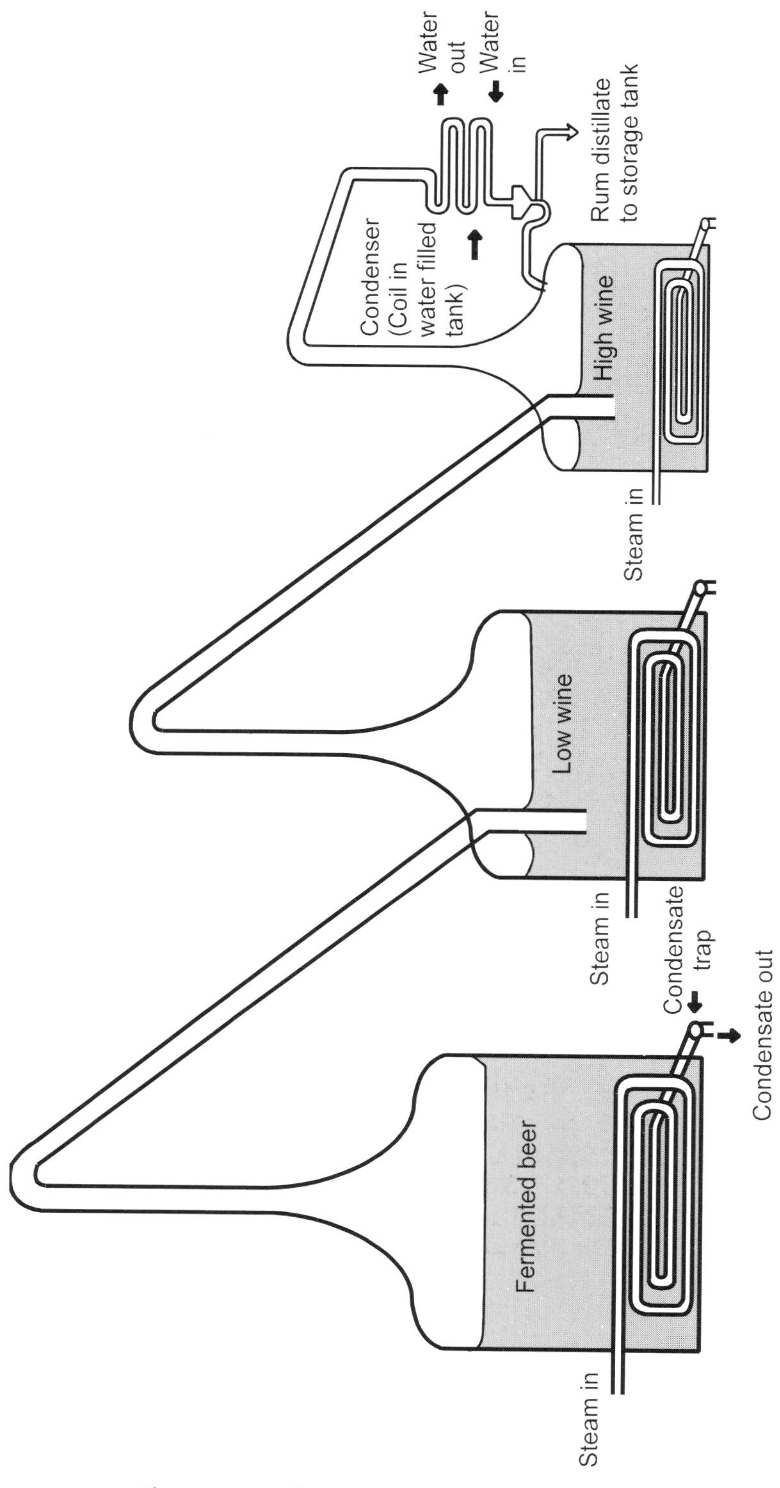

Figure 2. Triple pot still system for heavy rum production.

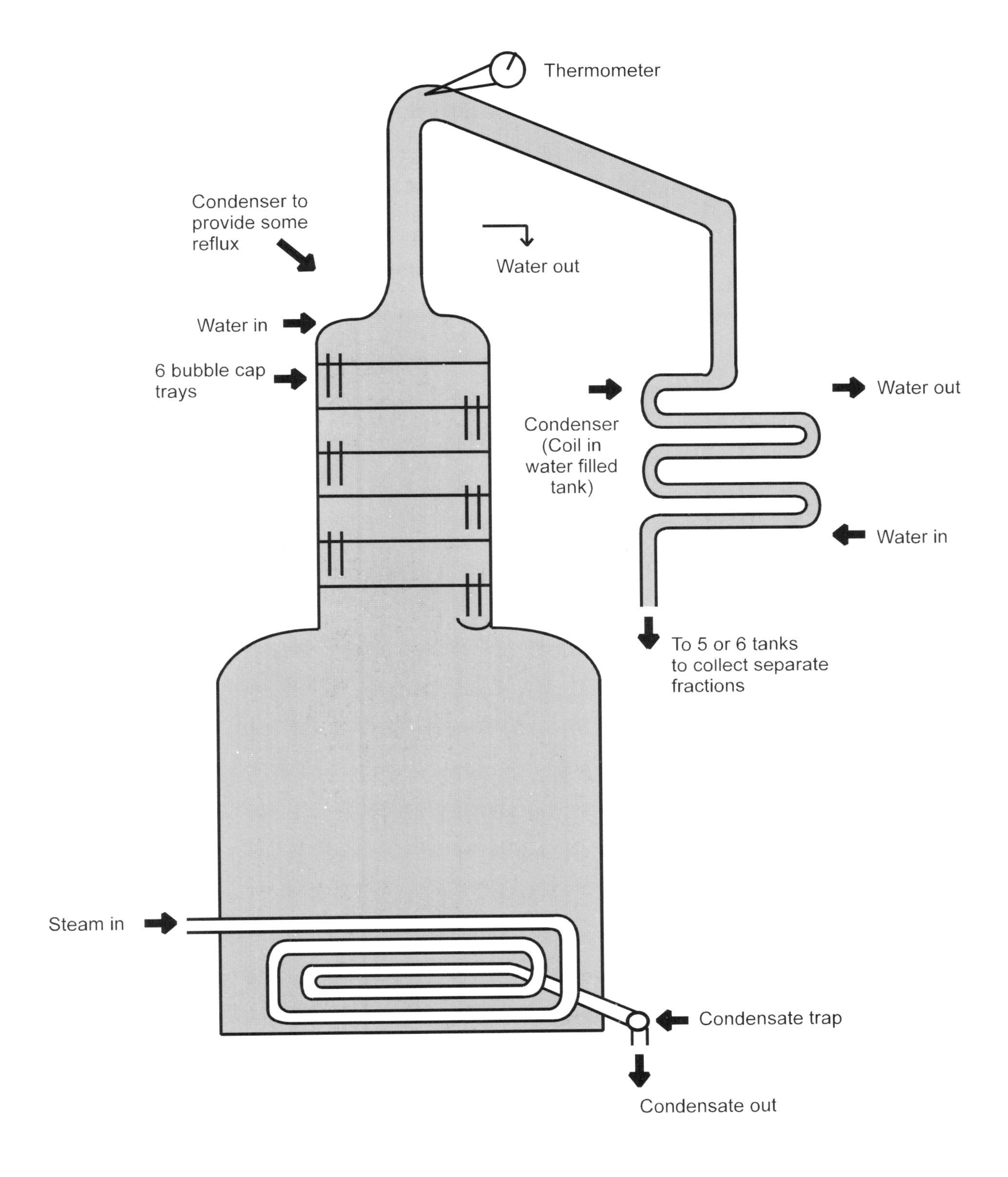

Figure 3. Pot still with bubble cap tray section.

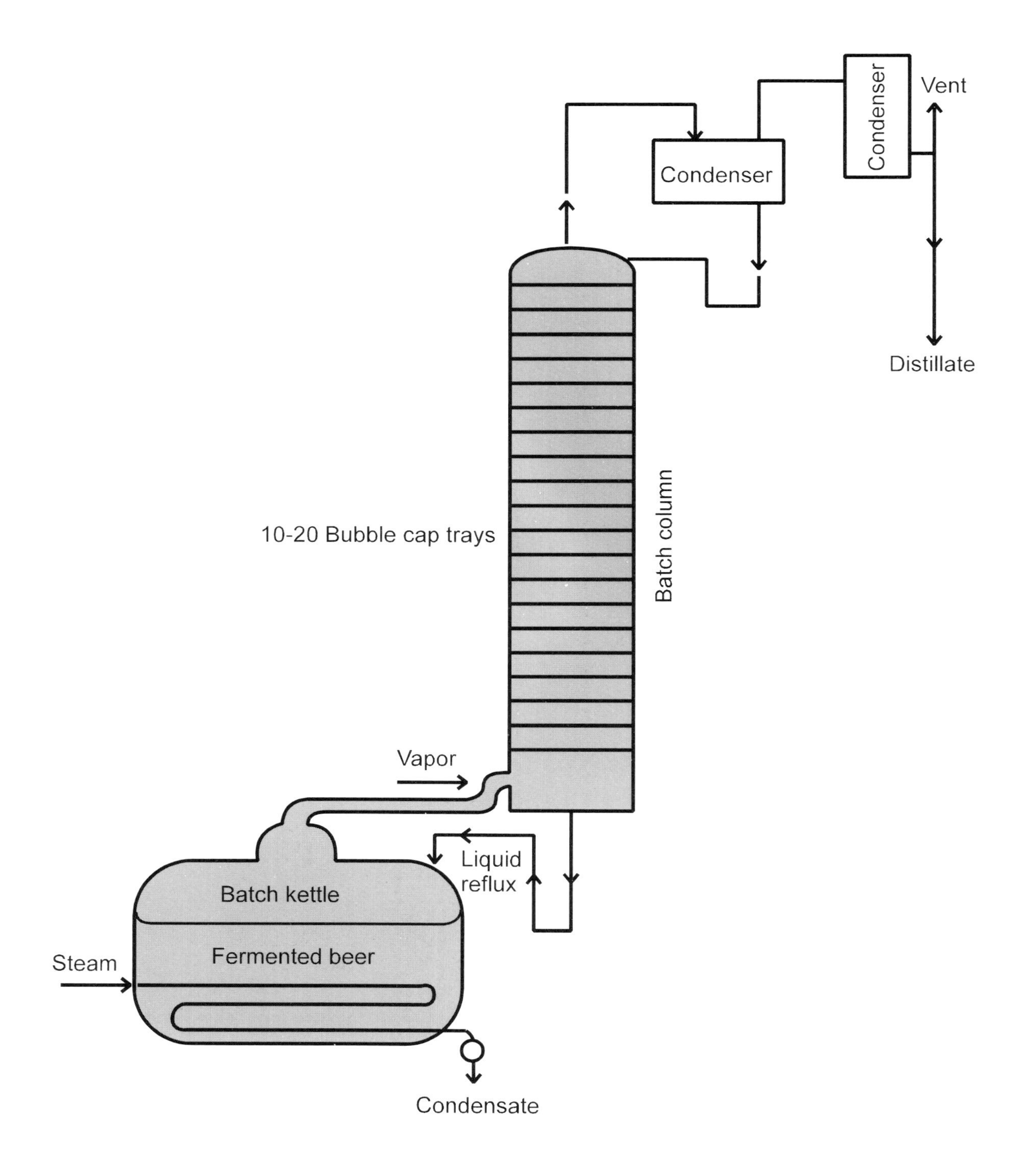

Figure 4. Batch still for rum production.

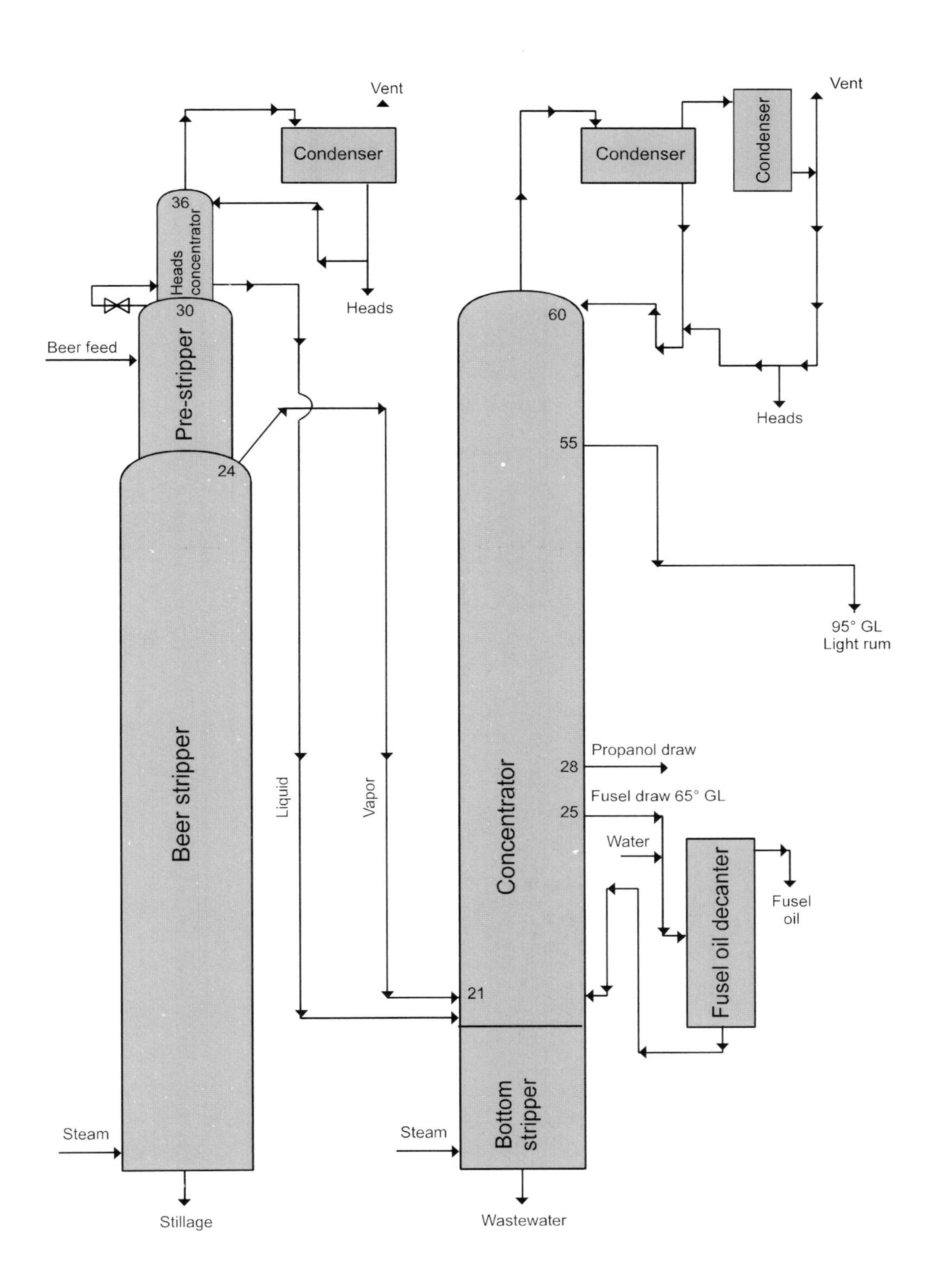

Figure 5. Modified Barbet beerstill for light rum production.

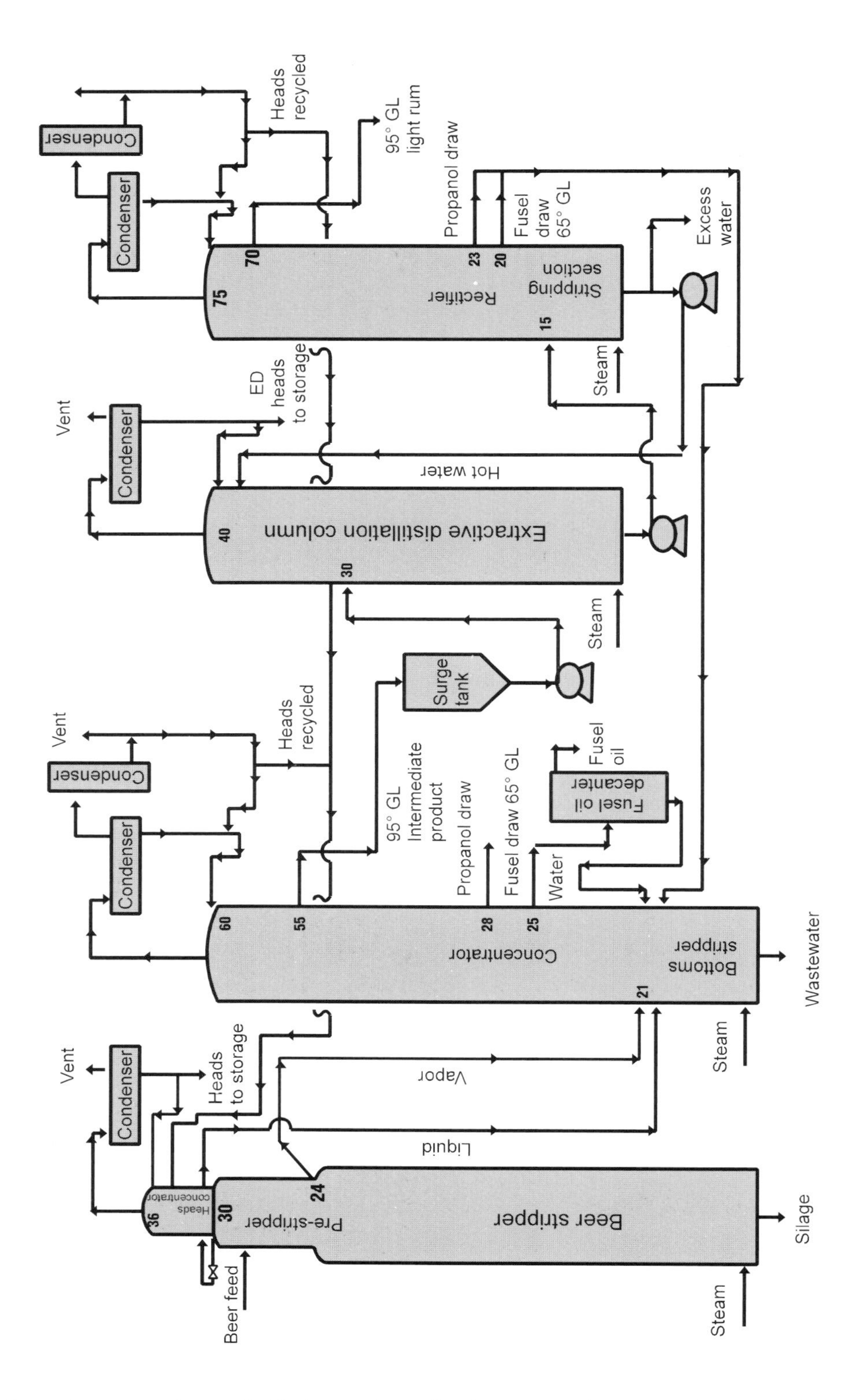

Figure 6. Four column distillation and rectification system for the production of light rum.

concentration of volatile organic acids. The fourth fraction was collected at a distilling temperature range of 78.5-85°C at a proof of about 90° GL. This fraction was characterized by the presence of most of the higher alcohols from the original continuous still distillate together with more esters, aldehydes and acids than in the previous fraction. The fifth and final fraction was collected over a distilling temperature range of 85-90°C and had a much lower proof than the other fractions (25-30° GL). This fraction was opalescent or turbid due to the presence of some of the highest boiling point esters and aldehydes which are more soluble in ethanol than in water.

Arroyo then recommended mixing the last four fractions in various proportions to make both light and heavy rums. He quoted an example of using 25% of the second fraction, 50% of the third fraction, 40% of the fourth fraction and 15% of the fifth fraction to make a light rum while using all the balances to make a heavy rum and avoid wastage.

In another combined process, Arroyo (1949b) suggested that the various congener sidestreams removed in the production of very light rums or neutral spirit by continuous distillation should be mixed, diluted with water and subjected to a fractional distillation almost identical to that used in the previous process. The distillate should be split into a similar set of five fractions with the first discarded, the third mixed with the light rum and the other three fractions used in various proportions for blending heavy rums. Arroyo claimed that these rums matured more rapidly than normally produced heavy rums when aged in oak barrels. As such there was a saving on warehouse space and other costs.

Rum maturation and blending

The normal and most reliable method of maturing rum to make it more suitable for consumption is by aging in oak barrels. Light rums with very low levels of congeners may require very little aging and may be acceptable after just a few weeks in barrels. In fact, many of the currently popular lighter rums are not aged at all. The heavy rums, however, tend to require much more aging to become palatable and may be held in barrels for five years or longer.

The barrels normally used for rum aging are obtained from the US, where government regulations specify that bourbon whiskey must be aged in new oak barrels thereby creating a large supply of once-used barrels. The bourbon barrels are charred on the inside and are used for aging rum either with the char intact or after its removal, depending on the preference of individual distillery.

Oak wood has a strong influence on the maturation and ultimate flavor of the rum, largely due to the extraction of compounds such as tannins and calcium salts and colorings such as quercitine. The type and quantity of the extracted compounds depend largely on the length of time and temperature of aging, the proof of the rum, the type and former usage of the barrel and the amount of contact surface it offers. New oak barrels, or fresh, once-used bourbon barrels naturally have more extractable compounds than old barrels that have been used many times over several years. Thus, a light rum stored in new or once-used barrels for a year or two may emerge with much of the odor and taste of a bourbon whiskey, as the main characteristics of bourbon are derived from the oak wood. This odor and taste may or may not be desirable. A program should therefore be established to identify individual barrels and code them as to history of usage in order to select particular types or ages of barrels for particular rum products. For example, in countries requiring that rum be aged in wooden barrels for a minimum period of time (which may be 1-3 years), light rums should be aged in the oldest available barrels as these will contribute the least amount of flavor. Even then, it may be necessary to treat the aged product with activated charcoal to reduce the color and flavor before bottling.

The main reactions that take place during aging to make rums more palatable are esterification, condensation and oxidation. The formation of esters involves a reaction between acids and the ethanol or other alcohols in the rum, and can take a considerable amount of time. The presence of esters gives the rum a generally desirable fruity aroma. In the condensations, molecules such as aldehydes and alcohols may combine to form acetals. In oxidations that occur with air passing through the pores in the wood, ethanol may be oxidized to acetaldehyde, which in turn may be oxidized to acetic acid, to then undergo esterification to ethyl acetate.

The temperature at which rums are aged will greatly affect the rate of maturation. It is reported that the rate of maturation at 25°C may double at 35°C (Kampen, 1975). This means that rums aged in the tropics will tend to mature more rapidly than those aged in temperate climates unless the warehouses are heated. The higher temperatures do, however, increase the annual rate of evaporation loss of rum through the wood.

The wood contact surface area may be increased by adding some charred or toasted oak chips to the barrels. The chips are relatively inexpensive as they can be made from scrap wood and can provide a means of standardizing the rate of maturation if fresh supplies are used each time.

Some rum producers add fruit extracts, sugar and even artificial rum essences to their products. The use of artificial essences should, however, be avoided or kept to a minimum as the product will usually have a noticeably artificial character. This is because an essence containing possibly four or five chemical compounds cannot give the overall 'roundness' of odor and taste that comes from the dozens or even hundreds of different congeners present naturally in a good heavy rum.

Characteristics of the aged rums should be carefully checked prior to bottling. Some blending may be required to ensure consistency with rum bottled previously under the same brand labels. Caramel coloring may be added to adjust the color of the rum to a standard in order to compensate for color variations acquired from the barrels used in aging.

Disposal of rum distillery wastes

The disposal of molasses stillage is a major problem in most rum producing areas as the liquid waste has a high biological oxygen demand (BOD) and will cause serious pollution if discharged into rivers or other watercourses.

There are some limited uses of the stillage. It may, for example, be sprayed on unpaved roads in dry seasons to reduce dust. Similarly, if evaporated to a syrup referred to as 'condensed molasses solubles', it may be used as a dust suppressant and source of minerals in cattle feeds. In some countries, a dried molasses stillage powder is sold for cattle feeding.

Evaporated stillage syrup may be burned in boilers as a partial replacement for bunker-C fuel oil; but the energy yield barely covers the energy required for the evaporation. There have also been reports of the use of molasses stillage as a binder in the production of concrete blocks. Generally, however, molasses stillage must be subjected to waste treatment processes. It was reported by Szendrey (1983) that the Bacardi Corporation in Puerto Rico uses an anaerobic digestion system to treat stillage and to produce methane for use as a boiler fuel. Other stillage waste treatment systems include anerobic and aerobic lagoons.

References

Arroyo, R. 1941. The manufacture of rum. Sugar 36(12).

Arroyo, R. 1942. The manufacture of rum. Sugar 37(1), (5), (7).

Arroyo, R. 1945. US Patent 2,386,924. Production of heavy rums.

Arroyo, R. 1945. The production of heavy bodied rum. International Sugar Journal 40(11):34-39.

Arroyo, R. 1945. Studies on rum. Research Bulletin No. 5 (University of Puerto Rico, Agricultural Experiment Station, Rio Piedras, Puerto Rico), December.

Arroyo, R. 1947. The economics of rum production. International Sugar Journal 49:292-294.

Arroyo, R. 1947. The economics of rum production. International Sugar Journal 49:325-327.

Arroyo, R. 1948a. The production of straight light rums from blackstrap. International Sugar Journal 50:150-152.

Arroyo, R. 1948b. The flavour of rum - recent chromatographic research. International Sugar Journal 50:210.

Arroyo, R. 1948c. Simultaneous production of light and heavy rum. International Sugar Journal 50:289-291.

Arroyo, R. 1949a. The Arroyo fermentation process for alcohol and light rum from molasses. Sugar Journal 11(8):5-12.

Arroyo, R. 1949b. A new rum distillation process. Sugar 44 (7):34-36.

Arroyo, R. 1949c. Rum distillery yields and efficiencies-factors affecting them. International Sugar Journal 51:163-169, 189-191.

Arroyo, R. 1950. Advanced features in rum fermentation. International Sugar Journal 52:42-44.

Bureau of Alcohol Tobacco and Firearms. 1982. Regulations under the Federal Alcohol Administration Act. Washington, D.C.

Clutton, D.W. 1974. Rum. The Flavour Industry, November/December pages 286-288.

Ianson, P. 1971. Rum Manufacture. Process Biochemistry, July, pages 35-39.

Kampen, W.H. 1975. Technology of the rum industry. Sugar & Azucar 70(8):36-43.

Lehtonen, M. and M. Suomalainen. 1977. Rum. In: Alcoholic beverages (A.M. Rose, ed). Academic Press, London. pages 595-635.

Paturau, J.M. 1969. By-products of the cane sugar industry. Elsevier, Amsterdam.

Szendrey, L.M. 1983. The Bacardi Corporation digestion process for stabilizing rum distillery wastes and producing methane. In: Energy from Biomass & Wastes VII Symposium, Lake Buena Vista, Florida, Institute of Gas Technology, Chicago, Il. pages 767-794.

Chapter 17

Fuel ethanol production

P.W. Madson and D.A. Monceaux
KATZEN International, Inc., Cincinnati, Ohio, USA

History

Motor fuel grade ethanol (MFGE) is the fastest growing market for ethanol worldwide; and MFGE production dwarfs the combined total production of all other forms of ethanol. Fermentation ethanol, as fuel (and solvent), has experienced several cycles of growth and decline since the early 1800s. By 1860, production had reached more than 90 million gallons per year. In 1861 Congress imposed a tax of $2.08 per gallon. About that time, oil was found in Pennsylvania. Thus began the cycle of 'control' of fuel ethanol markets (and therefore production) by taxation policy and oil industry influence on government. Petroleum interests dominated the world fuel industry in the post-World War II era, until a major policy shift by Brazil in the 1970s led to an ethanol-fueled motor vehicle strategy, followed a decade later by the US (Morris, 1994, personal communication). As a result, the combined motor fuel ethanol production from fermentation in the Western Hemisphere exceeded 4.5 billion gallons per year in 1998.

In Central and South America the dominant MFGE feedstock is sugar either in the form of cane juice directly from crushed cane (autonomous distilleries), or from molasses (annexed distilleries). In North America, the dominant feedstock is starch from grain, with 90% coming from corn. Feedstock choice follows regional dominant agricultural output (Katzen, 1987).

Since the technology for producing MFGE from sugar sources is an abbreviated form of ethanol production from starch, which is in turn an abbreviated form of production from whole grain; this chapter will focus on MFGE production from whole grain as typically practiced in North America. This technology is generally known as dry milling (Raphael Katzen Associates, 1978).

Introduction

A comparative evaluation of potable ethanol and MFGE production processes reveals numerous similarities. As the MFGE industry began to develop, it looked to the distilled spirits industry for technology. In the US many early MFGE plants copied distillery processes, differing primarily in the addition of dehydration facilities copied from the industrial-grade ethanol industry (Madson and Murtagh, 1991). This generally ensured a plant capable of producing ethanol.

This technology strategy, coupled with a strong ethanol market during the 1980s, often resulted in positive cash flows.

This technology strategy continued until a downward trend in ethanol pricing revealed the single greatest difference between distillery and MFGE economics. Distilleries are traditionally operated for consistency in flavor and quality of product. Other factors such as yield, energy efficiency, labor, etc., while being important, did not dominate the economics. For the beverage distiller it is counterproductive to reduce cost of production at the expense of flavor and quality and, possibly, market share. Flavor and product consistency are so important that any benefit associated with a process change must be extremely high to offset the inherent market risk, which could be catastrophic if the product has to be aged for several years. This, combined with the price differential between distilled potable spirits and MFGE, caused a shift to new technology development and differentiation of the MFGE industry in order to survive periods of high grain cost and low MFGE prices.

The MFGE producer has minimum product quality-related constraints. MFGE specifications for water content, acidity, solids, etc. (as defined in ASTM D-4806) can be met while concurrently minimizing operating costs. MFGE producers have traditionally operated with narrow profit margins. The drop in ethanol price during the mid 1980s resulted in most of the beverage distillery technology-based MFGE producers ceasing operations. Many of these operations were labor- and energy-intensive and operated with poor yields (Madson, 1990).

The design of a successful MFGE facility requires a clear understanding of the economic sensitivities. Evaluation of dry milling operating costs revealed that feedstock costs comprise over 60% of the total (Hill *et al.,* 1986). Energy consumption, at one time the central focus of debate, has been reduced via a rapid development of technology to less than 40,000 BTUs per gallon of product, which is approaching the point of diminishing returns in cost trade-offs (Hill, 1991; personal communication). The key issues today are feedstock conversion efficiency, capital investment, 'on-stream' time, labor costs and user-friendliness.

Conversion efficiency (yield)

Most MFGE producers have little control over feedstock pricing beyond hedging strategies such as trading in futures. The producer's primary edge is to maximize yield. Prior to the major growth of the MFGE industry, the typical yield in the production of spirits and industrial ethanol was five wine gallons per bushel (measured as 190° US proof hydrous spirits). In MFGE terms, that represents 2.375 undenatured gallons per bushel (gpb). By the early 1980s, the newly-developed MFGE technologies had demonstrated (in top-performing plants) achieve-ments of 2.55 gpb in dry milling plants and 2.45 gpb in wet milling plants. More recently, some dry milling MFGE plants have achieved sustained yields of 2.8 gpb.

Because of the different industry reporting procedures, this discussion is based on the 'pay-to-pay' analysis. That is, yields are presented on the basis of unadjusted bushels of grain purchased and ethanol sold (undenatured basis) over a time period exceeding three months. This results in a market-based yield figure that refers directly to profit per bushel of grain purchased.

What brought about this remarkable yield increase? The major development in technology has occurred in the dry milling industry, primarily because of the variety of technologies tested and the broad-based experience from which to learn (Katzen *et al.,* 1992).

Beginning with the cooking step, it has become clear that the controlling factor in design of a cooking system is *not* the cooking of starch, but rather elimination of bacteria in order to achieve and maintain sterility throughout the process (Kemmerling, 1989). Because conditions needed to achieve sterility are differ-ent from conditions required to cook starch, other factors must be considered. Cooking must be conducted with minimum solubilization of potential fermentables in order to minimize adverse reactivity; yet all fermentables must be released during the liquefaction, saccharification and fermentation processes for complete conversion to ethanol. This includes the fermentable sugars embedded in the fiber matrix. Premature solubilization of potential fermentables risks side reactions that can result in unfermentable starch and sugar complexes because of high temperature and the presence of water and other reactive components. These reactions may be as simple as retrogradation of starch or as complex as reactions between amino acids and carbohydrates. An example of a mashing and cooking system that achieves maximum yield is illustrated in Figure 1.

Competing 'cooking' factors can be balanced by selecting a grind that allows minimum mobility of the sugars and starch within the grain particle matrix yet provides necessary hydration. This is followed by instantaneous 'jet cooking' in the absence of adverse catalysts. By proper design of the cooking flash-down to liquefaction temperature (including valve selection), the 'locked in' fermentables can be released for full access by liquefying and saccharifying enzymes. Further, the non-starch fermentables are released from the fiber matrix to become available to the yeast. By keeping the fermentables 'locked in' within the particle matrix until the liquefaction tank has been reached, maximum retention and availability of fermentable value is achieved.

The next critical step is liquefaction. By liquefying to minimum dextrose equivalence (DE) at high temperatures for short time periods, adverse reaction conditions that convert fermentables to non-fermentables are minimized. Little of the starch is converted, and is therefore protected from adverse reaction until fermentation conditions are reached.

The key to creating maximum availability of fermentable carbohydrates is to reach the outlet of the mash cooler with virtually sterile mash while providing minimum exposure of carbohydrate to adverse reactions. At the same time, the system must maximize downstream availability of fermentable carbohydrate embedded in the fiber matrix. Upon reaching the fermentation temperature, undesirable side-reactions in the mixture are minimal.

From the mash cooler forward, yield is strictly a function of enzymatic hydrolysis, fermentation technology, sterility and completion. High sustained yields, above 2.75 gpb, have been achieved with SSF (Simultaneous Saccharification and Fermentation) technology, which was developed in the 1970s to solve a fundamental problem in the conversion of cellulose to

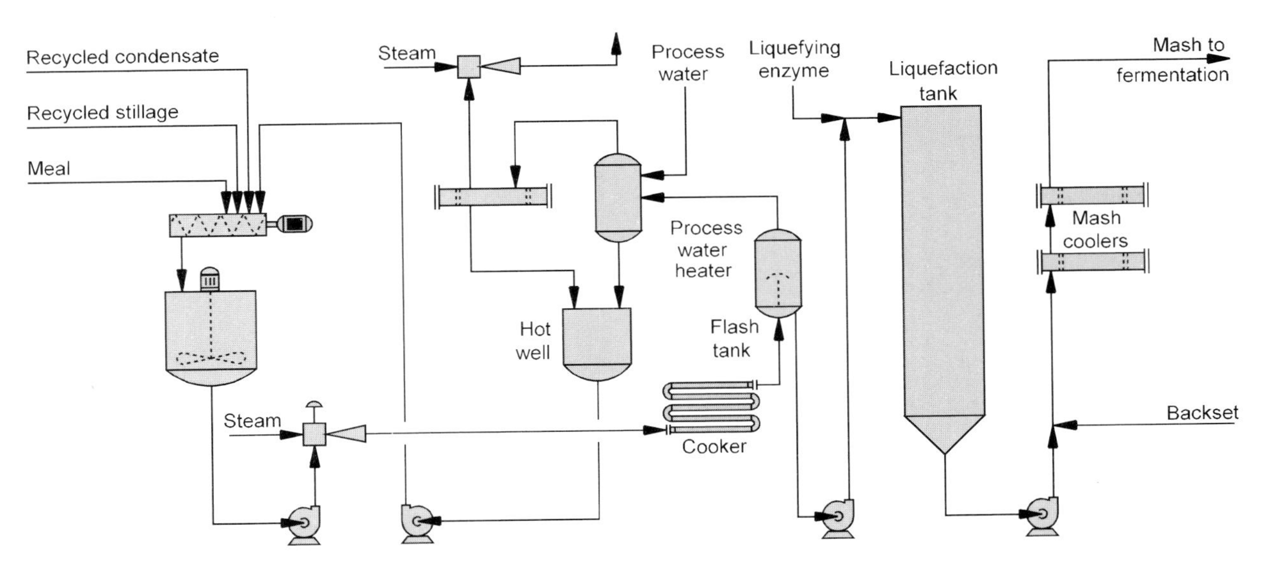

Figure 1. Mashing and cooking.

ethanol. The MFGE industry has advanced this technology to provide even higher yields by incorporating simultaneous yeast propagation (from active dry yeast) in the fermenter during initial saccharification. Thus SSYPF (Simultaneous Saccharification, Yeast Propagation and Fermentation) has become the low-cost, high-yielding technology of choice (Figure 2). Significant sugar concentrations do not develope in the fermenter, thus avoiding sugar inhibition of enzymatic hydrolysis. As a further consequence, bacterial growth is inhibited due to lack of substrate. Sugars are converted by yeast to ethanol as rapidly as they are produced. By proper maintenance of pH, nutrients and sterility, full conversion of available starch and sugars to ethanol is achieved. Any pH excursions below 4.2 at the end of fermentation correlate directly with losses in yield (Bowman and Geiger, 1984).

This high-yield SSYPF technology has been employed in four plants in North America for which long-term technical results have been reported (Katzen and Madson, 1991). South Point Ethanol of South Point, Ohio (64+ million gallons per year undenatured MFGE) was the first plant known to have achieved the 2.75 gpb sustained yield milestone with corn feedstock for more than one year of operation (Hill, 1991; personal communication). Reeve Agri-Energy Corporation of Garden City, Kansas, operates a 10 million gallons per year plant with yields from milo feedstock exceeding 2.75 gpb (Reeve and Conway, 1998; personal communication). Pound-Maker Agventures LTD of Saskatchewan, Canada, using wheat as feedstock, has achieved sustained yields equivalent to those of South Point Ethanol and Reeve Agri-Energy on a raw material starch and sugar basis (Wildeman and McCubbing, 1997; personal communication). More recently, Minnesota Energy Cooperative of Buffalo Lake, MN, in a plant producing 12 million gallons per year, has achieved sustained yields of 2.8 gallons of MFGE per bushel of corn (Robideaux and Johansen, 1999; personal communication). On a 'product sold' basis, the yield is 2.95 gallons of denatured MFGE per bushel.

It has been suggested that lower yields result in increased animal feed co-product production; and since the market price of the co-product, distillers dried grains with solubles (DDGS), is generally greater than that of grain, the production facility generates DDGS revenue to offset the losses in ethanol revenue. However, several issues need to be considered:

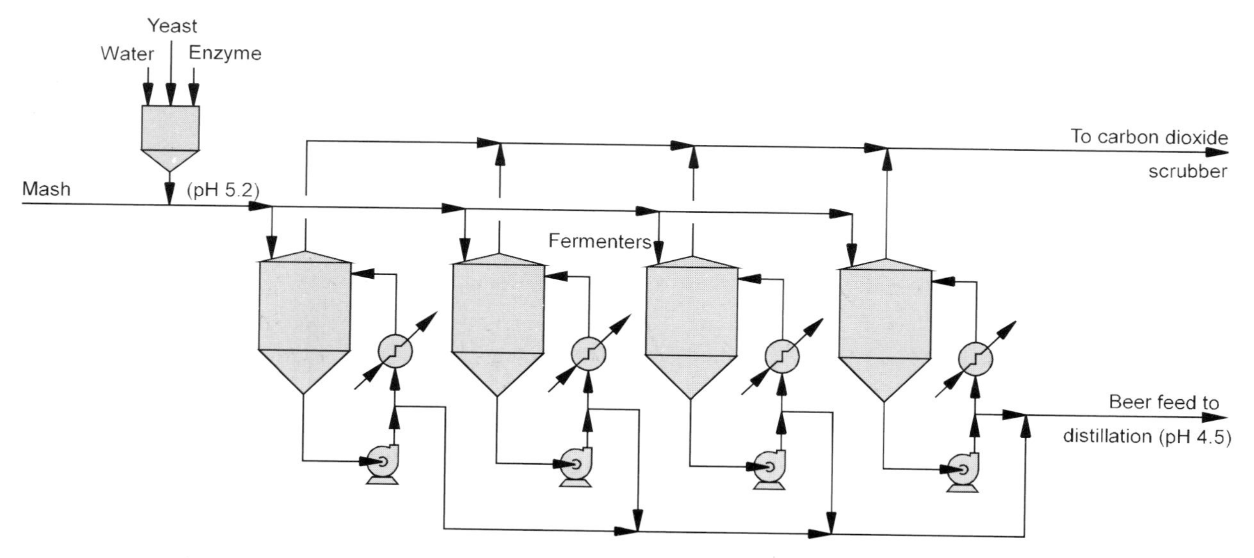

Figure 2. Simultaneous saccharification, yeast propagation and fermentation (SSYPF).

1. Every pound of starch or sugar not converted to ethanol must remain as starch, sugar, or be converted to a compound that does not involve the production of CO_2 (or other volatile by-product) in the requisite metabolic pathways. If not, production of one pound of DDGS will require consumption of two pounds of sugar, thereby negating the revenue trade-off.
2. If sugar or products of high-yield stoichiometric reactions pass directly through to DDGS, the soluble solids are recovered. Frequently, however, this results in complications in evaporator and dryer operations due to carbohydrate fouling and excess solubles syrup production. This can necessitate disposal of concentrated solubles syrup at significantly less than DDGS solids equivalent pricing. It can also result in a temporary decrease in production rate or may cause shutdown for cleaning
3. Most plants are limited in centrifugation, evaporation or dry-house capacity because these systems are the least productive from a return-on-investment perspective. Therefore higher ethanol yields maximize both plant production and productivity of the investment, since lower co-product production reduces demand on associated processing equipment.
4. For a plant of a given production capacity, increasing yield reduces the required size of most process equipment, with a corresponding decrease in capital investment. Higher yield, therefore, reduces debt service per unit of production.

What is this yield worth? If corn is priced at $2.50 per bushel, DDGS at $120.00 per ton and ethanol at $1.12 per undenatured gallon, the net result of a yield increase from 2.5 to 2.75 gpb is $0.145 per purchased bushel of grain ($0.053 per gallon of ethanol) in additional profit. This profit gain is after deduction of the $0.135 per bushel decrease in DDGS sales owing to reduction in carbohydrate pass-through to DDGS. Actual profit, however, will exceed $0.145 per bushel due to the efficiency value of increased production with no increase in fixed cost and little increase in variable cost.

What does it cost to achieve these yields? Fortunately, the investment and operating costs associated with this high-yield technology are lower than those of the common technologies available in the 1970s and early 1980s. High-yield milling, mashing, cooking, liquefaction and SSYPF technology represents one of those pleasant, but rare, situations in which it costs less to get more.

Cascade fermentation

Although there are substantial variations in the fermentation technologies applied in grain or starch conversion to ethanol, current operations can be described on a general basis. The broad divisions of technology are wet milling and dry milling. In wet milling, the major objective is to separate corn into a number of products such as starch, gluten, germ meal, germ oil, animal feed residue (gluten feed), dextrose, fructose, modified starches and a variety of specialty products. Production of MFGE from the lower grades of starch, or from all of the starch in an MFGE-dedicated plant, is an established conversion process in which a starch slurry is cooked, liquefied and saccharified prior to fermentation.

All fermentation systems in use today are continuous with respect to input and output. Fermentation of saccharified starch in the wet milling process is carried out by either simultaneous (SSF or SSYPF) or cascade processes. Figure 3 shows a typical system for cascaded saccharification and yeast propagation. Figure 4 shows the continuation of the cascade with pre-fermentation and fermentation. This cascade technology has been applied successfully to wet-milled starch feedstock. Application to whole corn dry milling operations has been carried out on a large scale, but has not yielded results equal to SSYPF technology.

Not only is less equipment required for SSYPF operation in dry milling plants, but the external

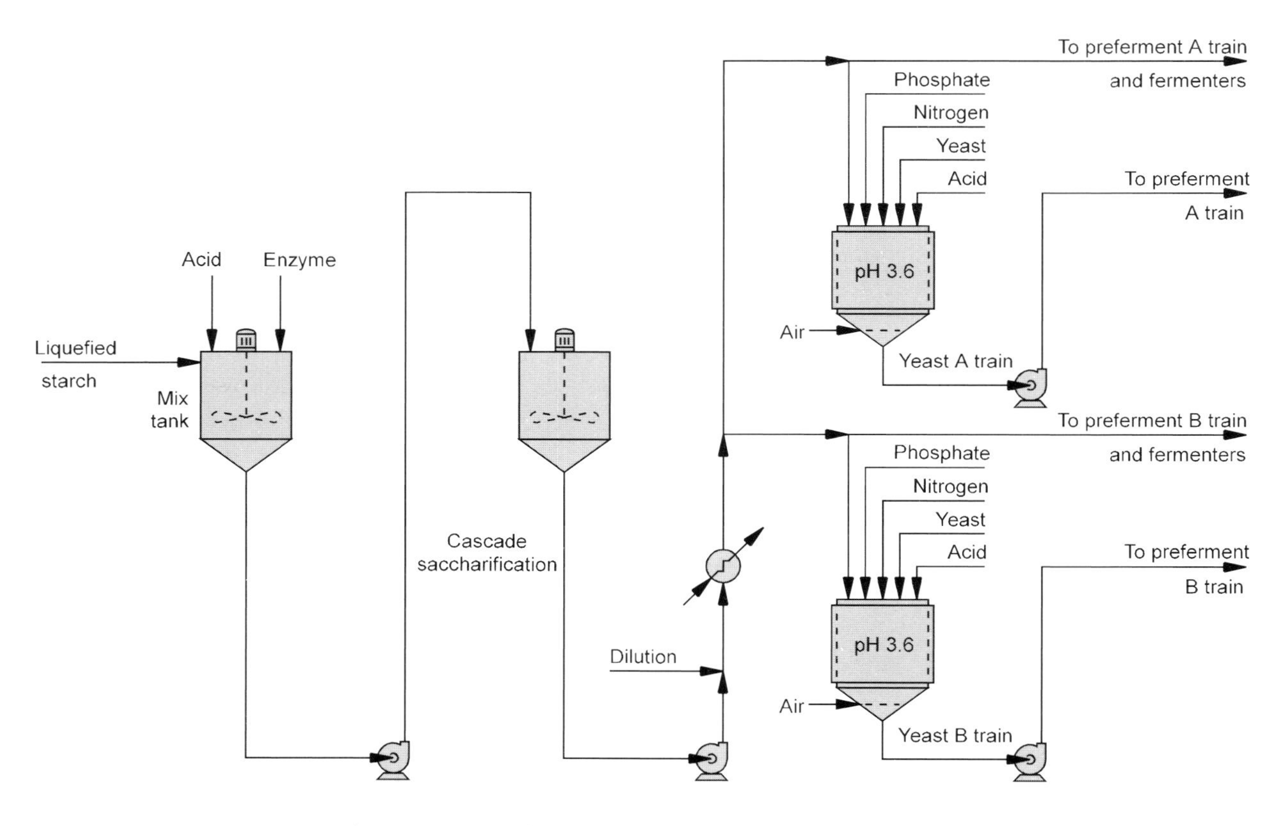

Figure 3. Cascade saccharification and yeast propagation.

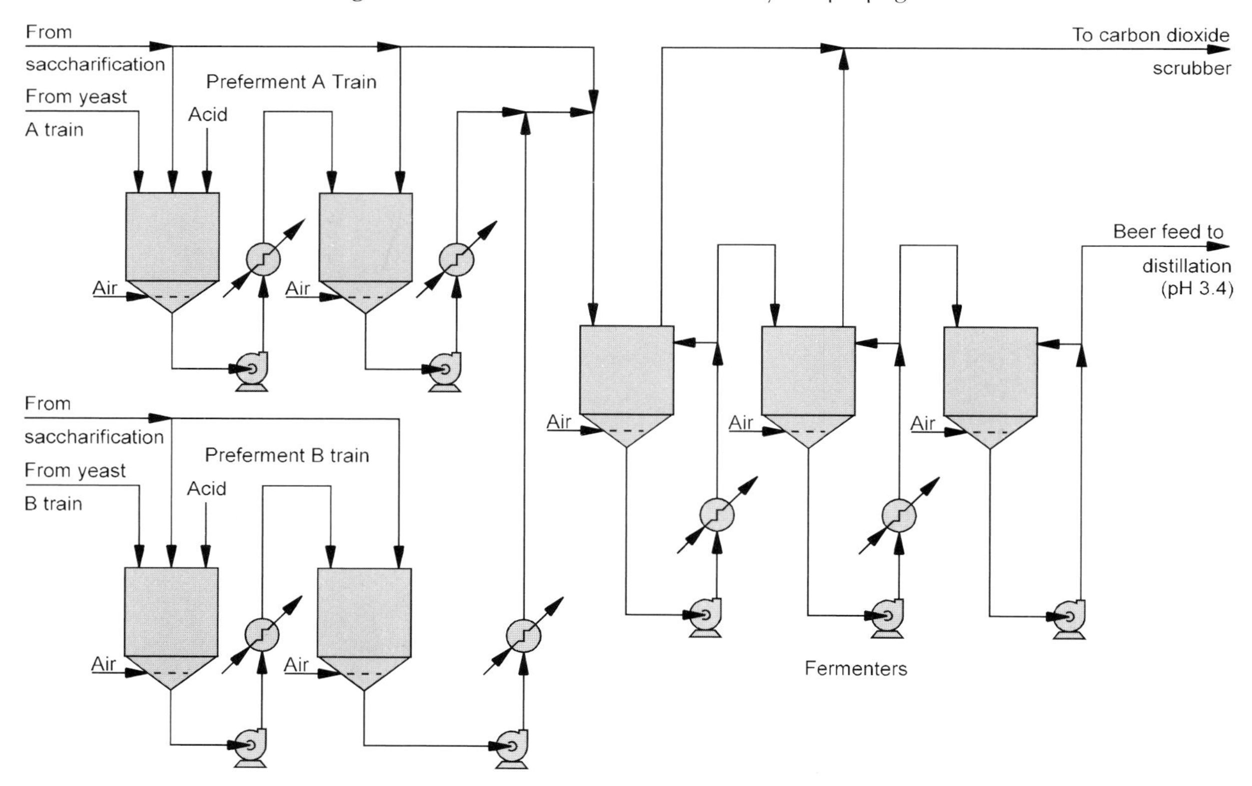

Figure 4. Cascade prefermentation and fermentation.

saccharification step of cascade systems, a major source of infections, is eliminated. Also, SSYPF, which starts at pH 5.2 and ends at pH 4.5, may be carried out in carbon steel fermenters. On the other hand, the cascade process requires maintenance of low pH to minimize bacterial infection. It operates at a pH near 3.5, thus requiring stainless steel construction and consequently higher investment.

Prior to the advent of fully-computerized fermenter control, including automated cleaning-in-place (CIP) systems, labor costs favored cascade operation. With today's automation and simplified design, labor costs associated with operating either fermentation technology are negligible.

Distillation and dehydration

Production of MFGE from fermented beers in the range of 8-10 wt.% ethanol has been carried out primarily by techniques similar to those found in the beverage spirits industry. However, the dehydration step, primarily conducted by ternary azeotropic distillation in the past (Figure 5), has been superceded by molecular sieve dehydration utilizing integrated pressure swing adsorption (PSA) technology (Figure 6), particularly in newer installations.

Considerable research efforts have been expended in the field of ternary azeotropic dehydration, utilizing agents such as benzene, cyclohexane, diethyl ether and n-pentane, to reduce the assumed high energy consumption of this process. However, in reality many systems have been operating with multistage distillation such that the dehydration process is operated entirely with recovered heat from the primary distillation system. Alternatively, recovered energy from dehydration is used to provide most of the energy for stripping and rectification. Commercial systems are producing MFGE from fermented beer with thermal energy consump-

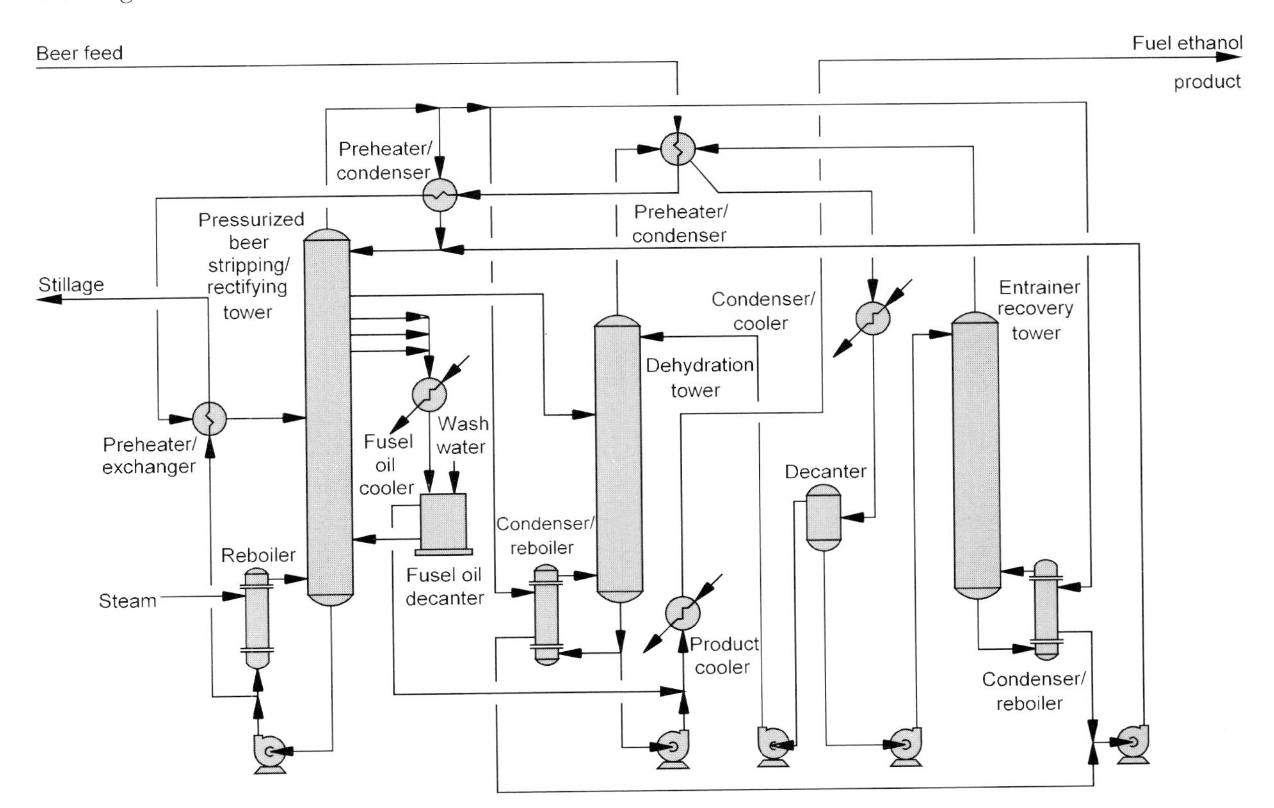

Figure 5. MFGE distillation and dehydration by ternary azeotrope.

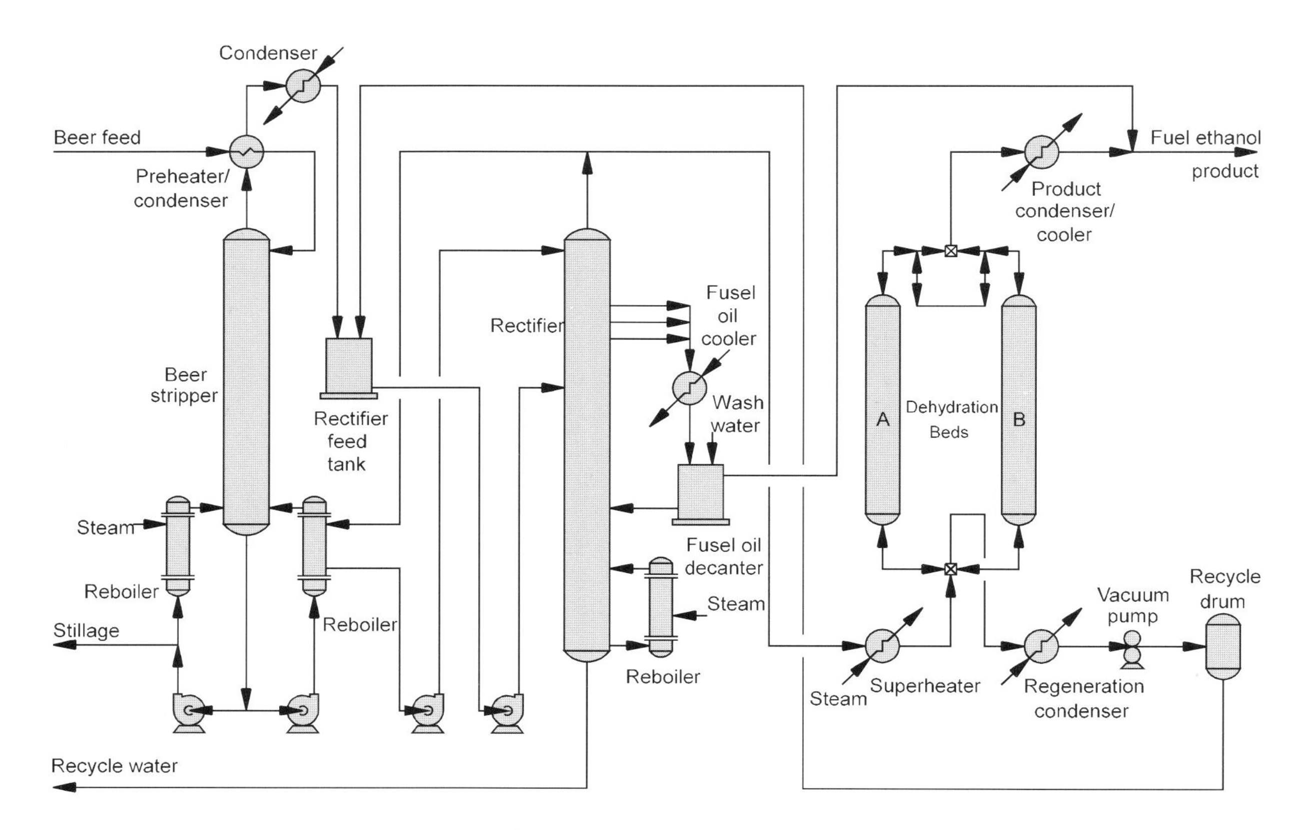

Figure 6. MFGE distillation and dehydration by pressure swing adsorption molecular sieve.

tion of 17,000 BTUs per gallon for combined distillation and dehydration. When molecular sieves are used in a well-integrated design, this operation consumes about 14,000 BTUs per gallon.

Molecular sieve dehydration

Molecular sieve adsorption technology for dehydrating MFGE has been actively developed since the late 1970s. Pressure swing adsorption is now the technology of choice for MFGE dehydration for new plants and retrofits. Once believed to apply only to small production facilities, a single-train molecular sieve unit has been in operation in Brazil since 1993 with an annualized capacity of more than 60 million gallons.

Molecular sieves are hard, granular substances, spherical or cylindrical extrudates manufactured from materials such as potassium aluminosilicates. They are graded according to the nominal diameter of the myriad of internal pores that provide access to the interstitial free volume found in the microcrystalline structure. A typical grade used in ethanol dehydration is Type 3Å. This designation means that the average diameter of the interstitial passageways is 3 Angstroms (Å). One Angstrom is a unit of measure equivalent to 10^{-8} centimeters). Thus, the passageways in the structure have a diameter that is of molecular scale. The water molecule has a mean diameter less than 3Å, while the ethanol molecule has a mean diameter greater than 3Å. In addition, the water molecules can be adsorbed onto the internal surface of the passageways in the molecular sieve structure. It is this combination of physical properties that make molecular sieves useful for the separation of mixtures of ethanol and water.

Water molecules can invade the inner structure of the molecular sieve beads and be adsorbed thereon, while the ethanol molecules are too large and pass out of the vessel leaving the water behind. Thus, dehydration of MFGE from sub-azeotropic concentrations is possible. It should be noted that this sieving process works to separate ethanol-water mixtures in either the liquid or vapor phase. Process details of course differ for vapor and liquid mixtures.

The earlier systems for such dehydration, particularly in the liquid phase, required hot gas regeneration to displace the water from the beads. The molecular sieve beads rapidly deteriorated due to excessive thermal shock. With a half life for the beads on the order of six months in the liquid systems, operating costs were high.

Application of vapor phase pressure swing, vacuum purge adsorption (PSA) technology for MFGE dehydration matured in the 1980s. With PSA technology, molecular sieve beads are regenerated by recycling a portion of the superheated, anhydrous ethanol vapors to one bed under vacuum while the other bed is producing anhydrous ethanol vapor under pressure. With this milder regeneration condition, molecular-sieve bead life is extended to several years. In some cases molecular sieve beads have operated for more than ten years with no appreciable deterioration. This results in insignificant adsorbent replacement expense and reduced overall operating costs.

With an automated operation, the vapor feed to the molecular sieve system can be taken directly from a pressurized rectifier, with the reprocessing of the hydrous regeneration stream in the same rectifier. In this way, recovery of the ethanol from the molecular sieve regeneration system adds less than 1% to the distillation energy consumption.

One of the major shortcomings of the early molecular sieve systems was the high maintenance cost of the compressor used to pressurize feed vapor from atmospheric pressure rectifiers. This problem was initially solved by feeding a liquid spirit to a pressurized vaporizer, thereby eliminating the compressor. However, higher energy consumption resulted. These problems have been overcome by new pressure-cascaded distillation systems integrated with the molecular sieve beds as shown in Figure 6. The rectifier is maintained at a pressure sufficient to economically operate a reboiler to provide energy to the atmospheric pressure beer stripper. Approximately two-thirds of the overhead rectifier vapor is used to provide this reboiler energy. The remaining ethanol spirit vapor is passed through a steam-heated superheater and then to the molecular sieve beds for dehydration. The thermal energy content of the resultant superheated anhydrous ethanol vapor can be recovered in an auxiliary reboiler on the beer stripper. The condensed anhydrous ethanol vapors (MFGE) are then cooled and passed to storage. The recovered ethanol and water from the regeneration phase of the pressure swing adsorption cycle is recycled to an appropriate feed point in the rectifier for recovery.

Animal feed co-products

The DDGS co-product operation for dry milling plants recovers the residual non-starch materials in the stillage from the beer stripper. This is accomplished by combining centrifuged solids (wet cake) with concentrated solubles (syrup) from the evaporated thin stillage after centrifugal separation, as shown in Figures 7 and 8. Recycle of dried DDGS acts as a base for blending the wet cake and syrup to yield a material suitable for operation of dryers, which may be either hot-gas or steam-tube type. The DDGS product meets the commercial specification of 26-30% crude protein. However, the more efficient dry milling plants produce DDGS containing 30-32% protein and 8-9% 'fat' as well as vitamin B complex from yeast propagated in the fermentation operation. This is a high-value, triple concentrate of the corn protein and oil, plus the incremental value of the propagated yeast.

In the wet milling industry, the primary animal feed co-product is corn gluten meal (CGM), a high value, low volume product containing 60% protein, but no appreciable fat or oil. The

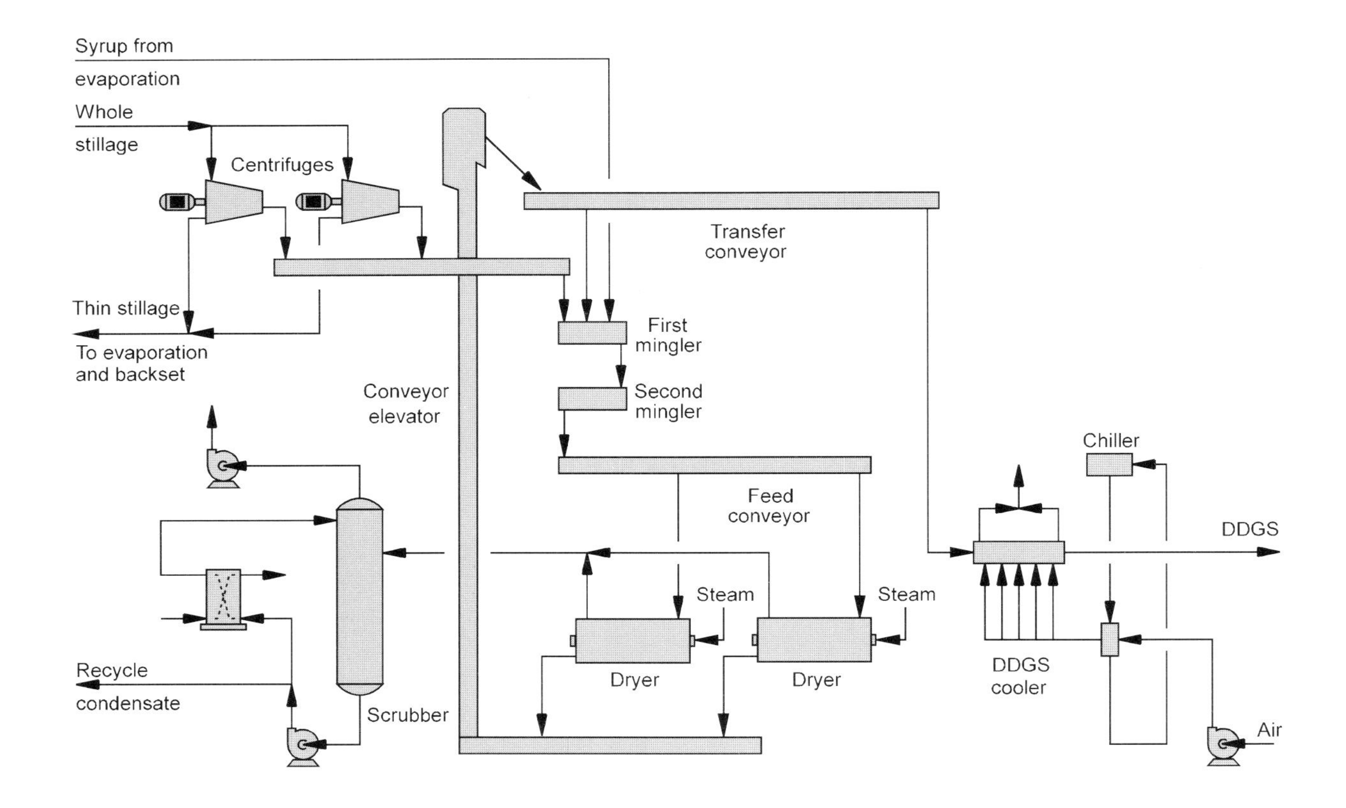

Figure 7. Centrifugation and drying.

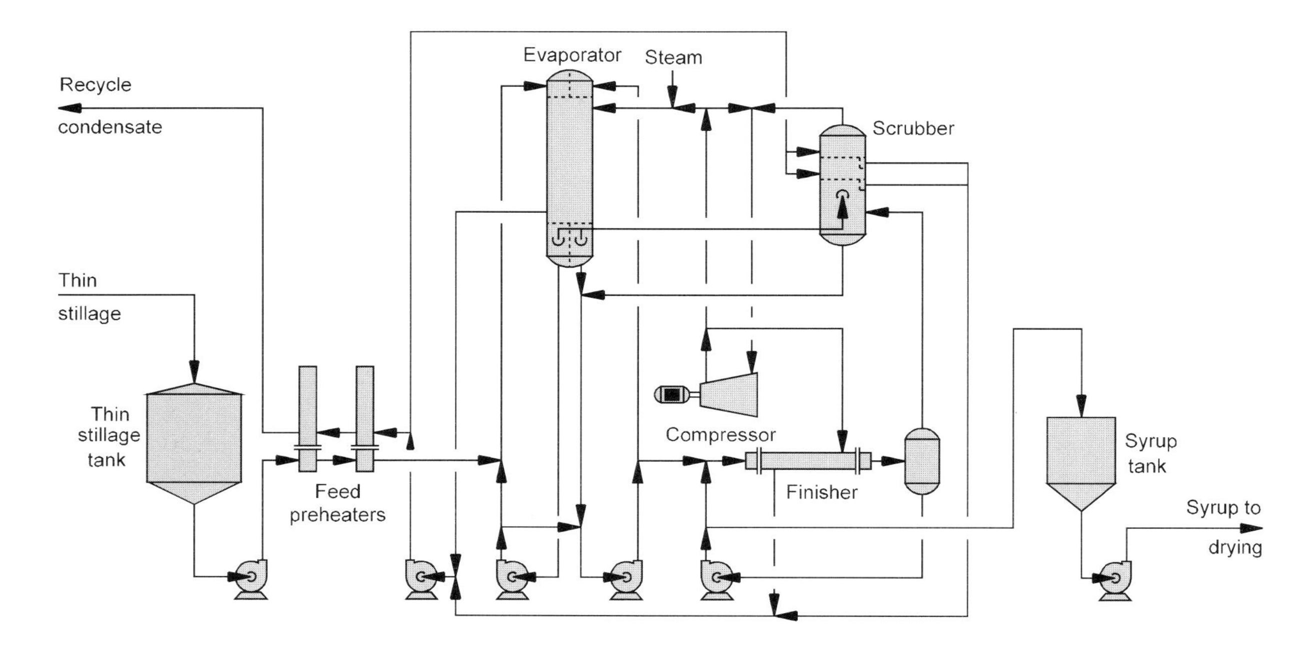

Figure 8. Evaporation.

secondary animal feed co-product is corn gluten feed (CGF), which is a concentrate of the residual fiber and liquid containing about 20% protein and very little fat or oil. This is a relatively low grade product compared to DDGS.

Energy use

Confusion abounds regarding energy use and efficiency of production of MFGE from corn. Some publications refer to very high steam usage, which is related to older practices in production of potable ethanol of various types from grain. Those operations hark back to an age when energy was cheap and potable ethanol had a high value. However, modern MFGE plants are designed with a high degree of efficiency with respect to energy consumption.

In wet milling operations, much of the energy use is charged to products other than ethanol. The ethanol operation itself requires only about 30,000 BTUs per gallon.

Dry milling operations charge all of the energy consumed to MFGE production, although much of this energy use is associated with production of the DDGS co-product. As an example of the efficiency of modern operations, thermal energy use of 39,000 BTUs of fuel and electrical use of 1.15 kilowatt hours per gallon is common. These usages can be compared with the gross heating value of ethanol, which is 84,000 BTUs per gallon, or the net heating value of about 75,000 BTUs per gallon.

Investment

Facilities for production of MFGE from corn vary widely in size and technology base; and it is therefore difficult to set a standard investment factor. In addition to large-scale wet milling and dry milling facilities, there are smaller plants built for unique opportunities, such as low-cost waste feedstocks, or adjacent cattle feedlots that can utilize the stillage directly. In effect, investment for any particular facility must be developed on a site-specific basis for the conditions at hand.

Experience in developing improved and simpler technology over the last decade permits estimation of approximate costs for wet milling and dry milling facilities in view of the most advanced and efficient technology available. Investment relating to large dry milling plants (turnkey) approaches $2 per gallon of annual production capacity (undenatured basis). It should be noted that in general, wet milling MFGE production units require only about half the investment of a dry milling plant. This is due to the fact that an MFGE facility is simply added on to a wet milling plant. All other facilities, such as raw material handling, primary processing, waste processing and utilities are already in place and provide sufficient capacity for the adjacent ethanol facility.

At the other extreme, small farm-based ethanol plants with direct acquisition of feedstocks and access to cattle feedlots for direct consumption of wet stillage can be built to produce in the range of 2-10 million gallons per year for investments in the range of $1.50 to $2.00 per annual gallon. However, in special situations, using pre-owned equipment, the required investment may be less than $1 per annual gallon of capacity.

References

Bowman. L., and E. Geiger. 1984. Optimization of fermentation conditions for alcohol production. Biotechnology and Bio-engineering 26:1492.

Hill, L.L. *et al.* 1986. South Point Ethanol 60-Million-Gallon-per-Year Fuel-Ethanol Plant Final Technical Report. Prepared for: US Department of Energy, DOE/ID/12188.

Kemmerling, M.K. 1989. Effects of bacterial contamination on ethanol yield and downstream processing. International Conference on Alcohols and Chemicals from Biomass, Guadalajara, Mexico.

Katzen, R. 1987. Large-scale ethanol production update. Alcohol Fuels 1987, Cancun, Mexico.

Katzen, R. and P.W. Madson. 1991. Bio-engineering improvements in corn fermentation to ethanol. Corn-Derived Ethanol Conference, Peoria, Illinois, May 12-21.

Katzen, R., P.W. Madson and B.S. Shroff. 1992. Ethanol from corn. State-of-the-art technology and economics. AIChE Annual Meeting, Biotechnology for Fuels, Chemicals, and Materials, Session No. 154, Miami Beach, Florida, November 1-6.

Madson, P.W. 1990. Bio-ethanol experiences in the USA. Zuckerind 115 No. 12, pp.1045-1048.

Madson, P.W. and J.E. Murtagh. 1991. Fuel ethanol in USA: review of reasons for 75% failure rate of plants built. International Symposium on Alcohol Fuels, Firenze '91, Florence, Italy.

Raphael Katzen Associates. 1978. Grain Motor Fuel Alcohol Technical and Economic Assessment Study. Prepared for: US Department of Energy, HCP/J6639-01.

Chapter 18

Ethanol distillation: the fundamentals

R. Katzen, P.W. Madson and G.D. Moon, Jr
KATZEN International, Inc., Cincinnati, Ohio, USA

Fundamentals of a distilling system

Certain fundamental principles are common to all distilling systems. Modern distillation systems are multi-stage, continuous, countercurrent, vapor-liquid contacting systems that operate within the physical laws that state that different materials boil at different temperatures.

Represented in Figure 1 is a typical distillation tower that could be employed to separate an ideal mixture. Such a system would contain the following elements:

a. a feed composed of the two components to be separated,
b. a source of energy to drive the process (in most cases, this energy source is steam, either directly entering the base of the tower or transferring its energy to the tower contents through an indirect heat exchanger called a reboiler),
c. an overhead, purified product consisting primarily of the feed component with the lower boiling point,
d. a bottoms product containing the component of the feed possessing the higher boiling point,
e. an overhead heat exchanger (condenser), normally water-cooled, to condense the vapor resulting from the boiling created by the energy input. The overhead vapor, after condensation, is split into two streams. One stream is the overhead product; the other is the reflux which is returned to the top of the tower to supply the liquid downflow required in the upper portion of the tower.

The portion of the tower above the feed entry point is defined as the 'rectifying section' of the tower. The part of the tower below the feed entry point is referred to as the 'stripping section' of the tower.

The system shown in Figure 1 is typical for the separation of a two component feed consisting of ideal, or nearly ideal, components into a relatively pure, overhead product containing the lower boiling component and a bottoms product containing primarily the higher boiling component of the original feed.

If energy was cheap and the ethanol-water system was ideal, then this rather simple distillation system would suffice for the separation of the beer feed into a relatively pure ethanol overhead product and a bottoms product of stillage, cleanly stripped of its ethanol content. Unfortunately, the ethanol-water (beer) mixture is not an ideal system. The balance of

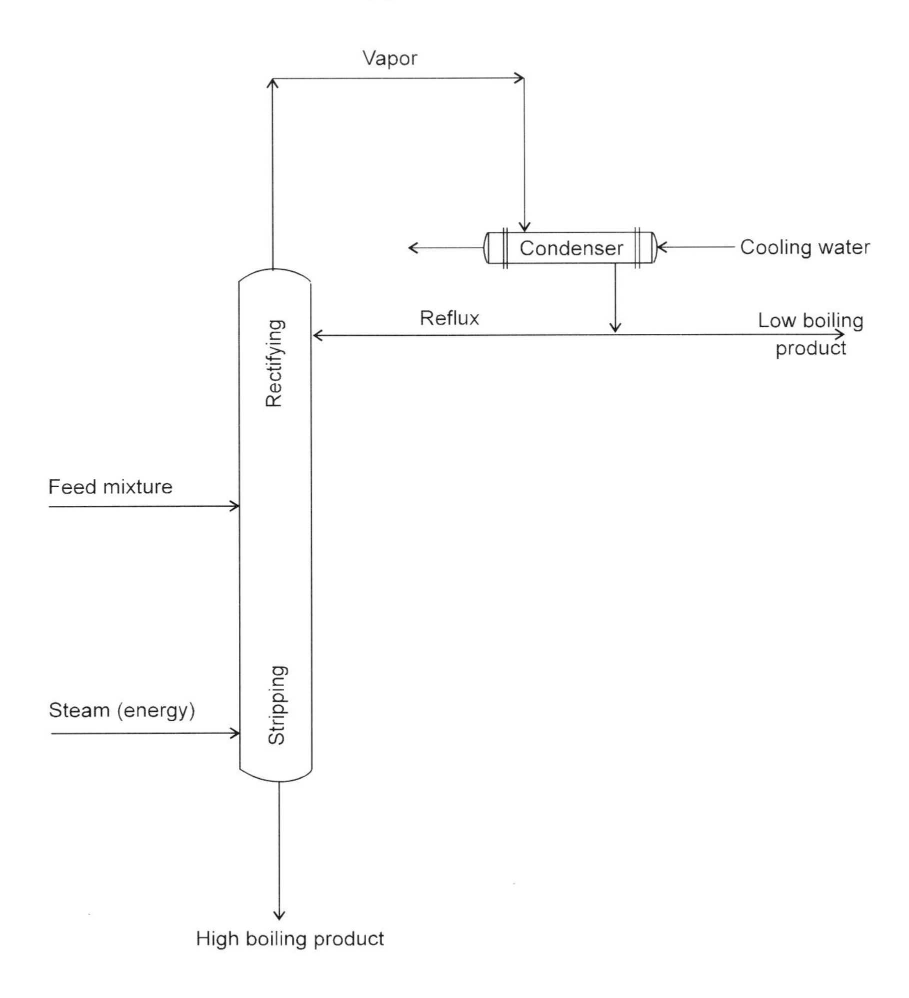

Figure 1. Ideal distillation system.

this chapter will be devoted to a description of the modifications required of the simple distillation system in order to make it effective for the separation of a very pure ethanol product, essentially free of its water content.

Figure 2 expands on Figure 1 by showing some additional features of a distillation tower. These are:

a. The highest temperature in the tower will occur at the base.
b. The temperature in the tower will regularly and progressively decrease from the bottom to the top of the tower.
c. The tower will have a number of similar, individual, internal components referred to as 'trays' (these may also be described as stages or contactors).
d. Vapor will rise up the tower and liquid will flow down the tower. The purpose of the tower internals (trays) is to allow intimate contact between rising vapors and descending liquids correlated separation of vapor and liquid.

Figure 3 shows a vapor-liquid equilibrium diagram for the ethanol-water system at atmospheric pressure. The diagram shows mole percent ethanol in the liquid (X axis) *vs* mole percent ethanol in the vapor (Y axis). The plot could also be made for volume percent in the liquid *vs* volume percent in the vapor and the equilibrium

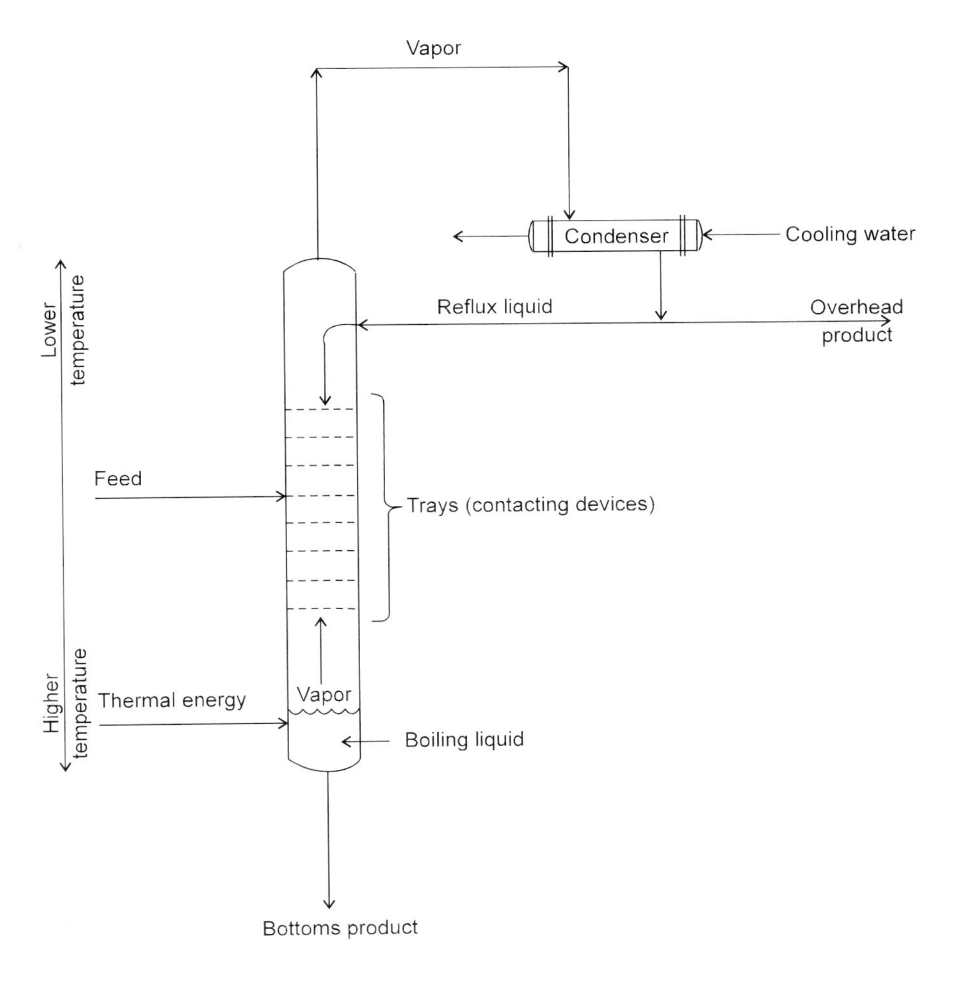

Figure 2. Typical distillation relationships.

curve would only be slightly displaced from that shown in Figure 3. Mole percent is generally used by engineers to analyze vapor/liquid separation systems because it relates directly to molecular interactions, which more closely describe the process occurring in a distillation system.

Analysis of the ethanol-water distillation system is mathematically straightforward when using molar quantities rather than the more common measurements of volume or weight. This is because of an energy balance principle called 'constant molal overflow'. Essentially, this principle states that the heat (energy) required to vaporize or condense a mole of ethanol is approximately equal to the heat (energy) required to vaporize or condense a mole of water; and is approximately equal to the heat (energy) required to vaporize or condense any mixture of the two. This relationship allows the tower to be analyzed by graphic techniques using straight lines. If constant molal overflow did not occur, then the tower analysis would become quite complex and would not lend itself easily to graphic analysis.

Referring to Figure 3, a 45° line is drawn from the compositions of the 0, 0-100% and 100%. This 45° line is useful for determining ranges of compositions that can be separated by distillation. Since the 45° line represents the potential points at which the concentration in the vapor equals the concentration in the liquid; it indicates those conditions under which distillation is

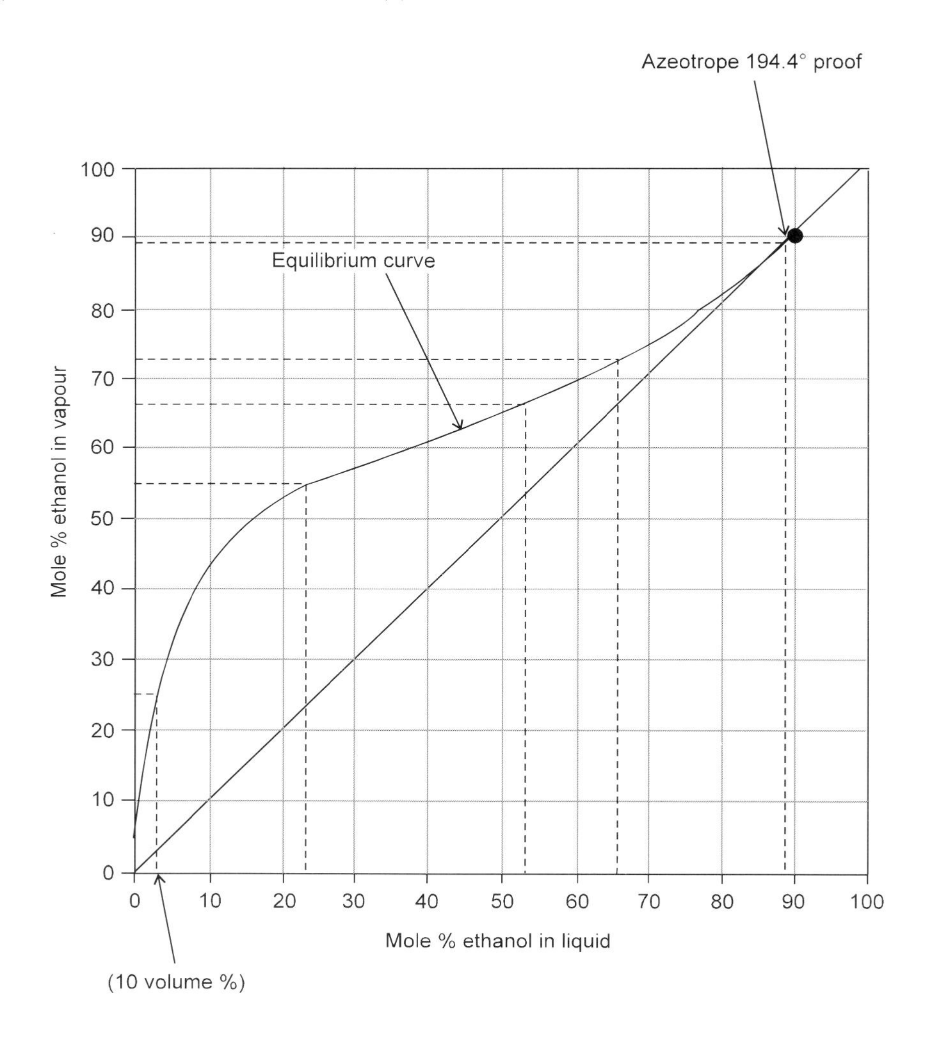

Figure 3. Vapor/liquid equilibrium for the ethanol/water system at atmospheric pressure.

impossible for performing the separation. If the equilibrium curve contacts the 45° line, an infinitely large distillation tower would be required to distill to that composition of vapor and liquid. Further, if the equilibrium curve crosses the 45° line, the mixture has formed an azeotrope. This means that even if the tower were infinitely large with an infinite amount of energy, it would be impossible to distill past that point by simple rectification.

Consider a very simple system consisting of a pot filled with a mixture of ethanol in water (a beer) containing 10 % by volume ethanol (3.3 mole %). This composition is identified in the lower left portion of Figure 3. A fire could be kindled under the pot, which would add thermal energy to the system. The pot would begin to boil and generate some vapor. If we gathered a small portion of the vapor initially generated and measured its ethanol content; we would find about 24 mole % ethanol (53 volume %). If we condense this vapor (note: there will be only a small amount of this vapor), boil it in a second pot and again collect a small amount of the first vapor generated, this second vapor would contain about 55 mole % (83 volume %) of ethanol (see Figure 3). If we should continue this simplified process to a third and fourth collection of small amounts of vapor; analysis would reveal that each successive portion of vapor would become richer in ethanol.

Thus we have created a series of steps by which we kept increasing the ethanol content of the analyzed sample, both liquid and vapor. Unfortunately, this oversimplified process is idealized; and practically speaking, is impossible. However if we had supplied our original pot with a continuous supply of ethanol-water feed and vapor generated in the first pot was continuously condensed and supplied to the second pot, etc. then the process becomes similar to the industrial distillation tower operation shown in Figure 2.

How far can this process be extended? Could we produce pure ethanol by continuously extending our process of boiling and reboiling? The answer is, no! We would finally reach a point in one of the downstream pots, where the vapor boiling off of the liquid was of the same composition as the liquid from which it was being generated. This unfortunate consequence limits our ability to produce anydrous ethanol from a dilute ethanol-water feed. What we finally encounter in our simp-lified process is the formation of an 'azeotrope'. This is a concentrated solution of ethanol and water that when boiled produces a vapor with a composition identical to the composition of the liquid solution from which it originated.

In summary then, we are limited in ethanol-water purification in any single multistage distillation tower to the production of azeotropic ethanol-water mixtures. These azeotropic solutions of ethanol and water are also known as constant boiling mixures (CBM) since the azeotropic liquid will have the same temperature as the azeotropic equilibrium vapor being boiled from itself. Without some sort of drastic process intervention, further ethanol purification becomes impossible. The question then becomes: What can we do to make it possible to produce anhydrous ethanol? Methods of doing so will be covered later in this chapter.

Figure 4 depicts the structure of the distillation process by dividing the vapor/liquid equilibrium information into three distinct zones of process and equipment requirements: stripping, rectifying and dehydration. This division is the basis for the design of equipment and systems to perform the distillation tasks.

Considerations in preliminary design

The engineer, given the assignment of designing a distillation tower, is faced with a number of fundamental considerations. These include:

a. What sort of contacting device should be employed? (e.g. trays or packing). If trays are chosen, what type will give the most intimate contact of vapor and liquid?
b. How much vapor is needed? How much liquid reflux is required? (What ratio of liquid: vapor is required?)
c. How much steam (energy) will be required?
d. What are the general dimensions of the distillation tower?

Distillation contactors

Trays are the most common contactor in use. What are the functions expected of tray contactors in the tower? Figure 5 depicts a single tray contactor in a distillation tower and shows the primary functions desired:

- Mixing rising vapor with a falling fluid.
- Allow for separation after mixing.
- Provide path for liquid to proceed down the tower.
- Provide path for liquid to proceed up the tower.

Figure 6 depicts a perforated tray contactor with certain accoutrements required to control the flow of liquid and vapor and to assure their intimate contact. Another type of tray contacting device, the disc-and-donut or baffle tray is shown in Figure 7. The characteristics of this type of contactor make it especially useful for distilling materials such as dry-milled grain beer, which would foul ordinary trays.

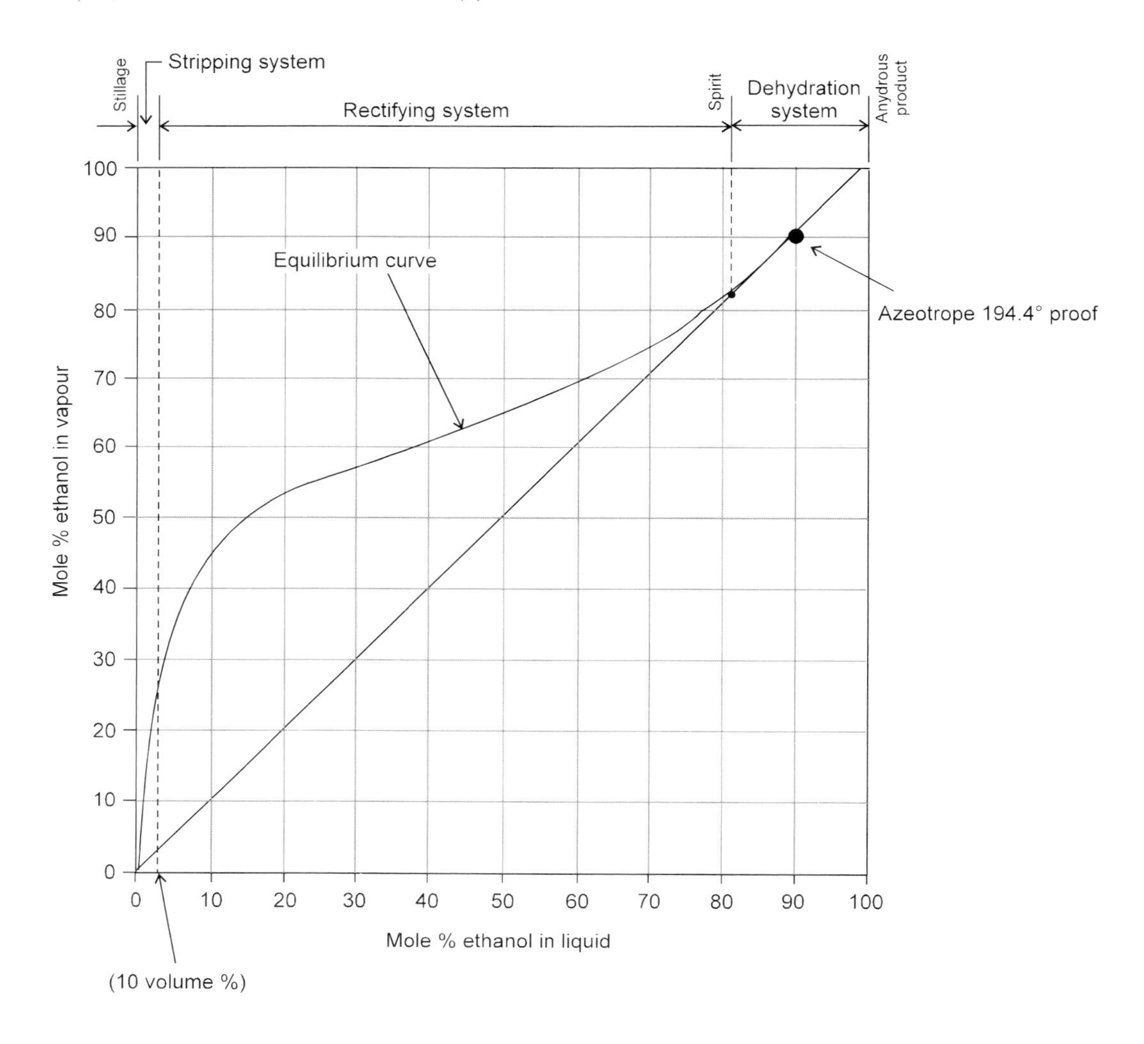

Figure 4. Structuring the distillation system strategy.

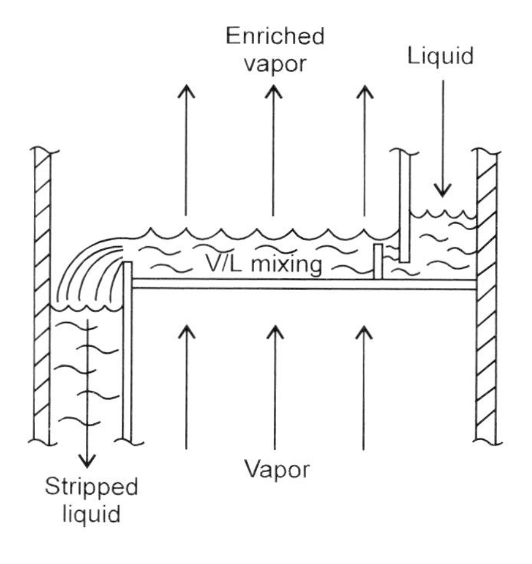

Figure 5. Distillation tray functions.

Energy analysis

In addition to the selection of the basic contacting device, the energy requirement must be established. This is accomplished by analyzing the vapor/liquid equilibrium data from Figure 4, for the liquid:vapor ratio to perform a continuous series of steps within the limits of the equilibrium curve. Table 1 demonstrates a simplified procedure to calculate the approximate energy requirement from the liquid:vapor ratio that will be employed in the tower design. Repetition of this type of calculation for different conditions produces a design chart like that shown in Figure 8 for the ethanol-water system. Such a graph is

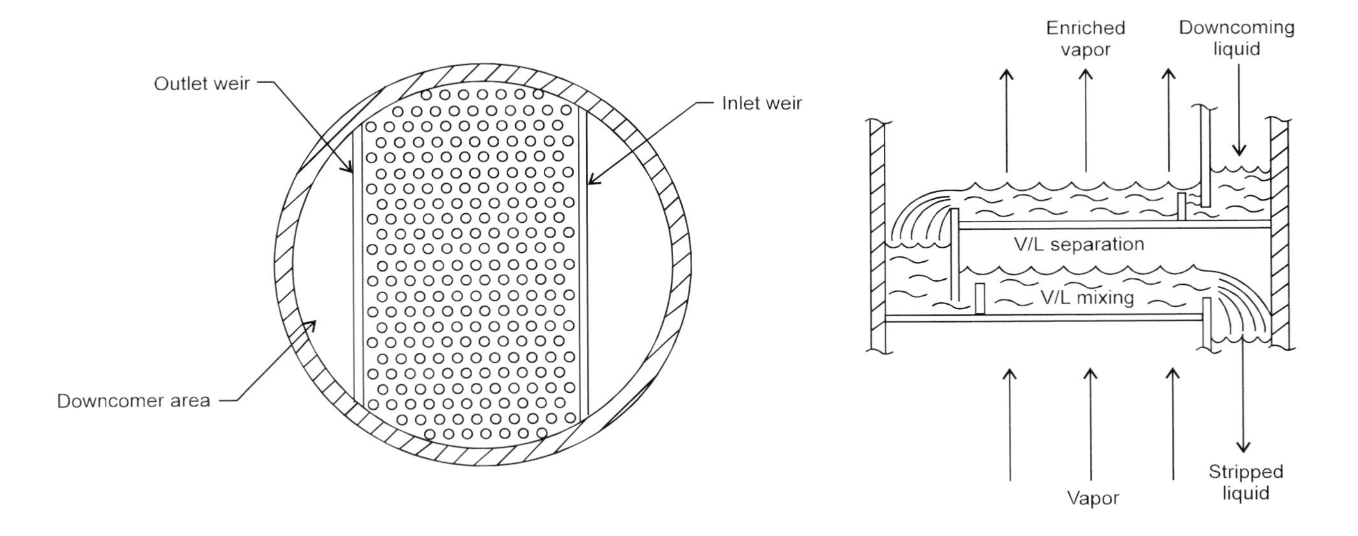

Figure 6. Perforated trays.

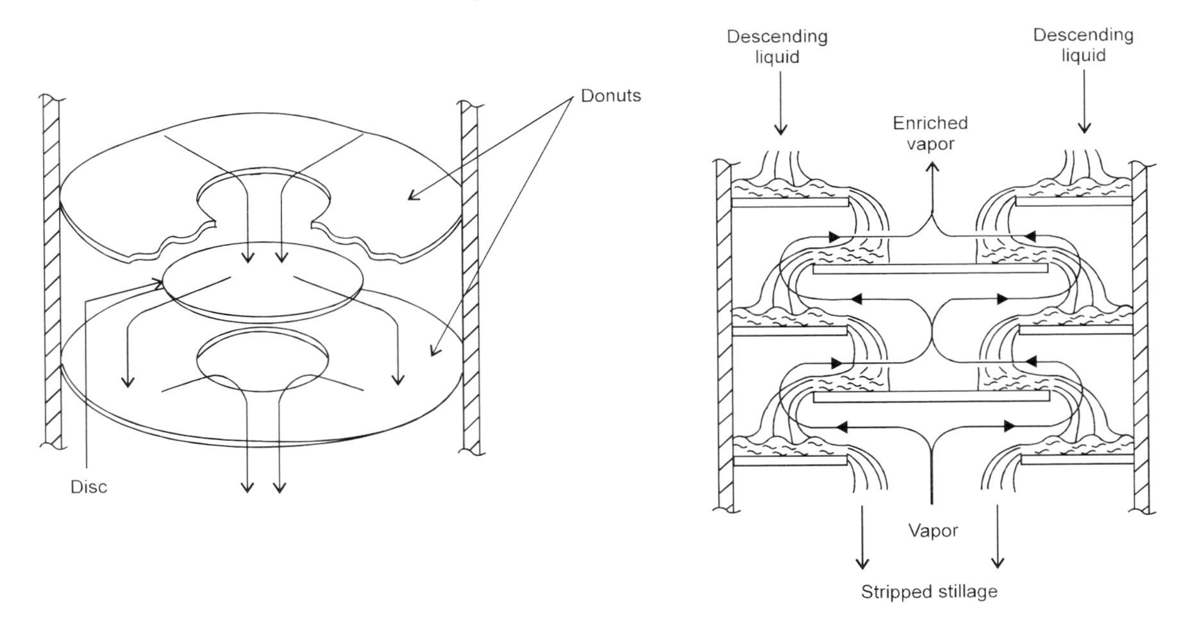

Figure 7. Disc-and-donut trays.

useful when calculations are needed to ascertain technical and economic feasibility and preliminary conditions for the design.

Figure 9 demonstrates how the liquid:vapor ratio, in connection with the number of stages (theoretically ideal trays) required for a specified separation between ethanol and water, is graphically determined. Note that the stages are constructed by drawing straight lines vertically and horizontally between the equilibrium curve (previously determined experimentally) and the operating lines. For an ethanol stripper/rectifier, there are two operating lines: one for the rectification section and one for the stripping section. The operating lines represent the locus of concentrations within the distillation tower of the passing liquid and vapor streams. The operating lines for a given tower are based on

Table 1. Simplified calculations for steam requirements for ethanol distillation.

Example 1. Calculate the steam requirement (lbs/gallon of product) for a 10% volume beer at 100 gpm (90 gpm water/10 gpm ethanol).

$L/V^* = 5.0$ (typical for a 10% volume beer) or $L = 5 \times V$

$$L = 90 \text{ gpm} \times \frac{500 \text{ lbs/hr}}{\text{gpm}} = 45{,}000 \text{ lbs/hr} = 5 \times V$$

Therefore, $V = 9{,}000$ lbs/hr (steam)

$$\text{And: } \frac{9{,}000 \text{ lbs/hr (steam)}}{10 \text{ gpm}} \times \frac{\text{hr}}{60 \text{ min}} = 15 \text{ lbs steam/gallon of product}$$

Example 2. Calculate the steam required for a 5% volume beer at 100 gpm (95 gpm water/5 gpm ethanol).

$L/V = 6.33$ (typical for a 5% volume beer) or $L = 6.33 \times V$

$$L = 95 \text{ gpm} \times \frac{500 \text{ lbs/hr}}{\text{gpm}} = 47{,}500 \text{ lbs/hr} = 6.33 \times V$$

Therefore, $V = 7{,}500$ lbs/hr (steam)

$$\text{And: } \frac{7{,}500 \text{ lbs/hr (steam)}}{5 \text{ gpm}} \times \frac{\text{hr}}{60 \text{ min}} = 25 \text{ lbs steam/gallon of product}$$

*L and V are liquid and vapor flow rates, respectively, expressed in lb-mole per hr.
Note: at base of column use simplifying assumption of water/steam. Therefore: $\frac{L \text{ (lb-mole/hr)}}{V \text{ (lb-mole/hr)}} = \frac{L \text{ (lbs/hr)}}{V \text{(lbs/hr)}}$

the energy input, as calculated and represented in Figure 8. Because of the principle of constant molal overflow, the operating lines can be represented as straight lines. If constant molal overflow was not valid for the ethanol/water distillation, then these lines would be curved to represent the changing ratio of liquid flow to vapor flow (in molar quantities) throughout the tower. The slope of the operating line (the ratio of liquid flow to vapor flow) is also called the internal reflux ratio. If the energy input to a tower is increased while the beer flow remains constant, the operating lines will move toward the 45° line, thus requiring fewer stages to conduct the distillation. Likewise if the energy input is reduced (lowering the internal reflux ratio), the operating lines will move toward the equilibrium curve, reducing the degree of separation achievable in each stage and therefore requiring more stages to conduct the distillation.

The calculations underlying the preparation of Figure 9 go beyond the scope and intent of this text, but have been included for continuity. The dashed lines represent the graphical solution to the design calculations for the number of theoretical stages required to accomplish a desired degree of separation of the feed components. Figure 9 is referred to as a McCabe-Thiele diagram. For further pursuit of this subject, refer to the classical distillation textbook by Robinson and Gilliland (1950).

Tower sizing

The goal of the design effort is to establish the size of the distillation tower required. Table 2 shows the basic procedure to determine the diameter required for the given distillation tower. Since all of the distillation 'work' is done by the

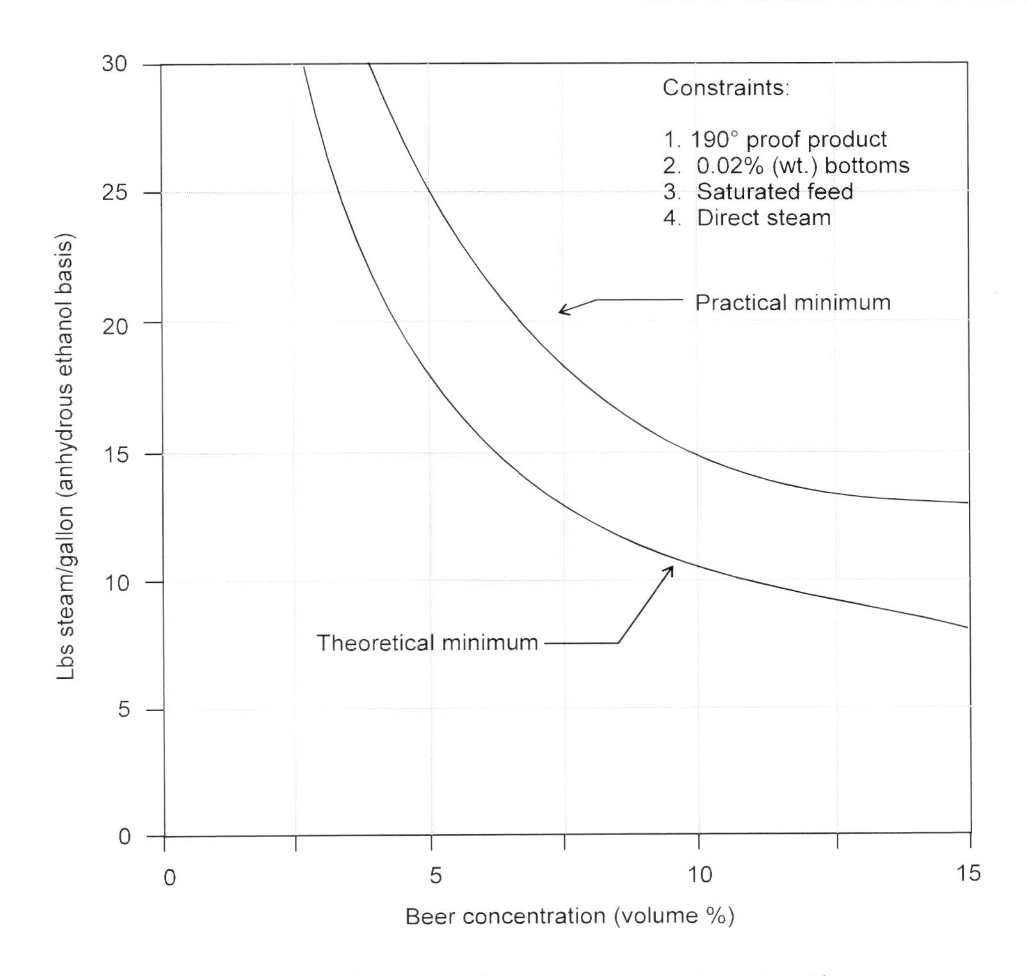

Figure 8. Steam requirements ethanol stripper/rectifier

trays, the tower is actually the 'container' to surround the vapor and liquid activity that is 'managed' by the trays. Tower diameter design is, therefore, actually the design of the necessary tray diameter for proper vapor/liquid interaction and movement.

The *f* factor (vapor loading) is an empirically determined factor that depends primarily upon tray type and spacing, fluid physical properties, froth stability and surface tension at the operating conditions of the system. The proper values for *f* are determined by field observations. In summary then, the *f* factor can be described as an adjusted velocity term (units are ft/sec) that when multiplied by the square root of the density ratio of liquid to vapor, will give the allowable vapor velocity in the empty tower shell, such that liquid entrainment and/or vapor phase pressure drop in the tower will not be exces-sive.

Excessive vapor velocity will first manifest itself by causing excessive liquid entrainment rising up the tower, causing loss of separation efficiency. Ultimately the excessive entrainment and pressure drop will cause tower flooding.

To achieve a well-balanced tower design, the foregoing analysis must be performed at each stage of the tower, from bottom to top. Composition changes, feed points, draws, etc., each can cause a different requirement. The tower must be examined to locate the limiting point.

Similar analyses, with empirically-observed performance coefficients, are applied to vapor passing through the trays and liquid, and to the

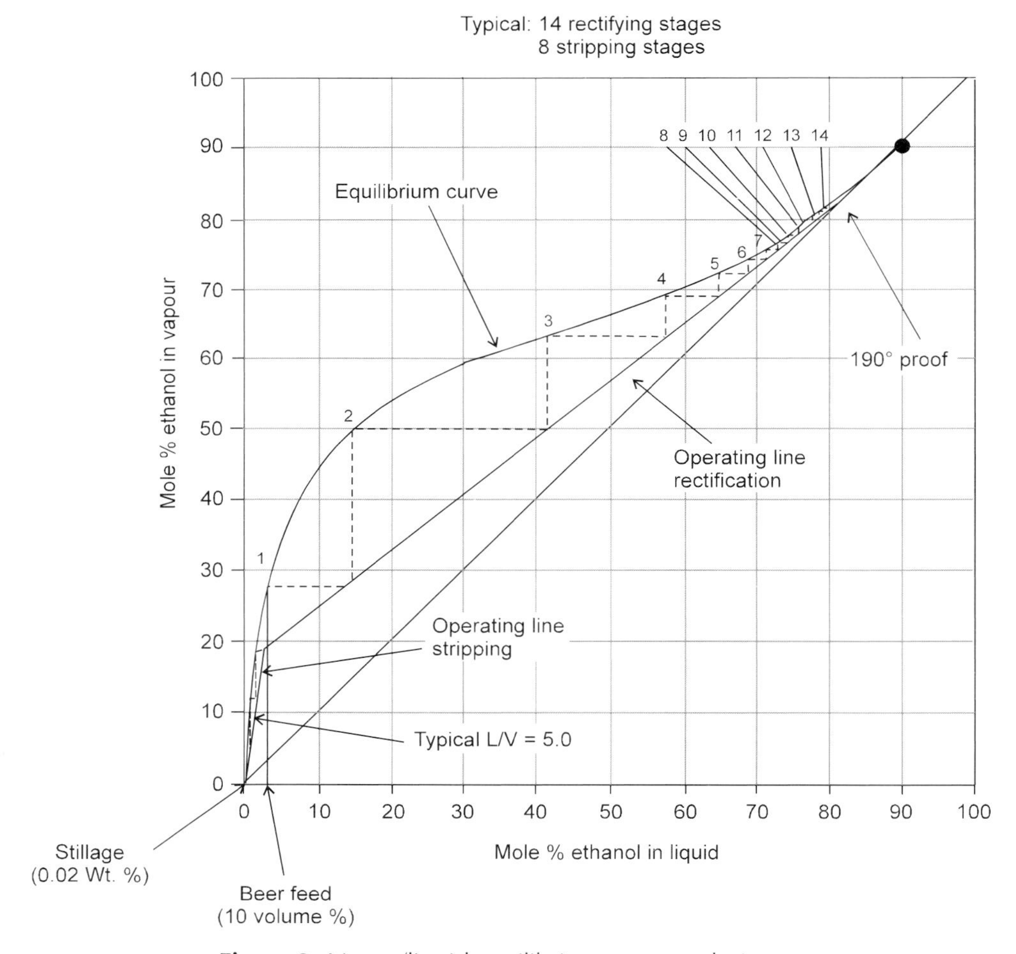

Figure 9. Vapor/liquid equilibrium stage analysis.

movement and control of liquid passing through downcomers and across the trays. These analytical procedures are beyond the scope of this text. Reference should be made to the aforementioned text by Robinson and Gilliland for further information.

Considerations in optimizing distillation system design

Optimizing the technical and economic design of distillation equipment and similar gas and vapor/liquid mass transfer systems involves a number of interrelated parameters. The positive/negative balance of a variety of contacting devices with different capacities and efficiencies for promoting vapor/liquid mass transfer must be taken into consideration. Along with the technical issues considered in such designs, economical operation is essential not only in the reduction of energy and other direct costs, but also in relation to investment and return on investment from the operation being considered. In this respect, distillation towers are not independent process-wise, as consideration must also be given to other auxiliaries such as reboilers, condensers, pumps, controls and related equipment.

Sizing towers

In determining optimum diameter and height of towers for distillation, absorption, stripping and similar mass transfer operations, design factors are affected by whether the installations will be

Table 2. Calculations for tower sizing (base of stripper).

Example 1. Calculate tower diameter required for a 10% volume beer at 100 gpm (15 lbs steam/gallon of product)

W (Vapor flow rate)	= 9,000 lbs/hr (steam)
P (Operating pressure at base)	= 1.34 ATM
M_{AVG} (Average MW of vapor)	= 18 lbs/lb-mole
ρ_L (Liquid mixture density	= 59.5 lbs/ft³ (227 °F)
T (Absolute operating temperature)	= 687 °R
D (Tower inside diameter (inches)	
f (Vapor loading factor)	= 0.05-0.3

Sizing equation:

$$D = 0.2085 \bullet \sqrt{\frac{W}{f \bullet \sqrt{\frac{P \bullet M_{AVG} \bullet \rho_L}{T}}}} = 0.2085 \bullet \sqrt{\frac{9000}{f \bullet \sqrt{\frac{1.34 \bullet 18 \bullet 59.5}{687}}}} = \frac{16.45}{\sqrt{f}}$$

Assuming f = 0.16 (specific to tray design and spacing), the tower diameter is:

$$D = \frac{16.45}{\sqrt{0.16}} = 41.125 \text{ inches}$$

Example 2 utilizes the following equation. Values for the terms are indicated in Example 1.

$$D = 0.2085 \bullet \sqrt{\frac{W}{f \bullet \sqrt{\frac{P \bullet M_{AVG} \bullet \rho_L}{T}}}} \quad \text{(Eq. 1)}$$

The final design equation can be derived beginning with the fundamental equation:

$$u = f \bullet \sqrt{\frac{\rho_L - \rho_V}{\rho_V}} \quad \text{(Eq. 2)}$$

Where u = average vapor velocity in empty tower shell (ft/sec)
ρ_L = liquid density (lbs/ft³)
ρ_V = vapor density (lbs/ft³)
f = tower vapor loading factor (ft/sec)

(Note: For most cases ρ_L is much greater than ρ_V, so that $\rho_L - \rho_V \cong \rho_L$. For example, water (steam) at 212°F and atmospheric pressure: ρ_L = 59.8 lbs/ft³ and ρ_V = 0.0373 lbs/ft³. Then $\rho_L - \rho_V$ = 59.7627 lbs/ft³ ≅ 59.8 lbs/ft³, which results in a negligible 0.06% error.)

Consequently,

$$u \cong f \bullet \sqrt{\frac{\rho_L}{\rho_V}} \quad \text{(Eq. 3)}$$

Imposing the equation of continuity: $W = A \bullet \rho_V \bullet u$ or $u = W/A \bullet \rho_V$

Where A = column cross-section area (ft²)
ρ_V = vapor density (lbs/ft³)
u = average vapor velocity in empty tower shell (ft/sec)
and W = vapor mass flow (lbs/sec)

Substituting for u in the equation above:

$$\frac{W}{A \bullet \rho_V} \cong f \bullet \sqrt{\frac{\rho_L}{\rho_V}} \quad \text{then} \quad \frac{W}{A} \cong f\sqrt{\rho_L \bullet \rho_v} \quad \text{or} \quad A \cong f\sqrt{\rho_L \bullet \rho_v}$$

Use the Ideal Gas Law to express the vapor density.

$$\rho_V = \frac{P \bullet M_{AVG}}{R \bullet T}$$ Then by substitution one obtains $$A = \frac{W}{f \bullet \sqrt{\frac{P \bullet M_{AVG} \bullet \rho_L}{R \bullet T}}}$$

Using the Universal Gas Constant R = 0.73 (ft^3)(atm)/(lb-mole)(°R) the equation becomes:

$$A = \frac{W}{1.17 \bullet f \bullet \sqrt{\frac{P \bullet M_{AVG} \bullet \rho_L}{T}}}$$ Now $$A = \frac{\pi \bullet D^2}{4} = 0.7854 \bullet D^2$$

and $$D^2 = \frac{W}{0.9192 \bullet f \bullet \sqrt{\frac{P \bullet M_{AVG} \bullet \rho_L}{T}}} = \frac{1.0879 \bullet W}{f \bullet \sqrt{\frac{P \bullet M_{AVG} \bullet \rho_L}{T}}}$$

Adjusting units: $$D = \frac{1.043 \bullet 12}{60} \bullet \sqrt{\frac{W}{f \bullet \sqrt{\frac{P \bullet M_{AVG} \bullet \rho_L}{T}}}}$$ Therefore, $$D(inches) = 0.2085 \bullet \sqrt{\frac{W}{f \bullet \sqrt{\frac{P \bullet M_{AVG} \bullet \rho_L}{T}}}}$$

---

indoors or outdoors. With indoor installations, building height limitations, as well as floor level accessibility, are an important factor in the design. Where there are height limitations, towers must be increased in diameter to provide for reduced tray spacing, which in turn will require lower vapor velocities. With outdoor installations, the 'sky is literally the limit', and refinery and petrochemical towers of 200 feet in height are not uncommon.

In either case, indoors or outdoors, the interrelated tower diameter and tray spacing are limited by allowable entrainment factors (*f* factors) (Katzen, 1955). If outdoors, tower heights and diameters must be related to maximum wind loading factors in the specific plant location and may be complicated by allowance for earthquake factors.

Tray and packing selection

Vapor/liquid contacting devices may be of two distinct types, namely packed or tray (staged) towers. In packed towers, the transfer of material between phases occurs continuously and differentially between vapor and liquid throughout the packed section height. By contrast, in tray towers, the vapor/liquid contact occurs on the individual trays by purposely interrupting down-flowing liquid using downcomers to conduct vapor-disengaged liquid from tray to tray and causing the vapor/liquid contact to occur between cross-flowing liquid on the tray with vapor flowing up through the tray. In other words, the vapor/liquid contact is intermittent from tray to tray, and is therefore referred to as being stagewise. Thus, for any given separation system, the degree of vapor/liquid contact will be greater with a greater height of the packed section, or in the case of tray towers, a greater number of trays used.

It is generally considered that packing-type internals may be used with relatively clean vapor and liquid systems where fouling is not a problem. Economics indicate that packing is applicable in small and modest sized towers. As the towers become larger, packing becomes complicated by the need for multiple liquid redistribution points to avoid potential vapor/liquid bypassing and reduction in efficiency. Structured packings (Fair *et al.*, 1990; Bravo *et al.*, 1985) are designed to minimize these problems by reducing the height requirement and controlling, to some extent, the distribution of liquid. However, high fabrication and specialized installation costs would indicate that these are applicable only for relatively low volume, high value product processing.

Trays of various types are predominant in vapor/liquid contacting operations, particularly on the very large scale encountered in the petroleum and petrochemical industries, in large scale operations of the chemical process industries and in the large scale plants of the motor fuel grade ethanol industry.

The venerable bubble cap tray, with a wide variety of cap sizes, designs and arrangements to maximize contact efficiency, has fallen out of favor during the past few decades because of the relatively high cost of manufacture and assembly. Valve trays of several types have taken over in operations requiring a relatively wide vapor handling capacity range (turndown). This has been extended by use of different weights of valves on the same tray. Specialty trays such as the Ripple, Turbogrid, tunnel cap and others designed to improve contact under certain specific circumstances have been used to a limited extent.

The long established perforated tray is a contacting tray into which a large number of regularly oriented and spaced small circular openings have been drilled or punched. These trays are commonly referred to as 'sieve trays' because of the original practice of putting the maximum number of holes in any given tray area. This original design produced a fairly inefficient operation at normal loading, and a very inefficient operation with decreased vapor loading. About 50 years ago, engineers began to suspect that the design approach had been in error, and that the hole area in the trays should be limited by the hole velocity loading factor to obtain maximum contact by frothing, as indicated in Figure 10.

The hole velocity loading factor (or perforation factor) is defined as the vapor velocity through the perforations adjusted by the square root of the vapor density at the specific tower location of a given perforated tray. With the parallel development of separation processes in the petroleum refining, chemical and ethanol industries, the modern approach has developed to what is now called 'perforated tray' design. The Fractionation Research Institute of the American Institute of Chemical Engineers diverted its efforts from bubble cap studies to perforated tray testing; and have established a basis for the design of perforated trays with high efficiency and wide capacity range (Raphael Katzen Associates, 1978).

Where foaming or tray fouling (caused by deposition of solid materials in the tower feed) can be an operational problem, novel designs such as the baffle tray may permit extended operating time between cleanings. Baffle trays may take a number of different forms. They can be as simple as appropriately spaced, unperforated, horizontal metal sheets covering as much as 50-70% of the tower cross sect-ional area; or they may take the form of a series of vertically spaced, alternating, solid disc-and-donut rings (see Figure 7). Towers up to 13ft in diameter are in operation using this simple disc-and-donut design concept.

Although system-specific data have been developed for each type of tray, it is difficult to correlate tray loading and efficiency data for a wide variety of trays on a quantitative basis. Each system must be evaluated based upon empirically-derived loading factors for vapor and liquid operations within the tower.

Auxiliaries

Energy input is of prime importance in tower design, particularly in ethanol stripping and rectification units. In aqueous and azeotrope-forming systems, direct steam injection has been common practice to maintain simplicity. However, current requirements to reduce the volume of waste going to pollution remediation facilities have minimized use of this simple steam injection technology to avoid the dilution effect of the steam being condensed and added to the stillage. Direct steam injection transfers both the energy and the water into the process. By imposing a heat exchanger (reboiler) between the steam and the process, only the energy is transferred into the tower. The condensate water is returned

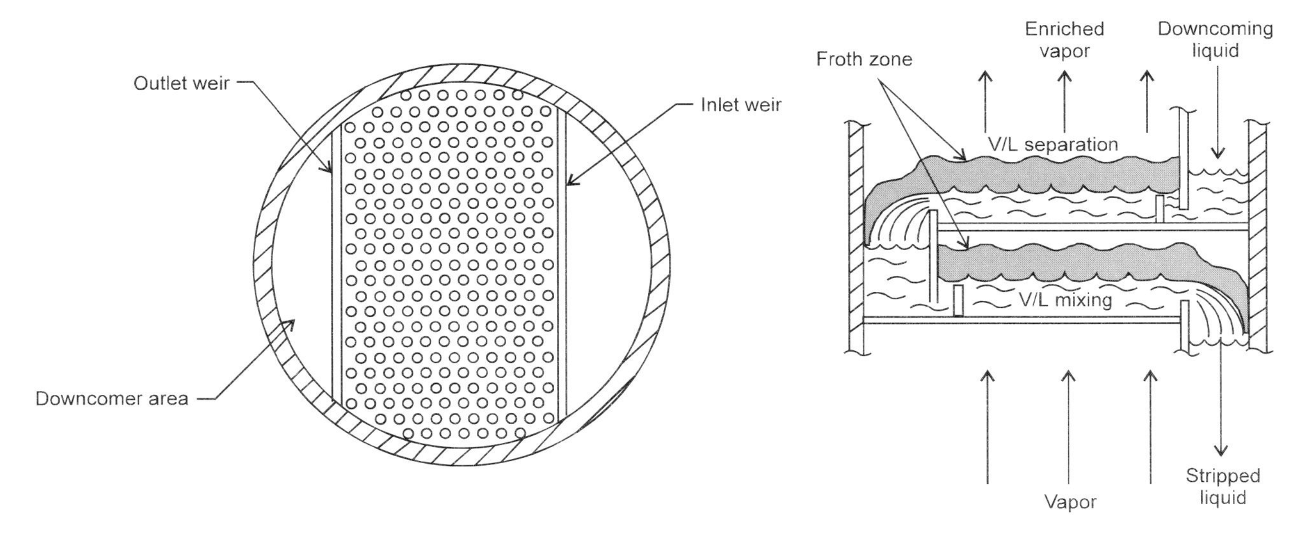

Figure 10. Perforated tray frothing.

in a closed-loop to the boiler, thus reducing the bottoms outflow from the process. Reboilers are thus growing in acceptance, and several types may be employed. Kettle and thermosyphon reboilers are preferred where fouling is not a problem. Where fouling can occur, high velocity, forced-circulation, flash heating reboilers are preferred. Figure 11 depicts the reboiler energy transfer by a forced-circulation reboiler as compared to Figure 1 which depicts direct steam injection.

Thermocompression injection of steam has also been utilized where low pressure vapors are produced from flash heat recovery installations and where higher pressure motive steam is also available.

Condenser design would appear to be simple. However, in many cases, water limitations require adapting condenser designs to the use of cooling tower water with limited temperature rise and minimal scale-forming tendencies. On the other hand, where water is extremely scarce, air-cooled condensers are used.

Energy conservation

The increasing cost of thermal energy, whether provided by natural gas, fuel oil, coal or biomass, is fostering an increased emphasis on heat recovery and a reduction in primary thermal energy usage (Fair, 1977; Petterson *et al.*, 1977; Mix *et al.*, 1978). Conventional bottoms-to-feed heat exchangers are now being supplemented with recovery of overhead vapor latent heat by preheating feed streams and other intermediate process streams. Techniques of multistage distillation (similar to multiple effect evaporation) are also practiced. Pressure-to-atmospheric, atmospheric-to-vacuum, or pressure-to-vacuum tower stages are utilized, with the thermal energy passing overhead from one tower to provide the reboiler heat for the next one. Two such stages are quite common and three stage systems have also been utilized (Katzen, 1980; Lynn *et al.*, 1986).

Furthermore, the modern technique of vapor recompression, commonly used in evaporation systems, is also being applied to distillation systems. Such a system can provide for compression of overhead vapors to a pressure and temperature suitable for use in reboiling a lower pressure stripping tower. However, the compression ratios required for such heat recovery may consume almost as much electrical energy as would be saved in thermal input. Alternative systems, using vapor recompression as an intermediate stage device in the distillation system, have also been proposed.

Control systems

Control systems can vary from manual control, through simple pneumatic control loops to fully automated distributed control (Martin *et al.*, 1970). High level computer control has facilitated the application of sophisticated control algorithms, providing more flexibility, reduced labor and higher efficiency with lower capital investment. Such systems, when properly adapted to a good process design, have proven more user-friendly than the control techniques utilized in the past.

Economic design

In integrating the technology discussed, the final analysis must be economic. Alternative systems must be compared on the basis of investment requirements, recovery efficiency and relative costs of operation. Thus, any heat exchangers installed for heat recovery must show a satisfactory return on the investment involved in their purchase and installation. In comparing alternative separation systems, the overall equipment costs must be compared against energy and other operating costs to determine

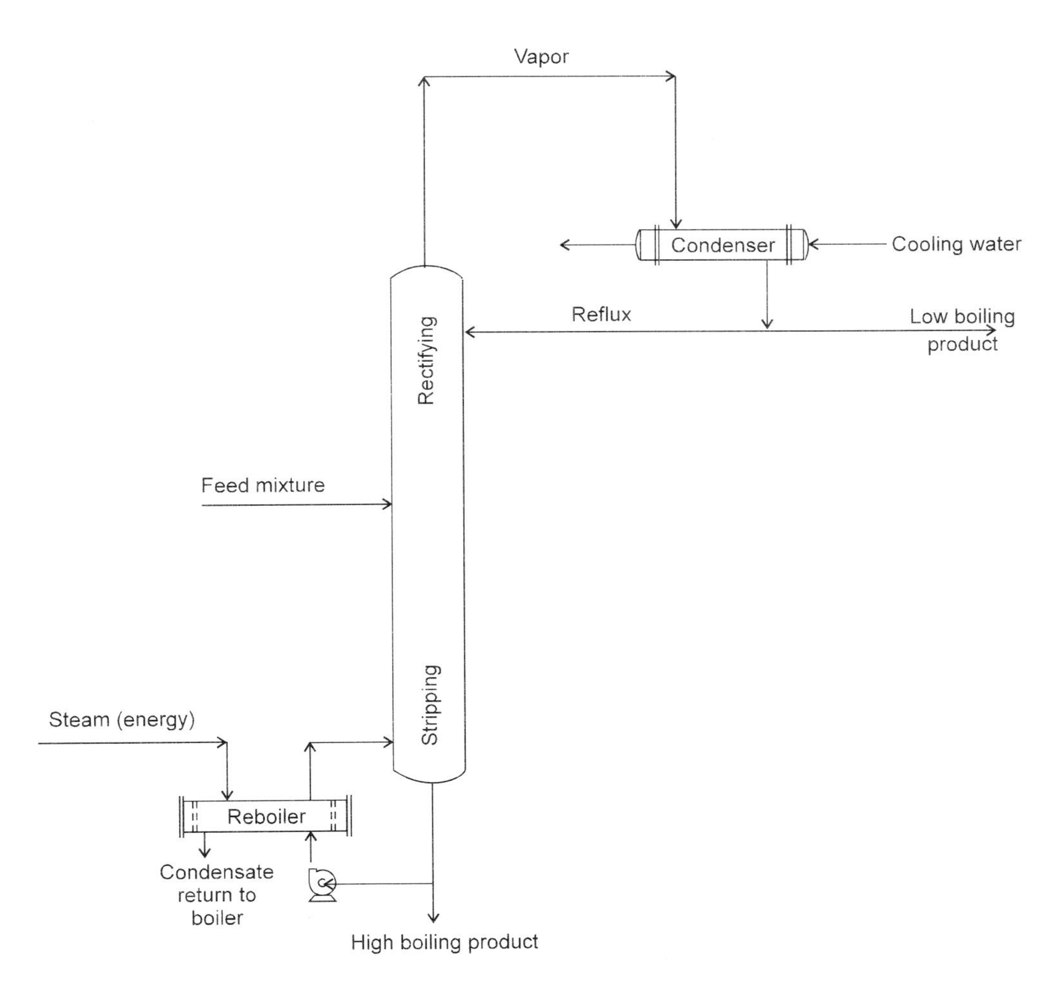

Figure 11. Energy transfer by a forced-circulation reboiler (See Figure 1 to compare with direct steam injection).

which system offers the best return. Modern computer-assisted designs incorporate economic evaluation factors so eco-nomic optimization can be determined rapidly.

Ethanol distillation/dehydration: specific systems technology

Proven industrial technologies are available for distillation of various grades of ethanol from grain, sugarcane, molasses and other feedstocks. Improvements have been made over the years, particularly during development of the motor fuel grade ethanol industry. In such installations, a key requirement is the mini-mization of total energy usage.

The operation that has been most subject to critical comment is the distillation process. Many relatively new 'authorities' in the field have based their criticism on technologies that go back 50-60 years, and have created an unwarranted condemnation of distillation as a viable process for low energy motor fuel grade ethanol production. Systems developed over the years will be described to show that much of such criticism is unwarranted and unjustified.

Production of industrial ethanol

Prior to the recent emphasis on motor fuel grade ethanol, the major ethanol product utilized worldwide was high purity, hydrous industrial ethanol, which is generally produced at a strength of 96° GL (192° US proof) (°GL = degress Gay Lussac = % by volume ethanol; US proof = 2 x % by volume ethanol). Efficient systems have been in commercial operation for many years for the production of such high grade ethanol from ethylene, grain, molasses and sulfite waste liquor. The basic distillation system is shown in Figure12.

In the case of synthetic ethanol (outside the scope of this publication), the beer stripping tower is not required and the refining system is a simple three tower unit, which achieves 98% recovery of the ethanol in the crude feed as a first grade product. The final product may contain less than 30 ppm total impurities and has a 'permanganate time' of more than 60 minutes.

For the production of industrial or beverage spirit products made by fermentation of grain, molasses or sulfite liquor, the system utilizes the full complement of equipment shown in Figure 12. The beer feed is preheated from the normal fermentation temperature in several stages, recovering low level and intermediate level heat from effluent streams and vapors in the process. This preheated beer is degassed and fed to the beer stripper, which has stripping trays below the beer feed point and several rectifying trays above it. The condensed high wines from the top of this tower are then fed to the extractive distillation tower, which may operate at a pressure in the order of 6-7 bars (87-101.5 psi). In this tower, most of the impurities are removed and carried overhead to be condensed as a low grade ethanol stream, from which a small purge of heads (acetaldehyde and other low boiling impurities) may be taken while the primary condensate flow is fed to the concentrating tower. The purified, diluted ethanol from the bottom of the extractive distillation tower is fed to the rectifying tower, which has an integral stripping section. In this tower, the high grade ethanol product, whether industrial or potable, is taken as a side draw from one of the upper trays. A small heads cut is removed from the overhead condensate. Fusel oils (mixtures of higher alcohols such as propyl, butyl, and amyl alcohols and their isomers, which are fermentation by-products or 'congeners') are drawn off at two points above the feed tray but below the product draw tray to avoid a buildup of fusel oil impurities in the rectifying tower. The overhead heads cut and the fusel oil draws are also sent to the concentrating tower.

It should be noted that the rectifying tower is heated by vapors from both the pressurized extractive distillation tower and the pressurized concentrating tower.

In the concentrating tower, the various streams of congener-containing draws are concentrated. A small heads draw is taken from

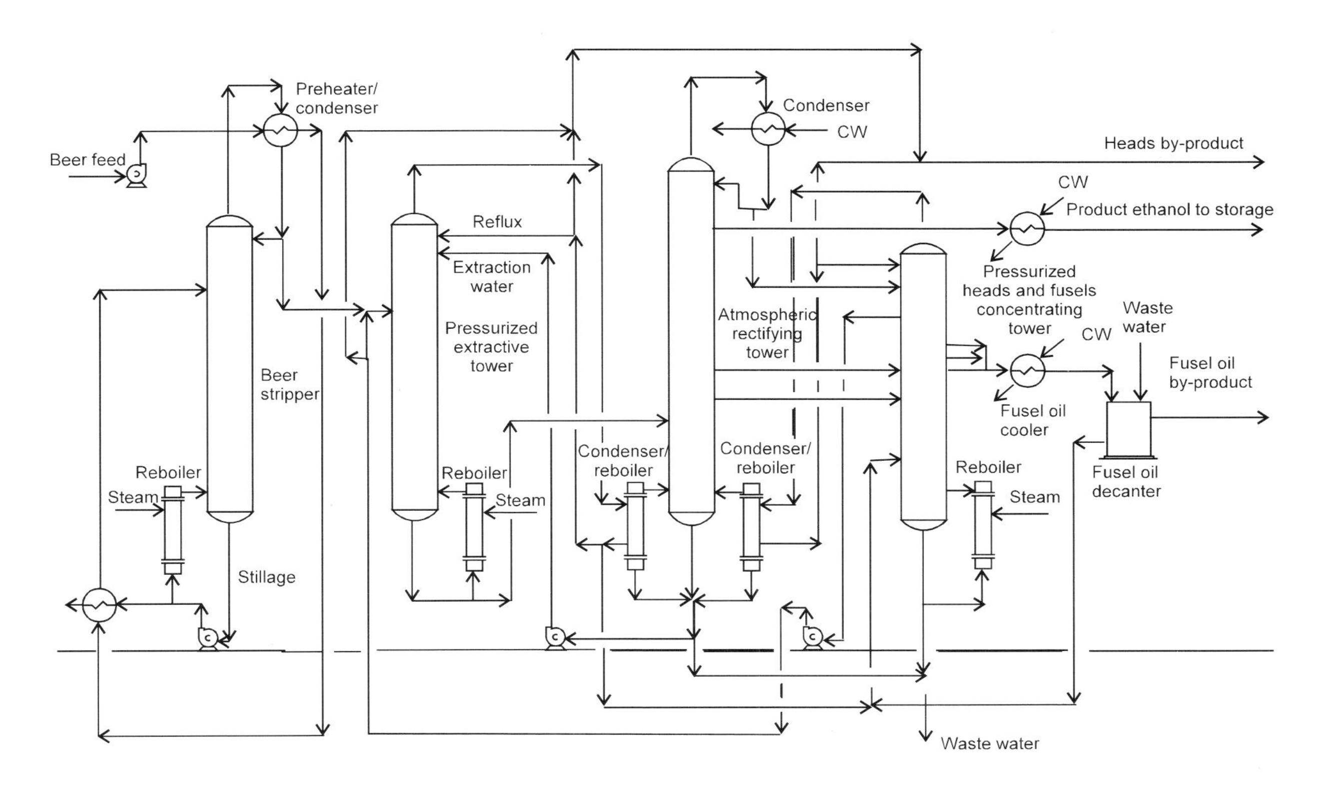

Figure 12. Low energy-consuming high grade hydrous ethanol distillation.

the overhead condensate, which contains the acetaldehyde fraction along with a small amount of the ethanol produced. This may be sold as a by-product or burned as fuel. A fusel oil side draw is taken at high fusel oil concentrations through a cooler to a washer. In the washer, water is utilized to separate the ethanol from the fusel oil, with the washings being recycled to the concentrating tower. The decanted fusel oil may be sold as a by-product. The ethanol recovered from the crude streams is taken as a side draw from the concentrating tower and fed back to the extractive distillation tower for re-purification and recovery of its ethanol content.

In an early version of this system, installed more than 50 years ago for the production of potable ethanol from grain and from molasses, all towers were operated at atmospheric pressure. However, installations made within the past 30 years utilize the multistage pressure system to reduce energy consumption to a level of about 60% of the all-atmospheric system.

The commercial installations utilizing the multistage pressure, or 'pressure cascading' technique operate with a steam consumption of 3.0-4.2 kg of steam/liter (25-35 lb/gallon) of 96° GL ethanol. This may be compared to about 6 kg of steam/liter for earlier conventional distillation systems.

Production of anhydrous ethanol

Systems have been designed and installed for production of extremely dry and very pure anhydrous ethanol for food and pharmaceutical use, primarily in aerosol preparations. These systems, as shown in Figure 13, yield ethanol containing less than 200 ppm of water (99.98° GL), less than 30 ppm of total impurities and more than 45 minutes permanganate time.

The two tower dehydrating system has been operated in two super-anhydrous plants in Canada, and was used to produce motor fuel grade ethanol (99.5° GL) in four installations in Cuba (prior to the advent of the Castro regime). The dehydrating tower and the entrainer-recovery tower are operated at atmospheric pressure. Thus, they may utilize either low pressure steam, hot condensate or hot waste streams from other parts of the ethanol process to minimize steam usage. To simplify equipment and minimize investment, a common condensing -decanting system is used for the two towers.

The entrainer used to remove water as a ternary (three component) azeotrope may be benzene, heptane (C_6-C_8 cut), cyclohexane, n-pentane, diethyl ether or other suitable azeotropic agents. The entrainer serves to create a three component azeotrope that boils at a temperature lower than any of the three individual components and lower than the ethanol/water binary (two component) azeotrope. Therefore the ternary mixture will pass overhead from the tower, carrying the water upward. Upon condensing, the mixture separates in a decanter into an entrainer-rich layer and a water-rich layer.

The hydrous ethanol feed enters the dehydrating tower near the top. The feed contacts the entrainer in the upper section of the tower. The three component mixture in this section of the tower seeks to form its azeotrope, but is deficient in water and contains more ethanol than the azeotrope composition. Therefore, the ethanol is rejected downward in the liquid and is withdrawn as an anhydrous product from the bottom of the tower. The water joins the entrainer, passing upward as vapor to form a mixture that is near the azeotrope composition for the three components. The condensed mixture separates into two layers in the decanter and the entrainer-rich layer is refluxed from the decanter back to the top of the tower. The aqueous layer is pumped from the decanter to the entrainer-recovery tower, in which the entrainer and ethanol are concentrated overhead in the condenser-decanter system. The

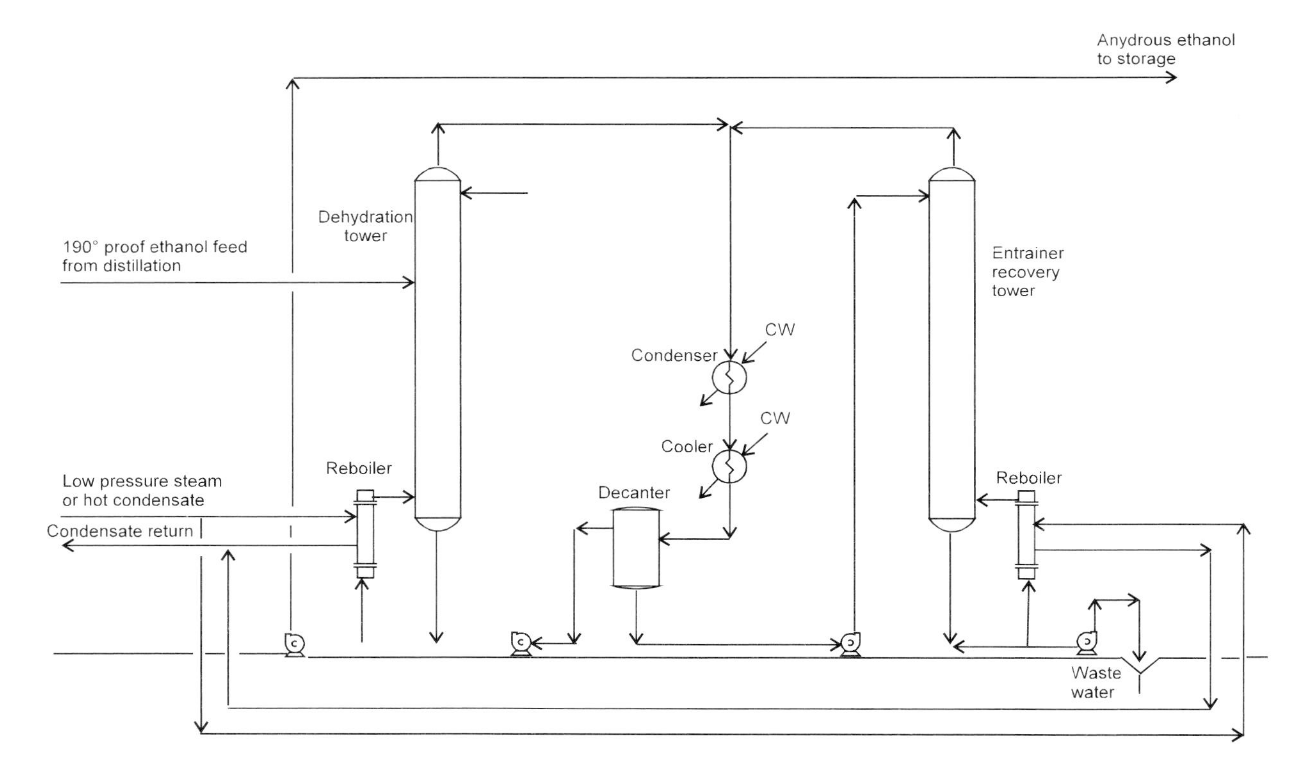

Figure 13. High grade anhydrous ethanol system.

stripped water, emerging from the base of the tower, may go to waste. If it has substantial ethanol content, it may be recycled to the beer well feeding the spirit unit, but this introduces the risk of traces of the entrainer in the hydrous ethanol which may not all be sent to the dehydration system. This system operates with a steam consumption of 1-1.5 kg/liter (8.3-12.5 lb/gallon) of anhydrous ethanol depending on the quality of product required. As indicated above, a major part of the equivalent steam energy can be provided by hot condensate and hot waste streams from the spirit unit.

Production of anhydrous motor fuel grade ethanol

In view of the growing demand for motor fuel grade ethanol (MFGE) in the US and other countries, a combination of key features in the improved systems described in Figures 12 and 13 has been used to maximize recovery of MFGE from fermented beer, while minimizing energy consumption. This system is shown in Figure 14 (US patent 4,217,178; Canadian Patent 876,620, 1980). Fermented beer feed is preheated in a multistage heat exchange sequence, varying somewhat in complexity with the size of the commercial facility. In effect, beer is preheated in a 'boot strapping' operation, which takes the lower level heat from the azeotrope vapors in the dehydrating system and then picks up heat from the excess of overhead vapors from the pressurized beer stripping and rectifying tower. Finally the beer is preheated in exchange with hot stillage from the same tower. This preheated beer, essentially at its saturation temperature, is de-gassed to remove residual CO_2 before it enters the pressurized beer stripper. This tower operates at a pressure of approximately 4 bars with heat provided by steam through a forced circulation reboiler. This is the only use of steam in this distillation system. The stillage leaving the base of this tower is partially cooled to the atmospheric boiling point via heat exchange with the preheated beer feed. This provides a stillage feed for vapor recompression

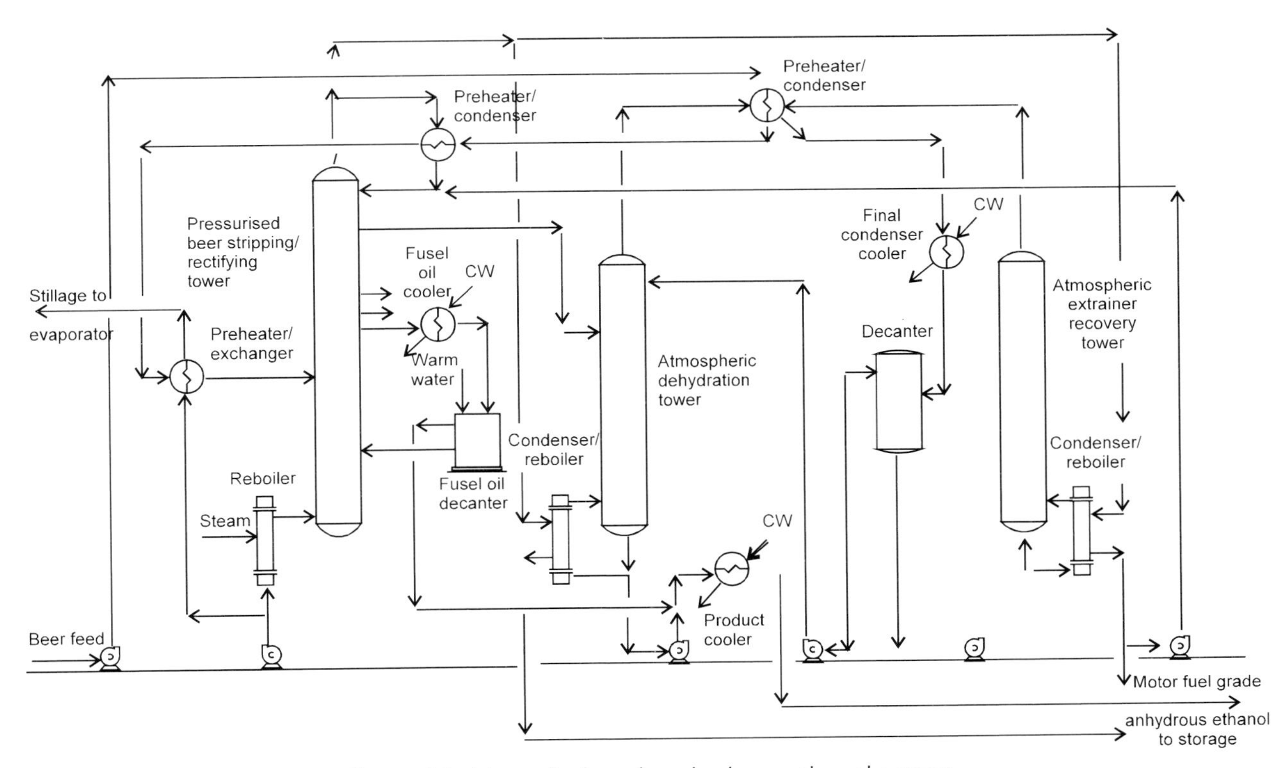

Figure 14. Motor fuel grade anhydrous ethanol system.

evaporation at an ideal temperature, requiring neither preheat nor flashing in the evaporator.

The ethanol stripped from the beer in the lower part of the beer tower is rectified to approximately 95° GL and taken as a side draw a few trays below the top of the rectifying section. The overhead vapors, under pressure, are used to boil up the atmospheric dehydration tower and the atmospheric entrainer recovery tower, as well as to provide preheat to the beer feed. The condensed overhead vapors are refluxed to the top of the pressurized beer tower, with a small draw of heads taken to avoid accumulations of the more volatile congeners such as acetaldehyde. The heads stream, amounting to less than 1% of ethanol production, can be burned as fuel in the plant boiler or sent directly into the MFGE final product, thus bypassing the dehydration system.

Side stream fusel oil draws are also taken from the rectifying section of the pressurized tower to a fusel oil decanter. The aqueous washings are returned to the beer stripping section of this tower; while the decanted, washed fusel oil is combined with the anhydrous ethanol plant. Fusel oil not only has a higher fuel value than ethanol, but serves as a blending agent between the ethanol and gasoline.

The 95° GL ethanol entering the atmospheric dehydration tower is dehydrated in the manner previously described. Steam consumption in this system, varying somewhat with the percentage of ethanol in the beer, is in the range of 1.8-2.5 kg/liter (15-21 lb/gallon).

References

Bravo, J.L. *et al.* 1985. Hydrocarbon Proc. Jan. p.91.

Katzen, R. 1955. Chem. Eng. Nov. p. 209.

Fair, J.R. 1977. Chem. Eng. Prog. Nov. p. 78.

Fair, J.R. *et al.* 1990. Chem. Eng. Prog. Jan. p. 19.

Katzen, R. 1980. Low energy distillation systems. Bio-Energy Conference, Atlanta, GA, April.

Kister, H.Z. *et al.* 1990. Chem. Eng. Prog. Sept. p. 63.

Lynn, S. *et al.* 1986. Ind. & Eng. Chem. 25:936.

Martin, R. L. *et al.* 1970. Hydrocarbon Proc. March 1970, p. 149.

Mix, T.J. *et al.* 1978. Chem. Eng. Prog. April p. 49.

Petterson, W.C. *et al.* 1977. Chem. Eng. Sept. p. 79.

Robinson and Gilliland. 1950. Elements of Fractional Distillation, McGraw Hill Book Co., Inc.

Chapter 19

Molecular sieve dehydrators. How they became the industry standard and how they work

R.L. Bibb Swain
Delta-T Corporation, Williamsburg, Virginia, USA

Introduction: the early days

Azeotropic distillation

The equipment used during the dawn of the fuel ethanol program was built using a strange combination of technologies borrowed from different industries. Most of the basic production expertise came from the beverage alcohol industry, but there was no need to remove all the water from beverage alcohol. To prevent phase separation when blended with gasoline, fuel-grade ethanol needed to be almost completely dry. Standard distillation still leaves over 4% water in the ethanol, so a process called 'azeotropic distillation' was used in all the early commercial ethanol plants to remove the final water from (dehydrate) the ethanol. Azeotropic distillation uses a third component, typically benzene or cyclohexane, to 'break' the azeotrope (the composition above which standard distillation becomes ineffective).

Azeotropic distillation systems tended to be quite expensive, difficult to operate and adjust, and consumed a significant amount of energy. Attempts were made to reduce the operating difficulties and energy consumption, but the two goals were incompatible. Multi-pressure techniques could reduce energy consumption, but at the expense of more difficult operation and considerably higher capital cost. A process upset could easily contaminate the ethanol product with benzene; and there was no way to totally protect plant workers from exposure to known or expected carcinogens.

Synthetic zeolytes and early molecular sieve dehydrators

An inventor named Skarstrom received a patent in 1957 for a device that utilized a synthetic zeolyte adsorbent that selectively removed water from air and some other gasses and vapors (Ruthven, 1984). A different zeolyte could even be used in virtually the same device to separate oxygen and nitrogen from air (Ruthven, 1984). These synthetic zeolytes became known as 'molecular sieves' due to the very precise pore size that enabled them to select and remove one molecule size from a bulk mixture containing molecules with a larger size or lower polarity. A properly designed system could dry air to a dew point of -100°F; and drying air became the first major application for molecular sieves. By the early 1980s, at least a dozen companies offered molecular sieve air dryers.

If molecular sieves could be used to dry air, why not ethanol? Experiments were under way at that time by several inventors and companies to prove efficacy of that concept. In the August 1981 issue of *Gasohol USA* (the only trade journal at the time), two companies advertised molecular sieve dehydrators that dried ethanol in the vapor phase. Ad-Pro Industries, Inc. of Houston, Texas, and Anhydrous Technology, Inc. from Sugarland, Texas, both offered small skid-mounted systems for drying up to 200 gallons per hour. Two other companies, Pall Pneumatics and Shroeder Farms, were marketing molecular sieve ethanol dehydrators that dried the ethanol in the liquid phase. The liquid phase systems used too much energy, had severe problems with desiccant fouling and finally were just not able to compete with the vapor phase units. The two companies eventually sold approximately a dozen of the small vapor phase dehydrators. Since the devices were relatively simple, many entrepreneurs built unauthorized copies for themselves and others.

The early vapor phase dehydrators had problems of their own. Aside from a general lack of reliability, the molecular sieve desiccant would deteriorate within a year; and all the dust that resulted would contaminate the ethanol product. The commercial molecular sieve desiccant was manufactured by mixing the molecular sieve crystalline powder (much like glass dust) with a clay binder, and forming the slurry into beads or extruded pellets about 1/8 inch in diameter. The finished product had the look and toughness of limestone. Any movement of a bed of the material resulted in abrasion between the particles. The resulting dust accumulated in the bed causing vapor channeling and product contamination. Bed movement could be caused by excessive temperature swings (thermal expansion and contraction), excessive pressure drop across the bed (fluidization) and local disturbance by high-speed vapor entering the bed during cycle transition (jetting). By 1987, Ad-Pro Industries and Anhydrous Technology, whose designs were subject to all these problems, were both out of business.

New technology becomes proven on large scale

In 1984, Delta-T Corporation was formed in Williamsburg, Virginia, to build and market a molecular sieve dehydrator design that had been developed over the past three years by Delta-T's founder. The new design eliminated the reliability and desiccant deterioration problems that plagued earlier designs. Delta-T built a 150-gallon/hr prototype at a small ethanol plant in Charles City, Virginia. After conducting a battery of tests to prove the concept and optimize the design, the prototype continued to produce ethanol commercially at the plant until the Virginia ethanol incentive expired many years later. Delta-T sold six dehydrators during its first eight months of operation, the largest of which was 800 gallons/hr.

In the meantime, larger ethanol plants continued to be built using azeotropic distillation dehydration technology. Molecular sieve dehydration was a relatively new technology with a somewhat troubled past; and Delta-T had no large reference unit to show a customer. In addition, the firms providing technology for the balance of the plant had their own azeotropic distillation technologies to sell the customer. They were quick to point out both the difficulties of getting uniform vapor flow through large beds and that the technology was not proven at a larger scale. Both were correct, but misleading.

Not able to sell a large dehydrator to an American ethanol plant, Delta-T formed a joint-venture with an Italian engineering firm to build a 30 million gallon/year dehydrator for the largest distillery in Europe, located in Sicily. Design specifications called for the feedstock ethanol to be as low as 160 proof (80% ethanol by volume), with a 199.4 minimum product proof. Performance guarantees were stringent, but Delta-T badly needed a large reference unit, so it contracted to provide the unit. Happily for all project participants, the new world-scale dehydrator worked just as promised, and Delta-T finally had proven the design at large scale.

Molecular sieve dehydrators had always offered the industry lower capital costs, lower operating costs, and greater personnel safety than azeotropic distillation. Now that a large unit was on line and performing well, the last obstacle to industry acceptance was overcome. At first, ethanol plant developers would get their plant technology from the older process design firms, but they would come directly to Delta-T for their molecular sieve dehydrator. When the process technology firms realized they would no longer be able to sell azeotropic distillation dehydration to their customers, they reluctantly adopted molecular sieve technology. Today, all new ethanol plants are built with molecular sieve dehydrators.

How does a molecular sieve dehydrator work?

Pressure swing adsorption

Most modern molecular sieve dehydrators use a process known in the industry as 'pressure swing adsorption' to remove water from a vaporized feed stream. We will concentrate herein on the removal of water from an ethanol vapor stream, but many other applications are possible with virtually the same device. The term 'pressure swing' refers to the fact that the dehydrator uses a relatively high pressure when water is being removed from the feed stream and a relatively low pressure when the molecular sieve desiccant is being regenerated (having water removed from the desiccant). Most commercial designs have two or more beds of desiccant and cycle the vapor flow through the beds to provide continuous operation. While one bed is on-line drying the feed vapor, another bed is being regenerated. In some designs one or more additional beds are being depressurized or repressurized in preparation for the next cycle (Le Van, 1996).

How does adsorption work?

The 'macroscopic' principles of adsorption are quite simple, but attempts to accurately explain how adsorption works often fail because the author gets bogged down in technical jargon. The basic characteristic of an adsorbent material is a stronger affinity for one type of atom or molecule than for the other types in the vapor stream. In the case of a molecular sieve ethanol dehydrator, we select an adsorbent with a strong affinity for water and little affinity for ethanol and the other impurities typically contained in the ethanol feed stream. As wet ethanol vapor passes through the bed, the desiccant adsorbs the water molecules but not the ethanol molecules. This process cannot continue indefinitely, because the desiccant has a finite capacity for water; and that capacity is affected by the operating temperature and pressure. Moisture capacity increases as pressure increases or temperature decreases, and vice versa. Since a pressure swing adsorption dehydrator operates at almost constant temperature (Ruthven, 1994), we can desorb the water adsorbed on the previous cycle by lowering the operating pressure and passing a purge vapor through the bed in the opposite direction to sweep the water molecules from the free space surrounding the desiccant beads.

The class of materials known as zeolites occurs naturally, but most commercial molecular sieves are man-made. Synthetic zeolites have a crystalline lattice structure that contains openings (pores) of a precise size, usually measured in angstroms (Å). Molecular sieves can be manufactured with different pore sizes by using different chemistry and manufacturing methods. A synthetic zeolite of type 3Å is used in most ethanol dehydrators, because the pores are 3Å in diameter while water molecules are 2.8Å and ethanol molecules are 4.4Å. Therefore, water molecules are strongly attracted into the pores but ethanol molecules are excluded. Many other adsorbent materials are available that have an affinity for water such as activated alumina.

However, other adsorbents have a wide range of pore sizes and therefore are much less selective than molecular sieves.

Water is so strongly attracted to type 3Å molecular sieve that for each pound adsorbed, 1,800 BTUs of heat are released. This effect is referred to as the heat of adsorption. When you remove that same pound of water during regeneration, you must supply 1,800 BTUs of heat (referred to as the heat of desorption). Thus, as feed vapor is dehydrated the bed of desiccant heats up; and as the bed is regenerated it cools down. Type 3Å molecular sieve is capable of adsorbing up to 22% of its weight in water, but pressure swing adsorption dehydrators operate at a much lower moisture loading and limit the dehydration period to prevent an excessive temperature swing.

Operational considerations

Energy needs, production capacities

By limiting the adsorption period, temperature swings are kept to a reasonable value and the heat of adsorption can be effectively stored in the bed of desiccant as sensible heat. The heat is then available to supply the necessary heat of desorption during the regeneration period. The ability to recover the heat of adsorption is why molecular sieve dehydrators are so energy efficient. Assuming the feed vapor comes directly from the rectifier column, the only energy used directly by the molecular sieve dehydrator is the steam required to superheat the vapor to bed operating temperatures (0.1 to 0.2 lbs of steam per gallon of anhydrous ethanol) and the electricity to operate the pumps (0.02-0.03 Kw hr per gallon). Many companies forget that additional energy is required to redistill the liquid that results during regeneration as well as the electricity to operate the cooling tower and air compressor. A properly designed system will have a totally inclusive energy consumption of about 4,000 BTUs per gallon of anhydrous ethanol, assuming the feed ethanol contains 5% water and the product is dried to 0.25% water. Additional energy is required if the feed ethanol is wetter or if the product is dryer.

Molecular sieve dehydrators can accommodate a wide range of specifications, and can be built for any desired production rate. Some are currently in service drying ethanol containing as much as 20% water and some can produce anhydrous ethanol with as little as 20 ppm water in the final product. The smallest units are designed for a product rate of about two gallons per hour (mostly for lab and test work), while the largest can produce 100 million gallons per year.

A typical operating cycle described

The following describes a typical operating cycle for a 2-bed dehydrator drying ethanol containing 5% water. We begin with Bed 1 on-line (drying the feed vapor) and Bed 2 being regenerated (see Figure 1). Wet ethanol vapor enters the top of Bed 1 under a modest pressure, passes downward through the bed and exits the bottom as dry vapor. 60-85% of the vapor leaves the system as anhydrous ethanol product; and the remainder is fed to Bed 2 as a regeneration purge stream. Bed 2 is maintained under a vacuum that shifts the equilibrium conditions so that the adsorbed water is desorbed. Operating pressures, temperatures and flow rates are adjusted so that regeneration is complete at the point in the cycle when the beds change position (valves are positioned to reverse the roles of the two beds). One bed typically stays on-line dehydrating the feed stream for 3-10 minutes before being regenerated. When the beds transition from dehydrate to regenerate mode, the pressure in the inlet header tends to sag, which can upset the rectifier column if provisions are not made to mitigate the pressure drop. After the beds switch, the bed that was being regenerated 'pressures up' to the dehydrating pressure and the vaccuum system lowers the pressure in the other bed to the regenerating pressure. The cycle repeats itself continuously.

Since one bed or the other is always receiving and drying the inlet vapor, the process seems to be continuous; but in reality it is more properly termed as 'cyclic batch'.

While a molecular sieve dehydrator can operate on a simple timed cycle, the best systems are managed by an 'intelligent' controller (computer or PLC) to provide greater operating safety, reliability and energy efficiency. Once adjusted to standard operating conditions, the molecular sieve dehydrator requires little operator interface and can tolerate reasonable variations in feed rate or quality without need for readjustment. Typical quality control measures include sampling the product, feed and regeneration liquid at regular intervals to assure the product meets specifications and that the system is operating efficiently.

References

LeVan, M.D. 1996. Fundamentals of adsorption proceedings of the fifth international conference on fundamentals of adsorption. Kluwer Academic Publishers, Boston, MA.

Ruthven, D.M. 1984. Principles of adsorption and adsorption processes. John Wiley & Sons, New York.

Ruthven, D.M. S. Farooq and K.S. Knaebel. 1994. Pressure swing adsorption. VCH Publishers, Inc., New York.

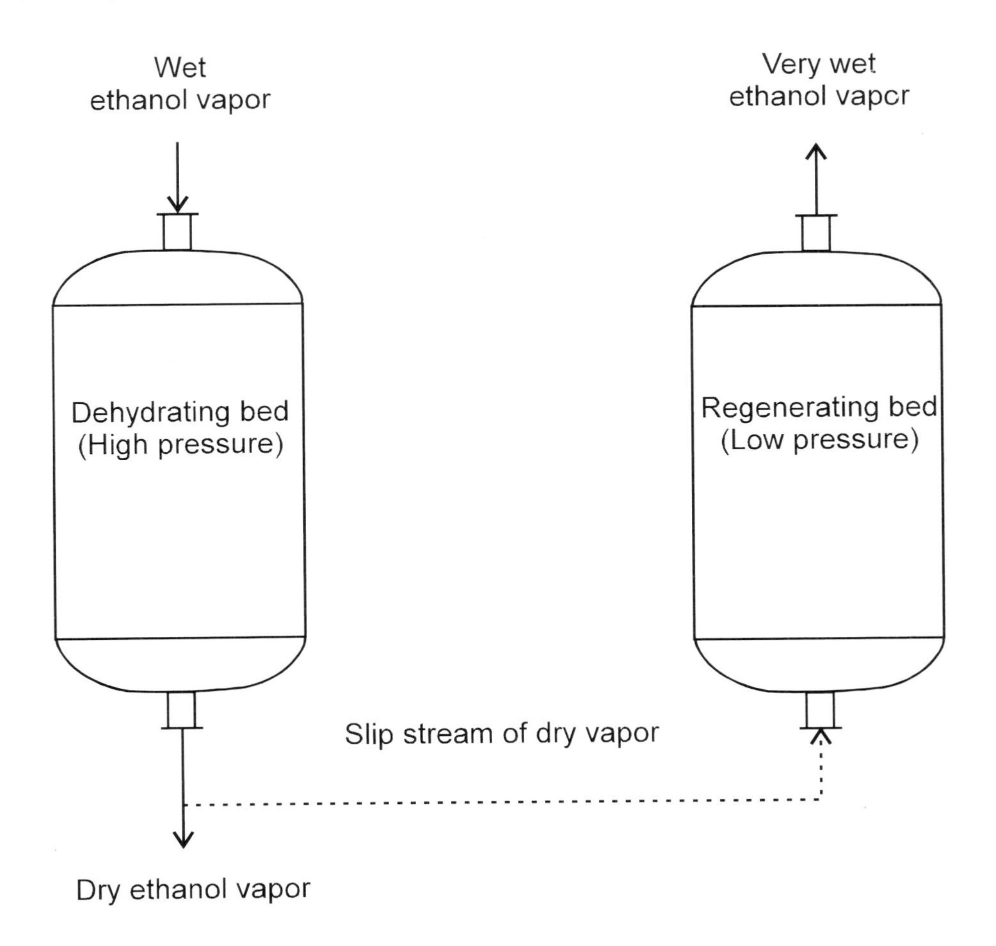

Figure 1. Operation of two bed molecular sieve dehydrator.

Chapter 20

Distillery quality control

S.A. Wright
Hiram Walker & Sons Ltd., Walkerville, Windsor, Ontario, Canadaario, Canada

Introduction

Quality can be described in many ways; but is essentially the conformance to established requirements, or a measure of the degree of excellence. Product quality is ultimately judged by the consumer, and therefore 'controls' must be in place to ensure that product quality lives up to the consumer's expectations. Quality control is the maintenance of quality within acceptable levels and tolerances. It is a comprehensive program put in place to conduct the inspections and tests necessary to maintain consistent production of a quality product at a consistently reasonable cost. Typical responsibilities of a quality control program include:

- establishment of specifications
- development and implementation of sampling plans and test methods
- recording and reporting of results
- problem troubleshooting

Specifications are established for raw materials, processes, components and finished products. Sampling and inspection schedules are created to maximize the significance and reliability of the test data. Testing methods are developed to determine that products and processes comply with established specifications within the given tolerances or limits. Testing protocols are either devised in-house utilizing staff expertise, or adopted from standard testing methods available from government agencies or trade associations (Kramer and Twigg, 1966). Test results are recorded and the necessary information reported using established communication channels in a manner that will facilitate necessary actions when required. Quality control personnel are frequently relied upon for input on troubleshooting during process upsets or quality problems because of their technical knowledge and familiarity with processes, specifications and test procedures.

This chapter does not expand on the principles and philosophies of total quality management systems such as ISO and TQM, but instead provides a 'work chest' of techniques and tools to be used in the day-to-day operation of a distillery quality control program. It presents an assortment of the traditional quality testing methods that are used in the industry today, along with some new and developing analytical techniques. Methods are described for each process operation in sequence, following the product flow through the plant; starting with receipt of raw materials and through to mashing, fermentation, distillation and co-product production. The distillery format most closely followed by the chapter is that of a whole grain beverage alcohol producer, while reference is made to other distillery formats where appropriate. Keep in mind that some of the methods described will require some amount of modification for proper application to the individual needs of the manufacturer.

Raw materials: sampling and inspection procedures

A wide variety of raw materials are used in the production of distilled alcohol. The starting materials for beverage alcohol production are mostly regulated by official standards of identity. For example, rum must be produced from sugarcane products; whisky production requires specific grain compositions; and brandy must be distilled from fruit fermentations. Production of non-beverage alcohol allows for a much larger selection of starting materials ranging from grains, tubers, fruits and other agricultural products to food processing by-products.

Quality checks must be in place to maintain consistent quality of the raw materials, to reduce the chances for downstream processing problems stemming from the use of substandard materials and to avoid potential product quality concerns. Typical quality testing used on the most common feedstock materials are described.

Sampling plans must be developed to ensure that samples drawn from incoming shipments are representative of the entire shipment. Composite samples are often required to overcome variations in the character of the material brought about during transport. The requirements for proper sampling protocols are dictated by the potential for inconsistency anticipated in the raw material shipments.

Corn, rye, barley and milo testing

Grain testing is conducted at the plant or designated facility in accordance with a standard grading procedure in a manner defined by the governing grain commission. The grade determined by the testing process, when conducted by a certified facility either on-site or off-site, can affect the negotiated price of the grain.

Aroma

A musty or sour odor usually indicates growth of spoilage organisms either due to improper storage after harvest or inadequate drying before storage. This can result in substandard alcohol yields; and in the case of beverage alcohol producers this opens the possibility for flavor defects in the finished product. Grain aroma is evaluated either by smelling the sample straight from the sample container; or if a more definitive evaluation is required, the grain is mixed with hot water and the hot mixture is then smelled in a glass snifter.

Weight

The weight of a standard measuring cylinder of specific volume filled with grain is determined and reported either as bushel weight (weight per 2,150.42 cubic inch Winchester bushel) or metric weight (g/L or other standard volume).

Color, foreign material and heat-damaged kernels

A standard sieving procedure is used to measure the amount of undersized grain in a standard sample size. The sample is inspected at this point for foreign materials and heat-damaged kernels, and these substances are weighed and recorded. Corn can be subject to heat damage if too much heat is applied during the forced-air drying procedures that are used on grains that have been harvested too wet. Heat-damage can result in an irreversible binding of the starch, making it unavailable to the distiller.

Moisture

High moisture content increases the chance for spoilage of the grain and can lead to the development of 'off' odors (sour, musty), which can be particularly troublesome to beverage distillers. While traditional methods for grain moisture testing make use of the reliable Model 919 Moisture Meter from Nuclear Enterprises Ltd. (Winnipeg, Canada), newer test methods are coming into common use. Recent developments in Near Infrared Spectroscopy (NIR) technology have opened many opportunities for NIR in grain quality testing, including moisture analysis. This and other applications for NIR in the distillery are discussed later in this chapter.

Barley malt

Malt is used in the production of whiskies as a source of amylolytic enzymes and as an important contributor of flavor. Malt has been replaced by microbial enzymes in all fuel alcohol plants; but remains mostly for its flavor characteristics in beverage distilleries. The use of microbial enzymes has a great many advantages over malt, as discussed elsewhere in this volume.

Whisky distillers each have their own specific quality requirements for malt. The malt supplier is accountable for assuring compliance with the quality specifications of the distiller. Typical malt specifications include bushel weight, moisture content, bacteria levels, α-amylase activity and total diastatic power. Testing methods for malt enzyme levels can be found in AOAC and ASBC standard methods manuals (AOAC, 1980; ASBC, 1992).

Grain milling

A consistent grind profile is important in helping maintain constant conditions in the cooker and in avoiding upsets in the spent grains operations. Irregularities in the size distribution of milled grains can lead to unevenly cooked mash, which can affect fermentation performance. In the spent grains dryers, inconsistent particle size can affect the loading of solids in the evaporators, which can lead to inconsistent evaporator operation.

Sieve testing

Milled grain samples are tested to determine particle size distribution by passing a measured weight of grain through a series of screens of progressively finer mesh size to separate the sample into size categories. Three or more sieve trays are used with standard US Series screen mesh sizes ranging typically from 0.0662 to 0.0165 inches. A mechanical sieve shaker provides a consistent shaking action for the test. After a timed shaking period, the sieves are removed from the shaker and the weight of grain material residing on each screen is determined. The weights are tabulated and a grind profile is reported. A typical grind profile for corn used in a continuous jet cooking system is shown in Table 1.

Table 1. A typical grind profile of corn.

Sieve number	*Screen opening (inches)*	*Quantity (grams)*
10	0.0662	10
20	0.0331	40
40	0.0165	30
Through 40	-	20

Fermentation quality control

Effective testing and monitoring of the fermenters is important to maintain fermentation efficiency and finished product quality. Table 2 is a list of the traditional quality control tests used for monitoring grain alcohol fermentations.

Balling test

The specific gravity of mash is measured using a Brix or 'Balling' hydrometer. Hydrometers of this type are calibrated against sucrose (% w/w) and are read directly in degrees Balling. Mash is tested at start of fermentation and at established intervals during fermentation. Finished fermenters are tested as well, as a measure of completion.

Table 2. Tests conducted in the mash before, during and after fermentation.

Mash to fermenter or fermenter set	*24 hour fermenter*	*Finished fermenter*
Balling	Balling	Balling
pH	pH	pH
acidity	acidity	acidity
temperature	temperature	residual carbohydrates
starch conversion (iodine test)	yeast count	alcohol
bacteria level	bacteria level	

Acidity/pH testing

Acidity/pH plays a central role in the mashing and fermenting operations. It affects enzyme activities during conversion of the starches in the mash, and influences yeast growth and congener formation in the fermenters. Low pH is a common cause of enzyme failure in mashing; and quick detection and correction of a low pH condition will minimize loss in yield. Low pH in the fermenter usually suggests a thriving population of bacteria, which, if left unchecked can disrupt both enzyme activities and yeast growth. Testing for both acid level and pH in the fermenter will provide more useful information than pH alone, as acid measurement is not affected by the pH buffering capacity of the mash. One should not underestimate the importance of acid monitoring. A record of acid level in the fermenter is a good indicator of fermenter quality and can be important in helping diagnose production problems. Acid testing is as easy as diluting a 10 ml sample of mash to approximately 50 ml with distilled water, adding a few drops of phenolphthalein indicator and titrating to endpoint (color change to pink) with 0.1 N sodium hydroxide.

Temperature monitoring

Proper control of process temperatures in distillery operations is essential to maintaining performance and quality. In grain distilleries, the mashing circuit requires a specific temperature sequence for proper cooking and gelatinization of the starches and for proper enzyme activities. In the fermenter, proper control of temperature is needed to keep the yeast at its required level of activity. Process temperatures are monitored and controlled automatically in most distillery operations; but periodic testing and calibration of the control equipment is an essential element of a quality control program.

Starch conversion testing

The effective conversion of grain starches to fermentable sugars is an important indicator of mashing performance, and a measure of this performance is an essential function of the quality control program. Starch is converted to sugars by two distinct enzymatic processes: liquefaction and saccharification. Liquefaction defines the 'thinning' of the mash by the actions of α-amylase on the cooked (soluble) starches. Thinning results from the enzyme randomly breaking the starch molecules into short-chain saccharides (dextrins). To easily test for completion of liquefaction, a few drops of 0.05 M iodine is added to a couple of milliliters of liquefied mash and checked for a color reaction. A deep blue color will indicate the presence of non-liquefied starch.

Saccharification is the enzymatic process of converting the dextrins into fermentable sugars (maltose or dextrose). Saccharification is not completed until well into the fermentation, and can only truly be measured by determining the residual level of unfermentable carbohydrate remaining at the end of fermentation. Starch conversion can best be assessed by monitoring the depletion of carbohydrates from the mash during the fermentation. A simple method for approximating carbohydrate levels in fermenting mash is with the Brix hydrometer. A hydrometer reading at the end of fermentation will indicate whether a fermentation has been completed. However, if a high reading is found on a finished fermenter, further testing will be required to establish whether the high reading was reporting unfermented sugars or unconverted starches. Methods for these measurements follow.

Carbohydrate testing

The measurement of specific sugars during fermentation can provide some useful information for quality control purposes. The depletion of sugars from the mash during fermentation can be monitored to measure starch conversion or fermentation performance. Finished fermenters can be tested for residual sugar content; and in instances where a high finished Balling is found, the sugar test will determine whether conversion or fermentation was at fault (high starch and low sugar indicates incomplete starch conversion; high sugar and low starch suggests fermentation problems).

Individual sugars can be measured by colorimetric test methods. Prepackaged test kits are available to simplify colorimetric measurement (Boehringer Mannheim GmbH, Germany). Automated measurement of sugars is available with the YSI autoanalyzer, which is widely used in the distilling industry (Yellow Springs Instrument Co.). Another common approach to sugar identification and measurement is high pressure liquid chromatography (HPLC).

HPLC applications

There are many quality control applications for HPLC in the distillery. HPLC is used for measuring sugars during fermentation, and can also be used to quantify organic acids, ethanol, glycerin and other by-products of the fermentation. A chromatography method is described in Table 3 for measuring carbohydrate and acid levels in a grain fermentation. The method uses a very basic chromatography system and requires a simple cleanup procedure for preparing the mash samples. Samples are prepared by first removing the coarse grain particles by filtration or centrifugation. The clear mash is then passed through a syringe-loaded 0.20 μm membrane filter, followed by a C18 solid-phase extraction cartridge (Sep Pak). The sample is then diluted with HPLC solvent or used at full strength.

Table 3. Chromatography conditions for HPLC analysis of fermentation mashes.

Column	Bio-Rad organic acid column HPX-87H
Pre-column	Bio-Rad Micro-Guard HPX-85H
Solvent	Sulfuric acid, 0.08 N in demineralized water, degassed
Solvent flow	0.06 ml/min.
Temperature	65°C
Injector	Waters UK6 (Millipore Corp. MA)
Pump	Waters Model 510
Detector	Waters Differential Refractomater R401

Microbiological testing

Proper control of the microbial activities in the fermenter is essential to maximizing the efficiency of alcohol production and in maintaining the quality of the distillate. Microbiological methods are used to test yeast quality and to monitor microbial contaminants in the raw materials, process equipment and manufacturing operations. Testing is done on the yeast prior to fermentation to determine purity and viability. During fermentation, yeast quality can be observed as an indicator of fermentation performance. Microbiological testing of spoilage organisms is an integral part of a plant sanitation program, which is needed to control microbial contaminants to manageable levels to maintain production efficiencies and product quality.

Testing yeast quality

Distillers yeast is either propagated in pure culture in the distillery or purchased in dry form (distillers active dry yeast) or as a compressed cake. When cultured yeast is propagated at the distillery, the yeast is tested at different stages in the process to ensure that quality specifications are met. Total yeast and bacteria counts are the most important indicators of yeast quality, but other characteristics such as percent viability, cell appearance and colony morphology are also important indicators of the quality of a yeast culture. When working with a yeast culture, one learns to recognize the features distinguishing healthy yeast from weak or unhealthy cultures.

If a commercial yeast preparation is used, quality is evaluated on arrival of the yeast at the plant and at specific intervals during storage. The yeast is sampled from a determined number of containers (usually bags or boxes) and yeast and bacteria counts are measured. Note is taken of the color, smell and texture of the yeast. A healthy yeast preparation is typically a creamy beige color and free from sour or musty odors. Compressed yeast should be of a texture dry enough to crumble slightly when prodded with a knife. If compressed yeast is found to be mushy or runny, it will most likely have developed a more intense odor, indicating it has been allowed to get too warm. This is harmful to the product, as it kills the yeast and causes it to lyse. Bacteria will quickly increase in number as the lysed yeast cells provide an enriched medium for growth. If the product is allowed to stay warm, the bacteria will quickly dominate and spoil the culture. The ideal storage temperature for compressed yeast is 4°C. Temperatures above 15°C should be avoided for storage beyond a few hours.

Active dry yeast is more shelf-stable and less prone to spoilage than compressed yeast. It can survive at room temperature for many months with little loss of activity. Stability is even better when stored at refrigerated temperatures. Active dry yeast is packaged in an inert atmosphere or under vacuum to extend shelf life. If a yeast is supposed to be vacuum packed, be sure to note the presence of a vacuum seal when opening the package for sampling. Also be aware that yeast activity will diminish more quickly once opened and exposed to air.

Active dry yeast is in the form of either small pellets (1-2 mm in diameter) or short thin strands. The uniformity of the yeast particles is important to quality. Presence of large irregular particles or powdery material in a sample could signal problems encountered during drying; and the result could be an inferior quality yeast. Microbiological testing will determine whether quality was affected.

Once the yeast is charged to the fermenter it is common practice to monitor performance of the culture during the fermentation. Sampling and testing at the start of fermentation, at 24 hrs and at the end of fermentation is common. Testing would include total yeast cell counts, percent viability and bacteria counts.

Test methods for yeast

Enumeration of yeast can be done by direct cell counting using a microscope or by a number of plating methods using appropriate culture media. Plating yeast does not lend itself well to fermentation monitoring, however. For yeast quality evaluations to have any practical value in the fermenter they must be determined quickly; and the incubation time required for results from plating, typically 36-48 hrs, is unacceptable. Direct yeast cell counting is the preferred method for fermenter quality control as it yields results in less than 15 minutes. Plating is reserved for other functions that can afford the necessary incubation period.

Direct counting procedure

Direct counting of yeast cells is accomplished using a microscope and a hemocytometer or equivalent counting chamber. A hemocytometer is specially ruled to provide a viewing area with a specific constant volume. By adding a liquid yeast sample to the chamber and counting the number of yeast cells within a ruled viewing area, a determination is made of the yeast cell numbers within a specific volume. The dimensions of the ruled area in the chamber and the corresponding liquid volume must be known to calculate the yeast concentration in standard units.

Methylene blue stain is used to distinguish between viable and dead cells. Methylene blue, a reducible dye, will stain dead cells blue while live cells remain colorless. The stain is absorbed by viable cells and is quickly reduced to its colorless form by being used as a hydrogen acceptor in metabolic reactions. Dead cells will also absorb the stain, but will remain blue. The stain is used at a concentration of 0.01g methylene blue powder per 100 ml distilled water.

A yeast sample or a sample of fermenting mash is diluted with water and then further diluted using methylene blue stain to a cell concentration that will allow for easy cell counting. The preparation is mixed well and held for five minutes. A drop of the mixture is applied to the counting chamber. The chamber is placed under the microscope at approximately 500X magnification and cells are counted in the ruled areas of the chamber. Counts of colorless and blue cells are recorded separately, if an indication of the proportion of viable cells is required. If only viable numbers are needed, then only the colorless cells are counted. The number of cells/ml = total cells in the central 25-square ruled chamber area x dilution factor x 10^4 (based on ruled area volume of 0.1 mm^3, or 0.0001 cm^3). The procedure and calculation of the number of cells per ml of sample are as follows:

Yeast count in a mash sample drawn from a 24 hr fermenter or a finished yeast propagator.

1. 1 ml sample is added to 9 ml tap water and mixed well.
2. 1 ml diluted sample is mixed well with 1 ml methylene blue solution, and held for 5 minutes then applied to the hemocytometer.
3. Counts are made of viable and dead cells in the four corner squares and the center square of the 25-square ruled area as indicated in Figure 1.
4. The five counts are totaled; and the final number is millions of cells per ml of sample. (i.e. if 120 cells are counted in the five squares then yeast count in the sample is 120 x 10^6/ml).

Plating yeast

Pour-plating is a standard microbiological procedure for measuring viable microorganisms. It has an advantage over direct yeast cell counting when large numbers of samples are being tested as plating can be done with less operator time and fatigue on large sample numbers than direct counting.

A sample is taken for testing using sterile methods and a measured amount is suspended in a sterile diluent (sterile water is acceptable). Further dilutions of the yeast suspension are made in sterile water; and a measured amount of the diluted suspension is mixed with molten agar medium in a Petri dish. When the agar has set, the plates are incubated at 32°C for 24-36 hrs and the number of colonies that have formed on the medium are then counted. Dilutions of the yeast should be made to the extent needed to produce between 30 and 300 yeast colonies on the Petri dish. If the colony concentration exceeds 300, the count will be depressed because of overcrowding; and if fewer than 30 colonies are formed, the statistical error becomes unacceptably high (Harrigan and McCance, 1976). Additional information on pour-plate methods can be found in most general microbiology textbooks.

When plating for yeast, it is necessary to use an agar medium that will allow for unrestricted growth of yeast while suppressing the growth of bacteria. Otherwise it becomes difficult to distinguish between yeast and bacteria colonies on the plate. Commonly used media for yeast plating in the distillery include potato dextrose agar (PDA) and wort agar. The low pH of these media is sufficient to suppress the growth of most bacteria while supporting the growth of yeast. Wort agar has a suitably low pH of 4.2. The pH of PDA is adjusted to 3.5-3.7 during preparation

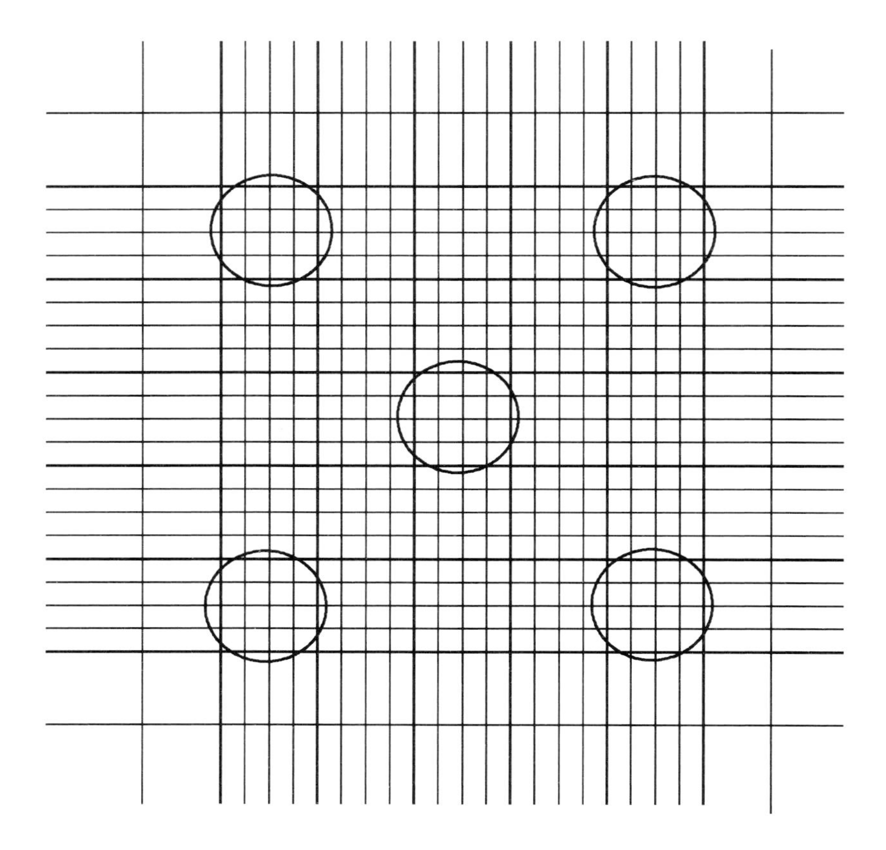

Figure 1. Ruled counting area of a hemacytometer used for direct counts of yeast.

by adding sterile tartaric or lactic acid to the molten agar immediately prior to pouring the plates.

Sanitation control programs

A properly implemented sanitation program is an essential component of a distillery quality control plan. A sanitation program can include the following functions:

1. Establishment of physical, chemical and bacteriological cleanliness standards covering all areas of production
2. Development and implementation of procedures for maintenance of sanitary standards
3. Monitoring of housekeeping and grounds-keeping activities
4. Monitoring of process and equipment cleanup operations
5. Development and maintenance of waste management operations
6. Training of personnel in proper sanitation practices
7. Inspection, record keeping and reporting responsibilities

The purpose of sanitary control in the distillery is to maintain conditions that will control spoilage organisms in the operations to manageable levels, i.e. to levels that will not impair production efficiencies or product quality. As a starting point in a control program, the production circuit is audited to identify the locations that pose the highest risk to sanitary operations. These might be areas that are hard to clean, or by design are an inherent source of contaminants. Such areas could include:

mash coolers
fermenter coolers
blanked off spur lines in mash piping (deadlegs)
valve bodies
fermenter vents and overflow piping
fermenter agitator seals
enzyme and yeast transfer lines

Such areas will require diligent monitoring and control. Proper control of these potential trouble spots alone will significantly raise the level of sanitation in the plant.

Sanitation inspection and testing methods

Surface testing

A variety of methods can be used for testing the sanitary quality of equipment surfaces. All methods require a means of removing the microorganisms from the test surface and a mechanism for counting them. Common test methods include:

1. *Surface swab/culturing.* A sterile moistened cotton swab is rubbed firmly across the test surface to collect a proportion of the microorganisms. The swab is submerged in a sterile diluent to suspend the microorganisms. The suspension is transferred to an agar medium to allow for colony growth on incubation.

2. *Agar contact testing.* A sterile plate or slide coated with a raised agar layer is pressed firmly onto the test surface. A proportion of the microorganisms on the area under test will be transferred to the agar surface and will grow into visible colonies on incubation.

3. *ATP bioluminescence.* This technology measures total ATP from biological residue (including microbial ATP) on test surfaces to estimate the potential for microbial contamination on the surface. Testing takes less than a minute to complete. The test surface is swabbed to remove microorganisms and other residues from the surface. The swab contents are analyzed in a hand-held instrument for total ATP content by measuring the amount of luminescence generated by reaction of the ATP with luciferase (firefly) enzyme. ATP level is a direct measure of the total amount of biological residue on the surface. This technology has gained widespread acceptance in the food industry as processors find it provides significant benefits over traditional methods of sanitation/cleanliness testing.

Product/mash sampling

Mash and other liquid samples are collected in sterile sampling containers provided with a secure fitting lid. Samples are taken at mash coolers, fermenters, mash transfer lines and any other location determined to be important for routine sanitary monitoring. Samples can be collected from valves where appropriate, or directly from the process streams and process vessels. Fermenters, for example, are commonly sampled by dipping the container directly into the mash. Samples can also be taken at the fermenter fill spout by holding the container in the mash stream.

When drawing a sample from a valve it is important to sanitize the valve first to prevent it from contaminating the sample. The valve body and adjoining sample line should be steamed or otherwise sanitized by liquid sanitizers prior to collecting the sample.

Enumeration of bacteria

The bacteria that create the most problems in distillery operations are species of lactobacilli, the so-called lactic acid bacteria. These Gram-positive bacilli are anaerobic or facultative anaerobes with a fermentative metabolism that produces high levels of acids, of which more than half is lactic acid. Lactobacilli begin to grow rapidly in the fermenter after about 24 hrs, at which point the oxygen has been mostly depleted and the aerobic bacteria have died. Lactic acid bacteria can be counted by plating and incubating the plates in a low oxygen atmosphere. A common growth medium for lactic acid bacteria is tomato juice agar (TJA). High recovery of lactic acid bacteria is achieved when TJA plates are incubated in an anaerobic jar or other comparable anaerobic system. Two other media used in the distillery for bacteria counts are Wallersteins' Nutrient (WLN) agar and Wallersteins' Differential (WLD) agar. WLD is used as a selective medium for bacteria and contains cyclohexamide to inhibit yeast and mold growth. When WLN is used, plates must be incubated anaerobically to prevent the growth of yeast. It is the opinion of the author that lactic acid bacteria grow to much the same extent on culture media whether they are incubated aerobically or anaerobically. It is noted by some laboratories that higher counts of lactic acid bacteria are detected on TJA than WLN or WLD; and others report that the best results are achieved from anaerobic incubation. In developing a quality control program for bacteria testing, different plating methods must be evaluated to find the one best suited to the specific conditions of the plant.

A test that has found application in some distillery laboratories for predicting contaminant levels in test samples without involving plating uses the following procedure:

1. Collect a sample in a sterile vessel (minimum 100 ml) using aseptic methods.
2. Aseptically withdraw a subsample from the vessel and measure total acidity by titration.
3. Incubate the remaining contents at 37°C for 24 hrs then retest total acid level.

The difference in the two acid values gives the amount of acid produced by bacteria during the incubation period; and this difference is an indicator of the relative amount of contamination in the sample when it was taken. The test cannot be used on fermenting mash samples because of interference caused by the yeast activity. If this method is to be used in place of standard plating techniques it will need to first be calibrated against the original plating method to establish a reasonable correlation.

Enzyme evaluation

Amyloglucosidase and α-amylase are the two enzymes most widely used in the distilling industry. Industrial enzymes are available as commercial products and most distillers purchase enzymes to meet their requirements. Some distilleries produce their own enzyme preparations to supplement or replace commercial products. There is a trend in the distilling industry today toward the use of nontraditional feedstocks for industrial or fuel ethanol production, and many of these materials have different enzyme requirements. Cellulases, cellobiase, beta-glucanase and lactase are finding use in some segments of the industry.

Quality control procedures are established to maintain the quality of the enzymes used at the plant. The primary concerns for enzyme quality from the distiller's perspective are the sanitary nature of the product and the level of enzyme activity. Commercial enzyme products can be scanned quickly for contaminant levels by placing a drop of the enzyme concentrate (or a dilution of the concentrate) on a slide for viewing under the microscope. If bacteria are visible and there is concern that the contaminant level might be unacceptably high, then a sample of the enzyme is drawn aseptically for serial dilution and plating on an appropriate growth medium. Usually a quick scan under the microscope is sufficient to determine whether the product is 'clean' or contaminated.

Amyloglucosidase activity testing

The 'strength' of an enzyme preparation is expressed in terms of its activity level, which is a determination of the amount of reaction that the enzyme is able to catalyze during a specific period of time under controlled conditions. An example of a test method for amyloglucosidase activity is given below. (This method was written for testing a whole-culture in-house enzyme preparation, and will require slight modification for use on commercial enzymes.)

Test method

Principle

This assay is based on the hydrolysis of soluble starch by the enzyme at a pH of 3.9 and 60°C for 60 minutes. The final reaction product, glucose, is measured at the end of the reaction time. Results are expressed in enzyme 'Units'. One amyloglucosidase Unit represents the amount of enzyme that will produce 1,000 mg glucose per ml from a soluble starch solution in one hour under specific test conditions.

Reagents

1. 4% starch solution: Slurry 4 g soluble starch (Improved Lintner method Soluble Starch, ASBC) with 30 ml distilled water and add to 60 ml boiling distilled water. Stir until starch is dissolved. Add 2 ml acetate buffer and bring to 100 ml.
2. Acetic acid-sodium acetate buffer, pH 3.9: 1.66% (w/v) sodium acetate trihydrate plus 3.8% (v/v) acetic acid.

Procedure

1. Add 4 ml starch solution to a 25 ml volumetric flask and adjust to 60°C in a water bath.
2. Coarse filter the enzyme sample and dilute 5 ml of filtrate to 500 ml with distilled water. Further dilution will be required when testing commercial enzymes.
3. Add 1 ml of diluted enzyme to the starch solution, mix well and place into the water bath for 60 minutes. Run another starch sample with no enzyme addition as a blank.
4. After 60 minutes, place both flasks into boiling water for 5 minutes to inactivate the enzyme.
5. Cool and bring both flasks up to volume with distilled water. Measure glucose concentration in the test flask. Run the blank as well to verify the absence of glucose in the original starch mixture.
6. Many methods can be used for measuring glucose. YSI sugar analyzers are widely used in this application (YSI Model 21, 27 or 2700, Yellow Springs Instrument Co.). Other methods include the Shaffer Somogyi test (AOAC, 1980), enzymatic/colorimetric methods (Boehringer Mannheim) and HPLC methods.
7. The glucose content per ml of sample is multiplied by the enzyme dilution factors (500/5, 25/1) to give the Unit count, expressed in g/ml.

Ethanol quality testing

Quality standards for beverage alcohol are dictated both by consumer expectations and by regulatory agencies. In the US, the standards for beverage alcohol quality testing and reporting are regulated by the Bureau of Alcohol, Tobacco and Firearms. The quality standards for fuel ethanol have been established by an independent organization, the American Society for Testing Materials (ASTM), which publishes standards, specifications and test methods for thousands of commercial products and materials. Although ASTM standards were created for use on a voluntary basis, most contracts between fuel ethanol producers and buyers will refer to ASTM standards for fuel ethanol quality.

Beverage alcohol test methods

Alcohol strength

Alcohol strength in a distilled spirit is determined by specific gravity, and can be measured by a variety of methods. In the US, distilled spirits alcohol content is reported in 'degrees proof' at 60°F (15°C), with proof representing twice the percent alcohol by volume. A whisky at 40% alcohol by volume in the US is reported as 80° proof. In Canada and the UK alcohol content is reported in percent alcohol by volume at 20°C. Specific gravity is traditionally measured by hydrometer or pycnometer, but today most strength testing is done using automated density meters (Anton-Parr and others).

Specific gravity is a measure of the density of a substance relative to the density of a reference substance (normally water), as expressed in the equation in Figure 2.

Measurement by hydrometer

Hydrometers operate on principles of liquid volume displacement by a submerged object. The volume displaced by a hydrometer is inversely related to the density of a liquid. A hydrometer is designed to measure the volume of liquid displaced by displaying the depth to which it will sink in the liquid when placed in an upright cylinder. To measure the specific gravity of an alcohol sample, a hydrometer cylinder is filled with sample and the temperature of the sample is taken using a calibrated thermometer. An alcohol hydrometer of the proper range is immersed in the product, and a reading is taken when the hydrometer has stabilized. Percent alcohol is then determined by reference to the appropriate chart. Different styles of hydrometers are used in different countries, depending on the measuring and reporting requirements of the country. In Canada and the UK, hydrometers are calibrated to 20°C. Hydrometers in the US are calibrated at 60°F (15°C). A selection of hydrometers from various countries is shown in Figure 3.

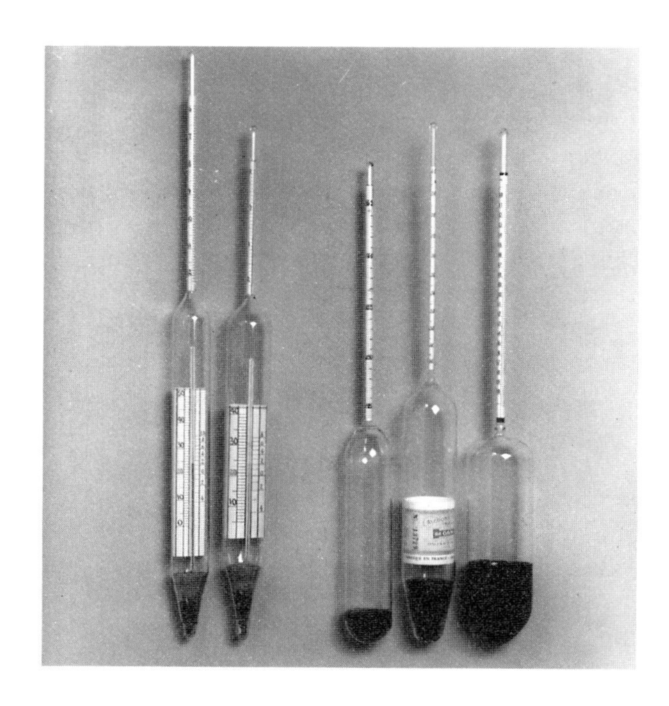

Figure 3. A selection of hydrometers. On the left, temperature compensated Brix or Balling hydrometers for monitoring fermentation progress. On the right, spirit hydrometers calibrated in specific gravity or proof (photo courtesy of Hiram Walker & Sons Ltd).

Measurement by pycnometer

A pycnometer measures the weight of a specific volume of an alcohol solution at a known temperature. This is compared to the weight of water under the same conditions to give the apparent specific gravity. Reference charts convert apparent specific gravity to percent alcohol. A pycnometer must first be calibrated by the user. Calibration is done by filling a thoroughly cleaned pycnometer with freshly distilled water and measuring the weight at a precise temperature. (The proper temperature for the weight determination may be specified

$$\text{Specific gravity} = \frac{\text{Weight of X mls of substance (at a specified temperature)}}{\text{Weight of X mls of water (at the same temperature)}}$$

Figure 2. Calculation of specific gravity.

by the reference table that will be used to calculate percent alcohol. If not, then it is best to select a temperature close to ambient, as this makes it easier to maintain the desired temperature when making the weight determination). The pycnometer is then drained and thoroughly dried (oven-dried overnight is best) and the dry weight of the pycnometer is measured. The test sample is then added to the pycnometer and the weight is measured at the same temperature used in the calibration. The weight of the empty pycnometer is subtracted from this weight and is also subtracted from the weight with water. The test sample weight is then divided by the water weight to give the specific gravity of the sample. Reference to the appropriate chart will determine percent alcohol or degrees proof, as required.

Measurement by density meter

Density meters are widely used in the industry today and have replaced hydrometers and pycnometers in most applications. The density meter determines density automatically by measuring the change in the oscillation period of an electronically-oscillated hollow tube when a substance is added to the tube. The oscillation period is affected by the mass of the sample; and because the volume of the oscillating tube is constant, and since mass = density x volume, the mass of the sample can be taken as proportional to its density.

The density meter is calibrated against water at a specific temperature to enable it to report specific gravity. To operate the instrument, a degassed sample of product is injected into the sample port, avoiding entrainment of air bubbles, and the specific gravity (or percent alcohol) is automatically displayed after a few minutes.

Gas chromatographic analysis of volatile components

The gas chromatograph (GC) is used to distinguish and quantify volatile components in a mixture of substances. It has numerous uses in the distillery for process monitoring and product quality testing. As a process tool, the GC can be used to profile the chemical constituents in the distillation columns. Determining the levels of different components at different draw points on the stills can aid in standardizing still operations and provide some assistance in troubleshooting.

Gas chromatography is used for finished product testing to quantify both desired congeners and unwanted contaminants in the spirits. Neutral beverage spirits demand a high level of purity, and in this case unwanted congeners such as diacetyl, acetaldehyde and fusel oils can be measured by GC. Other beverage alcohol products such as whiskies, rums, brandies and others, have unique congener compositions that can be profiled in part using the GC. Gas chromatography distinguishes and measures volatile compounds only. Although sample pre-treatment techniques can be used to convert certain nonvolatile constituents into volatile derivatives, the most important and largest volume flavor constituents in beverage spirits are already volatile and can be measured conveniently with no need for pre-treatment. The major flavor congeners include the fusel oils (n-propanol, isobutanol, active amyl, isoamyl alcohol), quantitatively the largest group, as well as ethyl acetate, acetaldehyde and methanol. Most distilled spirits producers test routinely for these components in their products.

Two examples of GC operating conditions for the analysis of beverage spirits are presented in Table 4. Method 1 uses a traditional packed glass chromatography column while Method 2 uses a capillary column. Samples are injected directly with no pre-treatment.

A large number of other flavor congeners can be measured using more specialized GC methods. Diacetyl, a contaminant congener with a butterscotch characteristic, can be measured using a more specialized electron capture detector. Major wood-derived congeners can be measured in matured spirits using gas chromatography or liquid chromatography methods. Fatty acids and fatty acid esters also contribute important flavor characteristics to matured spirits, and these too can be measured by GC.

Ethyl carbamate (urethane) in beverage alcohol products

Ethyl carbamate (EC), or urethane, is a natural by-product of fermentation and can be found in many fermented products, including spirits, wine and beer (Walker *et al.*, 1974; Joe *et al.*, 1976; Ough, 1976; Clegg *et al.*, 1988; Canas *et al.*, 1989). This compound is a known carcinogen. It first came to the attention of the distilled spirits industry in 1985, when Canadian regulators established EC limits for on all wines and spirits products sold in Canada. Since that time the wines and spirits industries in Canada, the US and many other countries have achieved reductions in EC levels in their products through investment in basic research, plant modifications and analytical testing.

The distilled spirits industry in the US set a target level of 125 ppb ethyl carbamate to be achieved for all new whisky spirits by January 1, 1989. In 1990 the US Food and Drug Administration (FDA) reported that US producers had met their objective (Food Chemical News, 1990), and today spirits and wine producers in the US and most other countries are continuing work to maintain lowest practical levels of ethyl carbamate in their products.

Many theories have been proposed on the sources of ethyl carbamate, the mechanisms for its formation and the steps needed to reduce its level in beverage alcohol products. Studies suggest different pathways to EC formation in different beverage types. It is accepted that EC is formed in grain-based spirits through the involvement of cyanide-based EC precursors originating in some cereal grains that are reacted during fermentation and distillation to form cyanates, which are then converted into EC in the spirit (Battaglia *et al.*, 1990).

Bourbon was identified early as having the highest potential for exceeding the EC target levels agreed to by US producers. The high barley content in bourbon mash (source of precursors) and the traditional bourbon distillation practices, which allowed a carryover of cyanate intermediates into the spirits, exposed bourbon to a greater potential for high EC than other styles of grain spirits. However, bourbon producers were successful in implementing design changes and operational changes which enabled all producers to reduce EC down to target levels.

Today, all countries exporting alcoholic beverages to Canada must meet the Canadian regulatory limits for ethyl carbamate (including 150 ppb for distilled spirits). Countries exporting to the US must meet the voluntary levels established by the US producers. Indications are clear that there is a global effort today for beverage alcohol producers to reduce EC to lowest technically feasible levels.

Table 4. Operating conditions for gas chromatographic analysis of neutral spirits and distilled beverage products for higher alcohols, ethyl acetate and acetaldehyde.

	Method 1	*Method 2*
Column	6' x 1/4" OD Glass 5% Carbowax 20 on Carbopak B	30 m x 53 mm DBWax capillary (megabore)
Oven program	65° to 92° @ 1.5°/min., hold 15 min.	35°C, hold 3 min., to 100°C @ 7.5°C/min., to 210°C @ 30°C/min.
Detector	FID @ 250°C	FID @ 300°C
Injector temperature	150°C	200°C
Carrier gas	Nitrogen	Helium

Ethyl carbamate testing

To honor their commitment to meet or exceed reductions to voluntary target levels for EC, bourbon distillers test their new whiskey spirits regularly, and report weekly average EC data to the US FDA. The preferred methods for EC testing use gas chromatography (GC) with one of a variety of GC detectors including flame ionization, ion trap, nitrogen selective/electrolytic conductivity and mass selective/G CMS (Walker *et al.*, 1974; Joe *et al.*, 1976; Ough, 1976; Clegg *et al.*, 1988; Canas *et al.*, 1989, Ayllot *et al.*, 1990; Canas *et al.*, 1988; Wright and Clegg, 1987; Battaglia *et al.*, 1990).

As with all GC procedures, proper sample preparation is the key to effective operation of the procedure. While unaged spirits and most aged 'high proof' spirits can be tested without needing sample purification/cleanup steps, other beverage types (cordials, coolers, wines and sherries) require pre-treatment to remove contaminants which would otherwise damage the analytical system. Pretreatment can be done by a variety of methods, using solvent/solvent extraction or solid-phase extraction procedures. Battaglia *et al.* (1990) gives a good review of sample pre-treatment methods used in EC testing. A method using GC-MS is outlined in Table 5. The procedure uses splitless injection onto a carbowax-type capillary column, with a mass selective detector operated in the Select Ion Monitoring mode.

Table 5. Operating conditions for GC-MS detection of ethyl carbamate in distilled beverages.

Instrument:	HP 6890 GC/5973MSD, 7683 Autoinjector
Column	J&W DB Wax 30 m x 0.25 mm ID x 0.25 µm
Carrier gas	Helium Initial flow: 1.0 mL/min. Purge flow: 26.6 mL/min. Time: 0.75 min. Total flow: 30.1 mL/min.
Injector	200°C Mode: splitless Volume: 1µL
Oven	55°C hold for 5 min. 10°C/min. to 175°C 30°C/min. to 235°C hold for 2 min.
Run time	21 minutes
MSD transfer line	240°C
MSD mode	SIM (initially use SCAN to identify EC retention time) Target ion: 62.0 m/z Qualifiers: 74.0 m/z, 89.0 m/z

Procedure:

1. Prepare 50, 100, and 200 ppb ethyl carbamate standards using 80° proof neutral spirits as the solvent.
2. Identify the EC peak using the 200 ppb standard (using SCAN mode and a NIST'98 spectra library).
3. Switch to SIM mode and construct a linear calibration curve.
4. Inject unknown and determine ppb ethyl carbamate.

Fuel ethanol standard specifications

Fuel ethanol producers maintain the quality of their product by standards specified in contractual agreements with their customers, the motor fuel blenders. Standard specifications and test methods have been established by the ASTM for fuel ethanol to guarantee the quality and safety of the product; and contract negotiations between fuel ethanol producers and buyers will usually stipulate ASTM Standard Specifications to determine the product quality.

ASTM is an organization dedicated to developing and publishing technical information and standards for a wide range of commodities and services. It is a voluntary organization that provides a forum for producers, users and consumers to develop standards that best meet their common interests. ASTM standard specifications are developed to establish a set of requirements to be satisfied by a product or material and to indicate the test methods necessary for determining whether these requirements are satisfied.

ASTM standard specifications

The standard specifications required for fuel ethanol are set out in the publication 'Standard Specification for Denatured Fuel Ethanol for Blending with Gasoline for Use as Automotive Spark-Ignition Engine Fuel' (ASTM, 1988). These specifications are summarized below.

Denaturants

The only denaturants shall be unleaded gasoline or rubber hydrocarbon solvent having a boiling point lower than 225°C. The minimum concentration of denaturants is 2% by volume.

Denatured fuel ethanol properties (at the time of blending with gasoline)

Water (max), mass %	1.25	
Nonvolatiles (max), mg/100 ml	5	
Chloride ion (max), mass % (mg/L)	40	(32)
Copper (max), mass (mg/L)	0.1	(0.03)
Acidity (as acetic acid) (max), mass % (mg/L)	0.007 (56)	
Appearance	Clear and bright	

Ethanol purity

The ethanol component excluding water must be a minimum of 98% by volume and cannot contain more than 0.5% by volume methanol or total ketones, or both.

Test methods

Standard test methods are described in ASTM publications for measuring water content, nonvolatile matter, acidity, chloride and copper. Test methods are published under given 'Designation' numbers in the Book of ASTM Standards reference series. Water content is tested using Karl Fischer reagent (Designation E-203 or D-1744). Nonvolatile matter is measured following evaporation of samples at 105°C (Designation D-1353). Acidity is determined by titration to a phenolphthalein end point (Designation D-1613). Chloride ion is measured using an ion selective electrode (Designation D-512, Method C). Copper is determined using atomic absorption spectrophotometry (Designation D-1688).

Maintenance of product quality in accordance with standard specifications is clearly an important quality control function. It is achieved by implementing suitable product sampling and testing protocols and by upholding the accuracy of the test results and maintaining good record keeping practices.

By-product and effluent management

Proper management of distillery by-products and effluents requires application of the same quality control principles as the other manufacturing processes. The same practices of standards development, sampling and testing plans, recording and reporting of data and process troubleshooting apply to these operations.

Distilleries are producers of high volume, high strength by-product streams. The de-alcoholized residue from the base of the beer still, called stillage, is a by-product that must be managed in a responsible manner to prevent it from being a serious liability to the distiller. Stillage represents as much as 90% of the original fermenter volume. Most grain distilleries are able to process this by-product through evaporation and drying operations to produce a valuable livestock feed known as distillers dried grains with solubles (DDGS). The evaporated liquid from these operations is condensed and usually discharged as a waste product.

Wastewater testing requirements

The test requirements for distillery wastewater monitoring are predicated by the discharge ordinances of the local wastewater authority. Limits are usually imposed for biochemical oxygen demand (BOD), chemical oxygen demands (COD), suspended solids, pH, temperature and oil/grease content. Standard test methods for wastewater analysis are outlined briefly in the sections that follow.

Biological and chemical oxygen demand

BOD is a measure of the amount of oxygen required for the decomposition of organic materials in a wastewater by microbial activities under controlled conditions. The test procedure is described in many publications, including the American Public Health Association manual, Standard Methods for the Examination of Water and Wastewater (Rand *et al.*, 1975). BOD testing requires five days for completion and this presents some disadvantages for routine use in the distillery laboratory. COD testing yields results much quicker, and can be performed conveniently using commercial disposable test kits. COD is a measure of the amount of oxygen consumed by a forced oxidative digestion of a (wastewater) material using a chemical oxidant. Some laboratories will establish a correlation between COD and BOD and will routinely use COD test results to approximate BOD.

Suspended solids/total solids

Total solids are determined by drying a measured volume of the wastewater sample to a constant weight at 103-105°C in a pre-dried, preweighed evaporation dish. Suspended solids are considered to be the non-filterable residue remaining on a filter medium following filtration of a wastewater sample. A sample is filtered through a Gooch crucible with a glass fiber filter disk (Gelman type A or equivalent), which has previously been rinsed, dried and weighed. The crucible and filter are dried at 103-105°C until constant weight is achieved (AOAC, 1980).

pH

Wastewater ordinances commonly specify pH limits to help protect the wastewater conveyance and treatment structures from corrosion and to help maintain a pH balance in the biological activities in the treatment plant. Distillery wastewaters are acidic in nature and will require adjustment with caustic to meet discharge pH specifications. Monitoring and control of wastewater pH is automated in many distilleries to ensure continuous pH compliance. Even with automated pH control, however, manual testing must still be done to verify the performance of the system.

Temperature

Temperature limits are established to protect the biological activities at the treatment plant against thermal shock from high volume, high temperature wastes. For wastewaters allowed to be discharged to surface waters, temperature is usually regulated because of the adverse effects of high temperature on oxygen solubility in the waste and on the biological activities in the receiving waters.

Grease and oil

Greasy substances have a very deleterious effect on water quality and can cause particular operational problems in wastewater treatment systems. A knowledge of the quantity of grease in a wastewater is helpful in avoiding or overcoming difficulties in treatment plant operations. Methods for measuring grease/oil in wastewaters are described in Standard Methods for the Examination of Water and Wastewater (Rand *et al.*, 1975).

Distillers dried grains analysis

Most grain distilleries recover their fermentation residues as distillers dried grains with solubles (DDGS). Many producers provide a guarantee on the composition of their products. Typical DDGS specifications will include composition of crude protein, fat, fiber, ash and moisture content. Table 6 shows typical compositional analysis of a corn-based DDGS (Sobolov *et al.*, 1988).

Table 6. Composition of corn distillers dried grains with solubles.

Component	*Percentage*
Moisture	12.0
Protein	27.0
Fat	8.0
Fiber	8.5
Ash	4.5

DDGS sample are tested regularly by the distiller to ensure that the quality is maintained within the guaranteed standards of composition. Testing is mostly performed in-house, but some producers prefer to send samples to contract testing laboratories. Conventional test methods used for standard compositional analysis of distillers dried grains include:

1. Protein determination by the Kjeldahl nitrogen method (AOAC, 1980).
2. Fat determination by hot ether extraction, evaporation of the solvent and measurement of the residue (AOAC, 1980).
3. Crude fiber determination on the defatted sample from the previous fat analysis. An AOAC method has the sample digested with acid then caustic, and the dried residue minus the ash weight is reported as fiber.
4. Ash measurement as the residue following sample heating at 315°C for 2 hrs.
5. Moisture measurement, which can be done either by drying to a constant weight in an oven, or using an automated moisture determination balance that operates an infrared heater in combination with a balance to directly determine percent moisture (produced by Ohaus, Cenco and others). Alternatively, moisture is determined by the AOAC approved Bidwell-Sterling method of distillation with toluene or benzene (AOAC, 1980).

Near infrared spectrometry (NIR)

Recent advances in near infrared spectroscopy are providing opportunities for automated testing of distillers grains and DDGS for component analysis. The traditional methods of testing, outlined above, are very time consuming and consequently tend to be run on composited samples collected over a week or more of production. There are many drawbacks to this style of sampling and testing. Composite testing will give a reading of the 'average' composition during the production period covered by the sample, and will not be representative any actual shipment of product. Composite testing produces historical data, because the product has usually left the plant by the time the test results show whether or not the product was in compliance with the specifications. There are clear advantages from a product quality perspective to running component testing on DDGS on a regular and timely basis, and NIR provides the ability to do this.

From an operations efficiency perspective, there is another advantage offered by the rapid testing capabilities of NIR. The component makeup of distillers dried grains can be considered to provide testament to the operational conditions and efficiencies in the plant. Mashing performance, fermentation efficiency and dryer house operations all have an impact on the composition of the DDGS. Levels of starch and fermentable sugars, as well as protein and fat levels can be used as evidence of an enzyme conversion problem in mashing or an incomplete fermentation. This concept of using NIR as a 'real time' diagnostic tool for distillery operations is still under development, and is being investigated by the author at the time of writing.

There are many other opportunities for NIR analysis in the distillery. NIR is being used at some plants for routine testing of incoming grains for starch and protein levels. NIR is very effective at measuring mash composition during fermentation; and can be used to monitor the progress of fermentation by testing for starch content, fermentable sugar level, ethanol concentration and total acidity simultaneously. With this information, one is able to detect problems in the fermenters as they are happening, and while there is still time to take corrective action. It also lends itself to establishing fermentation profiles to use as benchmarks for comparing production performance. In this manner, an NIR testing program can serve a valuable process control function. A fermentation profile of a whisky mash fermentation is presented in Figure 4 to demonstrate the type of information that can be generated using NIR

Near infrared spectrometry determines the composition of a substance by reading light reflected from, or transmitted through the substance at wavelengths in the near-infrared region of the electromagnetic spectrum (800-2,500 NM). Every constituent of a food or agricultural product exhibits typical patterns of absorption of light in the near infrared region. The NIR analyzer irradiates the sample over a broad range of near infrared wavelengths and then interprets the absorbance and reflectance information to predict the analytical values. As the advances in NIR technology continue to reduce the impact of NIR analyzer initial cost, and as the opportunities for NIR analysis in the distilling industry become more apparent, we will see a more prominent role for NIR in distillery quality control operations.

Mycotoxins

Mycotoxins are a wide group of toxic metabolites produced by a variety of molds, and demonstrate a range of health risks to susceptible animals and humans alike. Mycotoxins are considered to be unavoidable contaminants in susceptible food and feed crops, and it is not possible to predict their presence or to prevent entirely their occurrence by current agronomic practices (Wood , 1992). Grain alcohol distillers are aware of the potential risks imposed by mycotoxins on DDGS quality, and need to ensure that quality testing and control procedures are implemented to help protect the value of their all-important distillers co-products.

Of the hundreds of mycotoxins that are known to exist, those with largest agronomic and health risk and of most importance to distillers, include aflatoxins, fumonisins, DON and zearalenone. Aflatoxins are carcinogenic compounds produced by species of Aspergillus. Aflatoxins are found in corn, cottonseed, milo, soybeans, peanuts and tree nuts, but corn is the grain crop with the greatest potential for contamination with aflatoxins in the US (Jones, 1992). The incidence of aflatoxin in the US grain belt is dependent mostly on the prevailing weather during the preharvest and harvesting periods. Aflatoxin is the most prevalent mycotoxin in the US and because of its demonstrated carcinogenic effects, it presents the highest concern to distillers. The advisory limit for aflatoxin in feeds in 20 ppb.

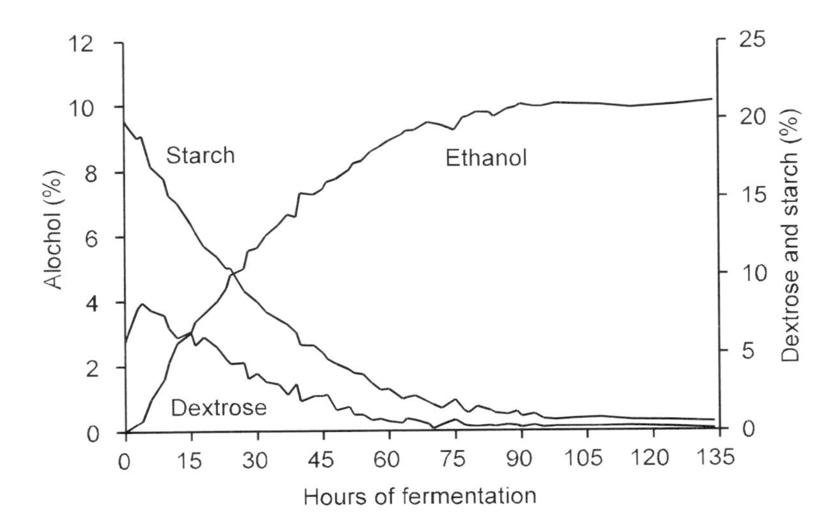

Figure 4. Whisky fermentation profile as determined by NIR analysis.

Fumonisins, deoxynivalenone (DON) and zearalenone are produced by species of Fusarium. Fumonisins and zearalenone affect mostly corn, while DON can contaminate corn and wheat. The effects of these toxins on susceptible animals (mostly horses and pigs) include feed refusal, vomiting, infertility and other toxin effects.

Many of the mycotoxins have high thermal stability and can withstand exposure to the process conditions in the distillery. Consequently there is a potential for mycotoxins to be processed through the distillery and concentrated in the dried grains.

Control strategies

Since it is not possible to prevent the occurrence of mycotoxins, the distiller needs to implement strategies to minimize the presence of mycotoxins in his grain supply. Strategies can include both selective purchasing and screening or testing. Purchase grains from geographic areas with a low incidence of contamination. By applying control at the 'supply chain' level, the risk of contaminated grains arriving at the gate is reduced. Spot testing/monitoring for a limited number of mycotoxins is common practice.

An effective monitoring program must include a good sampling plan and reliable analytical techniques. Mycotoxin contamination in grains is highly heterogeneous and precautions must be taken in sampling to obtain a reliable estimate of the concentration in a given lot. It is recommended that a minimum sample size of 5 lbs (of corn) be taken from every test lot to give confidence to the test result.

Aflatoxin screening

A traditional screening method used by distillers for aflatoxin prediction uses an ultraviolet light source to screen for fluorescence. The fluorescence is caused by a metabolite produced by the same fungus that produces aflatoxin, and therefore indicates the potential presence of the toxin. A sample of whole corn is spread thinly across a flat surface and scanned with a UV light. Kernels are observed for greenish-yellow fluorescence, with particular attention to cracked kernels and broken pieces. If fluorescence is detected, the lot is either rejected or the sample is re-tested using a more quantitative test

Semi-quantitative testing

Mycotoxin test kits are available for routine measurement of a variety of mycotoxins in grains. Test kits are sensitive, easy to use and reliable. Some can be performed in under 10 minutes. Many have been validated as conforming to the US Federal Grain Inspection Service (FGIS) standards.

Test kits can be used by distillers for grain testing and for monitoring toxin levels in DDGS. When selecting a test kit for DDGS testing it is important to validate the performance of the kit for the specific application. The test kit supplier can provide valuable assistance in test method development and performance validation. The performance and convenience of commercial mycotoxin test kits permits them to serve a valuable role in a mycotoxin control program.

References

AOAC. 1980. Official Methods of Analysis of the Association of Official Analytical Chemists. 13th Edition. Assoc. Off. Anal. Chem. Washington, DC.

ASBC. 1992. Methods of Analysis of the American Society of Brewing Chemists. Am. Soc. Brew. Chem. St. Paul, MN.

ASTM. 1988. Standard specification for denatured fuel ethanol for blending with gasolines for use as automotive spark-ignition engine fuel. Designation: D 4806-88. Book of ASTM Standards. ASTM, Philadelphia, PA.

Aylott, R.I., G.C. Cochrane, M.J. Leonard, L.S. Macdonald, W.M. Mackenzie, A.S. McNeish and D.A.Walker. 1990. Ethyl carbamate formation in grain based spirits. Part 1: post-distillation ethyl carbamate formation in maturing grain whisky. J. Inst. Brew. 96:213-221.

Battaglia, R., H.B.S. Conacher and B.D. Page. 1990. Ethyl carbamate (urethane) in alcoholic beverages and foods: A review. Food Additives and Contaminants 7(4):477-496.

Canas, B.J., D.C. Havery and F.L. Joe, Jr. 1988. Rapid gas chromatographic method for determining ethyl carbamate in alcoholic beverages with thermal energy analyzer detection. J. Assoc. Off. Anal. Chem. 71(3):509-511.

Canas, B.J., D.C. Havery, L.R. Robinson, M.P. Sullivan, F.L. Joe, Jr. and G.W. Diachenko. 1989. Ethyl carbamate levels in selected fermented foods and beverages. J. Assoc. Off. Anal. Chem. 72(6):873-876.

Clegg, B.S., R. Frank, B.D. Ripley, N.D. Chapman, H.E. Braun, M. Sobolov and S.A. Wright. 1988. Contamination of alcoholic products by trace quantities of ethyl carbamate (urethane). Bull. Environ. Contam. Toxicol. 41:832-837.

Food Chemical News. 1990. Urethane goals met, but aging effects cause uncertainty. CRC Press Inc. Washington, DC. March 26, 40-43.

Harrigan, W.F. and M.E. McCance. 1976. Laboratory Methods in Food and Dairy Microbiology. Academic Press, London, UK.

Joe, F.L., D.A. Kline, E.M. Miletta, J.A.G. Roach, E.L. Roseboro and T. Fazio. 1976. Determination of urethane in wines by gas-liquid chromatography and its confirmation by mass sprectrometry. J. Assoc. Off. Anal. Chem. 60(3):509-516.

Jones, J.M. 1992. Food Safety. Eagan Press, St. Paul, MN.

Kramer, A. and B.A. Twigg. 1966. Fundamentals of Quality Control for the Food Industry. The AVI Publishing Co. Inc., Westport, Conn.

Ough, C.S. 1976. Ethyl carbamate in fermented beverages and foods. I. Naturally occurring ethyl carbamate. J. Agric. Food. Chem. 24(2):323-328.

Rand, M.C., A.E. Greenberg and M.J. Taras. 1975. Standard methods for the examination of water and wastewater. American Public Health Association, Washington, DC.

Sobolov, M., D.M. Booth and R.G. Aldi. 1988. Whisky. In: Comprehensive Biotechnology. The Principles, Applications and Regulations of Biotechnology in Industry, Agriculture and Medicine (M. Moo-Young, ed) Pergamon Press, New York, NY.

Walker, G., W. Winterlin, H. Fonda and J. Seiber. 1974. Gas chromatographic analysis of urethane (ethyl carbamate) in wine. J. Agric. Food Chem. 22(6):944-947.

Wood, G.E. 1992. Mycotoxins in foods and feeds in the United States. J. Anim. Sci. 70:3941-3949.

Wright, S.A., and B.S. Clegg. 1987. A simplified gas chromatographic method for analysis of ethyl carbamate in distilled beverages. Poster presented at the 12th Annual Assoc. Off. Anal. Chem. Spring Training Workshop, Ottawa, Ont., Canada.

Chapter 21

Bacterial contaminants and their effects on alcohol production

C. Connolly
Alltech Inc., Nicholasville, KY, USA

Introduction

Bacterial contamination of alcohol production systems usually brings about formation of compounds such as organic acids that inhibit yeast function and/or metabolites that impart undesirable properties to the finished product. The "prevention is better than cure" approach is critical; and emphasis must be placed on maintaining conditions that minimize contamination. For this to be effective, it is necessary to recognize potential sources of contamination, know the most commonly encountered contaminants and understand the types of problems that result. Armed with this information, suitable measures may be put in place to minimize the likelihood of serious losses.

Sources of contamination

Bacterial contamination can arise from various points in an alcohol production process. Raw materials can act as a constant supply of contamination without regular and effective methods of control. Inefficient cleaning procedures allow residues that harbor bacteria to remain in tanks, pipelines and heat exchangers. Clean-in-place (CIP) systems using hot caustic detergents followed by treatment of vessels and pipelines with sanitizing agents are popular and widely used means of maintaining a clean and efficient plant. Such programs remove bacterial breeding grounds and reduce or eliminate the existing population of contaminants. Methods that measure the effectiveness of cleaning procedures have been described (Hammond, 1996). Pitching yeast, especially wet or caked yeast stored for extended periods or at temperatures above refrigeration, can act as a source of contamination. The addition of recycled thin stillage (backset), a common practice in fuel alcohol plant fermentations, can re-introduce viable bacteria. If the backset is returned prior to the jet cooker, contamination is less likely. The high temperatures employed during the cooking stage essentially sterilize the mash; but care must be taken to ensure that contamination is not re-introduced. Saccharification tanks and continuous yeast propagation systems can act as reservoirs for bacteria because the conditions are very well suited for microbial growth. If bacteria find their way into these systems and are not quickly eliminated, a constant inoculum of actively growing bacteria will be added to the fermenter along with the yeast, causing damaging effects during fermentation. Checks should be made at critical points to identify potential problems in the early stages and allow steps to prevent major losses. This is a particularly important concern in continuous fermentation systems where the fermenters are linked in series. One contaminated fermenter early in the process has the potential to contaminate all the others.

Commonly encountered bacterial contaminants

Bacterial contaminants of alcohol production systems include both Gram-positive and Gram-negative species. This classification system is based on response to a staining procedure developed by a Danish bacteriologist in 1884. Cells that possess a thick layer of a net-like structure called peptidoglycan on the exterior retain a purple dye and are termed Gram-positive (Figure 1). Bacteria with a thin layer of peptidoglycan between the outer membrane and the cell wall (gram-negative) are unable to retain the dye during washing with ethanol/acetone, and after exposure to a red counter stain appear pink when viewed under a microscope.

The Gram-positive lactic acid bacteria (LAB), because of their rapid growth rate and tolerance to high temperatures and low pH conditions, are the most troublesome group of bacterial contaminants found in breweries, distilleries and fuel alcohol plants (Chin and Ingledew, 1994). This group of bacteria, particularly species of the genera Lactobacillus and Pediococcus (mostly in brewing), are present in virtually every alcohol plant and can cause a variety of problems. Members of the genus Leuconostoc are commonly found in rum production processes (Murtagh, 1995) where molasses is used as the feedstock. In the manufacture of beverage alcohol, lactic acid bacteria can cause spoilage by producing off-flavors in the finished product. Examples are acetoin and diacetyl (undesirable above 0.15 ppm, Hough *et al.*, 1982), which will be discussed in more detail. The problems associated with bacterial contamination of fuel alcohol production processes are generally concerned with losses in ethanol yield. Narendranath *et al.* (1997) showed that the presence of lactobacilli at levels of 10^6 CFU/ml resulted in an approximate 2% reduction in ethanol yield from normal gravity wheat mashes.

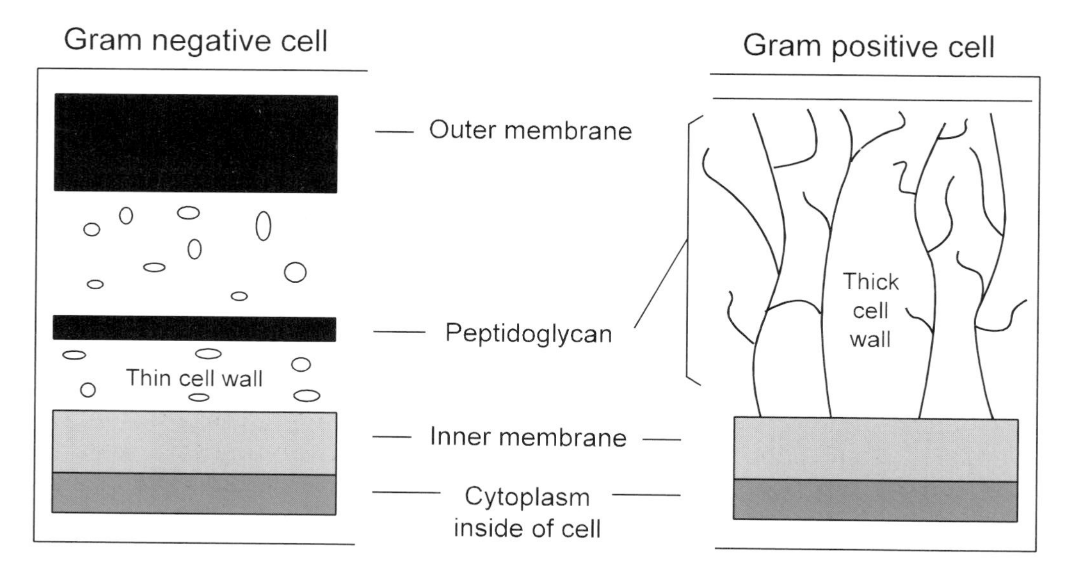

Figure 1. A schematic diagram of the outer structure of a bacterial cell.

Heavier inocula (10^9 CFU/ml) had a more pronounced effect, with reductions of 3.8–7.6% depending on the strain. Such reductions in ethanol yield and consequent revenue loss cannot be tolerated on an industrial scale. The negative effects were attributed to the diversion of carbohydrates for bacterial growth, competition with yeast for growth factors in the medium and the production of lactic acid. It was shown for the first time in this study that a direct relationship exists between the observed decrease in ethanol yield and the initial numbers of viable bacteria in the mash. In an earlier experiment, Makanjuola *et al.* (1992) reported a 17% loss in alcohol yield when lactic acid bacteria were present at 4.5 x 10^8 CFU/ml. They showed that acetic acid (a product of heterolactic fermentation) and lactic acid, when present at 0.5-9.0 g/L and 10-40 g/L, respectively, inhibited growth of *Saccharomyces cerevisiae*. It should be mentioned that ethanol is not the sole product of the alcohol fermentation process; but its yield and quality are usually most affected by the presence of contaminants. Co-products (e.g., distillers dried grains and carbon dioxide) are discussed elsewhere.

Bacterial cells are typically 10-20 times smaller than yeast cells and are easily distinguished microscopically. Another important difference between bacteria and yeast is the amount of time needed for cellular reproduction. Brewing and distilling yeast (*S. cerevisiae*) reproduce by forming a bud on the surface of the cell that eventually detaches, yielding a mother and a daughter cell. In contrast, bacteria reproduce by binary fission, a process whereby the cell extends longitudinally and a wall forms in the middle before the two daughter cells separate. Under optimal conditions, some bacteria reproduce in as little as 12-30 minutes. Even when conditions during fermentation are favorable, the same process can take up to 4 hrs for yeast. For this reason it is very important to prevent bacterial contaminants from entering the process, especially early in fermentation when yeast do not yet predominate and bacteria can proliferate without competition.

Lactic acid bacteria

This review will focus on the three genera most important in brewing and distilling, Lactobacillus, Leuconostoc and Pediococcus. Other contaminants include species of Lactococcus, whose importance lies within the confines of dairy processes; and bifidobacteria, which along with streptococci represent part of the normal microflora in the digestive tracts of humans and animals.

Lactic acid bacteria are Gram-positive, anaerobic, microaerophilic or aerotolerant, catalase negative rod or coccus-shaped cells that produce lactic acid as a major end product of carbohydrate metabolism. They are acid tolerant organisms and have complex nutritional requirements (Kandler and Weiss, 1986). Simple tests exist to differentiate the genera. For example, lactobacilli are rod-shaped cells while pediococci and Leuconostoc are round. Microscopic examination and study of fermentation profiles provide further differentiation. Complete identification of individual contaminants can prove difficult since there are many similarities among the species of each genus, especially in the case of Lactobacillus.

Metabolic pathways

The pathways for carbohydrate metabolism in lactic acid bacteria are strictly fermentative and the group can be roughly divided into three classes: obligately homofermentative, facultatively heterofermentative and obligately heterofermentative (Table 1).

In obligately homofermentative species lactic acid is the major end product of hexose (6-carbon sugar, e.g., glucose) metabolism, with two molecules formed from each molecule of sugar consumed. The glycolytic (Embden-Meyerhof-Parnas) pathway is used almost exclusively. Species of this class possess the enzyme aldolase but lack phosphoketolase, resulting in inability to ferment pentoses (5-carbon sugars) and gluconate.

Table 1. The three classes of lactic acid bacteria.

Type	*Examples*
Obligately homofermentative	*L. delbrueckii*, *L. acidophilus*, *P. damnosus*
Facultatively heterofermentative	*L. plantarum*, *L. casei*, *L. pentosus*, *L. sake*
Obligately heterofermentative	*L. brevis*, *L. buchneri*, *L. fermentum*

Facultatively heterofermentative species use glycolysis almost exclusively to produce lactic acid when hexoses are plentiful. These organisms possess both aldolase and an inducible phosphoketolase, allowing fermentation of pentoses and gluconate under certain conditions. The phosphoketolase pathway is repressed in the presence of glucose.

Obligately heterofermentative lactic acid bacteria produce a mixture of end products under normal conditions by using the phosphogluconate or phosphoketolase pathway to ferment all sugars. The end products include lactic acid, acetic acid, ethanol and carbon dioxide along with smaller amounts of formic and succinic acids. These organisms, except for a few strains (Kandler, 1983), can ferment pentoses and gluconate.

Most lactic acid bacteria, like yeast, are unable to ferment polysaccharides. An exception is *L. amylophilus*, which can degrade starch. Most are only able to metabolize smaller sugars such as glucose, fructose, maltose and sucrose. Figure 2 outlines the main pathways through which hexose sugars are catabolized and the various end products generated. In homofermentative species, the aldolase enzyme catalyzes breakdown of fructose 1,6-bisphosphate into two molecules of a 3-carbon compound (glyceraldehyde) that is further degraded to produce pyruvic acid and eventually lactic acid. The yield of lactic acid is stoichiometrically related (2:1) to the amount of hexose consumed. This pathway is common in Pediococcus, Streptococcus and some Lactobacillus species. The heterofermentative pathway is also found in some lactobacilli and in species of the genus Leuconostoc. The main difference between the homofermentative and heterofermentative pathways is the presence of a phosphoketolase enzyme in the latter, which is capable of converting xylulose-5-phosphate into glyceraldehyde-3-phosphate and the 2-carbon acetyl phosphate. As in homolactic fermentation, the triose phosphate moiety yields lactate via the formation of pyruvate. Depending on the oxidizing potential of the environment, the acetyl phosphate is converted into either ethanol or acetic acid. Ethanol is only formed in the absence of a suitable hydrogen acceptor; and so the commonly observed end products formed by the heterolactic fermentation of glucose under anaerobic conditions are equimolar amounts of lactic acid, acetic acid and carbon dioxide.

Lactic acid bacteria possessing a functional phosphoketolase pathway are capable of fermenting pentose sugars and some of their alcohols. Figure 3 illustrates how these sugars are incorporated into the pathway by conversion to xylulose-5-phosphate, which undergoes further breakdown (as outlined earlier). Obligately homofermentative lactic acid bacteria are unable to ferment such substrates due to the lack of a phosphoketolase enzyme, and hence are unable to form intermediates for further metabolism. Although employed to a lesser extent, some lactic acid bacteria use pathways for the conversion of pyruvate into other products. An example is diacetyl, which due to its characteristic butter-scotch flavor is a highly undesirable compound in finished products destined for human consumption.

Effects of lactic acid bacteria on yeast fermentations

Lactic acid bacteria affect fermentation mostly through production and excretion of compounds inhibitory to yeast. Scavenging of essential nutrients is also a problem; but the accumulation of organic acids over time presents a more serious threat. Lactic and acetic acids are toxic

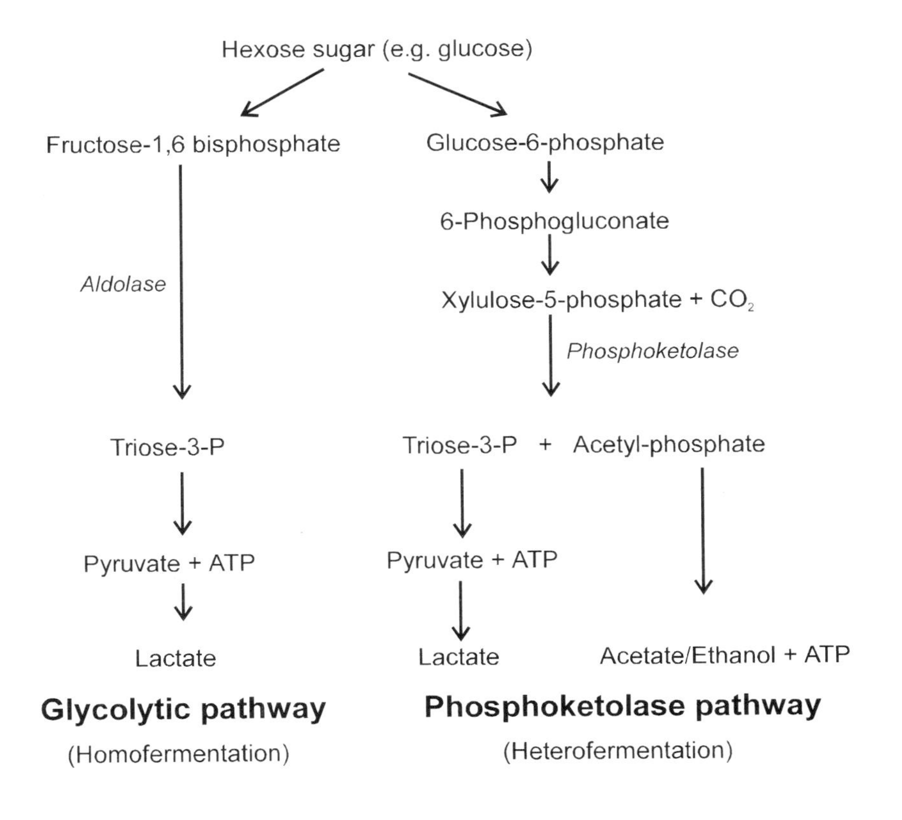

Figure 2. The main pathways of hexose fermentation in lactic acid bacteria (modified from Kandler, 1983).

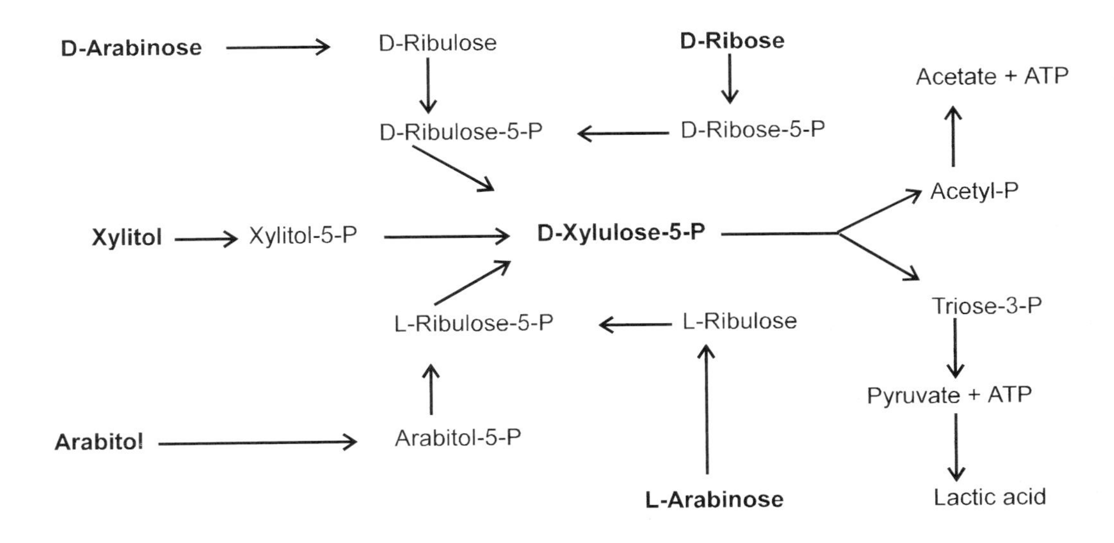

Figure 3. Pentose and pentitol fermentation in lactic acid bacteria (modified from Kandler *et al.*, 1983).

to yeast at 1-4% (w/v) and 0.05-0.9% (w/v), respectively (Makanjoula *et al.*, 1992). Since it is the undissociated form of an organic acid that is antimicrobial (Baird-Parker, 1980) and the pKa of acetic acid is higher than that of lactic acid, acetic acid is dissociated to a lesser extent in the fermentation mash and hence exhibits greater toxic effects. Acetic acid formed during heterolactic fermentation is not the sole cause of yeast inhibition (Huang *et al.*, 1996); but acetic acid and lactic acid act synergistically to inhibit yeast growth (Moon, 1983). Chin and Ingledew (1994) showed that lactic acid accumulation over five successive fermentations, each prepared with 50% laboratory backset, did not affect the yeast's ability to ferment sugars even when the concentration of lactic acid reached 14 g/L. It did, however, cause up to a 60% loss in yeast viability; and it was noted that this would present serious problems in continuous fermentation systems or if yeast were to be harvested for re-use. *S. cerevisiae* (the yeast normally used in alcohol production) also produces organic acids during the course of fermentation but in much smaller amounts and at sub-inhibitory concentrations.

Nutritional requirements of lactic acid bacteria

The nutritional needs of lactic acid bacteria are very similar to those of yeast, making them good competitors for essential nutrients and growth factors that may be present in limited quantities in fermentation mashes.

Nitrogen

Various amino acids and vitamins are necessary for growth of most lactic acid bacteria (Kandler and Weiss, 1986). Some homofermentative and heterofermentative species have protease enzymes that degrade proteins into useable forms of nitrogen; however most lactic acid bacteria are only weakly proteolytic (Priest, 1996) and rely on readily useable amino acids that can be directly incorporated into proteins or catabolized for energy. In mashes where available nitrogen is deficient, there can be strong competition between bacteria and yeast since both have absolute requirements for this essential nutrient. In many cases it is necessary to supplement the fermentation mash with assimilable nitrogen. The importance of nitrogen in relieving stuck fermentations has been extensively studied (Ingledew *et al.*, 1986; Thomas and Ingledew, 1990; O'Connor-Cox *et al.*, 1991; O'Connor-Cox and Ingledew, 1991; Jones and Ingledew, 1994).

Oxygen

Lactic acid bacteria are generally described as aerotolerant anaerobes. This means that they prefer oxygen-deficient environments (microaerophilic). They can also survive and grow where oxygen is completely absent, although usually at a slower rate than under aerobic conditions. The enzymes catalase and superoxide dismutase are usually absent. These are used by aerobic bacteria to dispose of toxic hydrogen peroxide and superoxide, respectively, both of which form during respiratory processes. However, *L. plantarum* and some species of Pediococcus can synthesize a catalase enzyme under certain conditions (Johnston and Delwiche, 1965) to degrade hydrogen peroxide. Iron and manganese ions are important for catalase activity and to neutralize the superoxide radical (O^-). Condon (1987) showed that some species contain NADH oxidases that are induced by the presence of oxygen. They convey aerotolerance to the organism by replenishing the supply of nicotinamide adenine dinucleotide (NAD) so that metabolic processes can continue. *S. cerevisiae* produces catalase and grows well under aerobic conditions, with oxygen being an absolute requirement for yeast growth in the early stages of fermentation. Oxygen is also needed for the synthesis of membrane sterols and unsaturated fatty acids (Ingledew, 1995). This represents another competitive advantage for lactic acid bacteria under the anaerobic conditions of yeast fermentation. It is important that every effort be made to make the conditions as suitable as possible for yeast, while at the same time removing competition in the form of bacterial contaminants for nutrients and other factors.

Metabolic by-products that affect flavor and product quality

The production of some compounds by contaminating bacteria can contribute to off-flavors in the final product; an effect referred to as sarcina (the old generic name assigned to all Gram-positive brewery contaminants). Knowing how off-flavors are formed and how to limit their production results in improved product quality.

A wide variety of substances can impart negative effects on flavor and taste profiles of beverage spirits. Acrolein (responsible for 'pepper gas'), commonly found in distilled alcohol products, is a water soluble, volatile liquid with a disagreeable odor. In an interesting study by Sobolov and Smiley (1959) it was concluded that acrolein is not formed directly from glycerol metabolism by contaminating bacteria. Instead, it may be formed chemically in the distillation process through the acid-catalyzed dehydration of ß-hydroxypropionaldehyde, an intermediate of glycerol metabolism (Figure 4). The tendency of acrolein to react with other congeners present in spirits to form stable complexes raises questions as to whether this compound itself is responsible for the negative effects on product flavor (Morrison, 1995).

Diacetyl, a ketone that smells like butterscotch, can develop during fermentation when pyruvate formed by glycolysis is converted into products other than the predominant lactic acid (Figure 5). It is closely related to 2,3-pentanedione, another important contributor to the flavor profile of the product. These compounds make up what is commonly referred to as the vicinal diketone content of beer. The flavor threshold of 2,3-pentanedione is 1 ppm (Fix, 1993) while that of diacetyl is about 10 times lower. Species of pediococci and some lactobacilli are the main lactic acid bacteria contaminants responsible for diacetyl production. Yeast produce diacetyl during the early stages of fermentation but it is removed by the action of various enzymes (Figure 6). It is for this reason that a 'resting period' is allowed at the end of beer fermentations.

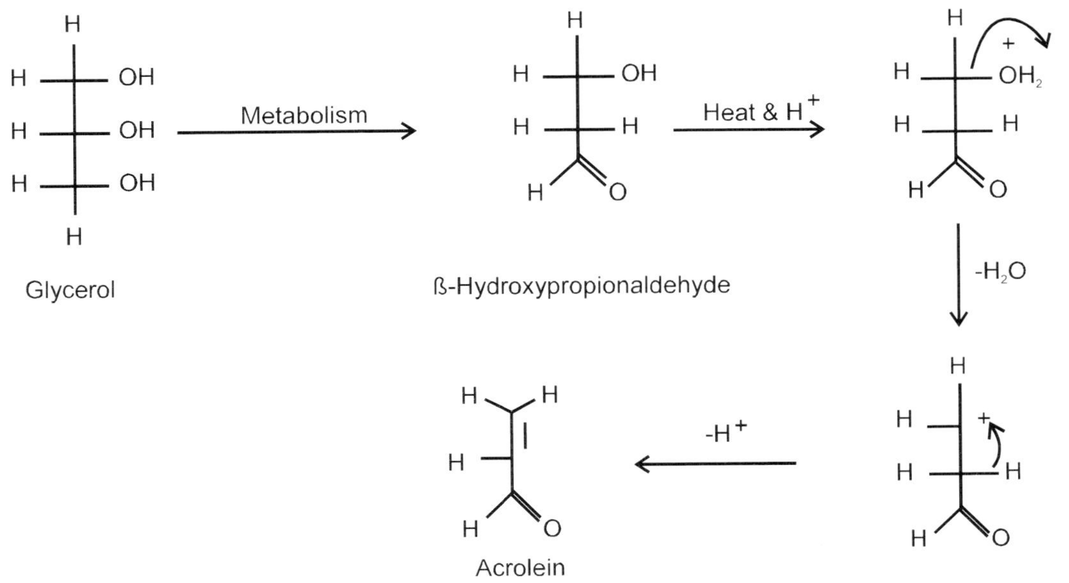

Figure 4. A possible reaction mechanism for the production of acrolein from glycerol (based on Sobolov and Smiley, 1959).

It is interesting to note the relationship between amino acids in the fermentation mash and the production of diacetyl and related compounds. Valine normally reacts with acetolactate to form various products thus removing some of the precursor for diacetyl production. Hence if valine is deficient in the mash, elevated levels of diacetyl are observed due to the availability of acetolactate (Fix, 1989). A similar correlation is found between final levels of 2,3-pentanedione where the precursor in this case is acetohydroxybutyrate and the essential amino acid is leucine.

Gram-negative bacterial contaminants

While Gram-positive lactic acid bacteria comprise the most important single group of bacterial contaminants, certain Gram-negative bacteria are also important in the context of brewing and distilling. The more commonly encountered of these are acetic acid bacteria, *Zymomonas mobilis*, and Enterobacteriaceae.

Acetic acid bacteria

The low threshold of yeast for acetic acid means that the lowest possible concentrations (<0.05% w/v) must be maintained in propagators and fermenters. The genera Acetobacter and Gluconobacter (previously Acetomonas) make up this group of 'vinegar bacteria' that produce acetic acid as the dominant end product of metabolism (Figure 7). Like lactobacilli, they are rod-shaped cells and are tolerant of acidic conditions. A very important difference, however, is the absolute requirement of acetic acid bacteria for oxygen. Members of the group are obligate aerobes, which means they are unable to survive and grow unless oxygen is present. This eliminates the potential to contaminate anaerobic fermentations. Their role as spoilage organisms is limited to aerobic processes in the plant although the possibility exists for development of at least some tolerance to carbon dioxide (Harper, 1980).

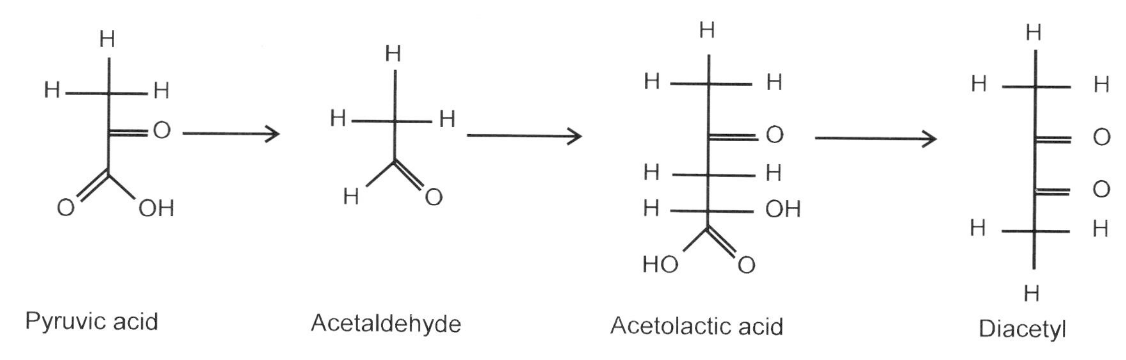

Figure 5. Pathway of diacetyl production from pyruvic acid.

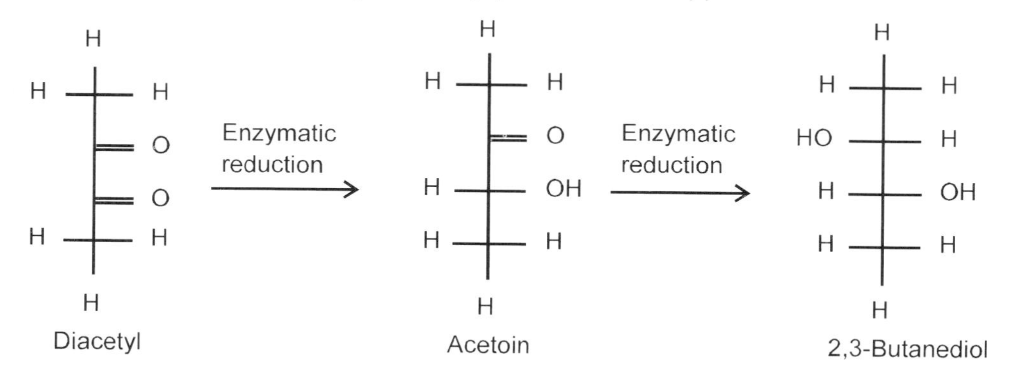

Figure 6. Removal of diacetyl by enzymatic reduction pathways in yeast.

The two most likely places to find acetic acid bacteria are the yeast propagation tank and the beer well. In the propagator, high agitation and aeration promote yeast growth. This also creates conditions ideal for the propagation of acetic acid bacteria; and in continuous propagation systems the problem may escalate over time. The medium in the propagation tank is not truly aerobic, especially when fermentation mash is used, because of its high dissolved solids and low dissolved oxygen content. However, acetic acid bacteria can grow in the headspace where agitation splashes contents around the tank thereby creating a suitable growth environment. In addition, ethanol produced by yeast under anaerobic conditions is an excellent substrate for these bacteria (particularly Acetobacter). Once the inoculum is pumped to the fermenter and an anaerobic system is established, acetic acid bacteria will usually die; but sufficient acetic acid to slow or inhibit yeast function may already be present. These bacteria are easier to control in batch propagation systems where the tank is emptied, cleaned and sanitized between runs. The addition of a microbiologically clean yeast inoculum to the propagator is also important; and routine checks should be performed to ensure that bacterial numbers are below specified levels.

Metabolism of acetic acid bacteria

Acetic acid bacteria use respiratory processes for the production of energy and can directly oxidize sugars and alcohols for this purpose. This can cause problems in the beer well where large amounts of alcohol are held for some time. If sufficient air (oxygen) is present in the beer and headspace, these bacteria, especially Acetobacter, will grow quite happily and convert ethanol into acetic acid and other products. While the negative effects of the acid on yeast are less important at this stage unless the yeast is to be recycled, there can still be substantial losses in alcohol yield.

Gluconobacter species prefer sugary environments such as those found in propagation tanks where there can be an abundant supply of glucose. With ethanol also present at this stage, both types of bacteria can cause serious losses in ethanol yield.

Acetobacter and Gluconobacter species are very similar morphologically with nearly identical cell shape, Gram stain reaction and motility characteristics. For this reason it is necessary to differentiate them using biochemical methods (Table 2). An important difference between Acetobacter and Gluconobacter species is the

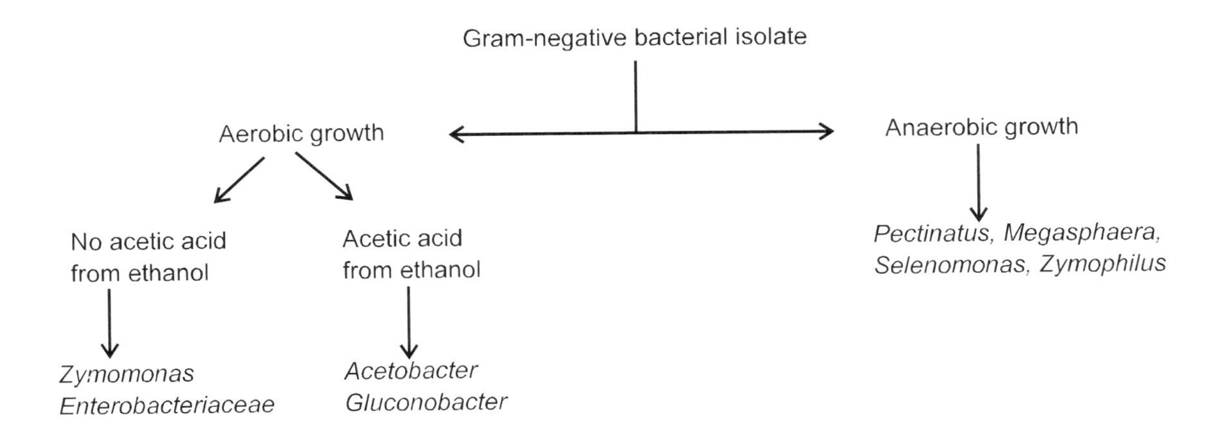

Figure 7. A diagnostic scheme for identification of Gram-negative bacterial contaminants (modified from Van Vuuren, 1996).

ability of the former to oxidize acetate completely to form carbon dioxide and water. This characteristic is often used to differentiate the two genera. Kulka *et al.* (1949) suggested that the ability of Acetobacter species to form giant colonies on wort agar could also be used as a means of differentiation.

Table 2. Some similarities and differences between genera of acetic acid bacteria.

Acetobacter	*Gluconobacter*
Gram-negative rod shaped cells	Gram-negative rod shaped cells
Carbohydrate metab-olism yields primarly acetic acid	Carbohydrate metab-olism yields primarly acetic acid
Peritrichous flagella if motile	Polar flagella if motile
Prefer alcohol substrates	Prefer sugar substrates
Oxidize acetate to CO_2 and H_2O	Do not oxidize acetate
HMP and TCA functional	HMP functional
Simple growth requirements	Complex growth requirements

Carbohydrate metabolism

Species of Acetobacter metabolize glucose in various ways. The hexose monophosphate pathway (HMP) and tricarboxylic acid cycle (TCA) predominate (King *et al.*, 1956), and lead to acetic acid as the main end product. Favorite sources of energy include ethanol, glycerol and lactate; and growth factor requirements depend on the carbon sources available (Rao and Stokes, 1953). The HMP and the Entner-Duodoroff pathway are used by Gluconobacter, but there is no functional TCA cycle due to the absence of essential enzymes such as isocitrate dehydrogenase. Glycolysis is very weak if present at all due to the lack of phosphofructokinase, an essential enzyme in the pathway (De Ley *et al.*, 1986). Carbon source preferences in decreasing order are mannitol, sorbitol, glycerol, fructose and glucose. The gluconate formed from glucose is not further metabolized, therefore acetic acid (formed from the HMP) is the main end product.

Nitrogen requirements

Acetobacter species are able to process simple nutrients to form all of the nitrogenous compounds needed for growth (De Ley *et al.*, 1986). For example, glutamate is formed from α-ketoglutarate (an intermediate in the TCA cycle) and ammonia. An enzyme-catalyzed transamination of glutamic acid with oxaloacetate yields aspartate, which in turn can be converted into alanine by ß-decarboxylation (Hough *et al.*, 1982). Since Gluconobacter species do not use the TCA cycle, they require a more complex range of nutrients for growth including the amino acids themselves or the oxo acids (carbon skeleton) from which they can be synthesized (Figure 8). In both cases, single amino acids cannot be used as sole sources of nitrogen and carbon (De Ley *et al.*, 1986) and there are no essential amino acids required by these bacteria.

Zymomonas mobilis

This contaminant is commonly found where molasses is used as a feedstock such as in rum production. It is facultatively anaerobic, fermenting sugars to produce ethanol and carbon dioxide (Figure 9). Some additional end products, including lactate and acetate, are also formed. *Z. mobilis* spoils beer mainly through the production of off-flavors like hydrogen sulfide, acetaldehyde and dimethyl sulfide (DMS). DMS actually imparts a pleasant character to beer when present at low levels (30 ppm), but gives an undesirable blackcurrant flavor above a threshold concentration of 80 ppm (Bamforth, 1997). Some strains of *Z. mobilis* are tolerant to levels of ethanol in the range of 10-13% by volume (Swings and De Ley, 1977) and their potential for use in industrial alcohol production has been the subject of many investigations

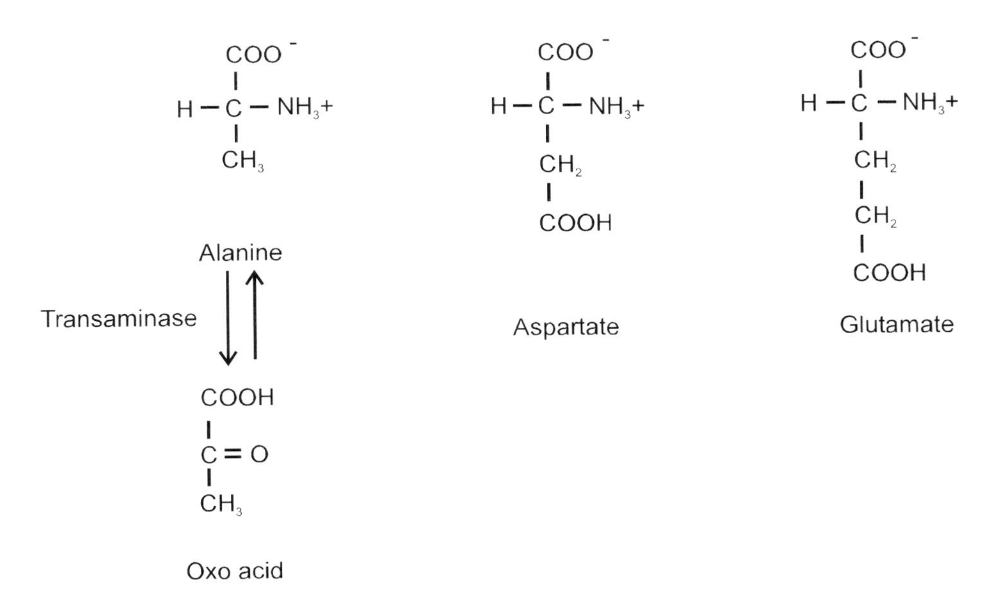

Figure 8. Examples of amino acid and oxo acid structure.

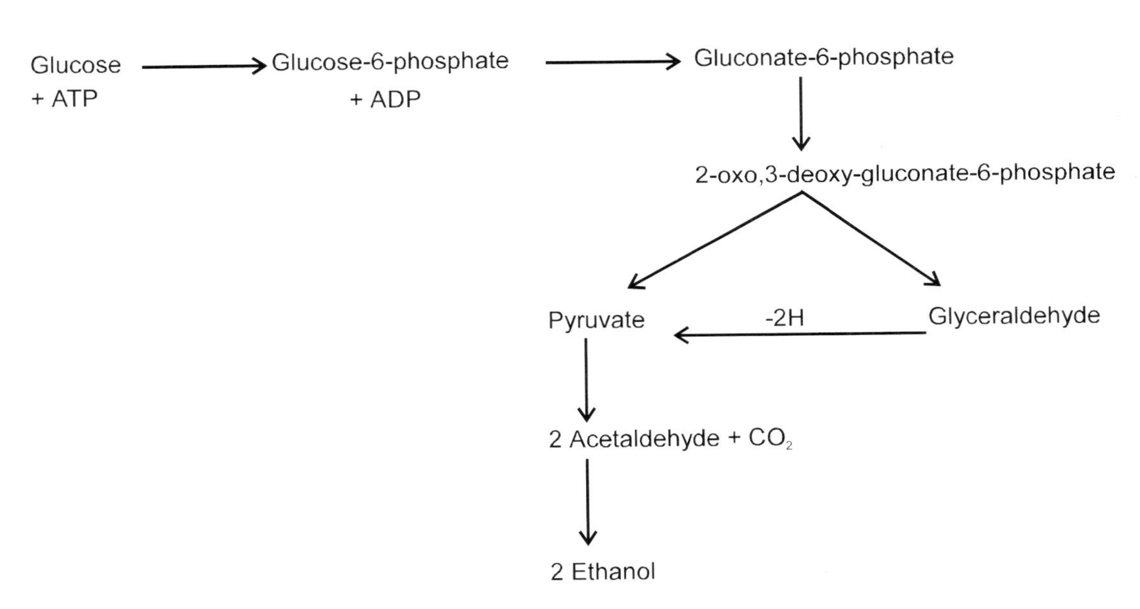

Figure 9. The Entner-Duodoroff pathway in Zymomonas (modified from Hough *et al.*, 1982).

(Amin and Allah, 1992; Gold *et al.*, 1992; Yanase *et al.*, 1995; Abate *et al.*, 1996; Gold, 1996).

Enterobacteriaceae

This group consists of many different genera of bacteria (e.g., Enterobacter, Citrobacter, Klebsiella, and Obesumbacterium), with only some having the potential to cause problems in alcohol production processes. Enterobacteriaceae are facultative anaerobes that produce a variety of end products from glucose through the use of different pathways, hence their association with the term 'mixed acid bacteria' (Ingledew, 1995; Figure 10). Lactic, acetic and succinic acids are commonly formed along with acetoin, 2,3-butanediol and ethanol. The relative amounts formed depend on the strain and the growth conditions (Van Vuuren, 1996; Hough *et al.*, 1982). They tend to be inhibited by low pH values and, with the exception of *O. proteus,* ethanol levels above 2% v/v (Hough *et al.*, 1982). Beer spoilage is mainly through production of off-flavors.

Isolation and identification of bacterial contaminants

Methods used for isolating bacterial contaminants involve inoculation on some kind of medium that will allow cells to grow and produce visible colonies. These media contain sources of nutrients, energy and minerals essential for the growth of the various microorganisms. Ingredients are usually added to inhibit other organisms that may be present or to promote the growth of potential isolates. For example, wort, beer and tomato juice tend to encourage the growth of lactobacilli during initial isolation; and cycloheximide (actidione) is added to inhibit yeast growth during bacterial analysis of fermentation samples. There are many publications dealing with this area and these should be consulted for a greater understanding of the principles and reasons underlying the use of different media. Reuter (1985) and Holzapfel (1992) discuss the use of culture media for the isolation of lactic acid bacteria. An in-depth review by Smith *et al.* (1987) explains why some media are considered better than others in brewing situations and provides recommendations together with formulations in most cases.

A wide variety of culture media is available and most of these are successful when used for their intended application (Table 3). Most media used for the isolation of the same kinds of contaminants are variations of similar formulations; and so the long list of potentially useful media can be shortened considerably. A major problem with isolating contaminants in this way is the need for the organism to grow outside its normal habitat before its presence can be detected. It is very difficult to create the necessary conditions in the laboratory that will allow for growth of all contaminants. There is no single medium suitable for enumeration of the total flora of fermenting mash, beer or brewing ingredients (Smith *et al.*, 1987). Therefore different media are often used when testing a single sample in order to increase the chances of recovery.

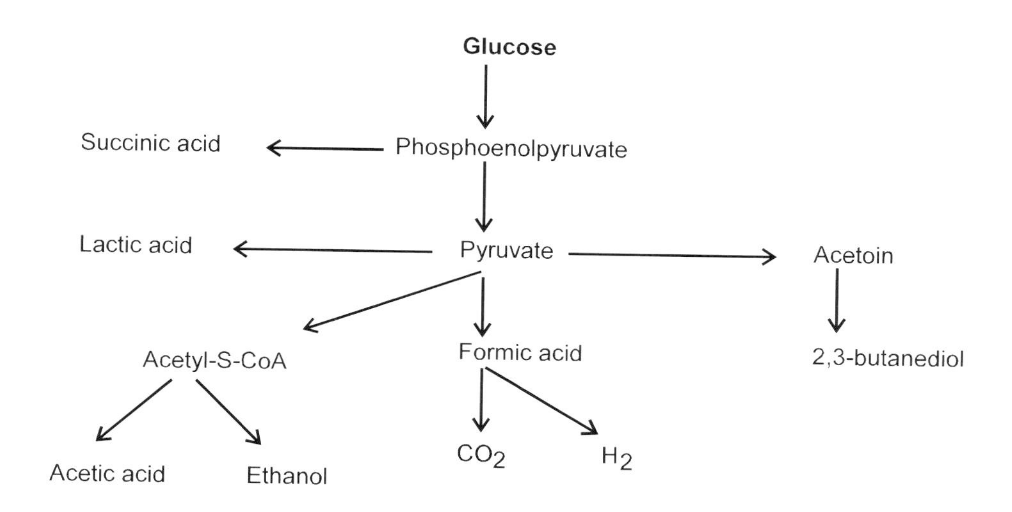

Figure 10. An outline of glucose metabolism by mixed acid bacteria (modified from Van Vuuren, 1996).

Incubation conditions can also play an important part in detecting contaminants. Lactic acid bacteria prefer anaerobic or at least CO_2 enriched atmospheres for growth during initial isolation. This can have the added effect of preventing growth of obligate aerobes, thus increasing selectivity. Conversely, media used for isolating contaminants of aerobic processes

Table 3. Some of the microbiological media used to isolate and enumerate bacterial contaminants found in alcohol production processes.

Medium	*Comments*
Raka Ray No. 3 agar	For isolation of lactic acid bacteria. Contains glucose, mannose and fructose for optimum recovery. Can be made selective by the addition of cycloheximide and 2-phenylethanol to inhibit yeast and Gram-negative bacteria, respectively.
Universal beer medium	Contains beer, tomato juice and peptonized milk to encourage growth of lactic acid bacteria.
Modified MRS agar	MRS with added maltose to encourage growth of lactic acid bacteria unable to ferment glucose.
VLB-S7 agar (Versuchs-und Lehranstalt für Brauerei)	Contains wort and maltose. Good for the selective isolation of lactic acid bacteria but needs extra incubation time because of slower growth rates.
Sucrose agar	General purpose medium that can be made selective for lactic acid bacteria by adding actidione, polymixin and 2-phenylethanol.
Carr's medium	Contains bromocresol green for the selective isolation of acetic acid bacteria. Distinguishes between Acetobacter and Gluconobacter species by ability to oxidize acetate.
Dextrose sorbitol mannitol agar	Contains cycloheximide to inhibit yeast and sodium desoxycholate to inhibit Gram-positive bacteria. For the selective isolation of acetic acid bacteria.
Acetic acid differential medium	Suppresses most brewery organisms except acetic acid producers. Contains ethanol, cycloheximide, and an indicator dye.
MYGP (malt extract, yeast extract, glucose and peptone)	General purpose medium useful for the cultivation of many brewery microorganisms (especially yeasts).
WLN (Wallerstein Laboratory Nutrient) agar	General non-selective medium containing bromocresol green indicator, which allows for differentiation based on colony appearance.
Lee's multidifferential medium (LMD)	General non-selective medium. Organisms that produce acid are detected by a color change of the indicator. May be made selective by addition of inhibitory agents.
MacConkey agar	Contains bile salts and neutral red indicator. Selective for enterobacteria.
Violet red bile	For selective isolation of enteric bacteria. Similar to MacConkey agar.
Sulfite agar	Contains sodium sulfite and lead acetate for the selective isolation of *Zymomonas* species.

are incubated under aerobic conditions. In most cases, the incubation temperature is 27-30°C, the optimum temperature required for growth of most contaminants. This temperature is also likely to be similar to temperatures in the process itself (yeast also likes this temperature range); and it is important as mentioned earlier to mimic the natural conditions as much as possible.

Practically all of the non-sporulating, Gram-positive, catalase-negative rod-shaped bacteria isolated from alcohol production plants belong to the genus Lactobacillus (Priest, 1996). There are a number of methods if further identification is required. These include serological tests (Nishikawa *et al.*, 1979) and the use of biochemical tests such the API 50 CHL diagnostic kit (bioMerieux). Most of the methods focus on identification of clinically important species (Gutteridge and Priest, 1996); but sometimes these methods or variations of them can be adapted for use elsewhere. Immunoassay analysis (Whiting *et al.*, 1992), conductance techniques (Kyriakides and Tiiurston, 1989) and protein fingerprinting (Dowhanick *et al.*, 1990) have all been evaluated and used for identification of bacteria in a brewing context. Efforts to develop new methods for fast reliable identification of bacteria are ongoing.

Early detection requires tests that can quickly and accurately identify microorganisms. A method using infrared spectroscopy, although very quick and easy to perform, was found to be unsuccessful due to lack of reproducibility (Lipkus *et al.*, 1990). Accuracy cannot be substituted for speed; and one without the other is useless. In complex systems such as fermentation processes where there are many types of microorganisms, methods are required that can differentiate among even closely related species. Applications of molecular biological techniques such as ribosomal RNA typing have resulted in more reliable methods than were previously available for the identification of all types of bacteria. Using these techniques it is possible to accurately detect to the species level very low levels of contaminating bacteria, sometimes as small as 1% of the total microbial population (Muyzer *et al.*, 1993). This represents a major development in the area of bacterial isolation and identification because it removes the need to culture the organism outside its natural habitat and therefore increases the chances for successfully identifying species that may otherwise go unnoticed. This methodology has been applied extensively in the food and beverage industries (Back *et al.*, 1996; Tilsala and Alatossava, 1997; Doyle *et al.*, 1995). In fuel alcohol plants, there tends to be a smaller variety of bacterial contaminants when compared to breweries because of the different processing conditions and the relatively fewer concerns with storage of the finished product. Usually, identification of the genus (e.g. Lactobacillus) is enough to know what type of action should be taken.

Treatment of bacterial contamination

Despite cleaning and sanitization procedures, bacteria may find their way into a process and will need to be removed as quickly as possible. In batch fermentation systems, the likelihood of a contamination problem (and difficulty of its subsequent removal) is greatly reduced compared to continuous processes. Emptying and cleaning the fermenter between runs removes residual material that could be a source of contamination for the fresh sterile mash. A fermenter in a continuous fermentation system cannot be emptied, cleaned and sanitized before filling without first isolating it from the rest of the process. This presents its own set of problems for the plant; but more importantly, the contamination is seldom confined to just one fermenter and it can take longer to get the whole process back on track. Methods of prevention are covered elsewhere and the discussion here will be confined to treatments commonly used to minimize the effects of contamination. These invariably involve the use of antimicrobial agents that selectively inhibit or kill bacteria while having little or no effect on yeast. For this reason antimicrobials can be added directly to the fermentation.

Antibiotics are compounds naturally produced by microorganisms for the purpose of inhibiting the growth of other microorganisms. In nature, this gives the producing organism an advantage for survival by removing competition for limited nutrients. 'Chemotherapeutic agent' is the term given to the synthetic analog of an antibiotic (Todar, 1996). Antibiotics have been used extensively in animal and human medicine for the treatment of bacterial infections (Brown, 1995). There are many different kinds of antibiotics; and they can be grouped according to their mechanism of action against bacteria.

Penicillin was the first antibiotic produced on an industrial scale, as a result of the efforts of Howard Florey, Ernst Chain and Norman Heatley during World War II (Quirke, 1998). A French medical student in 1896 had originally reported a substance produced by mold that could kill bacteria. In 1928 Alexander Fleming, the person most commonly associated with the discovery of penicillin, identified the producing organism as *Penicillium notatum*. Today, penicillin is produced by a related species, *Penicillium chrysogenum*, in a deep bed fermentation process (Brown, 1996). Most antibiotics have a narrow spectrum of activity, affecting only a specific group of bacteria, usually those found in biological systems similar to those of the producing organism. Penicillin belongs to the ß-lactam group of antibiotics and is effective against Gram-positive bacteria. It elicits its effects by preventing bacterial cell wall synthesis; and since the bacteria cannot reproduce without being able to produce new cells, they die. Because of the mode of action of penicillin, it has a more pronounced effect on actively growing popu-lations of bacteria and small concentrations are usually sufficient to inhibit growth. Typical dosage rates are generally unable to reduce the cell numbers of an existing population of bacteria because these have already formed cell walls and are less susceptible to attack. A penicillin preparation sold under the brand name Allpen, is commonly used in yeast fermentation processes to prevent the growth of lactic acid bacteria. It is usually effective at levels in the range of 0.5 to 2.0 ppm, although in practice higher doses may be required. This may be because penicillin is not very acid stable and is at least partially inactivated below pH 5.0 (Kelsall, 1995) although Ingledew (1998) has reported considerable antibacterial properties at pH values as low as 4.5.

Streptomycin, an aminoglycoside antibiotic produced by species of the genus Streptomyces, is another narrow spectrum antibacterial agent. It works mainly by creating pores in the cell wall of susceptible Gram-negative bacteria (Stratton, 1996). The end result is cellular death in a relatively short period of time. Combinations of antibiotics can sometimes exhibit synergistic effects, meaning that the mixture is more potent than either used separately. Penicillin/streptomycin combinations have been shown to act synergistically (Stratton, 1996) especially if the timing of the doses are staged, with the aminoglycoside added about 2 hrs before the ß-lactam. Streptomycin is most effective against aerobic Gram-negative bacteria, but it loses much of its inhibitory effect at pH values below neutrality (Amsterdam, 1996).

One of the most successful commercially-available antimicrobials is a wide-spectrum product. Marketed under the trade name Lactoside, this product contains a number of known antimicrobials that have been pH-stabilized. While they survive for effectiveness during fermentation, they are completely destroyed in distillation. All antibiotics used in the alcohol production industry are completely degraded during the distillation, so no residues remain in the distillers dried grains. This is an important concern at the present time with the imposition of bans on use of many antibiotic growth promoters in animal feeds, particularly virginiamycin (Spring, 1999).

References

Abate, C. 1996. Ethanol production by a mixed culture of flocculent strains of *Zymomonas mobilis* and *Saccharomyces spp*. Appl. Microbiol. and Biotech. 45:580-583.

Amin, G. and A.M.K. Allah. 1992. Byproducts formed during direct conversion of sugar beets to ethanol by *Zymomonas mobilis* in conventional submerged and solid-state fermentations. Biotechnology Letters 14:1187-1192.

Amsterdam, D. 1996. Susceptibility testing of antimicrobials in liquid media. *In:* Antibiotics in Laboratory Medicine (V. Lorian, ed). Maryland, Williams and Wilkins. pp. 52-111.

Back, W., I. Bohak, M. Ehrmann, W. Ludwig and K.H. Schleifer. 1996. Revival of the species *Lactobacillus lindneri* and the design of a species specific oligonucleotide probe. Systematic Appl. Microbiol. 19(3):322-325.

Baird-Parker, A.C. 1980. Organic acids. *In:* Microbial ecology of foods. (J. H. Silliker, R.P. Elliott, A.C. Baird-Parker, F.L. Bryan, J.H.B. Christian, D.S. Clark, J.C. Olson and T.A. Roberts, eds). London, Academic Press 1:126-134.

Bamforth, C. 1997. Brewing a better beer. Chemistry in Britain 33:37-39.

Brown, J. 1995. What is an antibiotic? http://falcon.cc.ukans.edu/~jbrown/antibiotic.html.

Brown, J. 1996. What is penicillin? http://falcon.cc.ukans.edu/~jbrown/penicillin.html.

Condon, S., 1987. Responses of lactic acid bacteria to oxygen. FEMS Microbiology Reviews 46:269-280.

Chin, P.M. and W.M. Ingledew. 1994. Effect of lactic acid bacteria on wheat mash fermentations prepared with laboratory backset. Enzyme Microbiol. and Tech. 16:311-317.

De Ley, J., M. Gillis and J. Swings. 1986. Family IV. Acetobacteraceae. Bergey's Manual of Systematic Bacteriology. Baltimore, Williams and Wilkins, 1:267-278.

Dowhanick, T., J. Sobczak, I. Russell and G. Stewart. 1990. The rapid identification by protein fingerprinting of yeast and bacterial brewery contaminants. Amer. Soc. of Brewing Chemists 48(2):75-79.

Doyle, L.M., J.O McInerney, J. Mooney, R. Powell, A. Haikara, and A.P. Moran. 1995. Sequence of the gene encoding the 16S rRNA of the beer spoilage organism *Megasphaera cerevisiae*. Industrial Microbiology 15(2):67-70.

Fix, G.J. 1989. Principles of brewing science. Brewers Publications, Boulder, Colorado.

Fix, G.J. 1993. Diacetyl: formation, reduction and control. Brewing Techniques. http://www.brewingtechniques.com/library/backissues/issue1.2/fix.html.

Gold, S.R., M.M. Meagher, R. Hutkins and T. Conway. 1992. Ethanol tolerance and carbohydrate metabolism in lactobacilli. Ind. Microbiol. 10:45-54.

Gold, R.S. 1996. Cloning and expression of the *Zymomonas mobilis* production of ethanol genes in *Lactobacillus casei*. Current Microbiology 33(4):256-260.

Gutteridge, C.S. and F.G. Priest. 1996. Methods for the rapid identification of micro-organisms. *In:* Brewing Microbiology (F.G. Priest and I. Campbell, eds). Chapman & Hall, London. pp. 237-270.

Hammond, J. 1996. Microbiological techniques to confirm CIP effectiveness. The Brewer, August. http://www.breworld.com/the brewer/9608/br1/html.

Harper, D.R. 1980. Microbial contamination of draught beer in public houses. Process Biochem. (Dec/Jan.). pp. 2-7.

Holzapfel, W.H. 1992. Culture media for non-sporulating Gram-positive food spoilage bacteria. International Journal of Food Microbiology 17:113-133.

Hough, J.S., D.E. Briggs, R. Stevens and T.W. Young. 1982. Microbial contamination in breweries. Malting and Brewing Science, Chapman & Hall, London 2:741-775.

Huang, Y. C., C.G. Edwards, J.C. Peterson and K.M. Haag. 1996. Relationship between sluggish fermentations and the antagonism of yeast by lactic acid bacteria. Amer. J. of Enology and Viticulture 47(1):1-10.

Ingledew, W.M., F.W. Sosulski and C.A. Magnus. 1986. An assessment of yeast foods and their utility in brewing and enology. J. Amer. Society of Brewing Chem. 44(4):166-170.

Ingledew, W. M. 1995. The biochemistry of alcohol production. *In:* The Alcohol Textbook (T.P. Lyons, D.R. Kelsall and J.E. Murtagh, eds). Nottingham University Press, Nottingham, UK. pp. 55-79.

Ingledew, W.M. 1998. Fuel Ethanol Workshop, Bryan & Bryan, Cotopaxi, Texas.

Johnston, M.A. and E.A. Delwiche. 1965. Distribution and characteristics of the catalases of Lactobacillaceae. Bacteriology 90(2):347-351.

Jones, A.M. and W.M. Ingledew. 1994. Fuel alcohol production: appraisal of nitrogenous yeast foods for very high gravity wheat mash fermentation. Proc. Biochem. 29:483-488.

Kandler, O. 1983. Carbohydrate metabolism in lactic acid bacteria. Antonie van Leeuwenhoek 49:209-224.

Kandler, O. and N. Weiss. 1986. Regular, nonsporing Gram-positive rods. Bergey's Manual of Systematic Bacteriology 2:1208-1234. Williams and Wilkins, Baltimore, Md.

Kelsall, D.R. 1995. The management of fermentations in the production of alcohol. *In:* The Alcohol Textbook (T.P. Lyons, D.R. Kelsall and J.E. Murtagh, eds). Nottingham University Press, Nottingham, UK. pp. 89-101.

King, T.E., E.H. Kawasaki and V.H. Cheldelin. 1956. Tricarboxylic acid cycle activity in *Acetobacter pasteurianum*. Bacteriology 72:418-421.

Kulka, D., J.M. Preston and T.K. Walker. 1949. Giant colonies of *Acetobacter* species as an aid to identification. Institute of Brewing 55:141-146.

Kyriakides, A.L. and P.A. Tiiurston. 1989. Conductance techniques for the detection of contaminants in beer. Tech. Ser. Soc. Applied Bacteriology. Oxford, Blackwell Scientific Publications 25:101-117.

Lipkus, A.H., K.K. Chittur, S.J. Vesper, J.B. Robinson and G.E. Pierce. 1990. Evaluation of infrared spectroscopy as a bacterial identification method. Indust. Microbiol. 6:71-75.

Makanjoula, D.B., A. Tymon and D.G. Springham. 1992. Some effects of lactic acid bacteria on laboratory scale yeast fermentations. Enzyme Microbiology and Technology 14:351-357.

Moon, N.J. 1983. Inhibition of the growth of acid tolerant yeasts by acetate, lactate and propionate and their synergistic mixtures. Appl. Bacteriology 55:453-460.

Morrison, J.A. 1995. The production of Canadian rye whisky. *In:* The Alcohol Textbook (T.P. Lyons, D.R. Kelsall and J.E. Murtagh, eds). Nottingham University Press, Nottingham, UK. pp. 69-192.

Murtagh, J.E. 1995. Molasses as a feedstock for alcohol production. *In:* The Alcohol Textbook (T.P. Lyons, D.R. Kelsall and J.E. Murtagh, eds). Nottingham University Press, Nottingham, UK. pp. 27-34.

Muyzer, G., E.C. De Wall, and A.G. Uitterlinden. 1993. Profiling of complex microbial populations by denaturing gradient gel electrophoresis analysis of polymerase chain reaction-amplified genes coding for 16S rRNA. Appl. and Environ. Microbiol. 59(3):695-700.

Narendranath, N.V., S.H. Hynes, K.C. Thomas and W.M. Ingledew. 1997. Effects of lactobacilli on yeast-catalyzed ethanol fermentations. Appl. and Environ. Microbiol. 63(11):4158-4163.

Nishikawa, N., M. Kohgo and T. Karakawa. 1979. Serological classification of brewer's and related yeasts with commercial factor sera for the detection of yeast contaminants in the brewing process. Fermentation Technology 57(4):364-368.

O'Connor-Cox, E.S.C., J. Paik and W.M. Ingledew. 1991. Improved ethanol yields through supplementation with excess assimilable nitrogen. Indust. Microbiol. 8:45-52.

O'Connor-Cox, E.S.C. and W.M. Ingledew. 1991. Alleviation of the effects of nitrogen limitation in high gravity worts through increased inoculation rates. Indust. Microbiol. 7:89-96.

Priest, F.G. 1996. Gram-positive brewery bacteria. *In:* Brewing Microbiology (F.G. Priest and I. Campbell, eds). Chapman & Hall, London. pp. 127-161.

Quirke, V.M. 1998. Howard Florey-Medicine Maker. Chemistry in Britain 34(10):35-38.

Rao, R.M.R. and J.L. Stokes. 1953. Nutrition of the acetic acid bacteria. Bacteriology 65:405-412.

Reuter, G. 1985. Elective and selective media for lactic acid bacteria. Intern. J. of Food Microbiol. 2:55-68.

Smith, C.E., G.P. Casey and W.M. Ingledew. 1987. The use and understanding of media used in brewing microbiology. Brewers Digest. pp. 12-16.

Sobolov, M. and K.L. Smiley. 1959. Metabolism of glycerol by an acrolein-forming *Lactobacillus*. Bacteriology 79:261-266.

Spring, P. 1999. The move away from antibiotic growth promoters in Europe. *In:* Biotechnology in the Feed Industry. Proc. 15th Annual Symposium (T.P Lyons and K.A. Jacques, eds). Nottingham University Press, Nottingham, UK. pp. 173-183.

Stratton, C.W. 1996. Mechanisms of action for antimicrobial agents: General principles and mechanisms for selected classes of antibiotics. *In:* Antibiotics in Laboratory Medicine (V. Lorian, eds). Williams and Wilkins, Baltimore, Md. pp. 579-603.

Swings, J. and J. De Ley. 1977. The biology of *Zymomonas*. Bacteriology Reviews 41:1-46.

Tilsala, T.A. and T. Alatossava. 1997. Development of oligonucleotide primers from the 16S-23\S rRNA intergenic sequences for identifying different dairy and probiotic lactic acid bacteria by PCR. Intern. J. of Food Microbiol. 35(1):49-56.

Thomas, K.C. and W.M. Ingledew. 1990. Fuel alcohol production: Effects of free amino nitrogen on fermentation of very high-gravity wheat mashes. Appl. and Environ. Microbiol. 56(7):2046-2050.

Todar, K. 1996. Bacteriology 330 Lecture Topics: Antimicrobial agents. http://www.bact.wisc.edu/bact330/lectureama.

Van Vuuren, H.J.J. 1996. Gram-negative spoilage bacteria. *In:* Brewing Microbiology (F.G. Priest and I. Campbell, eds). Chapman & Hall, London. pp. 163-191.

Whiting, M. M. Crichlow and W.M. Ingledew. 1992. Detection of *Pediococcus spp.* in brewing yeast by a rapid immunoassay. Appl. and Environ. Microbiol. 58(2):713-716.

Yanase, H., N. Kato, K. Tonomura, Y. Murooka and T. Imanaka. 1994. Strain improvement of *Zymomonas mobilis* for ethanol production. *In:* Recombinant Microbes for Industrial and Agricultural Applications, Vol. 19. Marcel Dekker Inc., New York, USA. pp. 723-739.

Chapter 22

Co-products from ethanol fermentation: alternatives for the future

P. Torre
Alltech Inc., Nicholasville, Kentucky, USA

Introduction

The technology utilized for conversion of plant material to alcohol is well-established. However, there is room for improvement, particularly in the areas of process technology and co-product utilization. Greater usage of by-products of alcohol fermentations will lead to enhanced profitability for both ethanol producers and the farmers who supply them with raw materials, ensuring success both for alcohol producers and for farmers. The types of co-products produced will depend on the feedstock and on the processing technology. Starch is consumed in the process of alcohol production, and protein and fiber are the major co-products. Protein has been investigated as both a concentrate for human and animal foods and as industrial material. Cereal fiber has in recent years been linked to dietary health effects.

Traditionally, the animal feed industry has provided the major outlet for ethanol fermentation by-products. However, new markets are needed to keep the industry viable (Hojilla-Evangelista *et al.,* 1992b). Products destined for the human and pet food markets are often of higher value than those destined for animal feed; and therefore additional invest-ment to purify and upgrade a product may be cost-effective (Gras and Simmonds, 1980). The types of co-products produced rely on a number of factors, including conversion technology, feedstock and milling.

Potential co-products from various feedstocks

Wheat

Industrial use of wheat has been limited because of its high value as a food commodity. Non-food uses of wheat have only been feasible during times when wheat prices have been low or if alternative methods of chemical production became too costly or difficult. For example, during World War II the demand for fuel and chemicals for the war effort was high; and industrial use of wheat for alcohol production increased because of shortages of fossil fuels and other cereal grains (Gagen, 1973). However, wheat may be the feedstock of choice in areas where, for example, supplies of corn are limited.

Protein

Wheat protein has found uses in construction materials, plastics, chewing gums, and pharmaceuticals (Gagen, 1973). Wheat has an approximate protein content of 10-13% by weight, of which 80-90% is made up of gluten (Weegels *et al.*, 1992). Gluten is an important additive in bakery and breakfast foods and in processed meat and fish products (Satterlee, 1981). Wheat gluten has a bland flavor, can absorb up to twice its weight in water and has elastic properties that enhance its value in baking, production of breakfast cereals and production of processed meat products. De-aminated gluten preparations have been used in fat-reduced dairy spreads as emulsion stabilizers. In a more recent application, Mohawk Canada Ltd. began marketing a high fiber food ingredient, Fibrotein™, a low fat cholesterol-free edible material derived from a by-product of ethanol fermentation from prairie spring wheat.

Fiber

Fiber, especially water-soluble or non-structural fiber, has been shown to have health benefits when included in human diets (Walker, 1974). The United States Food and Drug Agency has recognized the connection between dietary fiber and certain types of cancer and heart disease.

The majority of interest in fiber as a co-product of ethanol fermentation from wheat feedstock focuses on wheat bran. A Canadian company, Tkac and Timm Enterprises, Ltd., uses a patented process by which they are able to separate the bran and endosperm components of the wheat grain and then further segregate the bran layers (Tkac and Timm, Ltd., 1995). In the Tkac and Timm process, the wheat bran is removed from the kernel using friction and abrasion techniques. One of the strengths of the process is that it removes the bran prior to fermentation. Thus, the by-products have not undergone chemical alterations resulting from fermentation, and are closer to their original state than would otherwise be the case. Certain by-products from the Tkac and Timm process are in use in the European human food market. For example, the fiber product PrimAFibre™, which can be used in a variety of cereal-based foods, is marketed for human consumption.

Germ

Wheat germ contains high concentrations of protein and minerals, vitamins (including vitamin E), and increased percentages of unsaturated fatty acids in its oil compared to animal fats (Tsen, 1980). Commercial wheat germ contains small amounts of bran and endosperm. These properties suggest that wheat germ could be used to fortify bakery products; but baking quality is affected and processes must be modified if an acceptable commodity is to be produced. Wheat germ is particularly attractive as a protein supplement because of its lysine content.

Distillers dried grains (DDG), distillers dried solubles (DDS), and distillers dried grains with solubles (DDGS)

The major market for DDG, DDS, and DDGS is currently the animal feed market. Considerable research has been devoted to maximizing the value of these co-products in diets for ruminants and swine. In particular, DDGS has value in ruminant feeds as a source of rumen bypass protein. However, these products contain also high concentrations of dietary fiber and protein and have the potential for increased value as a flour supplement in baked goods. For example, Rasco and coworkers (1987) substituted 30% of the all-purpose flour in white bread, whole wheat bread and other baked goods with DDGS with positive results following evaluation by sensory panels. Dietary fiber and protein increased in the DDGS-enhanced products by 140-500% and 130-150%, respectively. The primary problem with incorporating DDGS into human food products has been smell and taste. However, in 1989 Rasco *et al.* patented a human food product produced from DDGS. The process developed by Rasco involved adjusting the pH of the stillage using organic and inorganic acids and neutralization with selected hydroxides or oxides before drying to a moisture content of 5-10% at a lower temperature to prevent adverse effects on color or flavor. The end product could be used in baked goods such as brownies, cookies, and pasta at concentrations ranging from 10-50%, depending on the product. Kim *et al.* (1989) found that wheat DDG was one of the more successful materials for use in extruded snack products.

Minor components

Phytate and derivatives

Phytate and its derivatives are found in wheat germ and bran (Murray *et al.,* 1987). Phytic acid has potential uses in medical applications, such as an imaging agent for organ scintigraphy and a contrasting agent for barium X-rays. Ability of phytic acids to chelate cations may suggest its use as a remedy to lead poisoning and to reduce calcium deposition. Phytate and its derivatives may also find applications in treatment of heart disease, certain cancers, and diabetes (Thompson, 1992). Phytate may also be useful in food preservation because of its antioxidant properties in oils (Graf, 1983).

Enzymes, enzyme inhibitors, and mycotoxins

Enzymes find a myriad of uses in the animal feed, medical and detergent industries. A variety of enzymes with documented uses are found in wheat and can be extracted (Murray *et al.,* 1987). For example, carboxypeptidase and phytase are found in wheat bran. Amylase inhibitors found in wheat may be useful to treat wheat to reduce sprouting during harvest. Enzymes and mycotoxins can also be biosynthesized from fungi growing on wheat grains. These compounds may have use in medicine and biotechnology for research processes. However, markets for such compounds, particularly mycotoxins in research processes are undeveloped.

Vitamin E

Vitamin E is a known antioxidant and has been linked to prevention of certain types of diseases such as cataracts, thrombotic disease, and certain cancers. It is also popular in certain cosmetics. Concentrations of 1.8-2.3 mg/100g have been reported in whole wheat grains (Thompson, 1992).

Cinnamates, ß-glucan, glycerides and lectins

Certain cinnamates found in wheat bran have good oxidation-reduction properties and the ability to absorb radiation; and therefore could be useful in products such as sunscreens. The cost effectiveness of extracting these compounds from wheat bran is uncertain.

Wheat bran contains ß-glucan, a hot water-insoluble polysaccharide. ß-glucan has been linked to control of heart disease and diabetes, as well as reduction in cholesterol. Wheat glycerides, contained in the germ, are used in cosmetic preparations such as creams, lotions, and lipsticks as an anti-irritant (Murray *et al.,* 1987). Since glycerides are effective in low concentrations, it is unclear whether demand may ever result in a significant market opportunity.

Carbohydrate-binding proteins known as lectins are found in the germ. Lectins have a broad range of medical and biochemical uses (Murray *et al.,* 1987). Lectins can be bound to other molecules, for example peroxidase, to increase their value. The major use for these lectins may be in diagnostic tests.

Corn: potential ethanol co-products

Products of corn wet milling for ethanol production have been outlined by Hayman *et al.* (1995). The corn kernel is presoaked and milled to produce starch, germ and fiber. The germ may be extracted for corn oil, perhaps the most valuable co-product of the process. The starch fraction is centrifuged and saccharified to produce gluten wet cake (the second most valuable co-product) and glucose, which may be fermented to produce ethanol. Distilling the ethanol leaves thin stillage, which when dewatered results in corn condensed distillers solubles. These solubles can be sprayed onto the corn fiber and fermented to produce corn gluten feed. Thus, wet milling procedures may have higher potential for value-added co-products.

Dry milling of corn involves breaking the kernel down to fine particles. The germ may be removed by aspiration or gravity procedures (Koseolu *et al.,* 1991). Pressing and solvent extraction may be used to remove oil from the germ. Protein concentrates may be obtained from the defatted germ meal by milling or alkali extraction-acid precipitation. Enzymes are used to convert the starch to glucose, and yeast is added to produce ethanol Wyman and Goodman, 1993).

A sequential extraction process for corn milling was developed to produce higher value co-products (Chang *et al.,* 1995). Oil yield and quality are enhanced by the process since the entire kernel is extracted. Yields are enhanced in different amounts depending on the source of corn (dent or high lysine) (Hojilla-Evangelista *et al.,* 1992a). The protein resulting from the product has a higher quality due to elimination of the need to steep in sulfur dioxide. Freeze-dried protein concentrates containing nearly 80% protein, compared to 60-62% for corn gluten meal, were produced. The protein was found to have a mild corn flavor, excellent emulsifying capacity and emulsion stability, and good heat stability (Myers *et al.,* 1994). The starch remaining from the process was of lower quality than that resulting from conventional wet milling, but still an excellent substrate for ethanol production. The cost of the process may result in consideration of protein as the main product and ethanol as the by-product (Chang *et al.,* 1995).

Corn protein concentrates

Corn protein concentrates have been the subject of considerable research (Sternberg *et al.,* 1980). However, for acceptance in food markets such concentrates will require unique characteristics. The ability of corn protein concentrates to complement oilseed proteins in terms of amino acid composition could be such a characteristic. For example, a corn protein concentrate (90% protein) isolated from corn gluten meal (Satterlee, 1981), because of its amino acid content, complemented other proten sources rich in lysine and tryptophan but low in methionine and cystine. Currently, markets for corn protein concentrates are undeveloped, although use in feed for infant animals and aquaculture could be profitable. Zein is the only corn protein currently used in non-food industrial applications. Zein possesses the ability to provide a tough coating that is resistant to water, grease, scuffing and microbial attack (Reiners *et al.,* 1973). Zein has been used in the production of packaging films, linoleum tiles, pill coatings, ink, and textile fibers.

Corn fiber

Corn fiber as a by-product of fermentation is presently used mostly in animal feeds. However, other opportunities may exist (Bothast, 1994). One possibility might be that the fibers could be used as feedstock for production of useful chemicals by microbial fermentation. This would require hydrolysis of the starch and hemicellulose fractions of the corn fiber to fermentable sugars. Corn bran flour was tested in the 1980s (Shukla, 1981) due to the increasing interest in health effects of dietary fiber. Corn bran has been used in nonfood applications as an extender and viscosifier for use in urea-formaldehyde plywood adhesives (Alexander, 1987).

Corn germ

Corn germ is a by-product of both dry and wet milling, although its properties may be altered slightly during wet milling (Nielsen *et al.,* 1973). Corn germ products may have high potential for use in human foods, although there are disadvantages compared to casein or soy isolates in terms of color and flavor (Nielsen *et al.*, 1973). Corn germ is 10-20% of the product generated during dry milling of corn. Defatted corn germ flour (DCGF) developed from dry milled corn germ may have use as a protein supplement in baked goods (Tsen, 1976). The product is high in protein, oil, minerals and vitamins; and possesses good emulsifying properties (Blessin *et al.*, 1979). Use of DCGF did not alter baking characteristics as long as acceptable levels were used (Tsen, 1980). The product may find use as a protein and mineral supplement in a number of foods such as cookies, bread and meat patties. As another example, Nielsen *et al.* (1979) produced a value-added flour product from wet milled corn germ that contained significant fiber, protein, lysine, and a good balance of other amino acids.

Gluten meal

Corn gluten meal produced by steeping corn kernels at the beginning of the wet milling process is a high protein product and may be one of the most valuable co-products of the wet milling process (Satterlee, 1981). However, a high unsaturated fatty acid content and sulfite levels (resulting from the steeping process) cause problems with odor and flavor. Various treatments have been attempted to produce more acceptable products (Wu *et al.*, 1994). Ethanol and carbon dioxide (CO_2), which would both be readily available in grain biorefineries, were shown to diminish the fermented flavor of corn gluten meal. Functional characteristics, texture, and flavor are negatively affected by supplementation with corn gluten meal, and therefore further work would be required to produce an acceptable product.

Corn oil

Most of the corn oil produced in the United States is a by-product of corn wet milling (Orthoefer and Sinram, 1987). Corn oil processing also generates a number of co-products, including wet gum which can be used with refining soapstock in animal feed or further processed to produce lecithin, a potentially valuable commodity for use as an emulsifier, antioxidant, nutrient or dispersing agent. Vegetable oil distillate, which is removed from the oil during the deodorization process, contains a number of chemicals including tocopherols, carotenoids, and flavor components that could be extracted for other value-added applications.

Distillers dried grains, distillers dried solubles, and distillers dried grains with solubles

Distillers grains from different sources vary in color, protein, fat, fiber, and taste; and must be selected carefully for use in value-added food products. Distillers grains from processing of high lysine corn may have the most utility for human foods, because of their higher protein quality (amino acid spectrum). DDGS from dry milling and fermentation of corn is chiefly used in the animal feed industry in the United States. However, use of the material in the human food industry could increase the value of the product to distilleries. In addition to the yeast cells from fermentation, DDGS contains all material present in the intact corn grain except the starch (Scheller, 1981). Protein and fiber are concentrated three-fold in DDGS compared to the original grain (Dong and Rasco, 1987). However, the amino acid balance of the product is problematic, with lysine being a limiting amino acid (Dong *et al.*, 1987). Distillers grains are allowed for use in food in the US as long as the original grains were fit for human consumption and the processing plants are approved for food manufacture. Examples of uses of distillers grains in human foods include incorporation into breads and other baked goods (Rasco *et al.*, 1987; O'Palka, 1987), canned products such as stew, chili, and hot dog sauce (Reddy *et al.*, 1986) puff-extruded products based on rice, potato, wheat, or corn flour doughs (Kim *et al.*, 1989) and spaghetti (Wu *et al.*, 1987). DDG, because of its high fiber content (one cup of DDG would supply the entire US daily requirement of fiber while 30 cups of corn flakes would be needed to supply an equal amount), could provide a significant source of dietary fiber if incorporated into human foods (San Buenventura *et al.,* 1987). Use of corn distillers grains can also cause problems with flavor, but these problems may be resolved in a similar manner as with corn gluten meal.

Minor components from stillage

Typically, 10-15 liters of stillage are produced for every liter of ethanol (Maiorella *et al.,* 1983). Most of the product enters the animal feed market or is recycled into the fermentation process to provide a source of nutrients for yeast cell growth. However, stillage may also have value as a fermentation medium for production of other products or may be extracted for isolation of minor components. Useful components contained in stillage include ethanol, acetic acid, propionic acid, glycerol, vitamins, alanine, valine, leucine, and proline (Dowd *et al.,* 1993). Glycerol in particular has a multiplicity of uses in the pharmaceutical, cosmetic, food, explosive, textile, and other industries (Julian *et al.,* 1990). At present, glycerol from plant sources is not competitive with glycerol produced from petrochemical products. However, as fossil fuels continue to deplete, plant sources of glycerol may become more viable, particularly as methods of recovery become more efficient and cost effective.

Stillage (or corn steep liquor) has been suggested as a medium for fermentative production of other useful products such as riboflavin, enzymes, and antibiotics (Maiorello *et al.,* 1983). Similarly, propionic acid is normally produced from petroleum feedstocks; and use of corn steep liquor may be an option as an abundant and inexpensive substrate for propionibacteria (Paik and Glatz, 1994). Similarly, corn wet milling by-products may in the future serve as fermentation substrates for organisms used to produce useful substances such as carotenoids, which are used as coloring agents to give poultry and aquaculture products a pigment acceptable to the consumer. For example, Hayman *et al.* (1995) found that use of thin stillage plus condensed distillers solubles and corn gluten feed gave the best accumulation of biomass and carotenoids. Another example would be pullulan, a biopolymer used as a film for coating and packaging food (Yuen, 1974). Currently, most pullulan is produced from petroleum products. An organism used for biosynthesis of pullulan (*Aureobasidium spp.*) was found to grow and produce pullulan on a media containing condensed distillers solubles (Leathers and Gupta, 1994). As with most applications for value-added co-products from alcohol fermentations, commercialization will depend on economic competitiveness with current sources, often petroleum products.

Oats: potential co-products of ethanol fermentation

Oats as a feedstock for alcohol fermentations have the advantages of higher starch yields on a per hectare basis (compared to wheat) as well as better growth on poor soil (Thomas and Ingledew, 1995). Oat yield per hectare is also higher than for barley or corn. Ethanol yield from oats (per hectare) is better than for wheat or barley, but poorer than for corn (Dale, 1991). Traditionally, oats have not been a feedstock for ethanol production. However, Thomas and Ingledew (1995), based on studies of processing oats into very high gravity (VHG) mashes, concluded that oat mash has an excellent potential as a feedstock for fuel ethanol production. The existence of a number of value-added products extracted from oat grains for use as nutraceuticals or functional foods suggests a market for products produced from by-products of fermentation of oats for ethanol. A patented process for fractionation of oats into endosperm and bran fractions (Burrows *et al.,* 1984), further refined (Collins and Paton, 1991; Collins and Paton, 1992), has the advantage of extracting also the ß-glucan component as well as the bran fraction. Via anion exchange procedures, Collins and Paton were able to recover other products with value-added potential including phenolic acids, alkaloids, fatty acids, organic acids and amino acids.

Oat protein

Oats have not been wet milled on an industrial scale. However, protein concentrates have been produced from oats via wet extraction techniques that possessed good amino acid profiles, bland flavoring and good emulsifying properties. The product was suggested for use in protein-fortified milk substitutes and breakfast beverages, prepared foods, breakfast cereals, baked goods and as a meat extender (Cluskey *et al.*, 1976). Others have attempted similar oat based protein concentrates (Wu *et al.*, 1973; Wu, 1990; Wu and Stringfellow, 1973). Oat based protein concentrates may have equal or greater nutritional value than other small grains, but are limiting in lysine, threonine, and methionine. Functional capacities of oat protein concentrates (foaming ability, fat-binding capacity, solubility, emulsifying properties) compared favorably with wheat gluten and soy protein isolate (Ma, 1983). Concentrated oat stillage from ethanol fermentation using either ground oats, groats, or oat flour resulted in distillers grains containing moderate protein levels and high dietary fiber, or oat flour distillers grains containing high protein levels and lower dietary fiber (Wu, 1990).

Oats are used by a Finnish alcohol producer to produce a high-protein oat flour in conjunction with ethanol production (Lapvetelainen and Aro, 1994). Once dietary fiber and starch are removed from the groat, a high protein oat flour is produced by dewatering the protein fraction, washing to remove undesirable flavors and spray drying. The oat product was found to have similar solubility and emulsifying properties compared to commercial soy protein concentrate. Oat protein has also been used in shampoos, due to its balanced amino acid content and reduced odor (Paton *et al.*, 1995).

Distillers dried grains, distillers dried solubles, and distillers dried grains with solubles

These products have the least potential use in human food products, probably due to their relatively high fiber content and low protein. However, the known markets for such products in the animal feed industry cannot be ignored.

Oat starch

Oat starch products have been developed for use in baby powders, antiperspirants, and other cosmetics due to concerns about the safety of talc. Other oat starch-based products include ingredients for lotions, creams, gels, shampoos, and sunscreens. Liposomes found in the groat may help movement of vitamins directly into the skin and act as a moisturizer. The lower quality starch component which has no value-added uses can be retained for use in ethanol fermentations.

Oat hulls

If the oats are dehulled before fermentation, the hulls may have use as a by-product. However, specialized fermentation technologies are required to compensate for the increased cellulose content of oat hulls. Industrial uses for oat hulls include manufacture of adhesive chemicals, abrasives filters and cavity preventers in the dental industry (Caldwell and Pomeranz, 1973).

Minor components: non-starch polysaccharides and phenolic acids

Oat gum, composed mainly of a ß-D-glucan (Wood *et al.*, 1989), accounts partially for the viscosity of cooked oatmeal and for emollient effects in cereal. Wet milling processes have been tested on a pilot plant scale for isolation of protein, starch, and gum fractions of the oat groat (Clark, 1972). ß-glucan is also found in the bran fraction of the oat groat. As noted for other feedstocks, ß-glucan has been investigated for effects in treating heart disease and diabetes, and in reducing cholesterol. Oat bran has been in demand for a number of years because of this cholesterol-lowering effect.

A number of phenolic acids have been identified in oat groats and hulls, including avenanthramides and hydroxycinnamic acids (Collins, 1986; Collins *et al.,* 1991). As was discussed for wheat, cinnamates have been noted for their redox potential and for use in sunscreens.

Barley fermentation co-products

Barley has been used as a feedstock for ethanol production in the US, Canada and Sweden. On average, barley contains 63-65% starch, 8-13% protein, 2-3% fat, 1-1.5% gums, 8-10% hemicellulose and 2-2.5% ash. Certain newly developed hull-less barley lines may be more suitable for ethanol fermentations due to increased ease of milling, mashing and fermenting (Ingledew *et al.,* 1995). Addition of ß-glucanase alleviates viscosity problems. Distillers grains remaining after fermentation of hull-less barley are comparable to wheat distillers grains. Because less starch is recovered from barley than from corn, by-product recovery becomes more essential to make ethanol production from barley profitable (Wu, 1985). Methodology suitable for wet milling to extract protein and starch from wheat should be adaptable to barley.

Barley protein

High protein, high lysine barley cultivars may have potential for production of protein concentrates as a by-product. Wu (1985) evaluated a process involving ultrafiltration and reverse osmosis for extracting a barley protein concentrate from barley fermentation stillage and concluded there was potential for use of barley distillers grains and centrifuged solids for use in human consumption.

Barley fiber

Similar to oats, barley may contain dietary fibers that are of benefit to human health (Salomonsson *et al.,* 1984). Certain varieties of barley cultivars have been found to contain elevated levels of ß-glucan comparable those in to oat bran (Aman and Graham, 1987). Certain insoluble fibers in barley (lignin and cellulose) have been linked to cancer prevention in rats (McIntosh *et al.,* 1993). Total dietary fiber in barley DDGS was found to be higher than in whole wheat, as well as DDGS from wheat, corn, or brown sorghum, and may be a potential source of dietary fiber for human consumption (San Buenaventura *et al.,* 1987).

Distillers dried grains, distillers dried solubles, and distillers dried grains with solubles

The flavor of barley distillers grains is a problem when considering uses in human food (Dawson *et al.,* 1987). However, defatting could partially alleviate these problems.

Minor components: enzymes, tocols and citric acid

Barley contains various enzymes such as amylase, amylase inhibitors, and oxalate oxidase that could be extracted for commercial use. The market for amylase in the distilling industry is known. Amylase inhibitors may have use in preventing sprouting in wheat. However, these markets await more commercially feasible extraction schemes.

Tocopherols and tocotrienols have known effects in lowering serum cholesterol and as antioxidants (Weber *et al.,* 1991; Burton and Traber, 1990). Milling and brewing may concentrate barley tocols, thus further enhancing their value as a co-product of barley ferm-entation.

Alcohol distillery wastewater (barley as a feedstock) has been tested as a substrate for fermentative production of citric acid by *Aspergillus niger* (Myung *et al.,* 1992). Treatment of the wastewater with Aspergillus resulted in a 2.4 g/L citric acid concentration and reduced the polluting potential of the wastewater by almost 40%.

Conclusions

The search for value-added by-products of the alcohol fermentation industry has most often been in response to issues of waste disposal from the process. The best example of this is the use of by-products considered to be waste by the alcohol industry as feed ingredients for livestock. However, maximizing the potential of co-product utilization will require identification of products that satisfy a need in other markets. As consumers become more concerned about the environment and depletion of fossil fuels, demand for bio-based, environmentally friendly products may increase. As new markets are identified, and as products produced synthetically from the petroleum industry become less profitable, alcohol producers may find it necessary to revise their processes so that alcohol becomes one of a stream of co-products to meet needs in the food, feed, industrial feedstock, and fiber product industries.

References

Alexander, R.J. 1987. Corn dry milling: Processes, products, and applications. *In*: Corn: Chemistry and Technology. (S.A. Watson and P.E. Ramstad, eds.) Amer. Assoc. Cereal Chem. p. 351-376.

Aman, P. and H. Graham. 1987. Analysis of total and insoluble mixed-linked (13)-ß-D glucans in barley and oats. J. Agric. Food. Chem. 35:704.

Blessin, C.W., W.L. Deatherage, J.F. Cavins, W.J. Garcia and G.E. Inglett. 1979. Preparation and properties of defatted flours from dry-milled yellow, white, and high lysine corn germ. Cereal Chem. 56:105-109.

Bothast, R.J. 1994. Genetically engineered microorganisms for the conversion of multiple substrates to ethanol. Proc. Corn Util. Conf. V., St. Louis, MO.

Burrows, V.D., R.G. Fulcher and D. Paton. 1984. US Patent No. 4,435,429.

Burton, G.W. and M.G. Traber. 1990. Vitamin E: antioxidant activity, biokinetics and avail-ability. Ann. Rev. Nutr. 10:357.

Caldwell, E.F. and Y. Pomeranz. 1973. Industrial uses of cereals, oats. *In*: Industrial Uses of Cereals. (Y. Pomeranz, ed.) St. Paul, MN. Amer. Assoc. Cereal Chem. p. 393-411.

Chang, D., M.P. Hojilla-Evangelista, L.A. Johnson and D.J. Myers. 1995. Economic-engineering assessment of sequential extraction processing of corn. Trans ASAE 38:1129-1138.

Clark, W.L. 1972. A new look at oats. The Agrologist, Nov.-Dec. pp. 8-11.

Cluskey, J.E., Y.V. Wu, G.E. Inglett and J.S. Wall. 1976. Oat protein concentrates for beverage fortification. J. Food Sci. 41: 799.

Collins, F.W. 1986. Oat phenolics: structure, occurrence and function. *In*: Oats, Chemistry and Technology. (F.H. Webster, ed.) St. Paul, MN: Amer. Assoc. Cereal Chem. 227-295.

Collins, F.W. and D. Paton. 1991. Recovery of values from cereal wastes. Can. Patent Applic. 2,013,190.

Collins, F.W. and D. Paton. 1992. Method of producing stable bran and flour products from cereal grains. U.S. Patent No. 5,169,660.

Collins, F.W., D.C. McLachlan and B.A. Blackwell. 1991. Oats phenolics: avenalumic acids, a new group of bound phenolic acids from oat groats and hulls. Cereal Chem. 68:184-189.

Dale, B.E. 1991. Ethanol production from cereal grains. Food Sci. Technol. Handbook of Cereal Sci. and Technol. (K.J. Lorenz and K. Kulp, eds.) New York: Marcel Dekker, Inc. 41:863-870.

Dawson, K.R., I. Eidet, J. O'Palka and L. Jackson. 1987. Barley neutral lipid changes during the fuel ethanol production process and product acceptability from the dried distillers grains. J. Food Sci. 52:1348-1352.

Dong, F.M. and B.A. Rasco. 1987. The neutral detergent fiber, acid detergent fiber, crude fiber and lignin contents of distillers dried grains with solubles. J. Food Sci. 52:403-405, 410.

Dong, F.M., B.A. Rasco and S.S. Gazzaz. 1987. A protein quality assessment of wheat and corn distillers dried grains with solubles. Cereal Chem. 64:327-332.

Dowd, M.K., P.J. Reilly and W.S. Trahanovsky. 1993. Low molecular weight organic composition of ethanol stillage from corn. Cereal Chem. 70:204-209.

Gagen, W.L. 1973. Industrial uses of wheat gluten, starch, millfeeds and other by-products. *In*: Industrial Uses of Cereals. Y. Pomeranz, ed. St. Paul, MN. Amer. Assc. Cereal Chem. p. 348-366.

Graf, E. 1983. Applications of phytic acid. J. Amer. Chem. Soc. 60:1861-1867.

Gras, P.W. and D.H. Simmonds. 1980. The utilization of protein-rich products from wheat carbohydrate separation processes. Food Technol. Australia 32:470-472.

Hayman, G.T., B.M. Mannarelli and T.D. Leathers. 1995. Production of carotenoids by *Phaffia rhodozyma* grown on media composed of corn wet milling co-products. J. Ind. Microbiol. 14:389-395.

Hojilla-Evangelista, M.P., D.J. Myers and L.A. Johnson. 1992a. Characterization of protein extracted from flaked, defatted, whole corn by the sequential extraction process. J. Amer. Oil. Chem. Soc. 69:199-204.

Hojilla-Evangelista, M.P., L.A. Johnson and D.J. Myers. 1992b. Sequential extraction processing of flaked whole corn: alternative corn fractionation technology for ethanol production. Cereal Chem. 43:643-647.

Ingledew, W.M., A.M. Jones, R.S. Bhatty and B.G. Rossnagel. 1995. Fuel alcohol production from hull-less barley. Cereal Chem. 72:147-150.

Julian, G.S., R.J. Bothast, L.H. Krull. 1990. Glycerol accumulation while recycling thin stillage in corn fermentations to ethanol. J. Ind. Microbiol. 5:391-394.

Kim, C.H., J.A. Maga and J.T. Martin. 1989. Properties of extruded dried distillers grains and flour blends. J. Food Process. Preserv. 13:219-231.

Koseolu, S.S., K.C. Rhee and E.W. Lusas. 1991. Membrane separations and applications in cereal processing. Cereal Foods World 36:376-383.

Lapvetelainen, A., and T. Aro. 1994. Protein composition and functionality of high protein oat flour derived from integrated starch ethanol process. Cereal Chem. 71:133-139.

Leathers, T.D. and S.C. Gupta. 1994. Production of pullulan from fuel ethanol by-products by *Aureobasidium sp.* Straw NRRL Y-12, 974. Biotechnol. Lett. 16:1163-1166.

Ma, C.Y. 1983. Chemical characterization and functionality assessment of protein concentrates from oats. Cereal Chem. 60:36-42.

Maiorella, B.L., H.W. Blanch and C.R. Wilke. 1983. Distillery effluent treatment and by product recovery. Process Biochem. 18:5-8.

McIntosh, G.H., L. Jorgensen and P. Royle. 1993. Insoluble dietary fiber-rich fractions from barley protects rats from intestinal cancers. R. Soc. Chem. (G.B.) 123 (Spec. Publ.): Chem. Abst. 119: 248800a.

Murray, E.D., M.A.H. Ismond, S.D. Arntfield and K.J. Shaykewich. 1987. Improved economics for agricultural resources through minor component recovery. University of Manitoba, 2nd edition.

Myers, D.J., M.P. Hojilla-Evangelista and L.A. Johnson. 1994. Functional properties of protein extracted from flaked, defatted, whole corn by ethanol/alkali during sequential extraction processing. J. Amer. Oil. Chem. Soc. 71:1201-1204.

Myung, G.S., H.A. Kab, G.L. Min and K.S. Seung. 1992. Treatment of alcoholic distillery wastes through citric acid fermentation. J. Korean Instit. Chem. Engineers 130:473-479.

Nielsen, H.C., G.E. Inglett, J.S. Wall and G.L. Donaldson. 1973. Corn germ protein isolate

– Preliminary studies on preparation and properties. Cereal Chem. 50:435-443.

Nielsen, H.C., J.S. Wall and G.E. Inglett. 1979. Flour containing protein and fiber made from wet-mill corn germ, with potential food use. Cereal Chem. 56:144-146.

O'Palka, J. 1987. Incorporation of dried distillers grains in baked products. Proc. Distillers Feed Conf. 42:47-54.

Orthoefer, F.T. and R.D. Sinram. 1987. Corn oil: composition, processing and utilization. *In*: Corn: Chemistry and Technology. (S.A. Watson and P.E. Ramstad, eds.) Amer. Assoc. Cereal Chem. Inc. p. 535-551.

Paik, H.D. and B.A. Glatz. 1994. Propionic-acid production by immobilized cells of a propionate-tolerant strain of *Propionibacterium acidipropionici*. Appl. Microbiol. Biotech. 42:22-27.

Paton, D., S. Bresciani, N.F. Han and J. Hart. 1995. Oats: chemistry, technology and potential uses in the cosmetic industry. Cosmetics and Toiletries 110:63-70.

Rasco, B.A. and W.J. McBurney. 1989. Human food product produced from dried distillers spent cereal grains and solubles. US Patent No. 4,828,846.

Rasco, B.A., S.E. Downey and F.M. Dong. 1987. Consumer acceptability of baked goods containing distillers dried grains with solubles from soft white winter wheat. Cereal Chem. 64:139-143.

Rasco, B.A., S.E. Downey, F.M. Dong and J. Ostrander. 1987. Consumer acceptability and color of deep-fried fish coated with wheat or corn distillers dried grains with solubles (DDGS). J. Food. Sci. 52:1506-1508.

Reddy, N.R., F.W. Cooler and M.D. Pierson. 1986. Sensory evaluation of canned meat-based foods supplemented with dried distillers grain flour. J. Food Qual. 9:233-242.

Reiners, R.A., J.S. Wall and G.E. Inglett. 1973. Corn proteins: potential for their industrial use. *In*: Industrial Uses of Cereal. (Y. Pomeranz, ed.) St. Paul, MN. Amer. Assoc. Cereal Chem. p. 285-302.

Salomonsson, A.C., O. Theander and E. Westerlund. 1984. Chemical characterization of some Swedish cereal whole meal and bran fractions. Swedish J. Agric. Res. 14:111-118.

San Buenaventura, M.L., F.M. Dong and B.A. Rasco. 1987. The total dietary fiber content of distillers dried grains with solubles. Cereal Chem. 64:135-136.

Satterlee, L.D. 1981. Proteins for use in foods. Food Technol. 35:53-70.

Scheller, W.A. 1981. Gasohol: The U.S. Experience. *In*: Cereals: A Renewable Resource. (Y. Pomeranz and L. Munck, eds.) St. Paul, MN. Amer. Assoc. Cereal Chem. p. 633-649.

Shukla, .T.P. 1981. Industrial uses of dry-milled corn products. *In*: Cereals: A Renewable Resource. (Y. Pomeranz and L. Munck, eds.) St. Paul, MN: Amer. Assoc. Cereal Chem. p. 633-649.

Sternberg, M., R.D. Phillips and L. Daley. 1980. Maize protein concentrate. *In*: Cereals for Food and Beverages, Recent Progress in Cereal Chemistry and Technology. (G.E. Inglett and L. Munck, eds.) NY Academic Press. p. 275-285.

Thomas, K.C. and W.M. Ingledew. 1995. Production of fuel alcohol from oats by fermentation. J. Ind. Microbiol. 15:125-130.

Thompson, L.V. 1992. Potential health benefits of whole grains and their components. Contemp. Nutr. Vol. 17.

Tkac, J.J. 1992. Process for removing bran layers from wheat kernels. US Patent No. 5,082,680.

Tsen, C.C. 1976. Regular and protein fortified cookies from composite flours. Cereal Foods World 21:633-640.

Tsen, C.C. 1980. Cereal germs used in bakery products: Chemistry and nutrition. *In*: Cereals for Food and Beverages, Recent Progress in Cereal Chemistry and Technology. (G.E. Inglett and L. Munck, eds.) NY Academic Press. p. 245-253.

Walker, A.R.P. 1974. Dietary fiber and the pattern of diseases. Ann. Intern. Med. 80:663.

Weber, F.E., V.K. Chaudhary and A.A. Qureshi. 1991. Suppression of cholesterol biosynthesis in hypercholesteremic subjects by tocotrienol of barley ingredients made from brewers grain. Cereal Foods World 36:680.

Weegels, P.L., J.P. Marseille and R.J. Hamer. 1992. Enzymes as a processing aid in the separation of wheat flour into starch and gluten. Starch 44:44-48.

Wood, P.J., J.W. Anderson, J.T. Braaten, N.A. Cave, F.W. Scott and C. Vachon. 1989. Physiological effects of ß-D-glucan rich fractions from oats. Cereal Foods World 34:878-882.

Wu, Y.V. 1985. Fractionation and characterization of protein rich material from barley after alcohol distillation. Cereal Foods World 30:540.

Wu, Y.V. 1987. Recovery of stillage soluble solids from hard and soft wheat by reverse osmosis and ultrafiltration. Cereal Chem. 64:260-264.

Wu, Y.V. 1990. Recovery of protein rich by-products from oat stillage after alcohol distillation. J. Agric. Food Chem. 38:588-592.

Wu, Y.V. 1994. Determination of neutral sugars in corn distillers dried grains, corn distillers dried solubles, and corn distillers dried grains with solubles. J. Agric. Food Chem. 42:723-726.

Wu, Y.V. and A.C. Stringfellow. 1973. Protein concentrates from oat flours by air classification of normal and high-protein varieties. Cereal Chem. 50:489.

Wu, Y.V., J.E. Cluskey, J.S. Wall, G.E. Inglett. 1973. Oat protein concentrates from a wet-milling process: composition and properties. Cereal Chem. 50:481-488.

Wyman, C.E., and B.J. Goodman. 1993. Biotechnology for production of fuels, chemicals, and materials from biomass. Appl. Biochem. Biotechnol. 39:39-59.

Yuen, S. 1974. Pullulan and its applications. Proc. Biochem. Nov. p. 7-9, 22.

The Alcohol Alphabet

A glossary of terms used in the ethanol-producing industries

Compiled by John E. Murtagh

Italic type denotes words (or minor variations of words) defined under separate entries.

A

Å Abbreviation for *Angstrom*.

Absolute ethanol A pharmaceutical term for *anhydrous ethanol*. It is generally defined as having less than 1% water.

Acetaldehyde Otherwise known as ethanal, acetic *aldehyde* or ethylaldehyde. A clear flammable liquid with a characteristic pungent odor. Chemical formula CH_3CHO. Boils at 21°C and freezes at -123.5°C. It is miscible in both ethanol and water. It has a narcotic effect on humans, and large doses may cause death by respiratory paralysis. It is a *congener* in the production of ethanol by *fermentation*, and is usually a major constituent of the *heads* fraction removed in *rectification*.

Acetic acid A colorless liquid with a pungent odor. Flammable at high concentrations. Chemical formula CH_3COOH. Acetic acid may be produced from *ethanol* by *Acetobacter bacteria* under aerobic conditions such as when a completed fermentation is agitated or aerated excessively.

Acetobacter A genus of *Gram-negative*, aerobic *bacteria* comprising ellipsoidal to rod-shaped cells as singles, pairs or chains. Otherwise known as acetic acid or vinegar bacteria, they are able to oxidize *ethanol* to acetic acid. They may be responsible for loss of yield in ethanol production if a *fermented mash* is agitated or aerated excessively.

Acetone Otherwise known as 2-propanone, dimethyl ketone or pyroacetic ether. It is a volatile, very-flammable liquid with a characteristic pungent 'mousey' odor and a sweetish taste. Chemical formula CH_3COCH_3. Boils at 56.5°C and freezes at -94°C. It is miscible with water, *ethanol* and most oils. Inhalation may cause headaches, and in large amounts, narcosis. It is a *congener* in the production of ethanol by *fermentation*, particularly from *molasses mashes*. It is also a by-product of the production of *butanol* by *anaerobic* fermentation using the *bacterium Clostridium acetobutylicum*. It tends to concentrate in the *heads* fraction in the *rectification* of *neutral spirit*.

Acid-acid process Term used in *starch* processing when *acid hydrolysis* is used to accomplish both the initial *liquefaction* and the final *saccharification* to simple *sugars*.

Acid-enzyme process Term used in *starch* processing when *acid hydrolysis* is used to accomplish the initial *liquefaction*, and an *enzyme* such as *amyloglucosidase* is used for the *saccharification* to simple *sugars*.

Acid hydrolysis The *hydrolysis* of a *polymer* by the use of acid. In the case of *starch* hydrolysis, acids may be used as an alternative to *enzymes* in either (or both) the *liquefaction* or *saccharification* processes.

Acid washing A process in which *yeast* recovered from a finished *fermentation* is acidified to reduce the level of *bacterial contamination* prior to recycling into a new fermentation.

Aguardiente An unaged alcoholic beverage produced in Central and South America by the *distillation* of *beer* derived from the *fermentation* of sugarcane juice or *molasses*. It is similar to crude *rum*.

Alcohol A member of a class of *organic* compounds containing carbon, hydrogen and oxygen. Considered to be *hydroxyl* derivatives of *hydrocarbons* produced by the replacement of one or more hydrogen atoms by one or more *hydroxyl* (OH) groups. Under the International Union of Pure and Applied Chemistry (IUPAC) naming system, the name given to an alcohol is derived from the parent hydrocarbon with the final 'e' changed to 'ol'. Thus *methane-methanol, ethane-ethanol* etc. The principal alcohol in fuel and beverage use is *ethanol* (otherwise known as *ethyl alcohol*.)

Alcohol Fuel Permit (AFP) A permit issued by the *Bureau of Alcohol, Tobacco and Firearms* allowing the holder to engage in the production of *ethanol* solely for fuel use.

Aldehyde A member of a class of *organic* compounds considered to be derived by the removal of hydrogen atoms from an *alcohol*. Aldehydes tend to be produced as *congeners*, or *by-products* of *fermentation*; and having a lower *boiling point* than *ethanol*, they tend to vaporize more readily and may accumulate as a *'heads'* fraction in the *distillation* process.

Alpha-amylase (α-amylase) An *enzyme* used in the *liquefaction* of *starch* in the grain *mashing* process prior to *saccharification* and *fermentation*. α-amylase *hydrolyzes* the long-chain *starch molecules* into short-chain *dextrins*. These are more suitable for subsequent saccharification by other enzymes to fermentable *glucose*. (α-amylase is an *endoenzyme* in that it works from the inside of the *amylose molecule*). In beverage alcohol production α-amylase may be derived from *malt* (sprouted barley), but in *fuel ethanol* production the enzyme is obtained solely as a *bacterial* product. The enzyme *molecule* contains a calcium atom, which is essential for its activity.

American Society for Testing and Materials (ASTM) A scientific and technical organization with headquarters in Philadelphia, PA established for 'the development of standards on characteristics and performance of materials, products and services and the promotion of related knowledge'. It sets voluntary consensus standards through committees representing producers and users. ASTM has published standards for *fuel ethanol* and *gasoline* that have been adopted by many states.

Amyl alcohol The principal constituent of *fusel oil*. Otherwise known as *pentanol*. Chemical formula $C_5H_{11}OH$. Eight *isomers* exist, the most common being primary isoamyl alcohol.

Amylase The name given to any *enzyme* that *hydrolyzes* (or breaks down) *amylose*, which is a major component of *starch*.

Amyloglucosidase An *enzyme*, also known as *glucoamylase*, which *hydrolyzes* amylose into its constituent *glucose* units. It is an *exoenzyme* in that it works from an outer end of the *molecule*. It is normally used in conjunction with *α-amylase* for the *liquefaction* and *saccharification* of *starch* in the grain *mashing* process, prior to *fermentation*. Amyloglucosidase is a product of fungal growth.

Amylopectin A major component of *starch* (together with *amylose*). The *molecule* is composed of large, branched chains of thousands of *glucose* units. The branching of the *molecule* distinguishes it from amylose and makes it less-easily *hydrolyzed* into its constituent *glucose* units for *fermentation*.

Amylose A major component of *starch* (together with *amylopectin*). The amylose *molecule* is composed of straight chains of hundreds of *glucose* units. In the grain-*mashing* process for *ethanol* production, amylose may first be broken down into short-chain *dextrins* by *a-amylase*, which are in turn broken down into single *glucose* units by *amyloglucosidase*.

Anaerobic Literally means 'without air'. The opposite of aerobic. For example, *yeast propagation* (or multiplication) is more rapid with aeration (i.e., under aerobic conditions), while yeast produces *ethanol* under anaerobic conditions of *fermentation*.

Anaerobic digestion Process of breaking down waste materials by *anaerobic bacterial* degradation. Normally accompanied by the production of *methane* gas.

Angstrom (Å) A unit of length equal to one hundred-millionth (10^{-8}) of a centimeter, used for measuring the diameter of chemical *molecules*. Thus, in *ethanol dehydration*, a *molecular sieve* material with holes 3 Angstroms in diameter may be used to separate water, which has a 2.5 Å diameter, from *ethanol*, which has a 4.5 Å diameter.

Anhydrous Literally means 'without water'. The term used for a substance that does not contain water. *Ethanol* for fuel use is commonly referred to as anhydrous, because it has had almost all of the water removed. See *absolute ethanol*.

Antibiotic A chemical substance produced by *microorganisms* with the capacity to inhibit growth or kill other microorganisms. The antibiotic most commonly used in *ethanol* production is *penicillin*.

Antifoam (or defoamer) A preparation composed of substances such as silicones, *organic* phosphates and *alcohols* that inhibits formation of bubbles in a liquid during agitation by reducing the surface tension. Antifoam may be used in *ethanol* production to control the development of foam in *fermenters*. Antifoam may also be used to control foam in *beer distillation* to increase the stripping capacity of a *column*.

Antiscalant (or scale inhibitor) A chemical compound or mixture of compounds added to water or *molasses beer* to reduce the incidence of scaling in *heat exchangers* or *distillation columns*. Usually, the principal ingredients of antiscalants are chelating compounds.

Arabinose A *pentose sugar* comprising a major constituent of *hemicellulose*. Chemical formula $C_5H_{10}O_5$. It is not fermented by normal *strains* of distillers *yeasts*. Also known as pectin sugar or gum sugar.

Archer Daniels Midland (ADM) Based in Decatur, IL. The largest US *fuel ethanol* producer and a major producer of *beverage* and *industrial alcohol*.

Atomic weight The relative mass of an atom based on a scale in which a specific carbon atom (carbon 12) is assigned a mass value of 12. Also known as relative atomic mass.

Azeotrope The term used to describe a constant boiling mixture. It is a mixture of two (or more) components with a lower *boiling point* than either component alone. For example water, which boils at 100°C, and *anhydrous ethanol*, which boils at 78.5°C, form a constant boiling mixture (azeotrope) at 78.15°C. The *vapor* of the mixture has the same composition as the liquid and therefore no further concentration can be achieved by normal *distillation*. Under normal pressures it contains approximately 97% by volume *ethanol* (194° *proof*). It is very expensive in terms of energy to attempt to reach 194° proof, so 190° proof is generally considered to be the practical, economic azeotrope limit for *fuel ethanol* distillation.

Azeotropic distillation A *distillation* process in which a liquid compound (*entrainer*) is added to the mixture to be separated to form an *azeotrope* with one or more of the components. Normally, the entrainer selected is easily separated from the component to be removed. For example, when *benzene* is used in azeotropic distillation to dehydrate *ethanol*, the *overhead condensate phase-separates* to yield a water-rich layer that can be withdrawn and a benzene-*ethanol* layer which is *refluxed*.

B

Backset Recycled *thin stillage*. It may be added to the *cooker* or to the *fermenter* and serves as a source of nutrients. It reduces the water required for *mashing* and reduces the volume of liquid residue to be evaporated. Improperly handled, it may be a major source of *bacterial contamination*.

Bacteria Any of a large group of microscopic plants constituting the class Schizomycetaceae having round, rod-like, spiral, or filamentous single-celled bodies that are often aggregated into colonies, are often motile by means of flagella, and reproduce by fission or by the formation of asexual resting spores. They may live in soil, water, organic matter or the live bodies of plants and animals. In *ethanol* production, bacteria are significant in that they compete with *yeast* to ferment the available *sugars* in a *mash* to products other than *ethanol* and cause losses in yield. However, some bacterial cultures may be added deliberately to *rum fermentation*s to help produce certain desired congeners. One genus of bacteria, *Zymomonas,* is being examined commercially for its ability to ferment sugars to *ethanol*.

Bacterial contamination The condition occurring when undesirable *bacteria* become established in a fermenting *mash* and reduce the *ethanol* yield. The *bacteria* use available *sugars* to produce various compounds (*congeners*), particularly acids, which may inhibit *yeast* activity. In severe situations, bacterial contamination may cause serious economic losses.

Balling (or Brix) A scale used to measure the *specific gravity* of a liquid in relation to that of a *solution* of *sugar* in water. Each unit on the scale is equivalent to 1% by weight of sugar. Thus a *mash* of 20° Balling has the same specific gravity as a 20% w/w sugar solution. The scale is frequently considered to indicate % dissolved solids in a liquid, although this is only true of solutions of pure sugar. Traditionally, the term 'Balling' has been used in grain distilleries, while 'Brix' has been used in sugar mills and *rum* or *molasses* alcohol distilleries. The measurement is accomplished by use of a Balling (or Brix) *hydrometer*.

Barbet time See Permanganate time.

Bargeload Generally refers to a river barge with a capacity of 10,000 *barrels* or 420,000 gallons.

Barrel A liquid measure equal to 42 US gallons or 5.6 cubic feet. Or, a wooden container used for the aging and maturation of alcoholic beverages. Barrels used for *whiskey* maturation are made of oak wood and have a capacity of about 52 US gallons. Barrels may only be used once for aging *bourbon whisky*, so there is a worldwide trade in used bourbon barrels for aging other alcoholic products such as Scotch whisky and *rum*.

Base losses The percentage of *ethanol* lost in the *stillage* at the base of a *beer stripping column* or a *rectifying column*. It is virtually impossible to achieve zero base losses, and it would be wasteful of steam. The base losses are generally monitored and controlled in an optimal range determined by steam costs, etc.

BATF Abbreviation for US Bureau of Alcohol, Tobacco and Firearms.

Batch cooking Cooking a set amount of grain *meal,* water and *backset* (if any) in a single vessel as a discontinuous operation. One or more batch cooks may be used to fill a single *batch fermenter.* Batch cooking is mainly confined to the *beverage alcohol* industry and smaller plants in the *fuel ethanol* industry.

Batch distillation Distilling batches of *beer* in a discontinuous operation. It is not commonly used in *fuel ethanol* production, but is used in the beverage *ethanol* industry particularly for the production of special types of heavily-flavored *distillates.*

Batch fermentation The fermentation of a set amount of *mash* in a single vessel in a discontinuous operation. In the *ethanol* production industries, batch *fermentation* predominates over *continuous fermentation.*

Beer The name given to the product of *fermentation.* In *grain alcohol* production beer may contain about 9-12% *ethanol.*

Beer preheater A heat-exchanging device used for heating *beer* before it enters a *beer still.* Usually, it also serves as the first *condenser* for the beer still, so that the *overhead vapors* heat the beer.

Beer still The *distillation* unit used for the initial removal of *ethanol* from finished *beer.* It generally consists of a *stripping section* that extracts the *ethanol* from the beer and a *concentrating or rectifying section,* which normally takes the *ethanol* up to 190^0 proof (95° *GL*). Beer stills may consist of a single tall *column,* or two or more columns standing side by side, linked by vapor pipes.

Beer stripping column See *Beer still.*

Beer well The holding vessel into which finished *beer* is transferred prior to *distillation.*

Benzene Colorless, flammable, aromatic *hydrocarbon* liquid. Chemical formula C_6H_6. Boils at 80.1°C and freezes at 5.4°C. Used as an *entrainer* for the *dehydration* of *ethanol* by *azeotropic distillation.* Known to be carcinogenic.

Beta-amylase (ß-amylase) An *enzyme* that *hydrolyses* the long-chain *amylose molecules* in *starch* into fermentable *maltose,* the *dimer* (or double-*molecule*) of *glucose.* (It is an *exoenzyme* in that it works from an outer end of the molecular chain. It is found in *malt* (sprouted barley) in association with *α-amylase.* With the advent of microbial *amyloglucosidase enzymes,* malt amylases are generally only used in the production of heavily-flavored *beverage alcohol.*

Betaglucan (ß-glucan) Gum-like *polymers* of ß-linked *glucose* as in *cellulose* instead of the α-linked glucose units as in *starch* (*amylose*). Betaglucan is commonly present in barley *mashes.* It is not broken down by *α-amylase* and causes foaming problems due to its viscous elastic nature.

Betaglucanase (ß-glucanase) An *enzyme* which *hydrolyses betaglucan.* It is frequently used in barley *mashing* to reduce foaming and viscosity problems.

Beverage alcohol Any form of distilled *ethanol* with or without *congeners* considered by law to be fit for human consumption. The laws in many countries confine beverage *alcohol* to that produced by *fermentation* as opposed to alcohol of synthetic origin.

Binary azeotrope An *azeotrope* or *constant boiling mixture* having two components such as *ethanol* and water.

Bio-ethanol See *fermentation ethanol.*

Bio-gas The gas produced by *anaerobic digestion* of wastes. It is mainly *methane.*

Biological oxygen demand (BOD) A measure of the oxygen-consuming capacity of *organic* matter contained in effluents. During decomposition, organic effluents have an oxygen requirement that may deplete the supply in a waterway and result in the death of fish and other aquatic life. BOD is normally measured on the basis of the weight of oxygen required per unit volume of an effluent.

Biomass Any renewable *organic* matter such as agricultural crops, crop waste residues, wood, animal and *municipal wastes*, aquatic plants, fungal growth, etc.

Blackstrap See *molasses*.

Blended whisky Defined by the *US Bureau of Alcohol, Tobacco and Firearms* as a mixture containing at least 20% of *straight whisky* on a *proof gallon* basis, together with other *whiskies* or *neutral spirits*. When a blend contains more than 51% of a straight whisky, it may be designated by that specific type, such as 'blended rye whisky'.

Blender Tax Credit A US federal income tax credit granted to blenders of *ethanol* and *gasoline* as an alternative to the *excise tax exemption*. It was originally introduced under the *Crude Oil Windfall Profits Tax Act of 1980*, at a level of 40 cents per gallon, and was subsequently raised to 50 cents by the *Surface Transportation Assistance Act of 1982*, and to 60 cents by the *Deficit Reduction Act of 1984*.

BOD (See *Biological Oxygen Demand*)

Boiler efficiency The thermal efficiency of a boiler in terms of the usable energy output (in the form of steam) in relation to the energy input in the form of fuel. Boiler efficiencies are commonly in the 70-80% range.

Boiling point The temperature at which the transition occurs from the liquid to the gaseous phase. In a pure substance at a fixed pressure the boiling point does not vary.

Bourbon whisky Defined by the *BATF* as *whisky* produced at not more than 160° *proof* from a fermented *mash* of not less than 51% corn, and stored at not more than 125° proof in charred, new oak *barrels*. In practice, the bourbon whisky *mash bill* is frequently about 65% corn, 25% rye, and 10% barley malt.

Brandy Defined by the *BATF* as 'an alcoholic *distillate* from the fermented juice, *mash* or wine of fruit, or from the residue thereof, produced at less than 190° *proof*, in such a manner that the distillate possesses the taste, aroma and characteristics generally attributed to the product'.

British Thermal Unit (BTU) The amount of heat required to raise the temperature of one pound of water one degree Fahrenheit under defined pressure conditions. It is the standard unit for measuring heat energy in the US.

Brix See *Balling*.

BTU Abbreviation for *British Thermal Unit*.

BTX Abbreviation for the three related *octane enhancers, benzene, toluene* and *xylene*.

Bubble cap A contacting device used on some *distillation plates*. It consists of a cylindrical chimney set in a hole in the plate and covered by a dome-shaped cap, which deflects the *vapors* rising up the chimney to cause them to pass through the liquid layer on the plate.

Bureau of Alcohol, Tobacco and Firearms (BATF) An agency of the US Department of the Treasury entrusted with enforcing laws covering the production, distribution and use of alcohol, tobacco and firearms.

Bushel A unit of dry volume equal to 2150.42 cubic inches or 1.244 cubic feet. When used to measure grain, bushel weight depends on the type and condition of the grain. In the case of corn, bushel volume generally averages about 56 lbs by weight. This has led to the use of a *distillers bushel*, which represents 56 lbs of grain, regardless of type or volume.

Butanol (butyl alcohol) A minor constituent of *fusel oil*. Chemical formula C_4H_9OH. Four *isomers* exist. They are all colorless, toxic flammable liquids. n-Butanol may be produced as a *co-product* with acetone and *ethanol* by the *fermentation* of selected *carbohydrates* with the *anaerobic bacterium Clostridium acetobutylicum*. Butanols are used as *solvents* and chemical intermediates.

By-products Products that are secondary to the principal product of a process. In *ethanol* production, *carbon dioxide* and *distillers dried* *grains* are normally considered by-products; but in certain circumstances they may be viewed as *co-products*, in that they may contribute significantly to the overall process economics.

C

Carbohydrate Any of a group of compounds composed of carbon, hydrogen and oxygen including the *sugars, starches, dextrans* and *celluloses*. They are the most abundant class of *organic* compounds in nature, constituting approximately 75% of the dry weight of all vegetation.

Carbon dioxide A colorless non-flammable gas. Composition CO_2. It does not support human respiration; and in high concentrations it causes asphyxiation. It is approximately 1.5 times the weight of air, and tends to accumulate in floor drains, pits and in the bottoms of unventilated tanks. It is produced by various means, notably the combustion of fuels in an excess of air and is a *by-product* of *yeast fermentation*. It may be recovered from *fermentations* and compressed to a liquid or solid (dry ice).

Carbon monoxide A colorless, odorless, flammable gas ('coal gas'). Composition CO. It is poisonous if inhaled, as it combines with blood hemoglobin to prevent oxygen transfer. It is very slightly lighter than air. It is produced by the incomplete combustion of fuels with a limited oxygen supply, as in auto engines. It is a major component of urban air pollution, which can be reduced by blending an oxygen-bearing compound (or *oxygenate*) such as *ethanol* into *hydrocarbon* fuels.

Carbon steel A steel deriving its particular properties from its content of carbon. These steels may range from 'low-carbon', (having less than 0.25% carbon), to 'high-carbon'). Low carbon or 'mild' steel is easily hot-worked and rolled, and is used in steel plates and beams. It is easily welded, but is readily subject to rusting and other forms of *corrosion*.

Caribbean Basin Initiative (CBI) The program arising from the Caribbean Basin Economic Recovery Act passed by the US Congress in 1983 to encourage industrial development in the Caribbean islands and Central America. Under this program, *fuel ethanol* and other products from the region may enter the US without being subject to customs duties.

Cassava A root crop with a high *starch* content grown in the tropics and subtropical regions. Known in Brazil as *manioc*, it is used as an alternative to sugarcane as a *feedstock* for *ethanol* production. It is also processed for food as 'tapioca'.

CBI Abbreviation for *Caribbean Basin Initiative*.

CCC Abbreviation for *Commodity Credit Corporation*.

CDA Abbreviation for *completely denatured alcohol*.

Cell recycle The process of recovering *yeast* from fermented *beer* to return it to the starting vessel of a *continuous fermentation* or to a new vessel in a *batch fermentation* system. It may include an *acid washing* step to reduce *bacterial contamination*.

Cellulase An *enzyme* capable of *hydrolyzing* long-chain *cellulose molecules* into simple *sugars* or short-chain *polymers*.

Cellulose The principal *polysaccharide* in living plants. It forms the skeletal structure of the cell wall, hence the name. It is a *polymer* of *glucose* units coupled by ß-type linkages into chains of 2,000-4,000 units. Cellulose normally occurs with other polysaccharides and *hemicelluloses* derived from sugars such as *xylose, arabinose* and *mannose*.

Celsius (or centigrade) A temperature scale in which (at normal atmospheric pressure) water freezes at zero degrees and boils at 100 degrees.

Centrifugal pump A machine for moving liquids by accelerating them radially outwards by means of a rotating impeller contained within a casing. It is the most common and most versatile type of pump in normal industrial use. It has an advantage in that it does not force a *positive displacement* of liquids, so it may be used in systems where output must be throttled or completely closed by control valves. This feature makes centrifugal pumps unsuitable for moving heavy viscous liquids such as *molasses* and syrups.

Centrifuge A machine for separating insoluble liquids or solids from liquids by the application of centrifugal force. The two main types of centrifuges are filter and sedimenter. In filtering centrifuges, solids are retained in a rotating perforated basket while liquid passes through the perforations. In sedimenting centrifuges, the mixture is thrown against a solid-walled cylinder and the heavier solid particles collect against the wall for removal while the lighter liquid collects in the central part. Centrifuges are commonly used in *ethanol* plants for *yeast* recovery and in *stillage dewatering*.

Chemical oxygen demand (COD) A laboratory test to determine the oxygen requirements for the chemical digestion of effluents. It should be distinguished from the *BOD* test, which determines the total of both chemical and biological oxygen requirements. COD is measured in a rapid test that involves heating the effluent in the presence of an oxidizing agent such as potassium dichromate and then determining the amount of oxygen absorbed by the effluent.

Chlorine dioxide Chemical formula ClO_2. It is a strongly-oxidizing, yellow-to-reddish-yellow gas at room temperature. It has an unpleasant odor similar to that of chlorine and reminiscent of nitric acid. It is unstable in light. It reacts violently with *organic* materials and is easily detonated by sunlight or heat in concentrations greater than 10% at atmospheric pressure. Boils at 11°C and freezes at -59°C. Chlorine dioxide may be used as a sterilant, and may be produced *in situ* for sterilizing *yeast mashes* by addition of sodium chlorite *solution* in the presence of acids or chlorine (or hypochlorite solution). It is considerably more effective as a sterilant than straight chlorine.

Chromatography A method for separating a mixture of chemical compounds into individual components by selective distribution between two immiscible materials (or phases), one stationary and the other mobile. The phases are selected so that the mobile phase will carry the various components through the stationary (or solid) phase at differing rates to give separation. *Gas chromatography* may be used to separate *ethanol* from the other *congeners* produced in fermentation and to measure them quantitatively. *High performance liquid chromatography* may be used to follow *starch hydrolysis* in *grain alcohol* production.

CIP Abbreviation for *cleaning-in-place system*.

Citrus molasses A *by-product* of the citrus juice industry. Citrus residue, mainly peel, is treated with lime and then passed through a press. The press liquor is then evaporated to a viscous, dark brown *molasses* of about 72° *Brix*. Citrus molasses is similar to cane *blackstrap molasses,* having about 45% total *sugars*. However, it has more *protein* and a much lower ash content. Citrus molasses may be diluted and fermented for *ethanol* production, but it may need pretreatment to reduce the content of d-limonene (commonly referred to as 'citrus-stripper oil'), which tends to inhibit *yeast* growth.

Cleaning-in-place system (CIP) A system designed to permit process equipment to be cleaned without disconnecting or dismantling. A sophisticated automatic CIP system may provide a sequence of water flushing cycles, detergent washing and chemical sterilizing of equipment at the press of a button.

Closed receiver (See *Receiver.*)

CMS Abbreviation for *condensed molasses solubles.*

Coccus (plural: cocci) A type of *bacteria* with spherically-shaped cells. They may occur as single cells, clusters or long chains.

COD Abbreviation for *chemical oxygen demand.*

Column A vertical cylindrical vessel containing a series of perforated *plates* or other contact devices through which *vapors* may pass to effect a separation of liquid mixtures by *distillation.*

Co-mingled tank The term used for *fuel ethanol* tanks at refineries or pipeline terminals where two or more suppliers may share the same tank for storing *ethanol.*

Commodity Credit Corporation (CCC) An agency of the *USDA* established to stabilize and protect farm incomes and prices by maintaining balanced and adequate supplies of agricultural commodities.

Completely denatured alcohol (CDA) A term used by the *BATF* to describe *ethanol* made unfit for human consumption by addition of specified *denaturants* such as methyl isobutyl ketone, *kerosene* or *gasoline.*

Concentrating column A column where *hydrous ethanol* is concentrated to decrease water content. A *beer still* may consist of a *stripping column* in which dilute *ethanol* is removed from *beer* and a concentrating column, which receives *ethanol vapor* from the stripping column and concentrates it up to about 190° *proof* (95°GL). If provision is made in a concentrating column to remove impurities such as *fusel oil,* then it is more correctly a *rectifying column.*

Condensate Liquid condensed from *vapor* in a *condenser.*

Condensed molasses solubles (CMS) The term used to describe *molasses stillage* concentrated by *evaporation.* The molasses residue (after *fermentation* and *distillation*) may be concentrated to about 60° *Brix* (or approximately 60% solids) to be sold as a substitute for molasses in animal feeds as a caking agent and dust suppressant. It contains high concentrations of salts.

Condenser A *heat exchange* device connected to the *vapor* discharge pipe of a *column* to permit the vapor to be cooled and condensed to a liquid. Condensers are commonly cylindrical vessels containing tubes through which cooling water is passed.

Congeners Chemical compounds produced with *ethanol* in the *fermentation* process. They are frequently referred to as impurities. Common congeners are *methanol acetaldehyde, esters* (such as ethyl acetate) and *fusel oils* (*higher alcohols,* particularly *amyl alcohols.*) *Fermentation* conditions may be adjusted to control congener formation depending on the requirements for the end product.

Constant boiling mixture See *Azeotrope.*

Continuous cooker A system into which a *mash* of water, grain and *enzymes* may be fed continuously to be cooked and discharged to the *fermentation* system. Continuous cookers generally consist of a *slurry tank* connected by a pump to a steam jet heater, a holding vessel or lengths of piping (to provide some residence time at the cooking temperature), one or more flash vessels (to cool the cooked mash), a holding vessel for enzymatic *liquefaction* and a *heat exchanger* for final mash cooling. Continuous cookers are more common in *fuel ethanol* plants than in *beverage alcohol* plants.

Continuous fermentation A system into which cooked *mash* may be fed continuously to be fermented and then discharged to the *beer well* and *distillation* system. Continuous *fermentation* systems generally consist of a series of interconnected tanks sized to provide sufficient residence time for the *fermentation* to proceed to completion. The *ethanol* industry is divided on *fermentation* systems, with more gallonage produced by batch *fermentation* than by continuous *fermentation*.

Continuous distillation A process using specially-designed equipment to permit a *volatile* component such as *ethanol* to be separated by *distillation* from a continuous flow of an *aqueous solution* such as *beer*.

Control loop A portion of a process control system that includes a sensing device connected to a signal transmitter, a controller, another signal transmitter and an actuator. For example, a pressure control loop on a *distillation column* may have a pressure sensor connected through a controller to a steam flow regulating valve.

Cooker A device for heating a slurry of grain and water to a sufficiently-high temperature for sufficient time to release and *gelatinize* the *starch* in the grain and thereby render it susceptible to enzymatic *hydrolysis*. Live steam is normally used for heating the slurry; and pumps or agitators are used to ensure mixing and even heating. Cooking may be performed continuously or in a batch mode.

Cooling tower A tower or other type of structure where air (the heat receiver) circulates in direct or indirect contact with warmer water (the heat source) to cool the water. Cooling towers are used in *ethanol* production plants to recirculate cooling water and to minimize the amount of water used from wells, rivers or public sources.

Cooper A person who makes or repairs wooden *barrels*.

Cooperage A place where wooden *barrels* are made or repaired. Also used to refer to a supply of barrels (i.e., the product of the work of a *cooper*).

Co-products Where the economics of a production process depend on the value of more than just the primary product, the secondary products are referred to as co-products rather than as *by-products*. For example, *distillers dried grain* may be considered a co-product of the production of *ethanol* from *dry milled* grain.

Cordials See *liqueurs*.

Corn steep liquor The sulfurous liquid that has been used for steeping and softening corn prior to *wet milling*. It contains extracted nutrients and serves as a nutrient source in subsequent *fermentation* of the *starch* stream.

Corn whisky Defined by the *BATF* as *'whisky* produced at under 160° *proof* from a fermented *mash* containing not less than 80% corn grain'. If corn whisky is stored in oak *barrels*, the *BATF* stipulates that the proof should not be more than 125°, and that the barrels be used, or uncharred if new. Furthermore, the whisky cannot be subjected to any treatment with charred wood.

Corrosion The destruction, degradation or deterioration of material (generally metal) due to the reaction between the material and its environment. The reaction is generally chemical or electrochemical, but there are often important physical and mechanical factors in the corrosion process.

Corrosion inhibitor A substance that reduces *corrosion*. For example, corrosion-inhibiting compounds may be added to *fuel ethanol* (particularly *methanol*) to reduce corrosion of tanks and automobile engine parts.

Co-solvent A liquid required to keep another liquid in solution in a third liquid. For example, in the *Dupont Waiver* for the use of 5% *methanol* in *gasoline*, 2.5% *ethanol* (or a *higher alcohol*) is required as co-solvent to improve the miscibility of the *methanol* in the gasoline.

Crude Oil Windfall Profits Tax Act of 1980 US federal legislation that extended to 1992 the *excise tax exemption* for *ethanol-gasoline* blends granted under the *Energy Tax Act of 1978*. It also introduced a *blender tax credit* of 40 cents per gallon of *ethanol* as an alternative to the excise tax exemption.

Cyclohexane A colorless, flammable, alicyclic *hydrocarbon* liquid of chemical formula C_6H_{12}, that boils at 80.3°C, and freezes at 6.5°C. It is used as an alternative to *benzene* as an *entrainer* in the *dehydration* of *ethanol* by *azeotropic distillation*.

D

DDG Abbreviation for *distillers dried grain*.

DDGS Abbreviation for *distillers dried grain with solubles*.

DE Abbreviation for *dextrose equivalent*.

Deadleg A length of *mash* piping closed either temporarily by a valve, or permanently to leave a dormant pocket of *mash* that may become a source of *bacterial contamination*.

Dealer tank wagon (DTW) Used in reference to *fuel ethanol* and gasoline sales, it is generally of 7,000-8,000 gallons capacity.

Decanter Vessel used for the separation of two-phase liquids. In a *fusel oil* decanter, an upper fusel oil phase is separated from a lower aqueous *ethanol* phase. In a *benzene* column *reflux* decanter, the upper, mainly-benzene, phase is separated from the lower, mainly-water, phase.

Deficit Reduction Act of 1984 US federal legislation that increased the *excise tax exemption* on *ethanol-gasoline* blends from 5 cents (as set by the *Surface Transportation Assistance Act of 1982*) to 6 cents per gallon. It also increased the alternative *blender tax* credit to 60 cents per gallon of *ethanol*.

Defoamer See **Antifoam**.

Dehydration The process of removing water from a substance, particularly the removal of most of the remaining 5% of water from 190° *proof ethanol* in the production of *absolute* or *anhydrous ethanol*.

Demethylizing column Occasionally referred to as a supplementary column, it is a *fractionating column* used to remove *methanol* in the production of *neutral spirit* and is located after the *rectifying column*. The demethylizing column is heated indirectly via a *reboiler*. The impure spirit enters part way up the column and the *methanol* is removed in the *overhead vapor* together with some *ethanol*, while the bulk of the *ethanol* descends to be removed at the base of the column as a product relatively free of *methanol*.

Denaturant A substance added to *ethanol* to make it unfit for human consumption so that it is not subject to taxation as *beverage alcohol*. The *BATF* permits the use of 2-5% unleaded *gasoline* (or similar, specified substances) for use as denaturants for *fuel ethanol*. (See also *specially-denatured alcohol*)

Department of Energy (DOE) A department of the US federal government established in 1977 to consolidate energy-orientated programs and agencies. The department's mission includes the coordination and management of energy conservation, supply, information dissemination, regulation, research, development and demonstration. The department includes an *Office of Alcohol Fuels*.

Dephlegmator Name commonly used for the first of two or more *condensers* attached to the *overhead vapor* line of a *distillation column*. It literally means an entrained liquid separator.

Desiccant A substance that absorbs water and can be used for drying purposes. For example, *potassium aluminosilicate* is used as a desiccant in *molecular sieve* systems for *ethanol dehydration*.

Detergent package A combination of detergents and *corrosion inhibitors* normally added to *fuel ethanol* to impart a cleaning action to *ethanol-gasoline* blends and thereby reduce engine fuel injector blockages.

Dewatering The removal of water (or other liquids) from a solid material, particularly the use of screens or centrifuges in the initial separation of *thin stillage* (liquid) from the solids contained in *whole stillage* in *DDG* processing.

Dextran A non-fermentable, large, branched-chain *polymer* of *sucrose molecules* produced in *molasses* by *bacterial contamination* (mainly *Leuconostoc mesenteroides*). It gives a ropey appearance to molasses when stirred or poured, and reduces *ethanol* yield on *fermentation*.

Dextrin Short-chain *polymers* of *glucose molecules* produced by the partial *hydrolysis* of *starch* with *α-amylase* or acid in the initial stage of conversion to *fermentable sugars*.

Dextrose An alternative name for *glucose*.

Dextrose equivalent (DE) A measure of the degree of *hydrolysis* of *starch*. It is no longer considered as significant as previously in *ethanol* production with the more general acceptance of *simultaneous saccharification and fermentation*.

Diethyl ether The traditional anaesthetic *ether*, which is employed as an *entrainer* in some *azeotropic distillation* processes for *fuel ethanol dehydration* (as an alternative to the more commonly-used *benzene*). It is a colorless, flammable liquid that boils at 34.6°C and freezes at -117°C. Chemical formula $(C_2H_5)_2O$. It readily forms explosive mixtures with air.

Differential pressure cell (DP cell) A device for measuring the difference in pressure of liquid on either side of a restricting orifice. It is used in flow measurement. (See *Orifice meter*).

Dimer A compound produced by linking together two *molecules* of a simpler compound (or *monomer*). It is the simplest form of *polymer*. For example, *maltose* is a dimer composed of two linked *glucose* molecules.

Disaccharide A compound *sugar* that yields two *monosaccharide* units on *hydrolysis*. For example, *lactose* yields *glucose* and *galactose*, *sucrose* yields *glucose* and *fructose*, while *maltose* yields two *glucose* units.

Disc and donut column A *beer distillation column* patented by Raphael Katzen with *trays* alternately consisting of annular shelves with open centers (donuts) and discs (of a slightly larger diameter than the donut holes) placed centrally below the holes. The effect is to give a circular curtain of liquid descending from tray to tray, through which the rising *vapors* must pass. It has the advantage of being unaffected by the solids content of the *beer* feed or any *scaling* that may occur in a typical tray column.

Distilland Material to be distilled, such as *beer*.

Distillate The portion of a liquid (*distilland*) removed as a *vapor* and condensed during a *distillation* process.

Distillation The process by which the components of a liquid mixture are separated by differences in *boiling point* by boiling and recondensing the resultant *vapors*. In *ethanol* production, it is the primary means of separating *ethanol* from *aqueous solutions*.

Distilled spirits permit (DSP) A permit issued by the *BATF* allowing the holder to engage in the production or warehousing of undenatured *ethanol* for beverage, industrial or fuel use.

Distillers bushel 56 lbs of any grain, regardless of volume.

Distillers dried grain (DDG) The dried residual *by-product* of a grain *fermentation* process. It is high in protein, as most of the grain *starch* has been removed. It is used as an animal feed ingredient. By strict definition, DDG is produced only from the solids separated from *whole stillage* by *centrifuging* or screening. In practice, the term is commonly used to describe the entire dried stillage residue, making it synonymous with *DDGS*.

Distillers dried grain with solubles (DDGS) The product derived by separating the liquid portion (*thin stillage* or solubles) from grain *whole stillage* by screening or *centrifuging*, then *evaporating* it to a thick syrup and drying it together with the grain solids portion.

Distillers dried solubles The product derived from separating the liquid portion (*thin stillage*) from grain *whole stillage*, *evaporating* it to a thick syrup and then drying it to a fine powder.

Distillers feeds The *by-products* of *fermentation* of cereal grains. See *DDG* and *DDGS*.

Distillers Feeds Research Council An organization established in 1945 to fund and coordinate university research into the utilization of *DDG*.

Distillers wet grain (DWG) The wet grain residue separated from *whole stillage* by screening or *centrifuging*. It has a very limited storage life and is generally used only where cattle feedlots are located near *ethanol* production plants.

Distillery A building or premises where alcohol is distilled.

Distillery run barrels Recently-emptied, used *whiskey barrels* that have not been sorted to remove those with or without defects.

DMA 67Y A detergent produced by Dupont. It is added to *fuel ethanol* to ensure that the resultant blend with *gasoline* will have clean-burning characteristics similar to detergent *gasolines* supplied by major oil companies.

Downcomer (or downpipe) A device used in *distillation columns* to allow liquid to descend from one *plate* to another. Downcomers may take the form of one or more round, oval or rectangular section pipes, or chordal baffles. The downcomer usually has a weir-type seal at the bottom to prevent *vapors* passing upward.

DP cell. See *differential pressure cell*.

Dry degermination A process for the removal of germ from grain without the need for steeping and *wet milling*. It may involve some pretreatment of the grain to raise the moisture content before processing. It is used in Scotland for corn milling in *grain whisky* plants and is used in the production of corn flakes; but is not commonly used in the US *ethanol* production industry.

Dry milling In the *ethanol* production industry dry milling refers to the milling of whole dry grain where, in contrast to wet milling, no attempt is made to remove fractions such as germ and bran. It may be carried out with various types of equipment, including *hammer mills* and *roller mills*.

DSP Abbreviation for *Distilled Spirits Permit*.

DTW Abbreviation for *Dealer Tank Wagon*.

Dual-flow plate (or tray) A perforated *plate* similar to a *sieve plate*, without *downcomers*. The perforations are of such a size and open area that the liquid descends by *weeping* through the holes.

Dunder A Caribbean synonym for *vinasse* or *molasses stillage*. It is commonly used to refer to vinasse that has been stored for some time to allow bacterial development prior to being used as *backset* in the production of heavily-flavored *rums*.

Dupont waiver The waiver received by Dupont from the *EPA* in 1985 for a blend of *gasoline* containing a maximum of 5% *methanol* plus a minimum of 2.5% *ethanol* or other approved *co-solvent*, together with an approved proprietary *corrosion inhibitor*.

DWG Abbreviation for *distillers wet grain*.

E

ED Abbreviation for *extractive distillation*.

Effect In the context of *evaporators*, the term is used to describe one vessel in a series. For example a quadruple effect *evaporator* consists of four linked vessels.

EITC Abbreviation for *energy investment tax credit*.

Endoenzyme An *enzyme* acting on internal portions of a large *polymeric molecule* rather than around the periphery. For example, the *α-amylase enzyme hydrolyzes* linkages within *amylose* and *amylopectin molecules*. In contrast, *amyloglucosidase* acts as an *exoenzyme* by only hydrolyzing the outermost linkages.

Energy investment tax credit (EITC) A 10% income tax credit for investments in equipment for the production of fuels such as *ethanol* from *biomass*. This credit was originally introduced under the *Energy Tax Act of 1978*, modified and extended under the *Crude Oil Windfall Profits Tax Act of 1980* and finally phased out in 1987 under the *Tax-Reform Act of 1986*.

Energy Policy Act of 1992 US federal legislation that amended the provisions of the Omnibus Reconciliation Act of 1990 to allow the proration of the *excise tax exemption* of 5.4 cents per *gallon* on *ethanol gasoline* blends that use less than the standard content of 10% *ethanol*. This was to meet the requirements of the Clean Air Amendments Act that required that oxygenate blends with 2.1% or 2.7% oxygen be used in *gasolines* in designated areas suffering from severe air pollution. (These oxygen rates correspond to *ethanol* usages of 6% and 7.7%, respectively.)

Energy Security Act of 1980 US federal legislation that supported the emerging *fuel ethanol* industry by establishing an independent *Office of Alcohol Fuels* within the *Department of Energy* and authorized programs of *loan guarantees*, price guarantees and purchase agreements with fuel *ethanol* producers.

Energy Tax Act of 1978 US federal legislation that instituted the first *excise tax exemption* for *gasoline* blended with 10% *fermentation ethanol*. It exempted the blends from the tax of 4 cents per gallon. It also created an *energy investment tax credit (EITC)* of 10% that applied to equipment for converting *biomass* to *ethanol* in addition to the standard 10% investment tax credit.

Entrainer A substance used to assist in the *dehydration* of *ethanol* by *azeotropic distillation*. Examples are *benzene*, *cyclohexane* and *pentane*. (See *azeotropic distillation*.)

Environmental Protection Agency (EPA) A US government agency established in 1970. EPA responsibilities include the regulation of fuels and fuel additives, including *ethanol gasoline* blends.

Enzymatic hydrolysis The *hydrolysis* of a *polymer* by the use of *enzymes*. In the case of *starch* hydrolysis, an *α-amylase enzyme* may be used in the initial hydrolysis to achieve *liquefaction* and an *amyloglucosidase enzyme* may be used to complete the hydrolytic *saccharification* to *fermentable sugars*.

Enzyme Any of a class of complex proteinaceous substances (such as amylases and lactases) produced by living organisms that catalyze chemical reactions without being destroyed. Enzymes may act outside the producing organism and maybe used in industrial processes such as saccharification.

EPA Abbreviation for *Environmental Protection Agency*.

Ester The product derived by the reaction of an acid with an *alcohol* or other *organic* compound having *hydroxyl* groups. For example, *ethyl acetate* is an ester produced by reacting *acetic acid* with *ethanol*. Esters tend to accumulate in *distillation* in the *heads* at the top of the column.

ETBE Abbreviation for *ethyl tertiary butyl ether*.

Ethanol Otherwise known as *ethyl alcohol*, *alcohol*, grain-spirit or *neutral spirit*, etc. A clear, colorless, flammable *oxygenated hydrocarbon*. Chemical formula: C_2H_5OH. It has a *boiling point* of 78.5°C in the *anhydrous* state. However, it forms a *binary azeotrope* with water with a boiling point of 78.15°C at a composition of 95.57% by weight *ethanol*.

Ether One of a class of *organic* compounds in which an oxygen atom is interposed between two carbon atoms in the *molecular* structure. Ethers may be derived from *alcohols* by the elimination of water. Ethers such as *diethyl ether* and *isopropyl ether* may be used as *entrainers* in the *dehydration* of *ethanol* by *azeotropic distillation*. Other ethers such as *MTBE* and *ETBE* may be used as *octane enhancers* in *gasoline*. Ethers are dangerous fire and explosion hazards. When exposed to air they form peroxides that may detonate on heating.

Ethyl acetate Chemical formula $CH_3COOC_2H_5$. A clear, volatile, flammable liquid with a characteristic fruity odor. It has a pleasant, sweet taste when diluted. Boils at 77°C and freezes at -83°C. It is produced by the reaction of *ethanol* with *acetic acid* and is a major component of the *esters* that give *rum* its characteristic odor.

Ethyl alcohol See *Ethanol*.

Ethyl carbamate Otherwise known as *urethane* or carbamic acid ethyl ether. A carcinogenic compound produced by heating urea with *ethanol* under pressure. Chemical formula $NH_2COOC_2H_5$. It is very soluble in *ethanol* or water. Traces of ethyl carbamate may be formed in the *distillation* of *beers* produced with the use of urea as a *yeast* nutrient. It does not present a significant hazard if such *ethanol* is only used for fuel purposes, but may present problems in the production of *whiskies*.

Ethylene glycol A slightly viscous, sweet-tasting, poisonous liquid. Chemical formula CH_2OHCH_2OH. It has a *boiling point* of 197.6°C, and is considerably *hygroscopic*. It is commonly used as an antifreeze. It may also be used as an *extractant* in *extractive distillation* processes for the *dehydration* of *ethanol*. (See *extractive distillation*.)

Ethyl tertiary butyl ether (ETBE) A colorless, flammable, *oxygenated hydrocarbon*. Chemical formula $C_2H_5OC_4H_9$. It may be produced from *ethanol* and tertiary *butanol (TBA)* or isobutylene. It is of similar structure to *methyl tertiary butyl ether (MTBE)*, having similar *octane-enhancing* properties, but has a significantly lower effective *Reid vapor pressure* in blending with *gasoline*.

Evaporation The process by which a substance in the liquid state is converted into the *vapor* state.

Evaporator A device used to evaporate part or all of a liquid from a *solution*. In the case of grain *stillage* processing, evaporators may be used to concentrate screened *thin stillage* to a syrup of about 35% solids, which may then be fed to a dryer. Evaporators are normally operated under vacuum (to obtain a lower *boiling point* of the liquid) and may consist of a series of interlinked vessels or *effects* operating under differing degrees of vacuum.

Excise tax exemption When *fermentation ethanol* is blended at a 10% concentration with *gasoline*, it is effectively exempted from the US federal excise tax at the rate of 60 cents per gallon of *ethanol* used or 6 cents per gallon of blend. This comes from the fact that the *ethanol* itself is exempted from the entire 9 cents per gallon of tax, and each of the 9 gallons of *ethanol* required to make a 10% blend is exempted from 5.66 of the 9 cents per gallon tax. This exemption was first introduced under the *Energy Tax Act of 1978* at a rate of 4 cents per gallon (the entirety of the tax at that time). It was increased to 5 cents per gallon under the *Surface Transportation Assistance Act of 1982* when the excise tax was raised to 9 cents. It was increased to the current effective level of 6 cents per gallon under the *Deficit Reduction Act of 1984*. It was scheduled to expire in 1992, but was extended to September 30, 1993 under the *Surface Transportation and Uniform Relocation Assistance Act of 1987*.

Extractant A substance such as *ethylene glycol* or *glycerol* be used in *extractive distillation* processes for the *dehydration* of *ethanol*.

Extractive distillation A process where an *extractant* is added to a mixture being distilled to change the *volatility* of one or more components. The less *volatile* mixture will then descend in a *continuous distillation column* while the more volatile components may be removed in the condensed *overhead vapors*. In the use of extractive distillation in the *dehydration* of *ethanol*, liquid extractants such as *ethylene glycol*, or *glycerol*, may be used. Salts such as potassium and sodium acetates may also be used alone in molten form or in mixtures with glycerol, etc. *Anhydrous ethanol* is recovered in the overhead *condensate* while the water combines with the extractant to emerge from the bottom of the column. (This is the reverse of the situation in *azeotropic distillation*.) The extractant is then separated from the water in another column (or an evaporator) and is recycled.

In the production of *neutral spirit*, light *rums* or *whiskies*, extractive distillation may be used to remove *fusel oils* and some other *congeners* in the condensed *overhead vapors*. In this instance the extractant is water, as some of the congeners with lower *volatility* than *ethanol* in a concentrated state may have higher volatility than *ethanol* when diluted with water and therefore rise up the extractive distillation column while the *ethanol* descends.

Exoenzyme An *enzyme* restricted to acting on the outer end of large *polymeric molecules* and cleaves molecules one by one. (See *Endoenzyme*.)

F

Facultative anaerobe Term used to describe a microorganism (such as a *yeast*) that is essentially aerobic (or air-requiring), but can also thrive under *anaerobic* (or air-free) conditions.

Fahrenheit scale A temperature scale in which the *boiling point* of water is 212°F and the freezing point is 32°F. (The zero point was originally established as the lowest point obtainable with a mixture of equal weights of snow and common salt.)

Farmers Home Administration (FmHA) A division of the *US Department of Agriculture* which among other activities is empowered to make *loan guarantees* for the establishment of *ethanol* production plants.

Federal Excise Tax Exemption See *Excise tax exemption*.

Feed plate (or feed tray) The *plate* or *tray* onto which the *distilland* (liquid to be distilled) is introduced in a *distillation column*. In theory, it is the point in a column above which enrichment or concentration occurs and below which stripping occurs.

Feedstock The raw material used in a process. For example, corn, *molasses, whey*, etc. may be used as feedstocks for *ethanol* production.

Fermentable sugars Simple *sugars* such as *glucose* and *fructose* that can be converted into *ethanol* by *fermentation* with *yeast*. They may be derived by the *hydrolysis* of *starch* or *cellulose feedstocks* or obtained from other sources.

Fermentation The enzymatic transformation by *microorganisms* of *organic* compounds such as *sugars*. It is usually accompanied by the evolution of gas as in the *fermentation* of *glucose* into *ethanol* and *carbon dioxide*.

Fermentation efficiency The measure of the actual output of a *fermentation* product such as *ethanol* in relation to the *theoretical yield*.

Fermentation ethanol The term used to distinguish *ethanol* produced by *fermentation* from *synthetic ethanol* produced from ethylene, etc. The difference is significant in that only *fermentation ethanol* qualifies for US federal and state *excise tax exemptions* for automotive fuel use. Furthermore, only *fermentation ethanol* may be used for beverage purposes.

Fermenter The vessel in which *mash fermentation* takes place. The vessel may be fabricated from steel, fiber glass, etc. and is normally fitted with an internal or external cooling system for controlling temperature of the fermenting mash.

FFV Abbreviation for *Flexible Fueled Vehicle.*

Flame arrester A device installed on the vapor vents of *alcohol* storage tanks and *distillation columns* to prevent entry of flames that might cause an explosion. It normally contains fine metal meshes with holes large enough to allow vapors to escape, but too small to allow flames to pass in the opposite direction.

Flash cooling The rapid cooling achieved when a hot liquid is subjected to a sudden pressure drop to reduce its *boiling point.*

Flash point The minimum temperature at which a combustible liquid will ignite when a flame is introduced. It depends on the *volatility* of the liquid to provide sufficient *vapor* for combustion. For example, *anhydrous ethanol* has a flash point of 51°F, while 90^0 *proof* ethanol has a flash point of 78°F.

Flexible fueled vehicle (FFV) A vehicle designed to operate on a variety of fuels such as *methanol, ethanol* or *gasoline* alone or in combinations without requiring major adjustments.

Flocculation The aggregation or coalescing of fine suspended particles or bodies into loose clusters or lumps. The flocculation characteristics of *yeast* or other *microorganisms* may be important features in their recovery for *cell recycling.*

Flowmeter A device for measuring the rate of flow of a liquid.

FmHA Abbreviation for the *Farmers Home Administration.*

Fluidized bed combustion boiler A boiler in which the coal or other fuel particles are kept in suspension by a rising column of gas rather than resting on a conventional grate. The system gives greater heat transfer and higher potential combustion efficiency. The system has the advantage that limestone may be added to the coal fuel to absorb sulfur gases to reduce emissions.

Fossil fuel Any naturally-occurring fuel of an *organic* nature that originated in a past geologic age such as coal, crude oil or natural gas.

Fractional distillation A process of separating mixtures such as *ethanol* and water by boiling and drawing off the condensed *vapors* from different levels of the *distillation column.*

Fructose A fermentable *monosaccharide*(simple *sugar*) of the chemical formula $C_6H_{12}O_6$. Its chemical structure is similar to that of *glucose,* but it is sweeter to the taste. It may be produced from *glucose* by enzymatic *isomerization,* as in the production of *high fructose corn syrup (HFCS).*

Fuel grade ethanol See *motor fuel grade ethanol.*

Fuel ethanol Usually denotes *anhydrous ethanol* that has been *denatured* by addition of 2-5% unleaded *gasoline* and is intended for use as an automotive fuel in blends with *gasoline.*

Fungible Literally means 'interchangeable in trade'. Commonly used to denote products suitable for transmission by pipeline. *Ethanol* is not considered fungible in this sense because it would absorb any water accumulating in pockets in a pipeline.

Fusel oil Term used to describe the *higher alcohols,* generally the various forms of *propanol, butanol* and *amyl alcohol* that are *congeners* or *by-products* of *ethanol fermentation.* Normally, predominantly isoamyl alcohol. Their presence in alcoholic beverages is known to be a cause of headaches and hangovers. The fusel oils have higher *boiling points* than *ethanol* and are generally removed in the *distillation* process to avoid accumulation in the *rectifier.* They may be subsequently added back into the *anhydrous* product for *motor fuel grade ethanol.*

Fusel oil decanter A device used to separate accumulations of *fusel oil* from *ethanol* based on the fact that fusel oil has a lower miscibility in water than *ethanol* and can therefore be removed by dilution with water. It generally consists of a tank with windows or sight glasses to permit the operator to observe and control the separation.

G

Galactose A *monosaccharide* of chemical formula $C_6H_{12}O_6$ that along with *glucose* is a constituent of the *disaccharide lactose*. It is an *isomer* of *glucose*, but is less-readily fermented by *yeasts* to *ethanol*.

Gas chromatography (GC) A technique for separating chemical substances in which the sample is carried by an inert gas stream through a tube (or column) packed with a finely-divided solid material. The various components in the sample pass through the column at differing velocities and emerge from the column at distinct intervals to be measured by devices such as a flame ionization detector or a thermal conductivity detector. (Where the solid material in the column is pretreated with a liquid to achieve the component separation, the process may be referred to as *gas liquid chromatography*) The technique may be used for separating and measuring the amount of *ethanol* and the various *by-products* formed in *fermentation*.

Gas liquid chromatography See *gas chromatography*.

Gasohol (or gasahol) A trade name registered by the Nebraska Agricultural Products Industrial Utilization Committee, later renamed the *Nebraska Gasohol Committee*. (The committee was responsible for laying the groundwork for the development of the present day US *fuel ethanol* industry). The trade name, in either spelling, covers a blend of *anhydrous* ethanol 'derived from agricultural products' with *gasoline* (not necessarily unleaded). The committee has freely granted permission for commercial use of the trade name provided it is not used for blends containing *alcohols* other than *ethanol*.

Gasoline A volatile, flammable, liquid *hydrocarbon* mixture suitable for use as a fuel in internal combustion engines. Normally consists of a blend of several products from natural gas and *petroleum* refining together with anti-knock agents and other additives. It is a complex mixture of hundreds of different hydrocarbons, generally in the range of 4-12 carbon atoms per *molecule*. The components may have *boiling points* ranging mainly between 30 and 200°C with blends being adjusted to altitude, season and legal requirements.

Gasoline extender The term used to describe *ethanol* when it is simply used as a partial replacement for *gasoline* without any consideration for its value as an *octane enhancer* or oxygenate.

Gay Lussac (GL) The name given to a scale of the concentration of *ethanol* in mixtures with water where each degree is equal to 1% by volume (i.e., 1° GL is equivalent to 2° US *proof*). It takes the name from the French chemistry pioneer, Joseph-Louis Gay Lussac.

Gay Lussac equation The equation for the *fermentation* of *sugar* by *yeast* to *carbon dioxide* and *ethanol*, established by the French chemist Joseph-Louis Gay Lussac in 1815: $C_6H_{12}O_6 \rightarrow 2CO_2 + 2C_2H_5OH$. (See *Stoichiometric yield*).

GC Abbreviation for *gas chromatograph*.

Gear pump A *positive displacement* rotary pump containing two intermeshed gear wheels in a suitable casing. The counter rotation of the gears draws fluid between the gear teeth on one side and discharges it on the other side. It is commonly used for viscous liquids such as *molasses* and *stillage* syrup.

Gelatinization In reference to the cooking of starchy *feedstocks*, gelatinization is the stage in which the *starch* granules absorb water and lose their individual crystalline structure to become a viscous liquid gel. Gelatinization is significant in that it is the preliminary process necessary to render *starch* susceptible to *enzymatic hydrolysis* for conversion to *fermentable sugars*.

Gin Defined by the *BATF* as 'a product obtained by original *distillation* from *mash*, or by redistillation of distilled spirits, or by mixing *neutral spirits* with or over juniper berries and other aromatics, or with or over other extracts derived from infusions, percolations or maceration of such materials, and includes mixtures of gin and neutral spirits. It shall derive its main characteristic flavor from juniper berries and be bottled at not less than 80^0 *proof*. Gin produced exclusively by original distillation or redistillation may be further designated as distilled'.

GL Abbreviation for *Gay Lussac*.

GLC Abbreviation for *gas-liquid chromatography*.

Glucoamylase. An *enzyme* that *hydrolyses starch* into its constituent *glucose* units. (See *Amyloglucosidase*.)

Glucan. See *Betaglucan*.

Glucanase An *enzyme* that *hydrolyses* glucan. (See *Betaglucanase*.)

Glucose A *fermentable sugar* otherwise referred to as *dextrose*. It is a *monosaccharide* and has the formula $C_6H_{12}O_6$. Glucose is the ultimate product in the *hydrolysis* of *starch* and *cellulose*, which are both *polymers* of hundreds or thousands of *glucose* units.

Glucose isomerase An *enzyme* that converts *glucose* into its *isomer, fructose*. It is used in the production of *high fructose corn syrup (HFCS)*.

Glucosidase See *Amyloglucosidase* or *Glucoamylase*.

Glycerol (or glycerine) A clear, colorless, viscous, sweet-tasting liquid belonging to the *alcohol* family of *organic* compounds. It has a chemical formula of $CH_2OHCHOHCH_2OH$, having three *hydroxyl* (OH) groups. It is a *by-product* of alcoholic *fermentations* of *sugars*. It is *hygroscopic* and may be used as an *extractant* in the *dehydration* of *ethanol*.

Grain alcohol Term used to distinguish *ethanol* produced from grain from *ethanol* produced from other *feedstocks*, or from *wood alcohol (methanol)*.

Grain whisky The term used in Scotland to distinguish the bland, nearly-neutral, continuous *distillation* product that forms the base of blended Scotch whisky from the *malt whiskies* that contribute most of the flavor in the blend. Grain whisky is usually produced from corn or barley.

Grain sorghum Otherwise known as *milo*, a sorghum grown for grain production, as distinct from *sweet sorghum* grown for the *sugar* content of its stem. It may be used as a *feedstock* for *ethanol* production.

Gram-negative See *Gram stain*.

Gram-positive See *Gram stain*.

Gram stain A widely-used microbiological staining technique that aids in the identification and characterization of *bacteria*. It was devised by Danish physician, Hans Christian Gram. Bacteria are described as *Gram-positive* if their cell walls absorb the stain and *Gram-negative* if they do not. The technique is particularly useful in examining *bacterial contaminants* in *fermentations*, as most Gram-positives are susceptible to control by *penicillin*.

H

Hammer mill A type of impact mill or crusher in which materials such as cereal grains are reduced in size by hammers revolving rapidly in a vertical plane within a steel casing. A screen with numerous holes of a selected diameter is installed in the casing to control the size of particles produced. It is commonly used for grinding corn as a *fermentation* feedstock.

Heads Term used to describe the impurities produced in *ethanol fermentations* (*congeners*) that have lower *boiling points* than *ethanol*. They include *methanol* and *aldehydes*.

Heads concentrating column A *distillation column* used to concentrate *heads* removed in the production of *neutral spirit*, light *rums* and *whiskies*.

Heat exchanger Device used to transfer heat from a fluid on one side of a barrier to another fluid flowing on the other side of the barrier. Common forms include shell-and-tube heat exchangers and plate-type heat exchangers. Various types of heat exchangers are used in *ethanol* plants for *mash* cooling, indirect steam heating (*reboilers*), *overhead vapor* condensing, etc.

Heat of condensation The heat given up when a *vapor* condenses to a liquid at its *boiling point*.

Heat of vaporization The heat input required to change a liquid at its *boiling point* to a *vapor* at the same temperature.

Hemicellulose Term used to describe non-cellulosic *polysaccharide* components of plant cell walls. The most common hemicelluloses are composed of *polymers* of *xylose* (a 5-carbon sugar, or *pentose*) together with a uronic acid (a sugar acid), *arabinose* (another pentose) and *mannose* (a *hexose*). Hemicelluloses have no chemical relationship to *cellulose*. They frequently surround the cellulose fibers and increase their bonding and tensile strength. The presence of hemicelluloses (and *lignins*) in close association with cellulose tends to impede the extraction and *hydrolysis* of wood cellulose to *sugars* for *ethanol* production.

Hexose A class of *monosaccharides* (simple *sugars*) containing six carbon atoms in the *molecule*. They have the formula $C_6H_{12}O_6$. Common examples are *glucose*, *fructose* and *galactose*.

HFCS Abbreviation for *high fructose corn syrup*.

Hiag process A process developed in Germany in the 1930s for the *dehydration* of *ethanol* by *extractive distillation* using a mixture of sodium and potassium acetates as the *extractant*.

High boilers Term used to describe the impurities produced in *ethanol fermentation* (*congeners*) that have higher *boiling points* than *ethanol*. Otherwise referred to as *tails*, they include the *higher alcohols*, *propanols*, *butanols* and *amyl alcohols* or *fusel oils*. In the context of *gasoline*, the term refers to components with *boiling points* considerably above the mid-range.

Higher alcohols *Alcohols* having more than two carbon atoms. They exist in various isomeric forms. As the number of carbon atoms increases, so does the number of *isomers*, but at a greater rate. The lower members of this group, namely *propanol*, *butanol* and *amyl alcohol* are major constituents of *fusel oil*.

High fructose corn syrup (HFCS) A product in which a large percentage of the *glucose* derived from *starch hydrolysis* has been converted into its sweeter-tasting *isomer fructose* by use of *enzymes*. It is frequently used as a substitute for cane or beet *sugar* as a sweetener in soft drinks, etc. It is a seasonal alternative to *ethanol* production in some corn processing plants.

High performance liquid chromatography (HPLC) An advanced form of liquid *chromatography* used for the separation of complex mixtures of non-volatile materials for analytical purposes. It involves very small particles packed in columns through which the liquids flow and high pressures to increase the rate of flow and shorten the time required to achieve the separation. The process is coupled with other devices to identify the constituents. It may be used in the analysis of the *sugars* and *dextrins* produced by *hydrolysis* of *starch*.

High test molasses (HTM) See *molasses*.

Hogshead A wooden *barrel* with a capacity of approximately 66 *US gallons* (250 *liters*) used in Scotland for aging *whisky*. It is usually constructed from *shooked bourbon whiskey* barrels of 52 gallons capacity by using additional staves and larger heads and hoops.

HPLC Abbreviation for *high performance liquid chromatography*.

HTM Abbreviation for *high test molasses*.

Hydrocarbon A member of a class of *organic* chemical compounds containing only hydrogen and carbon. It should be clearly differentiated from *carbohydrates*, which contain oxygen as well as hydrogen and carbon. The main natural sources of hydrocarbons are *petroleum*, coal, natural gas and bitumens.

Hydrolysis Literally means the breakdown, destruction or alteration of a chemical substance by water. In the case of *starch* and other *polymers* of *glucose*, a *molecule* of water is divided between two adjacent *glucose* units, in order to cleave the linkage. For example, *maltose* ($C_{12}H_{22}O_{11}$), which contains two *glucose* rings, requires the addition of a *molecule* of water (H_2O) to yield two separate *glucose molecules* ($C_6H_{12}O_6$). The hydrolysis may be accomplished with the use of acids or *enzymes*.

Hydrometer An instrument measuring the density, *specific gravity* or other similar characteristics of liquids. It is generally comprised of a long-stemmed glass tube with a weighted bottom which floats at different levels in liquids of different densities. The reading is taken at the meniscus (where the calibrated stem emerges from the liquid). The liquid temperature is normally determined when taking a reading; and reference is made to hydrometer tables to obtain a correction to a standard temperature. A *proof* hydrometer measures the content of *ethanol* in a mixture with water. A *Brix* or *Balling* hydrometer measures on a scale equivalent to the percentage of *sugar* by weight in an *aqueous solution*.

Hydroselection column Synonym for *extractive distillation column*.

Hydrous ethanol Term used for *ethanol* that has not been subjected to *dehydration*. It may refer to any mixture of *ethanol* and water, but frequently is used to denote *ethanol* at a concentration of about 190-192^0 *proof*, close to the *azeotropic* point.

Hygroscopic Term used to describe a substance with the property of absorbing moisture from the air. *Anhydrous ethanol* is hygroscopic, and its exposure to moist air should therefore be minimized.

Hydroxyl group A combination of one atom of oxygen and one atom of hydrogen (OH) forming an essential part of any *alcohol*. The spare atomic bond of the group is linked to a carbon atom as in *ethanol* (C_2H_5OH) or *methanol* (CH_3OH).

I

IDRB Abbreviation for *Industrial Development Revenue Bonds*.

Imperial gallon A measure of volume in the British system defined in 1824 as the volume occupied by 10 pounds of water at 62°F and 30 inches of barometric pressure. It is the equivalent of 1.2 US gallons, or 4.546 liters.

Industrial alcohol Denotes any *ethanol* intended for industrial uses such as *solvents, extractants*, antifreezes and intermediates in the synthesis of innumerable organic chemicals. The term covers *ethanol* of both *synthetic* and *fermentation* origin of a wide range of qualities and *proofs*, with or without *denaturants*.

Industrial development revenue bonds (IDRB) Debt incurred through local industrial development authorities in numerous US states. Such bonds were a popular way to finance *fuel ethanol* plants in the past because the interest on the bonds was not subject to income tax and the bonds could be sold with interest rates substantially below the prime rate.

Inoculum The portion of a culture of *yeast* (or *bacteria*) used to start a new culture or a *fermentation.*

Inulin A storage *polysaccharide* found in the roots and tubers of various plants, particularly Jerusalem artichokes. It consists of chains of an average of 30 *fructose* units. It is only slightly soluble in cold water, but dissolves readily in hot water. Unlike *starch,* it does not give a color reaction with iodine.

Inulinase An *enzyme* capable of *hydrolyzing inulin* to its component *fructose* units.

IPE Abbreviation for *isopropyl ether.*

Isoamyl alcohol The principal *alcohol* in *fusel oil.* It is an *isomer* of *pentanol,* of composition $C_5H_{11}OH$. It is a colorless liquid with a pungent taste and disagreeable odor. Boils at 132°C and freezes at -117.2°C. It is only slightly soluble in water but miscible with *ethanol.* It may be recovered by fractionation of *fusel oil,* and has a wide range of uses in *organic* synthesis, pharmaceuticals, photographic chemicals and as a *solvent* for fats, etc. The *vapors* are poisonous and at low concentrations may cause headaches and dizziness.

Isomer One of a series of two or more *molecules* with the same number and kind of atoms and hence the same *molecular weight,* but differing in respect to the arrangement or configuration of the atoms. For instance, *glucose* and *fructose* have the formula $C_6H_{12}O_6$, but different molecular structures.

Isomerase An *enzyme* that can convert a compound into an isomeric form. For instance, the *glucose isomerase enzyme* converts *glucose* into its sweeter-tasting *isomer fructose* in the production of *high fructose corn syrup (HFCS).*

Isomerization The process of converting a chemical compound into its *isomer* such as converting *glucose* to *fructose* in the production of *high fructose corn syrup (HFCS).* In *petroleum* refining, the term describes methods used to convert straight-chain to branched-chain *hydrocarbons,* or acyclic to aromatic hydrocarbons to increase their suitability for high *octane gasoline.*

Isopropyl ether (IPE) An ether (otherwise known as di-isopropyl ether) used in some *fuel ethanol* plants as an *entrainer* in the *dehydration* process as an alternative to *benzene,* etc. It is a colorless, volatile liquid with chemical formula $(CH_3)2CHOCH(CH_3)_2$, which boils at 67.5°C and freezes at -60°C. It readily forms explosive mixtures with air. Inhalation of vapors may cause narcosis and unconsciousness.

J

Jet cooker An apparatus for the *continuous cooking* of grain *mashes* in which the mash is pumped past a jet of steam that instantly heats the mash to *gelatinize* the *starch.*

Jobber A *gasoline* wholesaler. Some jobbers may operate under contract to one or more major oil companies, distributing *gasoline* to branded gas stations. Other jobbers may operate independently, buying *gasoline* from various sources for distribution to unbranded retail outlets. Independent jobbers are frequently major buyers and blenders of *fuel ethanol.*

K

Karl Fischer titration A method to chemically determine the amount of water present in a sample of *ethanol* and/or other substances. When correctly practiced, the method provides an extremely accurate measurement of very small quantities of water in *ethanol* even if *gasoline denaturant* is present. (See *titration*).

Kerosene One of the three permissible *denaturants* for *fuel ethanol* as specified in *BATF* regulations. It is a refined *petroleum* fraction used as a fuel for heating, cooking and for jet engines. It has a *boiling point* range somewhat higher than that of *gasoline*, generally between 180°C and 290°C.

Kjeldahl method An analytical method for the determination of nitrogen in *organic* compounds. As nitrogen is an essential element in *protein*, the method may be used for determining protein in such materials as *distillers dried grains*. Kjeldahl protein, or crude protein, is Kjeldahl N x 6.25.

Kluyveromyces fragilis (or marxianus) A *lactose*-fermenting *yeast* used in the production of *ethanol* from cheese *whey*.

Kubierschky process The first patented process for the continuous *dehydration* of *ethanol* with *benzene*. With relatively minor variations, the process developed in 1914 based on Young's earlier batch process is still used today in many *fuel ethanol* plants.

L

Lactase An *enzyme* that *hydrolyzes lactose* into *glucose* and *galactose*. This hydrolysis allows lactose-containing *feedstocks* such as cheese *whey* to be fermented by the common *Saccharomyces cerevisiae yeasts*. A principal source of lactase is the yeast *Kluyveromyces fragilis*, which can also directly ferment lactose to *ethanol*.

Lactic acid The *organic* acid produced in the *fermentation* of carbohydrates by *Lactobacillus bacteria*. Its production is the principal reason for loss of yield in contaminated *ethanol fermentations*. Pure lactic acid is a colorless, odorless, *hygroscopic*, syrupy liquid with a *boiling point* of 122°C and a formula $CH_3CHOHCOOH$.

Lactobacillus A genus of *bacteria* that produce *lactic acid* as a major product in the *fermentation* of *carbohydrates*. Lactobacilli are found extensively in fermenting food products such as souring milk and in grain dust. They are the principal cause of loss of yield in *ethanol fermentations*. Otherwise referred to as lactic acid bacteria, they are generally *Gram-positive* and controllable with *penicillin* and certain other *antibiotics*.

Lactose The principal *sugar* in milk and cheese *whey*. It may be fermented by suitable *yeasts* to *ethanol*. It is a *disaccharide* readily *hydrolysed* to its two components, *glucose* and *galactose*. Lactose has the formula $C_{12}H_{22}O_{11}$.

Lag phase Applied to *yeast propagation*, lag phase refers to the initial period in which the *yeast inoculum* becomes adapted to the *mash* prior to the rapid increase in cell numbers referred to as the *logarithmic phase*.

Latent heat The quantity of energy absorbed or released when a substance undergoes a change of state, i.e., from a solid to a liquid (melting) or from a liquid to a *vapor*, or vice versa. No change in temperature is involved. For example, water requires a large amount of latent heat (measured in BTUs or calories) to convert from the liquid to vapor state (steam) at 100°C, while steam releases the latent heat again on condensing back to the liquid state.

Lead phase-out The reduction in the amount of lead (normally in the form of *tetraethyl lead*) that could be used as an *octane enhancer* in leaded *gasoline* in the US. The reduction, which went from 1.1 grams per gallon prior to July 1, 1985 to less than 0.1 grams per gallon after January 1, 1986 caused an increase in demand for alternative octane enhancers such as *ethanol* and *MTBE*.

Light whisky Defined by the *BATF* as *whisky* produced at more than 160^0 *proof* and stored in used or uncharred oak *barrels*.

Lignin A *polymeric*, non-carbohydrate constituent of wood that functions as a support and plastic binder for *cellulose* fibers. It may comprise 15-30% of wood and can only be separated from the cellulose and *hemicellulose* components by chemical reaction at high temperatures. Its presence in wood is a major barrier to the *hydrolysis* of cellulose to *sugars* for *fermentation* purposes.

Lignocellulose Woody materials made up largely of *lignin*, *cellulose* and *hemicelluloses*. The chemical bonding between the constituents renders it resistant to *hydrolysis*.

Lime A white alkaline powder composed of calcium oxide. It is added to grain *mash* to adjust the *pH* and to provide calcium ions in order to prevent the inactivation of the *α-amylase enzyme molecule* by the loss of its essential calcium atom.

Liquefaction Conversion of a solid substance to the liquid state. In reference to *starch*, it is the stage in the *cooking* and *saccharification* process in which *gelatinized starch* is partially *hydrolyzed* by *α-amylase* (or occasionally by an acid) to give soluble *dextrins*. This converts the *starch mash* into a free-flowing liquid.

Liqueurs and cordials Defined by the *BATF* as 'products obtained by mixing or redistilling distilled spirits with or over fruits, flowers, plants or pure juices therefrom, or any other natural flavoring materials, or with extracts derived from infusions, percolation or maceration of such materials, and containing *sugar, dextrose* or levulose, or a combination thereof, in an amount not less than 2.5% by weight of the finished product'.

Liter A metric measurement of volume defined as the equivalent of 1000 cubic centimeters or 0.2642 US gallons.

Loan guarantee A procedure by which the *DOE* and *FmHA* separately encouraged the construction of *fuel ethanol* plants by acting as guarantors for up to 90% of the amount of loans made by banks and other lenders to approved borrowers.

Logarithmic phase Applied to *yeast propagation*, it refers to the period in which cell numbers increase at an exponential rate after the initial *lag phase*.

Low boilers In reference to *ethanol distillation*, the term is applied to the *congeners*, or *fermentation by-products*, which boil at a lower temperature than *ethanol*. More commonly referred to as *heads*, these compounds are principally *aldehydes* and *methanol*.

LPA Abbreviation for *liters* of pure *alcohol*.

M

Macromolecule A giant *molecule* in which there is a large number of one or more relatively simple structural units or *monomers*.

Malt Barley grains that have been steeped in water and then allowed to germinate. The germination is normally halted by drying the grains when the sprouts are about the same length as the grains. At this stage, the malt (or 'malted barley') contains considerable amounts of *α- and ß-amylase enzymes* that *saccharify* the barley *starch* and other additional *starch* in a *mash* to yield *fermentable sugars*. (In Scotland, the drying may be done by exposing the malt to a flow of peat smoke to impart a smoky odor to the malt.) Malt is used in *whisky* production, mainly for its contribution to product flavor. In *fuel ethanol* production the necessary saccharifying *enzymes* are normally derived from microbial sources.

Malt whisky In the US malt whisky is defined by the *BATF* as a *whisky* produced at less than 160^0 proof from a fermented *mash* containing at least 51% malted barley, and stored at under

125^0 proof in charred, new oak *barrels*. In Scotland, malt whiskies are made from a 100% malted barley mash and may be aged in previously-used oak barrels. Malt whiskies may be mixed with *grain whiskies* to impart much of the characteristic flavor of blended Scotch whisky.

Maltase An *enzyme* capable of *hydrolyzing maltose sugar molecules* into their two component *glucose* units. Occasionally the name is loosely applied to the *amyloglucosidase enzyme*, which *hydrolyzes polysaccharides* to *glucose*.

Maltose A *fermentable sugar* which is a *dimer* (or *disaccharide*) of *glucose* in that it is comprised of two linked *glucose* units. It is the normal end product of *starch saccharification* by the *ß-amylase enzyme* in *malt*.

Manioc See *cassava*.

Mannose A fermentable 6-carbon *sugar* (or *hexose*) which is an *isomer* of *glucose*. Chemical formula $C_6H_{12}O_6$. It is a constituent of *hemicellulose*.

Mash A mixture of milled grain or other fermentable *carbohydrate* in water used in the production of *ethanol*. The term may be used at any stage from the initial mixing of the *feedstock* in water prior to any cooking and *saccharification* through to the completion of *fermentation*, when it becomes referred to as *beer*.

Mash bill The percentages of different types of grains used in the preparation of a *mash* in *beverage alcohol* production. For example, a typical bourbon whisky mash bill may consist of 65% corn, 25% rye and 10% barley *malt*.

McCabe-Thiele diagram A graphic method for calculation of the number of *theoretical plates (or trays)* required in a *distillation column* to achieve a desired separation of two components.

Meal The floury or granular product resulting from milling or grinding of cereal grains.

Mechanical vapor recompression (MVR) A method used in *evaporation* in which the water vapor leaving the *evaporator* is recompressed and recycled to heat the same vessel. This means that mechanical energy is used rather than heat energy. The recompressor may be operated by electricity or by a steam turbine (if high pressure steam is available and there is a demand elsewhere for the low pressure turbine exhaust steam). Mechanical vapor recompression can greatly increase the steam usage efficiency, but the capital costs for equipment are also greater.

Metabolism The chemical processes in living cells by which energy is derived for vital processes, growth and activities.

Methane A colorless, odorless, tasteless, readily-combustible, asphyxiant, lighter-than-air gas. The first member of the paraffin (or alkane) series of *hydrocarbons*, it has a formula CH_4. It occurs in high proportions in natural gas, and is produced by decaying vegetation and other *organic* matter as in landfills and marshes. The methane produced in sealed landfills or from the *anaerobic digestion* of *thin stillage* and other wastewaters is used for boiler fuel in some *ethanol* plants.

Methane digester (or anaerobic digester) A system of covered tanks used for treatment of *organic* waste streams such as *thin stillage*. The system is initially inoculated with a suitable culture of *methane*-producing *bacteria* and operates on a continuous basis to generate methane for use as a boiler fuel.

Methanol (or methyl alcohol) A colorless poisonous liquid with essentially no odor and very little taste. It is the simplest *alcohol* and has the formula CH_3OH. It boils at 64.7°C. It is miscible with water and most *organic* liquids, including *gasoline*. It is extremely flammable, burning with a nearly invisible blue flame. It is a *congener* of *ethanol fermentations*. Having a lower *boiling point* than *ethanol*, it tends to be a major component of the *heads* stream on *distillation*. Due to its miscibility with *benzene*, its presence in a hydrous *ethanol* feed may reduce the efficiency of *dehydration* processes where benzene is used as an *entrainer*. *Methanol* is produced commercially by the catalyzed reaction of hydrogen and carbon monoxide. It was formerly derived from the destructive distillation of wood, which caused it to be known as *wood alcohol*. *Methanol* may be blended with *gasoline*, but requires a *co-solvent* such as *ethanol* or a *higher alcohol* to maintain it in *solution*. (See *Dupont waiver*. See *Demethylizing column*).

Methyl tertiary butyl ether (MTBE) A colorless, flammable, liquid *oxygenated hydrocarbon*. Chemical formula $(CH_3)_3COCH_3$. It contains 18.15% oxygen and has a *boiling point* of 55.2°C. It is produced by reacting *methanol* with *isobutylene*. It is used as an *octane enhancer* in *gasoline*.

MFGE Abbreviation for *motor fuel grade ethanol*.

Microorganism A collective term for microscopic organisms including *bacteria*, *yeasts*, viruses, algae and protozoa.

Milo (or millet, or grain sorghum) Much smaller than corn, this cereal grain has a similar *starch* content and yields almost the same amount of *ethanol* per *bushel* on *fermentation*. Milo is more drought-resistant than corn and is frequently grown in areas unsuited to corn. Milo *fermentations* may tend to foam more than corn fermentations and will produce a characteristic surface crust if not agitated.

Molar solution A solution of salts or other substances in which one *molecular weight* of the substance in grams (one *mole*) is dissolved in enough *solvent* to make up to one liter. Molarity (or molar concentration) is a measure of moles per *liter* of a solution and is indicated by the letter M as in 1M, 10M, etc.

Molasses The thick liquid remaining after *sucrose* has been removed from the mother liquor (of clarified concentrated cane or beet juice), in *sugar* manufacture. *Blackstrap* molasses is the syrup from which no more sugar may be removed economically. It has usually been subjected to at least three *evaporating* and *centrifuging* cycles to remove the crystalline sucrose. Its analysis varies considerably, depending on many factors including sugar mill equipment and operational efficiency; but it may contain approximately 45-60% *fermentable sugars* by weight and approximately 10% ash (or minerals). It is commonly used as an *ethanol feedstock* when prices are favorable. *High test molasses (HTM)* is not a true molasses as it is the mother liquor from which no crystalline sugar has been removed by centrifugation but which has been treated with acid to reduce crystallization. It may contain approximately 80% sugars by weight and is very low in ash. Typically HTM is only produced in years when the sugar price does not justify its recovery. It may be used as an *ethanol* feedstock when prices are favorable, and has the advantage over blackstrap of causing less *distillation column scaling*. However, it requires more nutrients for *fermentation*. (See also *citrus molasses*).

Mole A quantity of a chemical compound or element equal to its *molecular weight* in grams. By this definition, moles of different compounds or elements contain the same number of *molecules*. Thus in the *vapor* state, as ideal gases, moles of different compounds occupy the same volume at the same temperature and pressure. This leads *distillation* equipment designers to base their calculations on the mole percentages of different compounds in a mixture to be separated, such as *ethanol* and water, rather than the percentages by weight or volume.

Molecule The smallest unit into which a pure substance can be divided and still retain the composition and chemical properties of the substance. For example the formula for water is H_2O; and the smallest unit recognizable as water is the single molecule of two atoms of hydrogen linked to one atom of oxygen.

Molecular sieve A microporous substance composed of materials such as crystalline aluminosilicates belonging to a class known as *zeolites*. The size of the pores in the substance may vary with its chemical structure, being generally in the range of 3 to 10 *Angstrom* units in diameter. With material having a very precise pore size, it is possible to separate smaller *molecule*s from larger ones by a sieving action. For example, in *ethanol dehydration* with a *potassium aluminosilicate* material prepared with pores of a diameter of 3Å units, water *molecule*s with a diameter of 2.5 Å may be retained by adsorption within the pores while *ethanol molecules* of a diameter of 4 Å cannot enter and therefore flow around the material.

The term molecular sieve is frequently used loosely to describe the entire *ethanol* dehydration apparatus holding the beads of sieve material and includes the equipment and controls necessary to regenerate them when saturated with water.

Molecular weight The sum of the *atomic weights* of the atoms in a *molecule*. For example, water *molecule*s are composed of two atoms of hydrogen (atomic weight of one) and one atom of oxygen (atomic weight of 16) to give a total molecular weight of 18.

MON Abbreviation for *motor octane number*.

Monomer A single *molecule* of a substance of relatively low *molecular weight* and simple structure, which is capable of conversion to *polymers* by combination with other identical or similar *molecule*s. For example, *glucose* is a monomer, which by interlinking of *molecule*s can be built into large polymers such as the *amylose* and *amylopectin* contained in *starch*.

Monosaccharide A *sugar monomer*. Examples are *glucose*, *fructose* and *galactose*. These are the simplest forms of sugars, which are more readily fermented by *yeast*s than their *polymers* (or *polysaccharides*).

Mother yeasting A system of *yeast propagation* frequently used for *molasses fermentation*s in which the propagator is not emptied entirely when inoculating a *fermenter* and the portion retained is used for starting another *yeast* propagation cycle. By adding acid to maintain a low *pH* in the propagator it may be possible to repeat the process for numerous cycles without excessive *bacterial contamination*.

Motor fuel grade ethanol (MFGE) Refers to *anhydrous ethanol* prior to *denaturation* to *fuel ethanol*. The term is used to distinguish it from the various grades of *industrial* and *beverage alcohol*. MFGE is relatively crude, with considerable impurities, but conforms to legal *anhydrous* standards. It has not undergone the *rectification* required to make it suitable for industrial or beverage uses.

Motor octane number (MON) One of two methods commonly used to assess the *octane rating* of an automobile fuel. (See *Octane rating*.)

MSW Abbreviation for *municipal solid waste*.

MTBE Abbreviation for methyl tertiary butyl ether.

Multiple effect evaporator A system comprising a series of interlinked *evaporator* vessels (or *'effects'*) operating under increasing degrees of vacuum. As liquids boil at lower temperatures with increasing vacuum, it is possible to use the *latent heat* of the *vapors* from one vessel to heat the next (and so on through a series of up to about six vessels under increasing vacuums) with steam or other external source of heat only being introduced to the first vessel. This allows considerable economy in energy consumption. Multiple effect evaporators are commonly used for concentrating *thin stillage* solids into a syrup.

Municipal solid waste (MSW) Regular city trash or garbage. There have been numerous proposals and some pilot-scale attempts to produce *ethanol* from MSW by *hydrolyzing* and *fermenting* the *cellulosics* contained in the material. Such efforts have, however, generally run into problems due to the normal wide variability of MSW.

MVR Abbreviation for *mechanical vapor recompression*.

N

Nebraska Gasohol Committee A committee previously known as the Nebraska Agricultural Products Utilization Committee. Starting in 1973, it was responsible for initiating trials on the use of *ethanol-gasoline* blends, which led to the development of a nationwide *fuel ethanol* industry. It is the registered owner of the trade name *'Gasohol'* or *'Gasahol'*.

Net energy balance The amount of energy available from a fuel by combustion, less the amount of energy taken for its production. In the early 1980s a lot of emphasis was placed on the need for a net energy balance in the production of *fuel ethanol* to be used as a *gasoline extender*. The emphasis was later reduced when it became appreciated that a) much of the energy used in *ethanol* production may come from low quality sources such as coal, wood or bunker-C oil, which cannot be used as automobile fuels, and b) that *ethanol* has a value as an *octane enhancer* and oxygenate and not just as a *gasoline* extender.

Neutral spirit Defined by the *BATF* as 'distilled spirits produced from any material at or above 190^0 *proof*'. In practice, neutral spirit is purified, odorless, tasteless and colorless *ethanol* produced by *distillation* and *rectification* techniques that remove any significant amount of congeners. It is used in the production of beverages such as *vodka, gin, cordials* and cream *liqueurs*.

Nitrogen oxides Air-polluting gases contained in automobile emissions, which are regulated by the *EPA*. They comprise colorless nitrous oxide (N_2O) (otherwise know as dinitrogen monoxide or as the anaesthetic 'laughing gas'), colorless nitric oxide (NO) and the reddish-brown nitrogen dioxide (NO_2). Nitric oxide is very unstable and on exposure to air it is readily converted to nitrogen dioxide, which has an irritating odor and is very poisonous. It contributes to the brownish layer in the atmospheric pollution over some metropolitan areas. Other nitrogen oxides of less significance are nitrogen tetroxide (N_2O_4) and nitrogen pentoxide (N_2O_5). Nitrogen oxides are sometimes collectively referred to as NO_x (or Nox) where x represents any proportion of oxygen to nitrogen.

Normal solution A *solution* containing one equivalent weight of a dissolved substance per liter. One volume of a normal acid will neutralize an equal volume of a normal alkali (or vice versa). This principle is used in measuring the acidity in *fermentations* by *titration* with a standardized alkali. The 'normality' of solutions is indicated by the letter N, as in 0.1N, 2N, etc.

O

Occupational Safety and Health Administration (OSHA) An agency of the US Department of Labor with the mission of developing and promulgating occupational safety and health standards, developing and issuing regulations and conducting inspections and investigations to ensure their compliance.

Octane A flammable liquid *hydrocarbon* of chemical formula C_8H_{18} found in *petroleum*. One of the eighteen *isomers* of octane, 2,2,4-trimethylpentane is used as a standard in assessing the *octane rating* of fuels.

Octane enhancer Any substance such as *ethanol, methanol MTBE, ETBE, benzene, toluene, xylene,* etc., that will raise the *octane rating* when blended with *gasoline*.

Octane rating (or octane number) A laboratory assessment of a fuel's ability to resist self-ignition or 'knock' during combustion in a spark-ignition engine. A standardized-design, single-cylinder, four-stroke engine with a variable compression ratio is used to compare the knock resistance of a given fuel with that of reference fuels composed of varying proportions of two pure *hydrocarbons*. One component is the *octane isomer* 2,2,4-trimethylpentane (sometimes referred to as isooctane), which has a high resistance to knock and is given the arbitrary rating of 100. The other component, heptane, has a very low knock resistance and is given the base rating of zero. For fuels with a rating higher than 100 octane, the rating is obtained by determining how much *tetraethyl lead* needs to be added to pure isooctane to match its knock resistance.

The engine knock tests are performed under two sets of operating conditions, the so-called 'motor' or M method, and the 'research' or R method. The average of the two results is taken to be a good indicator of a fuel's performance in a typical automobile on the road. Hence, the use of the *(R+M)÷2* octane rating on labels on service pumps.

OFA Abbreviation for *Oxygenated Fuels Association*.

Office of Alcohol Fuels A division of the US *Department of Energy* charged with a wide range of activities to promote the development of the production and use of *alcohol* fuels. It was established by the *Energy Security Act of 1980*.

Oil Price Information Service (OPIS) A service based in Bethesda, Maryland that publishes a weekly bulletin of prices of *gasoline* and other *petroleum* products at various distribution terminals nationwide. Reference is sometimes made to OPIS prices for a particular terminal location and grade of *gasoline* when a precise basis is required in sales contracts where the price of *fuel ethanol* is to be related to that of *gasoline*.

Oligomer A *macromolecule* formed by the chemical union of two, three or four identical units known as *monomers*. The oligomers of two units are *dimers*, three units are *trimers* and four units are *tetramers*. (The prefix oligo- means 'few'.)

Oligosaccharide Short-chain *polymers* of simple *sugars* (or *monosaccharides*) generally considered to cover the range of 2-8 units. Short *dextrins* produced by *hydrolysis* of *starch* are included in this category.

Omnibus Reconciliation Act of 1990 US federal legislation that established a 'small-producer' tax credit of 10 cents per gallon on the first 15 million gallons of *fuel ethanol* produced annually by plants with a production capacity of under 30 million gallons per year. It also reduced the *excise tax exemption* (provided under the *Deficit Reduction Act of 1984*) for *gasoline* blends containing 10% *ethanol* from 6 to 5.4 cents per blended gallon and the alternative *blender tax credit* from 60 to 54 cents per gallon of *ethanol* used. The Act took effect on December 1, 1990 and extended the tax exemptions to September 30, 2000 and the tax credits to December 31, 2000.

OPIS Abbreviation for *Oil Price Information Service*.

Organic Adjective for chemical compounds containing carbon and hydrogen with or without oxygen, nitrogen or other elements.

Organoleptic testing The quality control process of checking samples of alcoholic products on the basis of odor and taste. It is normally performed by comparing samples of new production with older samples of acceptable quality that have been designated as standards.

Orifice meter An instrument used to measure fluid flow by recording the differential pressure across a restriction (or 'orifice plate') placed in the flow stream and the static or actual pressure acting on the system.

OSHA Abbreviation for *Occupational Safety and Health Administration*.

Overhead vapors The *vapors* emerging from the top of a *distillation column* that are conducted to a *condenser* system.

Oxygenated fuels Literally meaning any fuel substance containing oxygen, the term is commonly taken to cover *gasoline*-based fuels containing such oxygen-bearing compounds as *ethanol, methanol, MTBE, ETBE*, etc.

Oxygenated fuel tends to give a more complete combustion of its carbon to *carbon dioxide* (rather than *monoxide*) and thereby to reduce air pollution from exhaust emissions.

Oxygenated Fuels Association (OFA) An organization founded in 1983 by the leading US producers of *methanol* and *ethanol* to address the uses, growth and application of *alcohol* fuels as oxygenates and to develop industry standards. Its headquarters are located in Washington, DC.

P

Packed distillation column A column filled with a packing of ceramic, metal or other material designed to increase the surface area for contact between liquids and *vapors*.

PADD Abbreviation for *Petroleum Administration for Defense District*.

Pasteurization A method devised by Pasteur to partially sterilize a fluid by heating to a specific temperature for a specific length of time. Pasteurization does not greatly change chemical composition, but destroys undesirable *bacterial contaminants*. The practice may be applied to *molasses* or other liquid *fermentation feedstocks* to reduce the initial level of *bacterial contamination*.

Patent A certificate or grant by a government of an exclusive right with respect to an invention for a limited period of time. In the US the effective period is normally 17 years from the date of granting. Patents may cover processes, machines or other apparatus, methods of manufacture, composition of materials and designs. Patents are published and freely available and are required to provide sufficient information on an invention so that anyone 'reasonably versed in the art' would be able to implement it. Thus, patents can be very useful sources of technical information. Regardless of what may be stated in the title, summary or introduction, the legal essence of a patent is in the very precisely and narrowly-defined claims appearing at the end of the text. Generally, the patent examiner does not revise the summary of a patent to comply with the claims that are finally approved; so that the summaries published in various journals can be quite misleading as to the scope of the respective patents.

Penicillin The collective name for salts of a series of *antibiotic organic* acids produced by a number of Penicillium and Aspergillus molds that are active against most *Gram-positive bacteria* and some *Gram-negative cocci*. (See *Gram stain*) The commonest type of penicillin used in *ethanol fermentations* to control *bacterial contamination* is the potassium G form, otherwise known as benzyl penicillin potassium.

Pentane A colorless, highly-flammable *hydrocarbon* liquid with a pleasant odor. Chemical formula C_5H_{12}, *boiling point* 37.1°C, freezing point -129.7°C. It is used as an *entrainer* for the *dehydration* of *ethanol* by *azeotropic distillation*. It presents a severe explosion risk at low concentrations in air.

Pentanol See *amyl alcohol*.

Pentose General term for *sugars* with five carbon atoms per *molecule* such as *xylose* and *arabinose*, which are constituents of *hemicellulose*. Pentoses are not fermented by normal *strains* of distillers *yeasts*.

Permanganate (or Barbet) time A laboratory test used for assessing the quality of samples of *industrial* or *beverage alcohol*. It is the time required for an alcohol sample to decolorize a standard potassium permanganate solution. The time is an indication of the reducing (deoxidizing) power of the sample, and is considered to be a crude measure of the presence of congeners.

Petroleum A naturally-occurring complex *hydrocarbon*, which may be liquid (crude oil), gaseous (natural gas), solid (asphalt, tar or bitumen) or a combination of these forms.

Petroleum Administration for Defense District (PADD) A designation of regions in the US for the purposes of the *Department of Energy's* presentation of statistics on the usage of *petroleum*-based fuels including diesel fuel, *gasoline* and *gasoline-ethanol* blends. The term originated in World War II as 'Petroleum Administration for War Districts'. The boundary lines were drawn to reflect the natural logistical regions of the petroleum economy. Thus, the country is divided into 5 PADDs: No. 1 (east coast) includes primarily the area receiving most of its supply of crude oil by tanker, No. 2 (midwest) and No. 3 (Gulf coast) are defined in light of normal transport and supply operations, No. 4 (Rocky Mountains) is largely self-contained, No. 5 (west coast) is naturally isolated from the rest of the country.

PG Abbreviation for *proof gallon.*

pH A value measuring the acidity or alkalinity of an *aqueous solution*. Defined as the logarithm of the reciprocal of the hydrogen ion concentration of a solution. Pure water is used as the standard for arriving at pH because water *molecule*s dissociate into H and OH ions with recombination at such a rate that at 22°C there is a concentration of oppositely-charged ions of 1/10,000,000 or 10^{-7} *mole* per liter. This is defined as a pH of 7. Solutions with a pH of less than 7 are acidic, and greater than 7 are alkaline. As the pH scale is logarithmic, a solution with a pH of 5 has 10 times the acidity of a solution of pH 6. Control of pH is important in *ethanol* production both for obtaining optimal *enzymatic* activity and in controlling the growth of *bacterial contaminants*.

Phase separation The phenomenon of a separation of a liquid (or *vapor*) into two or more physically distinct and mechanically separable portions or layers. It is used to advantage in the *dehydration* of *ethanol* by *azeotropic distillation* with an *entrainer* such as *benzene*. When the *overhead vapors* are condensed, the liquid separates into two distinct layers or phases. The upper phase contains most of the benzene with some *ethanol* and a little water, while the lower contains most of the water with some *ethanol* and a little benzene. The lower layer is removed and redistilled in an entrainer recovery column to recover the benzene and *ethanol* and to discharge the bulk of the water.

Plate (or tray) A contacting device placed horizontally at intervals within a *distillation column*. Plates may be simple perforated discs with or without *downcomers* as in the *sieve plate* and the *dual flow plate*, or they may have *bubble caps, tunnel caps* or various types of floating valves to improve the contact between the rising *vapor* and descending liquid. Sieve plates and tunnel caps are the most common form of plates used in *ethanol* production facilities.

Plate (or tray) distillation column A *distillation column* with horizontally arranged contacting *plates* located at intervals up the column; as compared to a *packed column,* which is filled randomly with contacting devices.

Polymer A *macromolecule* formed by the chemical union of identical combining units known as *monomers*. While generally referring to a combination of at least five monomers, in many cases the number of monomers is quite large, such as the average of 3,500 *glucose* monomers in *cellulose*.

Polysaccharide A *polymer* composed of numerous *sugar monomers* or *monosaccharides*. Examples are *cellulose* and the *amylose* and *amylopectin* in *starch*. For *fermentation* purposes, polysaccharides must be subjected to hydrolysis to yield their component fermentable monosaccharides.

Positive displacement pump A class of pumps in which the displacement is accomplished mechanically. They may take various forms such as the reciprocating cylinder-piston, *gear* or lobe mechanisms. Generally, they are used to move relatively low volumes of fluid at high pressures. As they force a positive displacement, they may not be used in systems where the output may be throttled or closed off without incorporating a pressure relief or recycle system. Positive displacement pumps are frequently used for moving *molasses* and *stillage* syrups.

Potassium aluminosilicate A zeolite. (See *molecular sieve*).

Pot still A simple *batch distillation* unit used for the production of heavily-flavored *distillates* for beverage use. It consists of a tank (heated either by an internal steam coil or by an external fire) and an *overhead vapor* pipe leading to a *condenser.* It may be used in the production of heavily-flavored *rums* and *whiskies.*

Prefermenter See *yeast* propagator.

Proof A measure of the *absolute ethanol* content of a *distillate* containing *ethanol* and water. In the US system, each degree of proof is equal to 0.5% *ethanol* by volume, so that absolute *ethanol* is 200^0 proof. In the Imperial system 100 proof is equal to 57.06% *ethanol* by volume, or 48.24% by weight, while absolute *ethanol* is 75.25 over proof, or 175.25 proof.

Proof gallon (PG) The volume of a liquid that contains the equivalent of one gallon of *ethanol* at 100^0 *proof.*

Proof tables Books of tables compiled by the *BATF* and other organizations for the calculation of *proof* of *ethanol*-containing liquids with adjustments for varying temperatures.

Propagation The process of increasing numbers of organisms by natural reproduction. In the case of *yeast,* under *aerobic* conditions propagation occurs by budding to form new cells. (See *yeast propagator*).

Propanol (or propyl alcohol) A minor constituent of *fusel oil.* Chemical formula C_3H_7OH. It exists as either of two *isomers.* Both are colorless, toxic, flammable liquids with odors similar to that of *ethanol.*

Protein Any of a class of high *molecular weight polymer* compounds composed of a variety of linked amino acids. Proteins are an essential ingredient in the diets of animals and humans. *Yeast* dry matter is approximately 50% protein. Corn *DDG* generally contains over 27% crude protein.

PSI Abbreviation for pounds per square inch.

PSIG Abbreviation for pounds per square inch gauge.

R

Rack price Used in reference to *gasoline* and *fuel ethanol,* it is the wholesale price at the tank-truck loading terminal exclusive of any federal, state or local taxes.

Reboiler A device for supplying heat to a *distillation column* without introducing live steam. It generally consists of a shell-and-tube *heat exchanger* connected to the base of the column with liquid from the column entering inside the tubes to be heated indirectly by steam on the shell side.

Receiver A tank into which new *distillates* flow from the *still* for verifying *proof,* quantity and quality before transfer to shipping or storage tanks. Where the tank is sealed or has locks on the valves to meet government excise regulations for preventing unlawful access, it may be referred to as a 'closed receiver'.

Recoopered barrels Used wooden *barrels* that have been repaired to replace any cracked staves or heads to be suitable for reuse in the aging of alcoholic beverages.

Rectification The process of concentrating and purifying *ethanol* or other materials in a *rectifying column.* The process may be operated simultaneously with *beer distillation* with the rectifying column directly connected to a *beer stripping column.* In *beverage ethanol* production the rectification process may involve concentration and removal of *fusel oil, esters* and *heads.* In *fuel ethanol* production the process commonly only involves concentration and the removal of fusel oil. In some instances, where *benzene* is to be used as an *entrainer* in the subsequent *dehydration* process, the heads may also be removed in the rectification process.

Rectifying column (rectifier, rectification column or rectifying section) The portion of a *distillation column* above the *feed tray* in which rising *vapor* is enriched by interaction with a countercurrent descending stream of condensed vapor. In the case of *ethanol* production, the rectifying column may have valves at draw-off points on various trays to allow the removal of accumulated *fusel oil* and other *congeners.* In some instances the *ethanol* product is withdrawn from the *reflux* of the *overhead vapor condensate,* but in others the product may be drawn from one of the upper trays to permit the *heads* or *low boilers* to be purged from the reflux. (See *Concentrating column).*

Reflux The portion of the condensed *overhead vapors* returned to a *distillation column* to maintain the liquid-vapor equilibrium.

Reflux ratio The ratio of the amount of *condensate refluxed* to the amount withdrawn as product. Generally, the higher the reflux ratio the greater the degree of separation of the components in a *distillation* system.

Reid vapor pressure (Rvp) A measure of the *vapor pressure* in pounds per square inch of a sample of *gasoline* at 100°F. It is an indication of the *volatility* of a *gasoline.* The blending of *ethanol* with *gasoline* tends to increase Rvp while the blending of *MTBE,* and more particularly *ETBE,* tends to reduce Rvp.

Renewable Fuels Association (RFA) The Washington DC-based trade association for the US *fuel ethanol* industry. Its membership includes companies involved in the production, blending and marketing of *ethanol*-blended fuels.

Research octane number (RON) See *Octane rating.*

Reverse osmosis A technique used in water purification and wastewater and *stillage* treatment in which pressure is applied to the liquid in a suitable apparatus to force pure water through a membrane that does not allow the passage of dissolved ions.

RFA Abbreviation for *Renewable Fuels Association.*

(R+M)÷2 The formula for calculating *octane rating.*

Roller mill A mill for crushing or grinding grain or other solid material by passing it between two steel rollers. The rollers may be smooth, or serrated to shear the grain, and they may turn at differing speeds to increase the abrasion. Roller mills are suitable for small grains such as wheat, but do not perform as well as a *hammer mill* on corn.

RON Abbreviation for *research octane number.*

Rotameter A commonly-misused brand name of a type of *flowmeter.*

Rum Defined by the *BATF* as 'an alcoholic *distillate* from the fermented juice of sugarcane, sugarcane syrup, sugarcane *molasses,* or other sugarcane by-products, produced at less than 190° *proof,* in such a manner that the distillate possesses the taste, aroma and characteristics generally attributed to rum'. Unlike the specifications for *whiskies,* the *BATF* does not require that rum be aged in oak *barrels.* British regulations specify that rum be produced 'from sugarcane products in sugarcane-growing countries'.

Rvp Abbreviation for *Reid vapor pressure.*

Rye whisky Defined by the *BATF* as *whisky* produced at not more than 160° *proof,* from a fermented *mash* of not less that 51% rye and stored at not more than 125° proof in charred, new oak *barrels.*

S

Saccharification The process of converting a complex *carbohydrate* such as *starch* or *cellulose* into *fermentable sugars* such as *glucose* or *maltose*. It is essentially a *hydrolysis*. The process may be accomplished by the use of *enzymes* or acids.

Sacc' tank (or saccharification tank) A vessel where cooked *mash* is held for the *saccharification* process. Sacc' tanks are generally going out of use with the adoption of *simultaneous saccharification and fermentation* in the *fermenter*.

Saccharomyces A genus of unicellular *yeasts* of the family Saccharomycetaceae distinguished by the general absence of mycelium and by their facility to reproduce asexually by budding. This genus includes the species *Saccharomyces cerevisiae*, which is the *yeast* most commonly used by bakers, brewers, distillers and wine producers.

SBA Abbreviation for *Small Business Administration*.

Scale inhibitor See *Antiscalant*.

Scaling The precipitation of salts in a *distillation column* or *heat exchanger* which if uncontrolled may reduce capacity and eventually block the equipment and make it unusable. Scaling occurs because some salts such as calcium sulfate and oxalate are less soluble at high temperatures and/or in the presence of *ethanol*. Scaling problems are more common in *molasses beer* distillation, but may occur in grain beer distillation when conducted under pressure (i.e., at high temperatures).

Scrubber A device for the removal of entrained liquid droplets in a gas stream. Scrubbers may be used to recover *ethanol* from the *carbon dioxide* vented from *fermenters*.

SDA Abbreviation for *specially denatured alcohol*.

SG Abbreviation for *specific gravity*.

Shooked barrel (or shook) A used *bourbon whisky barrel* that has been dismantled to reduce the space requirements for transportation.

Sieve analysis A laboratory test made on grain *meal* to check that the milling process is being conducted correctly. The meal is added to the top of a stack of sieves with increasingly fine mesh sizes descending downwards. The sieve stack is vibrated for a standard time period and the weight percentage retained on each screen is determined. With *hammer mills*, the sieve analysis will generally show that the meal gradually becomes more coarse as the hammers wear and need turning or replacement.

Sieve plate (or sieve tray) A *distillation column plate* with perforations of precise number and size such that the ascending *vapor* passes through the plate vertically to mix with the descending liquid held on the plate. A sieve plate differs from a *dual flow plate* in that it is designed for the descending liquid to overflow down a *downcomer*, instead of *weeping* through the perforations.

Simultaneous saccharification and fermentation (SSF) A procedure in which *saccharification* of a cooked *starch mash* occurs in the *fermenter* (by addition of *enzymes*) simultaneously with the commencement of *fermentation* (by *yeast*). This procedure is replacing the traditional process taken from the whisky industry in which there is a specific holding stage for saccharification with *malt* or microbial *enzymes* (in a *sacc' tank*) before the mash goes to a fermenter.

Slurrying tank The vessel in which grain *meal* is mixed into a slurry with water before being pumped through a *continuous cooking* system.

Small Business Administration (SBA) A US federal government agency charged with encouraging and assisting the development of small businesses. It offers financial assistance such as direct loans, loan guarantees and direct participation loans for virtually any legitimate small business development. The SBA has assisted in the financing of many small *fuel ethanol* facilities.

Solute One or more substances dissolved in another substance (a *solvent*) to form a *solution*. The solute is uniformly dispersed in the solvent in the form of *molecules* (as with *sugars*) or ions (as with salts).

Solution A uniformly dispersed mixture at the *molecular* or ionic level of one or more substances (the *solute*) in one or more other substances (the *solvent*). *Ethanol* and water form a liquid-liquid solution, while sugar and water form a solid-liquid solution.

Solvent A substance capable of dissolving another substance (a *solute*) to form a uniformly-dispersed mixture (*solution*) at the *molecular* or ionic level.

Specially denatured alcohol (SDA) The term used to describe *ethanol* denatured with any formulation of compounds selected from a list approved by the *BATF*. The *denaturant* renders the *ethanol* unfit for beverage purposes without impairing its usefulness for other applications.

Specific gravity (SG) The ratio of the density of a material to the density of a standard reference material such as water at a specific temperature. For example, the specific gravity of *ethanol* is quoted as 0.7893 at 20/4°C. This means that a given volume of *ethanol* at 20°C weighs 0.7893 times the weight of the same volume of water at 4°C.

Spent sulfite liquor (SSL) See *sulfite waste liquor*.

Spirit whisky Defined by the *BATF* as a mixture of *neutral spirits* with at least 5% on a *proof gallon* basis of *whisky* or *straight whisky*.

SSF Abbreviation for *simultaneous saccharification and fermentation*.

SSL Abbreviation for *spent sulfite liquor*. (See *sulfite waste liquor*).

Starch A mixture of two *carbohydrate polymers* (*amylose* and *amylopectin*), both of which are composed of *glucose monomers* linked by glycosidic bonds. *Starch* is a principal energy storage product of photosynthesis and is found in most roots, tubers and cereal grains. *Starch* may be subjected to *hydrolysis* (*sacchar-ification*) to yield *dextrins* and *glucose*.

Still An apparatus used in *distillation* comprising a vessel in which the liquid is *vaporized* by heat and a cooling device in which the *vapor* is condensed. It may take the form of a batch or continuously operable unit.

Stillage The mixture of non-fermentable (or non-fermented) solids and water, which is the residue after removal of *ethanol* from a fermented *beer* by *distillation*. Stillage may be dried to recover the solid material (as *DDG* in the case of grain *feedstocks*).

Stoichiometric yield The yield of a product of a chemical reaction as calculated from the theoretical equation. For example, in the *Gay Lussac* equation for the *fermentation* of *glucose* to *ethanol* by *yeast*, $C_6H_{12}O_6 \rightarrow 2CO_2 + 2C_2H_5OH$, 100 parts (by weight) *glucose* should yield 48.89 parts *carbon dioxide* and 51.11 parts *ethanol*. Due to a variety of factors, this yield is not achieved in practice.

Stoichiometry The branch of chemistry that deals with the quantities of substances that enter into and are produced by chemical reactions.

Stover The dried stalks and leaves remaining from a crop, particularly corn, after the grain has been harvested. It is of interest as a potential source of *cellulose feedstock* for *ethanol* production.

Straight whisky Defined by the *BATF* as a product conforming to the requirements for the *whisky* designation, which has been stored in charred, new oak *barrels* for a period of at least two years. The straight whisky definition may include mixtures of straight whiskies of the same grain type, produced at the same *distillery*, all of which are not less than four years old.

Strain See *Yeast strain*.

Stripping column (or stripping section) The portion of a *distillation column* below the *feed tray* in which the descending liquid is progressively depleted of its *volatile* components by the introduction of heat at the base.

Sub-octane blending The practice of blending *ethanol* with selected *gasoline* components that have *octane ratings* below the commercially or legally acceptable level. It takes advantage of the fact that *ethanol* can enhance the octane rating of the mixture to bring it to an acceptable level.

Sucrose Common table *sugar*, derived from beet or cane sources. It is a *disaccharide* comprising a *monomer* of *glucose* linked to a monomer of *fructose*. It has a chemical formula $C_{12}H_{22}O_{11}$. It is directly fermentable by common *yeasts* to produce *ethanol*.

Sugar Any of a class of water-soluble, simple *carbohydrate*, crystalline compounds that vary widely in sweetness includeing the *mono-saccharides* and lower *oligosaccharides*. Sugars may be chemically reducing or non-reducing compounds and are typically optically active. Examples include the monosaccharides *glucose, fructose, mannose* and *xylose*, and the *disaccharides sucrose, maltose, lactose,* and the trisaccharides raffinose and maltotriose.

Sulfite waste liquor (SWL) An effluent produced in the sulfite pulping process used in some paper mills. It partly consists of a dilute solution of *sugars* produced by the *acid hydrolysis* of *cellulose*. It may be used as a *feedstock* for the production of *ethanol* by *fermentation* using selected *yeast* strains after stripping out the sulfite (or sulfur dioxide) with steam.

Supplementary column See *Demethylizing column*.

Surface Transportation and Uniform Relocation Assistance Act of 1987. US federal legislation which extended to September 30, 1993 the *excise tax exemption* for *ethanol-gasoline* blends, as originally granted by the *Energy Tax Act of 1978* and subsequently increased by the *Surface Transportation Assistance Act of 1982* and the *Deficit Reduction Act of 1984*.

Surface Transportation Assistance Act of 1982 US federal legislation that increased the *excise tax exemption* originally granted under the *Energy Tax Act of 1978* from 4 to 5 cents per gallon of *ethanol-gasoline* blend when the tax on *gasoline* was raised to 9 cents per gallon. It also increased the alternative *blender tax credit* to 50 cents per gallon of *ethanol*.

Sweet sorghum A plant of the species *Sorghum bicolor*, closely related to *milo or grain sorghum*, which is cultivated primarily for the sweet juice in its stem. It is widely used for cattle fodder and silage. The juice contains *sucrose*, which may be extracted by crushing with modified sugar mill equipment. The juice may be used in the fermentative production of *ethanol*. The plant has an advantage over sugarcane in that it can be grown over a much wider range of climatic regions.

SWL Abbreviation for *sulfite waste liquor*.

Synthetic ethanol *Ethanol* produced by any of several synthetic processes such as the catalytic hydration of ethylene, the sulfuric acid hydration of ethylene and the Fischer-Tropsch process, in which it is a major by-product of the synthesis of *methanol* by catalytically reacting *carbon dioxide* and hydrogen. Synthetic *ethanol* is chemically identical to *fermentation ethanol*, but does not qualify for US federal or state incentives for blending with *gasoline* and may not be used in the production of alcoholic beverages.

T

Tafia An unaged Caribbean or South American alcoholic beverage produced by *batch distillation* of *beers* obtained by the *fermentation* of sugarcane juice or *molasses*. It is similar to *aguardiente* and *rum*.

Tails See *high boilers*.

TBA Abbreviation for tertiary *butyl alcohol*. See *butanol*.

Tequila Defined by the *BATF* as 'an alcoholic *distillate* from a fermented *mash* derived principally from *Agave tequilana Weber* ('blue' variety) with or without additional fermentable substances, distilled in such a manner that the distillate possesses the taste, aroma and characteristics normally attributed to tequila. It is a distinctive product of Mexico, manufactured in that country in compliance with the laws regulating the manufacture of tequila for consumption in that country'. Tequila may be matured in wooden containers to acquire a light gold color.

Ternary azeotrope An *azeotrope* or *constant boiling mixture* made up of three components. For example, a mixture of 74% volume *benzene*, 18.5% *ethanol* and 7.5% of water forms an azeotrope boiling at 64.9°C.

Tetramer A *macromolecule* produced by linking four identical units known as *monomers*. (See *oligomer)*.

Tetraethyl lead (or lead tetraethyl) An organo-metallic compound used as an *octane enhancer* in *gasoline*. Its use is regulated by the *EPA* to control air pollution. If added to *gasoline* on its own, tetraethyl lead would leave combustion residues of lead and lead oxide on engine cylinder walls. To prevent accumulation of these deposits, ethylene dibromide and dichloride are also added to convert the lead to *volatile* halides before release into the atmosphere.

Theoretical plate (or tray) A *distillation column plate* or *tray* that produces perfect *distillation* i.e. it would yield the same difference in composition between liquid and *vapor* as that normally existing between the liquid and the *vapor* when in equilibrium. A distillation column giving the same separation as 10 simple theoretical distillations is said to have 10 theoretical plates.

Theoretical yield See *stoichiometric yield*.

Thermal efficiency The ratio of the energy output of a process to the energy input.

Thermal vapor recompression A method using the same recompression and recirculation principles as in *mechanical vapor recompression* (to enhance the efficiency of *evaporation)* except that recompression is performed by passing steam through a venturi jet on the vapor line.

Thermophilic Literally meaning 'heat loving', it refers to *microorganisms* such as *bacteria* that thrive at warm temperatures, generally in the range of 35-60°C. Lactobacillus is an example of a thermophilic bacterium that multiplies rapidly at around 55°C. Appreciation of this fact has been a reason why the use of *sacc' tanks* has fallen into disfavor.

Thin stillage The liquid portion of *stillage* separated from the solids by screening or *centrifuging*. It contains suspended fine particles and dissolved material. It is normally sent to an *evaporator* to be concentrated to a thick syrup and then dried with the solids portion to give *DDGS*.

Titration A method of volumetrically determining the concentration of a substance in *solution* by adding a standard solution of known volume and strength until the reaction between the two is completed, usually as indicated by a change in color due to an added chemical indicator. For instance, it is common practice to perform a titration with 0.1 *Normal* sodium hydroxide solution using phenol-phthalein as indicator to determine the acidity of a *mash* or *beer*.

Toluene An aromatic *hydrocarbon* compound that can be used as an *octane enhancer* in *gasoline*. It is a colorless, flammable liquid with a benzene-like odor. Chemical formula $C_6H_5CH_3$. It boils at 110.7°C and freezes at -94.5°C. It is derived from the catalytic reforming of *gasoline* or the *distillation* of coal tar light oil. It is known to be carcinogenic.

Total sugars as invert (TSAI) A simple crude analytical measure of reducing *sugars* in *molasses*.

Transglucosidase An *enzyme* with the opposite effect of *amyloglucosidase* in that it brings about a *polymerization* of *glucose* into *polysaccharides.* It may be present as an impurity in commercial *enzyme* preparations.

Trimer A *macromolecule* produced by the linking three identical units known as *monomers.* (See *oligomer).*

Tray See *plate.*

TSAI Abbreviation for *total sugars as invert.*

Tunnel cap A contacting device used occasionally on *distillation plates.* It is essentially an elongated *bubble cap*, consisting of a chimney portion and a domed cover that deflects the *vapors* rising through the chimney, causing them to pass down through the liquid layer on the *plate.* Tunnel caps may be made from interlocking metal channels to form an integral part of the plate.

U

Upgrading A term used loosely to mean the *dehydration* of *hydrous ethanol.*

Ultrafiltration A process for the separation of colloidal or very fine solid materials, or large dissolved *molecules*, by filtration through microporous or semi-permeable membranes. The process may be used for removal of *protein* from cheese *whey* prior to *fermentation.*

United States Department of Agriculture (USDA) A federal government department with the mission to improve and maintain farm income, to develop and expand markets for agricultural products, etc. Through such agencies as the *Farm Home Administration (FmHA)* and the *Commodity Credit Corporation (CCC)* the USDA has been involved in assisting the development of the *fuel ethanol* industry.

Urethane See *Ethyl carbamate.*

USDA Abbreviation for *United States Department of Agriculture.*

US gallon A measure of 231 cubic inches liquid at 60°F. It is the equivalent of 3.785 *liters* in the metric system or 5/6 of an *Imperial gallon.*

V

Vacuum distillation A process that takes advantage of the fact that liquids boil at lower temperature under reduced pressure. Thus, *distillation* under vacuum may be used for substances that otherwise have a relatively high *boiling point* such as *ethylene glycol* (which may be used in the *dehydration* of *ethanol*). The process may also be used in the distillation of *ethanol*-water mixtures, as the *azeotrope* is formed at a higher *proof* under lower pressures. Vacuum distillation may be used in the production of *rum* or *ethanol* from molasses to reduce the incidence of *scaling* in the *column* that occurs at high temperatures.

Vacuum fermentation A process of operating a *fermentation* under vacuum, so that the *ethanol* or other product is *vaporized* and removed as it is formed to avoid its concentration becoming inhibitory to the *yeast.* In a patented variation known as the Vacuferm process, instead of maintaining the entire *fermenter* under vacuum, the fermenting *beer* is circulated through a vacuum chamber to flash off the *ethanol* before returning the beer to the fermenter.

Vapor A dispersion in air of *molecules* of a substance that is liquid or solid in its normal state at standard temperature and pressure. An example is water vapor or steam.

Vaporization (or Volatilization) The conversion of a chemical substance from a liquid or solid state to a *vapor* or gaseous state by the application of heat, by the reduction of pressure or by a combination of these processes.

Vapor pressure The saturation pressure exerted by *vapors* when in equilibrium with their liquid or solid forms.

Vent condenser The final *condenser* in a series of two or more connected to the *overhead vapor line* of a *distillation column*. As the final condenser, it is the one from which non-condensed (or non-condensable) gases and *vapors* are vented to the atmosphere or to a *scrubber* system.

Vinasse The term sometimes applied to the *stillage* of *molasses*, grape juice or other liquid *ethanol feedstocks*.

Vodka Currently defined by the *BATF* as '*neutral spirit* so distilled, or so treated after *distillation* with charcoal or other materials as to be without distinctive character, aroma, taste or color'. It should be noted that the charcoal treatment is now optional, whereas in earlier *BATF* regulations it was mandatory and the minimum time of treatment and amounts of fresh charcoal to be used were specified.

Volatile Adjective to describe a solid or liquid that readily converts to the *vapor* state.

Volatility The tendency of a solid or liquid to pass into the *vapor* state at a given temperature. With automotive fuels, the volatility is determined by measuring the *Reid vapor pressure (Rvp)*.

Volatilization See *Vaporization*.

W

Wash A British synonym for distillers *beer*.

Wet milling A process in which corn is first steeped (or soaked) in water containing sulfur dioxide. This softens the kernels and loosens the hulls. (The liquid when separated is known as *corn steep liquor*.) The grain is then degermed by abrasion and liquid separation. Oil is extracted from the germ, while the remainder of the kernel is ground in impact fiber mills and passed through stationary screens to separate the *starch* and gluten from the fibrous portion. The heavier *starch* is then separated from the gluten by use of *centrifuges*. The *starch* portion may then be processed into commercial *starch*, or it may be used as a *feedstock* for production of *ethanol* or *HFCS*.

Weeping The condition when droplets of liquid fall through the holes of a *sieve plate* in a *distillation column*. It may be caused by a) steam flow that is too low, or b) too low a liquid flow to maintain a level on the plate, or c) having a tilted plate such that liquid depth is uneven.

Whey The serum or watery part of milk separated from the curd in the process of making cheese. It contains *lactose* and may be used as a *feedstock* for *ethanol* production.

Whisky Defined by the *BATF* as 'an alcoholic *distillate* from a *mash* of grain, produced at less than 190° *proof*, in such as manner that the distillate possesses the taste, aroma and characteristics generally attributed to whisky'. With the exception of *corn whisky*, it should be stored in oak barrels.

Whole stillage The entire stillage emerging from a *distillation* unit before any removal of solids by screening or *centrifuging*.

Wine gallon A *US gallon* of liquid measure as distinct from a *proof gallon*.

Wood alcohol See *methanol*.

Wort A brewery term used occasionally in distilleries. It refers to an unfermented *mash*, particularly if produced from a liquid *feedstock* or if the solids have been removed to yield a relatively clear, free-flowing liquid.

X

Xylene One of a group of three isomeric, aromatic *hydrocarbons* having a formula of $C_6H_4(CH_3)_2$. Commercial xylene is normally a mixture of the o- m- and p- *isomers* with a *boiling point* range of 137-145°C. It is a clear, flammable liquid. Derived from catalytic reforming of napthas or the distillation of coal tar, it may be used as an *octane enhancer* in *gasoline*.

Xylose A *pentose* (or 5-carbon sugar) derived from the *hydrolysis* of *hemicellulose*. Chemical formula $C_5H_{10}O_5$. It is not fermented by normal strains of distillers *yeasts*.

Y

Yeast Any of certain unicellular fungi, generally members of the class Ascomycetaceae (a few are members of the class Basidiomycetaceae). Many yeasts are capable of producing *ethanol* and *carbon dioxide* by the *anaerobic fermentation* of sugars. Yeasts are composed of approximately 50% *protein* and are a rich nutritional source of B vitamins.

Yeast autolysis The disintegration of *yeast* cells by the action of their own *enzymes*.

Yeast cream *Yeast* concentrated by centrifuging or decanting from the contents of a propagator or fermenter to be used as *inoculum* in another *fermenter*.

Yeast propagator (or prefermenter) A tank used for the propagation or development of a *yeast* culture prior to transfer to a *fermenter*. It is normally fitted with aeration, agitation and cooling devices and is designed for ease of cleaning and sterilization.

Yeast recycle See *cell recycle*.

Yeast strain A pure culture of *yeast* derived from a single isolation. Strains may be specially selected for certain characteristics such as the ability to efficiently produce and tolerate high levels of *ethanol*.

Z

Zeolite See *Molecular sieve*.

Zymomonas A genus of the Pseudomonadaceae family of *bacteria* which are characterized by being *Gram-negative* and non-spore-forming. The genus Zymomonas is distinguished by its *fermentation* of *sugar* to *ethanol*. The principal species being examined commercially for *fuel ethanol* production is *Zymomonas mobilis*. It is, however, considered an undesirable contaminant in *beverage alcohol fermentations* in that it tends to produce hydrogen sulfide from sulfur compounds in the *mash*, particularly that derived from *molasses*.